Fundamental
Astronomy

Springer

Berlin
Heidelberg
New York
Hong Kong
London
Milan
Paris
Tokyo

Physics and Astronomy  ONLINE LIBRARY

http://www.springer.de/phys/

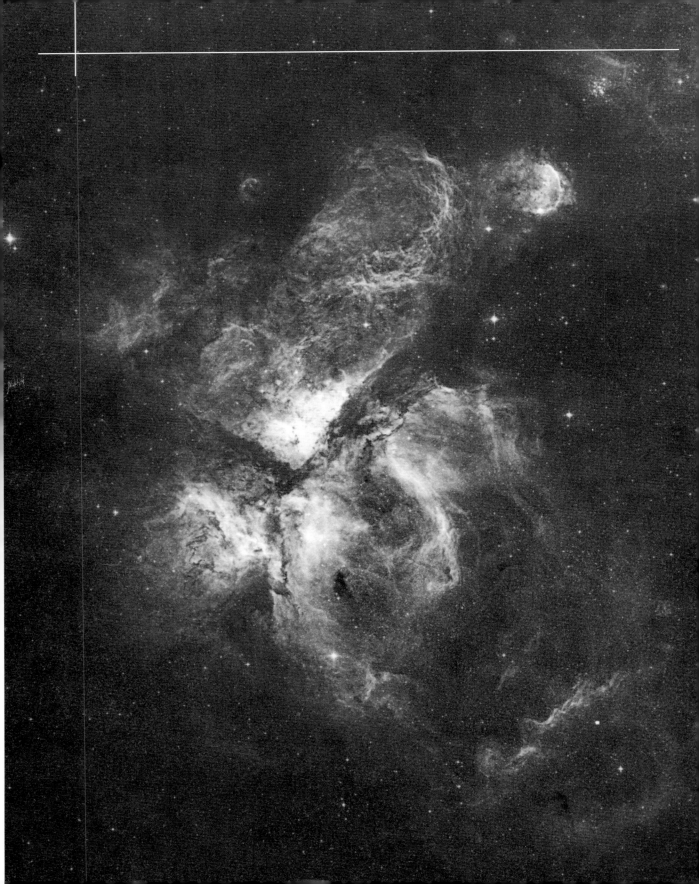

H. Karttunen
P. Kröger
H. Oja
M. Poutanen
K. J. Donner (Eds.)

Fundamental Astronomy

Fourth Edition
With 411 Illustrations
Including 36 Colour Plates
and 73 Exercises with Solutions

 Springer

Dr. *Hannu Karttunen*
University of Turku, Tuorla Observatory,
Väisäläntie 20, 21500 Piikkiö, Finland

Dr. *Pekka Kröger*
Isonniitynkatu 9 C 9, 00520 Helsinki, Finland

Dr. *Heikki Oja*
Observatory and University of Helsinki,
Tähtitorninmäki, 00014 Helsinki, Finland

Dr. *Markku Poutanen*
Finnish Geodetic Institute,
Geodeetinrinne 2, 02430 Masala, Finland

Dr. *Karl Johan Donner*
Finnish Geodetic Institute
Ilmalankatu 1 a, 00240 Helsinki, Finland

ISBN 3-540-00179-4   4th Edition
Springer-Verlag Berlin Heidelberg New York

ISBN 3-540-60936-9   3rd Edition
Springer-Verlag Berlin Heidelberg New York

Library of Congress Cataloging-in-Publication Data.
Tähtitieteen perusteet. English. Fundamental astronomy /
H. Karttunen . . . [et al.] eds. – 4th ed. p. cm.
Includes bibliographical references and index.
ISBN 3540001794 (alk. paper)
1. Astronomy I. Karttunen, Hannu. II. Title.
QB43.2 .T2613 2003   520–dc21   2002042569

Cover picture: The 8.1 m Gemini North telescope on Mauna Kea,
Hawaii, was set in operation in 1999. Its twin, Gemini South, was
dedicated in 2000. (Fig. 3.18a in this volume)

Frontispiece: The η Carinae nebula, NGC3372, is a giant HII region
in the Carina spiral arm of our galaxy at a distance of 8000 light-years.
(Photograph European Southern Observatory)

Title of original Finnish edition:
Tähtitieteen perusteet (Ursan julkaisuja 56)
© Tähtitieteellinen yhdistys Ursa Helsinki 1984, 1995

Sources for the illustrations are given in the captions and more fully
at the end of the book. Most of the uncredited illustrations are
© Ursa Astronomical Association, Raatimiehenkatu 3A2,
00140 Helsinki, Finland

Springer-Verlag Berlin Heidelberg New York
a member of BertelsmannSpringer Science+Business Media GmbH

http://www.springer.de

© Springer-Verlag Berlin Heidelberg 1987, 1994, 1996, 2003
Printed in Germany

The use of general descriptive names, registered names, trademarks,
etc. in this publication does not imply, even in the absence of a specific
statement, that such names are exempt from the relevant protective
laws and regulations and therefore free for general use.

Data conversion: LE-TeX, Jelonek, Schmidt & Vöckler GbR, Leipzig
Cover design: Erich Kirchner, Heidelberg
Layout: Schreiber VIS, Seeheim
Printing and binding: Universitätsdruckerei Stürtz, Würzburg

Printed on acid-free paper
SPIN: 10760987   55/3141/ba   5 4 3 2 1 0

Preface to the Fourth Edition

While editing the first version of this book we could hardly imagine that it would keep us busy for decades. Since then, several revised Finnish and English editions have appeared. Due to different production techniques, the Finnish and English versions diverged to different evolutionary tracks. This fourth edition brings these tracks together again. Now, finally, we have both of them in a similar computer readable form, which will make them easier to update. The conversion has really exhausted us; we can only hope that while updating the text and fixing the old errors we have not incorporated too many new ones.

The main difference to the previous editions is the restructuring of some chapters. Earlier chapters 5 (Radiation Mechanisms) and 6 (Temperatures) have been combined, and part of the material of the appendices has been absorbed to appropriate places. Also the chapters on the solar system and cosmology have been more extensively revised. In other chapters there are several small revisions. Many of the tables in Appendix C have also been updated.

Several people have sent us their comments. We are grateful to all of them, and we hope that our keen readers will continue to send us suggestions for improving this book as well as inform us on possible errors. Comments can be sent e.g. by e-mail to Hannu.Karttunen@astro.utu.fi.

During the years we have worked with the book, we have been impressed by the high professional standards of the people at Springer. With a few of them we have had the pleasure of communicating through mail or e-mail, but there are many others that have remained unknown to us but who have made the success of this work possible. It is a great pleasure to be able to express our gratitude to all of them.

Helsinki
February 2003

The Editors

Preface to the First Edition

The main purpose of this book is to serve as a university textbook for a first course in astronomy. However, we believe that the audience will also include many serious amateurs, who often find the popular texts too trivial. The lack of a good handbook for amateurs has become a problem lately, as more and more people are buying personal computers and need exact, but comprehensible, mathematical formalism for their programs. The reader of this book is assumed to have only a standard high-school knowledge of mathematics and physics (as they are taught in Finland); everything more advanced is usually derived step by step from simple basic principles. The mathematical background needed includes plane trigonometry, basic differential and integral calculus, and (only in the chapter dealing with celestial mechanics) some vector calculus. Some mathematical concepts the reader may not be familiar with are briefly explained in the appendices or can be understood by studying the numerous exercises and examples. However, most of the book can be read with very little knowledge of mathematics, and even if the reader skips the mathematically more involved sections, (s)he should get a good overview of the field of astronomy.

This book has evolved in the course of many years and through the work of several authors and editors. The first version consisted of lecture notes by one of the editors (Oja). These were later modified and augmented by the other editors and authors. Hannu Karttunen wrote the chapters on spherical astronomy and celestial mechanics; Vilppu Piirola added parts to the chapter on observational instruments, and Göran Sandell wrote the part about radio astronomy; chapters on magnitudes, radiation mechanisms and temperature were rewritten by the editors; Markku Poutanen wrote the chapter on the solar system; Juhani Kyröläinen expanded the chapter on stellar spectra; Timo Rahunen rewrote most of the chapters on stellar structure and evolution; Ilkka Tuominen revised the chapter on the Sun; Kalevi Mattila wrote the chapter on interstellar matter; Tapio Markkanen wrote the chapters on star clusters and the Milky Way; Karl Johan Donner wrote the major part of the chapter on galaxies; Mauri Valtonen wrote parts of the galaxy chapter, and, in collaboration with Pekka Teerikorpi, the chapter on cosmology. Finally, the resulting, somewhat inhomogeneous, material was made consistent by the editors.

The English text was written by the editors, who translated parts of the original Finnish text, and rewrote other parts, updating the text and correcting errors found in the original edition. The parts of text set in smaller print are less important material that may still be of interest to the reader.

For the illustrations, we received help from Veikko Sinkkonen, Mirva Vuori and several observatories and individuals mentioned in the figure captions. In the practical work, we were assisted by Arja Kyröläinen and Merja Karsma. A part of the translation was read and corrected by Brian Skiff. We want to express our warmest thanks to all of them.

Financial support was given by the Finnish Ministry of Education and Suomalaisen kirjallisuuden edistämisvarojen valtuuskunta (a foundation promoting Finnish literature), to whom we express our gratitude.

Helsinki   *The Editors*
June 1987

Contents

The Virgo cluster of galaxies. Each galaxy contains hundreds of billions of stars
(Photograph National Optical Astronomy Observatories)

1. Introduction

1.1 The Role of Astronomy

On a dark, cloudless night, at a distant location far away from the city lights, the starry sky can be seen in all its splendour (Fig. 1.1). It is easy to understand how these thousands of lights in the sky have affected people throughout the ages. After the *Sun*, necessary to all life, the *Moon*, governing the night sky and continuously changing its phases, is the most conspicuous object in the sky. The *stars* seem to stay fixed. Only some relatively bright objects, the *planets*, move with respect to the stars.

The phenomena of the sky aroused people's interest a long time ago. The *Cro Magnon people* made bone engravings 30,000 years ago, which may depict the phases of the Moon. These calendars are the oldest astronomical documents, 25,000 years older than writing.

Agriculture required a good knowledge of the seasons. Religious rituals and prognostication were based on the locations of the celestial bodies. Thus time reck-

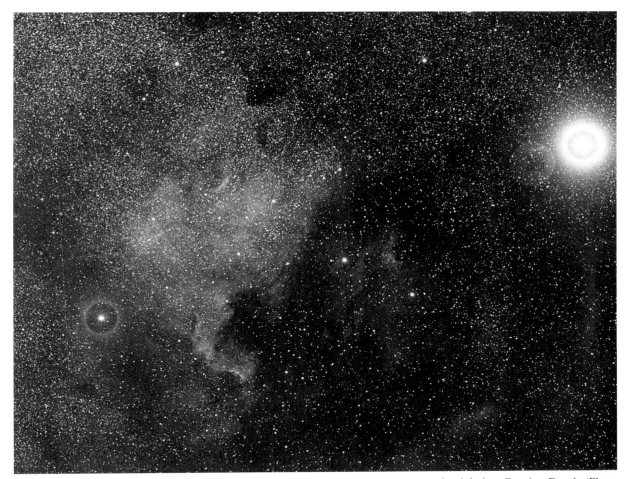

Fig. 1.1. The North America nebula in the constellation of Cygnus. The brightest star on the right is α Cygni or Deneb. (Photo M. Poutanen and H. Virtanen)

oning became more and more accurate, and people learned to calculate the movements of celestial bodies in advance.

During the rapid development of seafaring, when voyages extended farther and farther from home ports, position determination presented a problem for which astronomy offered a practical solution. Solving these problems of navigation were the most important tasks of astronomy in the 17th and 18th centuries, when the first precise tables on the movements of the planets and on other celestial phenomena were published. The basis for these developments was the discovery of the laws governing the motions of the planets by *Copernicus*, *Tycho Brahe*, *Kepler*, *Galilei* and *Newton*.

Astronomical research has changed man's view of the world from geocentric, anthropocentric conceptions to the modern view of a vast universe where man and the Earth play an insignificant role. Astronomy has taught us the real scale of the nature surrounding us.

Modern astronomy is fundamental science, motivated mainly by man's curiosity, his wish to know more about Nature and the Universe. Astronomy has a central role in forming a scientific view of the world. "A scientific view of the world" means a model of the universe based on observations, thoroughly tested theories and logical reasoning. Observations are always the ultimate test of a model: if the model does not fit the observations, it has to be changed, and this process must not be limited by any philosophical, political or religious conceptions or beliefs.

1.2 Astronomical Objects of Research

Modern astronomy explores the whole Universe and its different forms of matter and energy. Astronomers study the contents of the Universe from the level of elementary particles and molecules (with masses of 10^{-30} kg) to the largest superclusters of galaxies (with masses of 10^{50} kg).

Astronomy can be divided into different branches in several ways. The division can be made according to either the methods or the objects of research.

The Earth (Fig. 1.3) is of interest to astronomy for many reasons. Nearly all observations must be made through the atmosphere, and the phenomena of the upper atmosphere and magnetosphere reflect the state of interplanetary space. The Earth is also the most important object of comparison for planetologists.

The Moon is still studied by astronomical methods, although spacecraft and astronauts have visited its surface and brought samples back to the Earth. To amateur astronomers, the Moon is an interesting and easy object for observations.

In the study of the planets of the solar system, the situation in the 1980's was the same as in lunar

Fig. 1.2. Although space probes and satellites have gathered remarkable new information, a great majority of astronomical observations is still Earth-based. The most important observatories are usually located at high altitudes far from densely populated areas. One such observatory area, containing several European telescopes, is on La Palma, which belongs to the Canary Islands

Fig. 1.3. The Earth as seen from the Moon. The picture was taken on the last Apollo flight in December, 1972. (Photo NASA)

exploration 20 years earlier: the surfaces of the planets and their moons have been mapped by fly-bys of spacecraft or by orbiters, and spacecraft have soft-landed on Mars and Venus. This kind of exploration has tremendously added to our knowledge of the conditions on the planets. Continuous monitoring of the planets, however, can still only be made from the Earth, and many bodies in the solar system still await their spacecraft.

The Solar System is governed by the Sun, which produces energy in its centre by nuclear fusion. The Sun is our nearest star, and its study lends insight into conditions on other stars.

Some thousands of stars can be seen with the naked eye, but even a small telescope reveals millions of them. Stars can be classified according to their observed characteristics. A majority are like the Sun; we call them *main sequence stars*. However, some stars are much larger, *giants* or *supergiants*, and some are much smaller, *white dwarfs*. Different types of stars represent different stages of stellar evolution. Most stars are components of *binary* or *multiple systems*, many are *variable*: their brightness is not constant.

Among the newest objects studied by astronomers are the *compact stars*: *neutron stars* and *black holes*. In them, matter has been so greatly compressed and the gravitational field is so strong that Einstein's general theory of relativity must be used to describe matter and space.

Stars are points of light in an otherwise seemingly empty space. Yet interstellar space is not empty, but contains large clouds of *atoms*, *molecules*, *elementary particles* and *dust*. New matter is injected into interstellar space by erupting and exploding stars; at other places, new stars are formed from contracting interstellar clouds.

Stars are not evenly distributed in space, but form concentrations, *clusters of stars*. These consist of stars born near each other, and in some cases, remaining together for billions of years.

The largest concentration of stars in the sky is the *Milky Way*. It is a massive stellar system, a *galaxy*, consisting of over 200 billion stars. All the stars visible to the naked eye belong to the Milky Way. Light travels across our galaxy in 100,000 years.

The Milky Way is not the only galaxy, but one of almost innumerable others. Galaxies often form *clusters of galaxies*, and these clusters can be clumped together into *superclusters*. Galaxies are seen at all distances as far away as our observations reach. Still further out we see *quasars* – the light of the most distant quasars we see now was emitted when the Universe was one-tenth of its present age.

The largest object studied by astronomers is the whole Universe. *Cosmology*, once the domain of theologicians and philosophers, has become the subject of physical theories and concrete astronomical observations.

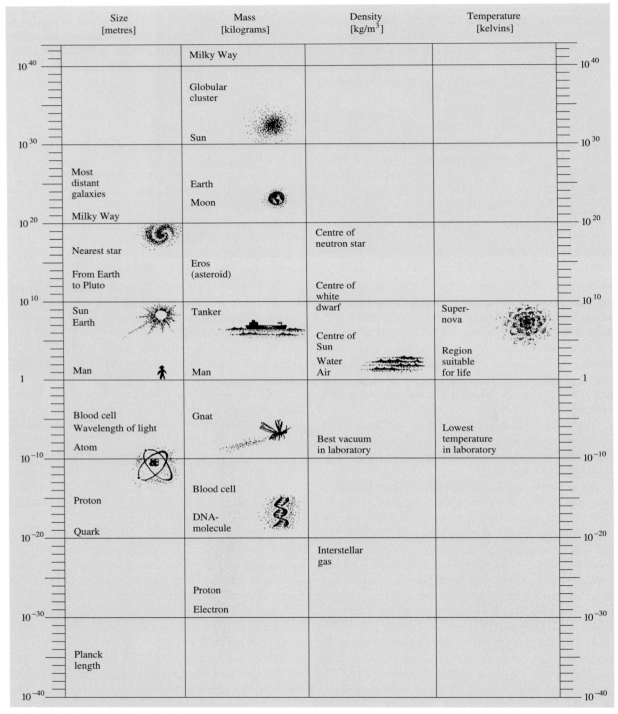

Fig. 1.4. The dimensions of the Universe

Table 1.1. The share of different branches of astronomy in *Astronomy and Astrophysics Abstracts* for the first half of 1998 (and for comparison 1980). This index service contains short abstracts of all astronomical articles published during the half-year covered by the book

Branch	Pages	%	(% in 1980)
Astronomical instruments and techniques	90	6	(6)
Positional astronomy, celestial mechanics	33	2	(4)
Space research	17	1	(2)
Theoretical astrophysics	180	13	(10)
Sun	117	8	(8)
Earth	51	4	(5)
Planetary system	133	9	(16)
Stars	253	17	(19)
Interstellar matter, nebulae	88	6	(7)
Radio sources, X-ray sources, cosmic rays	64	5	(9)
Stellar systems, galaxy, extragalactic objects, cosmology	424	29	(14)

Among the different branches of research, *spherical*, or positional, *astronomy* studies the coordinate systems on the celestial sphere, their changes and the apparent places of celestial bodies in the sky. *Celestial mechanics* studies the movements of bodies in the solar system, in stellar systems and among the galaxies and clusters of galaxies. *Astrophysics* is concerned with the physical properties of celestial objects; it employs methods of modern physics. It thus has a central position in almost all branches of astronomy (Table 1.1).

Astronomy can be divided into different areas according to the wavelength used in observations. We can speak of radio, infrared, optical, ultraviolet, X-ray or gamma astronomy, depending on which wavelengths of the electromagnetic spectrum are used. In the future, neutrinos and gravitational waves may also be observed.

1.3 The Scale of the Universe

The masses and sizes of astronomical objects are usually enormously large. But to understand their properties, the smallest parts of matter, molecules, atoms and elementary particles, must be studied. The densities, temperatures and magnetic fields in the Universe vary within much larger limits than can be reached in laboratories on the Earth.

Fig. 1.5. The globular cluster M13. There are over a million stars in the cluster. (Photo Palomar Observatory)

Fig. 1.6. The Large Magellanic Cloud, our nearest neighbour galaxy. (Photo National Optical Astronomy Observatories, Cerro Tololo Inter-American Observatory)

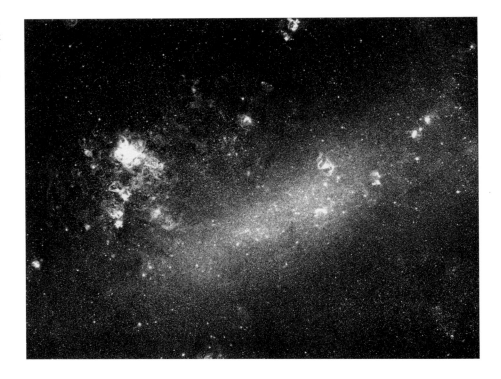

The greatest natural density met on the Earth is $22{,}500 \, \mathrm{kg \, m^{-3}}$ (osmium), while in neutron stars densities of the order of $10^{18} \, \mathrm{kg \, m^{-3}}$ are possible. The density in the best vacuum achieved on the Earth is only $10^{-9} \, \mathrm{kg \, m^{-3}}$, but in interstellar space the density of the gas may be $10^{-21} \, \mathrm{kg \, m^{-3}}$ or even less. Modern accelerators can give particles energies of the order of 10^{12} electron volts (eV). Cosmic rays coming from the sky may have energies of over 10^{20} eV.

It has taken man a long time to grasp the vast dimensions of space. Already *Hipparchos* in the second century B.C. obtained a reasonably correct value for the distance of the Moon. The scale of the solar system was established together with the heliocentric system in the 17th century. The first measurements of stellar distances were made in the 1830's, and the distances to the galaxies were determined only in the 1920's.

We can get some kind of picture of the distances involved (Fig. 1.4) by considering the time required for light to travel from a source to the retina of the human eye. It takes 8 minutes for light to travel from the Sun, $5\frac{1}{2}$ hours from Pluto and 4 years from the nearest star. We cannot see the centre of the Milky Way, but the many globular clusters around the Milky Way are at approximately similar distances. It takes about 20,000 years for the light from the globular cluster of Fig. 1.5 to reach the Earth. It takes 150,000 years to travel the distance from the nearest galaxy, the Magellanic Cloud seen on the southern sky (Fig. 1.6). The photons that we see now started their voyage when Neanderthal Man lived on the Earth. The light coming from the *Andromeda Galaxy* in the northern sky originated 2 million years ago. Around the same time the first actual human using tools, *Homo habilis*, appeared. The most distant objects known, the quasars, are so far away that their radiation, seen on the Earth now, was emitted long before the Sun or the Earth were born.

2. Spherical Astronomy

Spherical astronomy is a science studying astronomical coordinate frames, directions and apparent motions of celestial objects, determination of position from astronomical observations, observational errors, etc. We shall concentrate mainly on astronomical coordinates, apparent motions of stars and time reckoning. Also, some of the most important star catalogues will be introduced.

For simplicity we will assume that the observer is always on the northern hemisphere. Although all definitions and equations are easily generalized for both hemispheres, this might be unnecessarily confusing. In spherical astronomy all angles are usually expressed in degrees; we will also use degrees unless otherwise mentioned.

2.1 Spherical Trigonometry

For the coordinate transformations of spherical astronomy, we need some mathematical tools, which we present now.

If a plane passes through the centre of a sphere, it will split the sphere into two identical hemispheres along a circle called a *great circle* (Fig. 2.1). A line perpendicular to the plane and passing through the centre of the sphere intersects the sphere at the *poles P and P'*. If a sphere is intersected by a plane not containing the centre, the intersection curve is a *small circle*. There is exactly one great circle passing through two given points Q and Q' on a sphere (unless these points are an-

tipodal, in which case all circles passing through both of them are great circles). The arc QQ' of this great circle is the shortest path on the surface of the sphere between these points.

A *spherical triangle* is not just any three-cornered figure lying on a sphere; its sides must be arcs of great circles. The spherical triangle ABC in Fig. 2.2 has the arcs AB, BC and AC as its sides. If the radius of sphere is r, the length of the arc AB is

$$|AB| = rc , \quad [c] = \text{rad} ,$$

where c is the angle subtended by the arc AB as seen from the centre. This angle is called the *central angle* of the side AB. Because lengths of sides and central

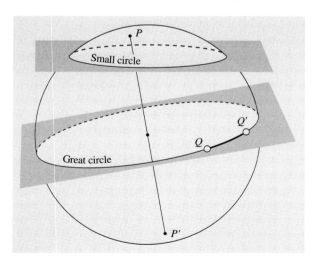

Fig. 2.1. A great circle is the intersection of a sphere and a plane passing through its centre. P and P' are the poles of the great circle. The shortest path from Q to Q' follows the great circle

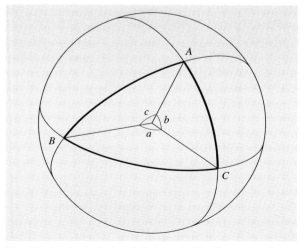

Fig. 2.2. A spherical triangle is bounded by three arcs of great circles, AB, BC and CA. The corresponding central angles are c, a, and b

angles correspond to each other in a unique way, it is customary to give the central angles instead of the sides. In this way, the radius of the sphere does not enter into the equations of spherical trigonometry. An angle of a spherical triangle can be defined as the angle between the tangents of the two sides meeting at a vertex, or as the dihedral angle between the planes intersecting the sphere along these two sides. We denote the angles of a spherical triangle by capital letters (A, B, C) and the opposing sides, or, more correctly, the corresponding central angles, by lowercase letters (a, b, c).

The sum of the angles of a spherical triangle is always greater than 180 degrees; the excess

$$E = A + B + C - 180° \qquad (2.1)$$

is called the *spherical excess*. It is not a constant, but depends on the triangle. Unlike in plane geometry, it is not enough to know two of the angles to determine the third one. The area of a spherical triangle is related to the spherical excess in a very simple way:

$$\text{Area} = Er^2, \quad [E] = \text{rad} . \qquad (2.2)$$

This shows that the spherical excess equals the solid angle in steradians (see Appendix A.1), subtended by the triangle as seen from the centre.

To prove (2.2), we extend all sides of the triangle Δ to great circles (Fig. 2.3). These great circles will form another triangle Δ', congruent with Δ but antipodal to it. If the angle A is expressed in radians, the area of the slice $S(A)$ bounded by the two sides of A (the shaded area in Fig. 2.3) is obviously $2A/2\pi = A/\pi$ times the area of the sphere, $4\pi r^2$. Similarly, the slices $S(B)$ and $S(C)$ cover fractions B/π and C/π of the whole sphere.

Together, the three slices cover the whole surface of the sphere, the equal triangles Δ and Δ' belonging to every slice, and each point outside the triangles, to exactly one slice. Thus the area of the slices $S(A)$, $S(B)$ and $S(C)$ equals the area of the sphere plus four times the area of Δ, $\mathcal{A}(\Delta)$:

$$\frac{A + B + C}{\pi} 4\pi r^2 = 4\pi r^2 + 4\mathcal{A}(\Delta) ,$$

whence

$$\mathcal{A}(\Delta) = (A + B + C - \pi)r^2 = Er^2 .$$

As in the case of plane triangles, we can derive relationships between the sides and angles of spherical triangles. The easiest way to do this is by inspecting certain coordinate transformations.

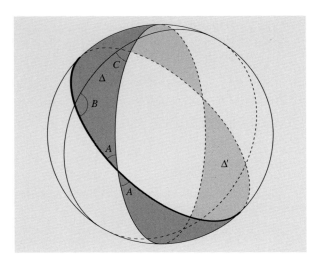

Fig. 2.3. If the sides of a spherical triangle are extended all the way around the sphere, they form another triangle Δ', antipodal and equal to the original triangle Δ. The shaded area is the slice $S(A)$

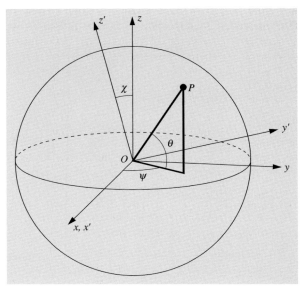

Fig. 2.4. The location of a point P on the surface of a unit sphere can be expressed by rectangular xyz coordinates or by two angles, ψ and θ. The $x'y'z'$ frame is obtained by rotating the xyz frame around its x axis by an angle χ

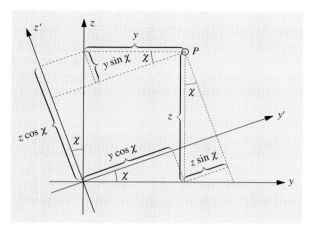

Fig. 2.5. The coordinates of the point P in the rotated frame are $x' = x$, $y' = y \cos \chi + z \sin \chi$, $z' = z \cos \chi - y \sin \chi$

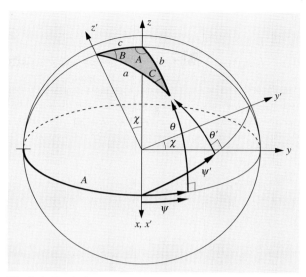

Fig. 2.6. To derive triangulation formulas for the spherical triangle ABC, the spherical coordinates ψ, θ, ψ' and θ' of the vertex C are expressed in terms of the sides and angles of the triangle

Suppose we have two rectangular coordinate frames $Oxyz$ and $Ox'y'z'$ (Fig. 2.4), such that the $x'y'z'$ frame is obtained from the xyz frame by rotating it around the x axis by an angle χ.

The position of a point P on a unit sphere is uniquely determined by giving two angles. The angle ψ is measured counterclockwise from the positive x axis along the xy plane; the other angle θ tells the angular distance from the xy plane. In an analogous way, we can define the angles ψ' and θ', which give the position of the point P in the $x'y'z'$ frame. The rectangular coordinates of the point P as functions of these angles are:

$$
\begin{aligned}
x &= \cos \psi \cos \theta \,, & x' &= \cos \psi' \cos \theta' \,, \\
y &= \sin \psi \cos \theta \,, & y' &= \sin \psi' \cos \theta', & (2.3) \\
z &= \sin \theta, & z' &= \sin \theta'.
\end{aligned}
$$

We also know that the dashed coordinates are obtained from the undashed ones by a rotation in the yz plane (Fig. 2.5):

$$
\begin{aligned}
x' &= x \,, \\
y' &= y \cos \chi + z \sin \chi \,, & (2.4) \\
z' &= -y \sin \chi + z \cos \chi \,.
\end{aligned}
$$

By substituting the expressions of the rectangular coordinates (2.3) into (2.4), we have

$$
\begin{aligned}
\cos \psi' \cos \theta' &= \cos \psi \cos \theta \,, \\
\sin \psi' \cos \theta' &= \sin \psi \cos \theta \cos \chi + \sin \theta \sin \chi \,, & (2.5) \\
\sin \theta' &= -\sin \psi \cos \theta \sin \chi + \sin \theta \cos \chi \,.
\end{aligned}
$$

In fact, these equations are quite sufficient for all coordinate transformations we may encounter. However, we shall also derive the usual equations for spherical triangles. To do this, we set up the coordinate frames in a suitable way (Fig. 2.6). The z axis points towards the vertex A and the z' axis, towards B. Now the vertex C corresponds to the point P in Fig. 2.4. The angles ψ, θ, ψ', θ' and χ can be expressed in terms of the angles and sides of the spherical triangle:

$$
\begin{aligned}
\psi &= A - 90° \,, & \theta &= 90° - b \,, \\
\psi' &= 90° - B \,, & \theta' &= 90° - a \,, & \chi = c \,. & (2.6)
\end{aligned}
$$

Substitution into (2.5) gives

$$
\begin{aligned}
&\cos(90° - B) \cos(90° - a) \\
&\quad = \cos(A - 90°) \cos(90° - b) \,,
\end{aligned}
$$

$$
\begin{aligned}
&\sin(90° - B) \cos(90° - a) \\
&\quad = \sin(A - 90°) \cos(90° - b) \cos c \\
&\qquad + \sin(90° - b) \sin c \,,
\end{aligned}
$$

$$
\begin{aligned}
&\sin(90° - a) \\
&\quad = -\sin(A - 90°) \cos(90° - b) \sin c \\
&\qquad + \sin(90° - b) \cos c \,,
\end{aligned}
$$

or

$$\sin B \sin a = \sin A \sin b ,$$
$$\cos B \sin a = -\cos A \sin b \cos c + \cos b \sin c , \quad (2.7)$$
$$\cos a = \cos A \sin b \sin c + \cos b \cos c .$$

Equations for other sides and angles are obtained by cyclic permutations of the sides a, b, c and the angles A, B, C. For instance, the first equation also yields

$$\sin C \sin b = \sin B \sin c ,$$
$$\sin A \sin c = \sin C \sin a .$$

All these variations of the *sine formula* can be written in an easily remembered form:

$$\frac{\sin a}{\sin A} = \frac{\sin b}{\sin B} = \frac{\sin c}{\sin C} . \quad (2.8)$$

If we take the limit, letting the sides a, b and c shrink to zero, the spherical triangle becomes a plane triangle. If all angles are expressed in radians, we have approximately

$$\sin a \approx a , \quad \cos a \approx 1 - \frac{1}{2}a^2 .$$

Substituting these approximations into the sine formula, we get the familiar sine formula of plane geometry:

$$\frac{a}{\sin A} = \frac{b}{\sin B} = \frac{c}{\sin C} .$$

The second equation in (2.7) is the *sine-cosine formula*, and the corresponding plane formula is a trivial one:

$$c = b \cos A + a \cos B .$$

This is obtained by substituting the approximations of sine and cosine into the sine-cosine formula and ignoring all quadratic and higher-order terms. In the same way we can use the third equation in (2.7), the *cosine formula*, to derive the planar cosine formula:

$$a^2 = b^2 + c^2 - 2bc \cos A .$$

2.2 The Earth

A position on the Earth is usually given by two spherical coordinates (although in some calculations rectangular or other coordinates may be more convenient). If neces-

sary, also a third coordinate, e. g. the distance from the centre, can be used.

The reference plane is the *equatorial plane*, perpendicular to the rotation axis and intersecting the surface of the Earth along the *equator*. Small circles parallel to the equator are called *parallels of latitude*. Semicircles from pole to pole are *meridians*. The geographical *longitude* is the angle between the meridian and the zero meridian passing through Greenwich Observatory. We shall use positive values for longitudes east of Greenwich and negative values west of Greenwich. Sign convention, however, varies, and negative longitudes are not used in maps; so it is usually better to say explicitly whether the longitude is east or west of Greenwich.

The *latitude* is usually supposed to mean the *geographical latitude*, which is the angle between the plumb line and the equatorial plane. The latitude is positive in the northern hemisphere and negative in the southern one. The geographical latitude can be determined by astronomical observations (Fig. 2.7): the altitude of the celestial pole measured from the hori-

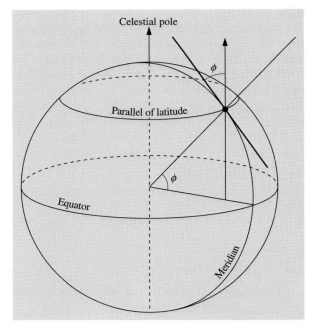

Fig. 2.7. The latitude ϕ is obtained by measuring the altitude of the celestial pole. The celestial pole can be imagined as a point at an infinite distance in the direction of the Earth's rotation axis

zon equals the geographical latitude. (The celestial pole is the intersection of the rotation axis of the Earth and the infinitely distant celestial sphere; we shall return to these concepts a little later.)

Because the Earth is rotating, it is slightly flattened. The exact shape is rather complicated, but for most purposes it can by approximated by an oblate spheroid, the short axis of which coincides with the rotation axis (Sect. 7.5). In 1979 the International Union of Geodesy and Geophysics (IUGG) adopted the Geodetic Reference System 1980 (GRS-80), which is used when global reference frames fixed to the Earth are defined. The GRS-80 reference ellipsoid has the following dimensions:

equatorial radius	$a = 6,378,137$ m,
polar radius	$b = 6,356,752$ m,
flattening	$f = (a-b)/a$
	$= 1/298.25722210.$

The shape defined by the surface of the oceans, called the *geoid*, differs from this spheroid at most by about 100 m.

The angle between the equator and the normal to the ellipsoid approximating the true Earth is called the *geodetic latitude*. Because the surface of a liquid (like an ocean) is perpendicular to the plumb line, the geodetic and geographical latitudes are practically the same.

Because of the flattening, the plumb line does not point to the centre of the Earth except at the poles and on the equator. An angle corresponding to the ordinary spherical coordinate (the angle between the equator and the line from the centre to a point on the surface), the *geocentric latitude* ϕ' is therefore a little smaller than the geographic latitude ϕ (Fig. 2.8).

We now derive an equation between the geographic latitude ϕ and geocentric latitude ϕ', assuming the Earth is an oblate spheroid and the geographic and geodesic latitudes are equal. The equation of the meridional ellipse is

$$\frac{x^2}{a^2} + \frac{y^2}{b^2} = 1 \ .$$

The direction of the normal to the ellipse at a point (x, y) is given by

$$\tan \phi = -\frac{dx}{dy} = \frac{a^2}{b^2} \frac{y}{x} \ .$$

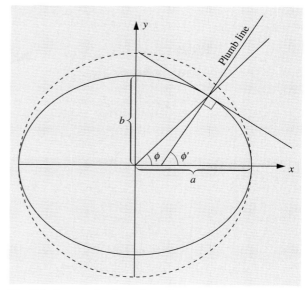

Fig. 2.8. Due to the flattening of the Earth, the geographic latitude ϕ and geocentric latitude ϕ' are different

The geocentric latitude is obtained from

$$\tan \phi' = y/x \ .$$

Hence

$$\tan \phi' = \frac{b^2}{a^2} \tan \phi = (1 - e^2) \tan \phi \ , \qquad (2.9)$$

where

$$e = \sqrt{1 - b^2/a^2}$$

is the eccentricity of the ellipse. The difference $\Delta\phi = \phi - \phi'$ has a maximum $11.5'$ at the latitude $45°$.

Since the coordinates of celestial bodies in astronomical almanacs are given with respect to the centre of the Earth, the coordinates of nearby objects must be corrected for the difference in the position of the observer, if high accuracy is required. This means that one has to calculate the *topocentric* coordinates, centered at the observer. The easiest way to do this is to use rectangular coordinates of the object and the observer (Example 2.5).

One arc minute along a meridian is called a *nautical mile*. Since the radius of curvature varies with latitude, the length of the nautical mile so defined would depend on the latitude. Therefore one nautical mile has been defined to be equal to one minute of arc at $\phi = 45°$, whence 1 nautical mile $= 1852$ m.

2.3 The Celestial Sphere

The ancient universe was confined within a finite spherical shell. The stars were fixed to this shell and thus were all equidistant from the Earth, which was at the centre of the spherical universe. This simple model is still in many ways as useful as it was in antiquity: it helps us to easily understand the diurnal and annual motions of stars, and, more important, to predict these motions in a relatively simple way. Therefore we will assume for the time being that all the stars are located on the surface of an enormous sphere and that we are at its centre. Because the radius of this celestial sphere is practically infinite, we can neglect the effects due to the changing position of the observer, caused by the rotation and orbital motion of the Earth. These effects will be considered later in Sects. 2.9 and 2.10.

Since the distances of the stars are ignored, we need only two coordinates to specify their directions. Each coordinate frame has some fixed reference plane passing through the centre of the celestial sphere and dividing the sphere into two hemispheres along a great circle. One of the coordinates indicates the angular distance from this reference plane. There is exactly one great circle going through the object and intersecting this plane perpendicularly; the second coordinate gives the angle between that point of intersection and some fixed direction.

2.4 The Horizontal System

The most natural coordinate frame from the observer's point of view is the *horizontal frame* (Fig. 2.9). Its reference plane is the tangent plane of the Earth passing through the observer; this horizontal plane intersects the celestial sphere along the *horizon*. The point just above the observer is called the *zenith* and the antipodal point below the observer is the *nadir*. (These two points are the poles corresponding to the horizon.) Great circles through the zenith are called *verticals*. All verticals intersect the horizon perpendicularly.

By observing the motion of a star over the course of a night, an observer finds out that it follows a track like one of those in Fig. 2.9. Stars rise in the east, reach their highest point, or *culminate*, on the vertical NZS, and set in the west. The vertical NZS is called the *meridian*. North and south directions are defined as the intersections of the meridian and the horizon.

One of the horizontal coordinates is the *altitude* or *elevation*, a, which is measured from the horizon along the vertical passing through the object. The altitude lies in the range $[-90°, +90°]$; it is positive for objects above the horizon and negative for the objects below the horizon. The *zenith distance*, or the angle between

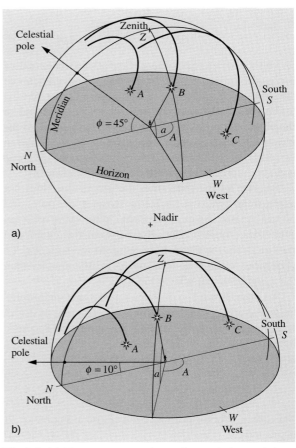

Fig. 2.9. (a) The apparent motions of stars during a night as seen from latitude $\phi = 45°$. **(b)** The same stars seen from latitude $\phi = 10°$

the object and the zenith, is obviously

$$z = 90° - a .\qquad(2.10)$$

The second coordinate is the *azimuth*, A; it is the angular distance of the vertical of the object from some fixed direction. Unfortunately, in different contexts, different fixed directions are used; thus it is always advisable to check which definition is employed. The azimuth is usually measured from the north or south, and though clockwise is the preferred direction, counterclockwise measurements are also occasionally made. In this book we have adopted a fairly common astronomical convention, measuring the azimuth *clockwise* from the *south*. Its values are usually normalized between 0° and 360°.

In Fig. 2.9a we can see the altitude and azimuth of a star B at some instant. As the star moves along its daily track, both of its coordinates will change. Another difficulty with this coordinate frame is its local character. In Fig. 2.9b we have the same stars, but the observer is now further south. We can see that the coordinates of the same star at the same moment are different for different observers. Since the horizontal coordinates are time and position dependent, they cannot be used, for instance, in star catalogues.

2.5 The Equatorial System

The direction of the rotation axis of the Earth remains almost constant and so does the equatorial plane perpendicular to this axis. Therefore the equatorial plane is a suitable reference plane for a coordinate frame that has to be independent of time and the position of the observer.

The intersection of the celestial sphere and the equatorial plane is a great circle, which is called the *equator of the celestial sphere*. The north pole of the celestial sphere is one of the poles corresponding to this great circle. It is also the point in the northern sky where the extension of the Earth's rotational axis meets the celestial sphere. The celestial north pole is at a distance of about one degree (which is equivalent to two full moons) from the moderately bright star Polaris. The meridian always passes through the north pole; it is divided by the pole into north and south meridians.

Fig. 2.10. At night, stars seem to revolve around the celestial pole. The altitude of the pole from the horizon equals the latitude of the observer. (Photo Pekka Parviainen)

The angular separation of a star from the equatorial plane is not affected by the rotation of the Earth. This angle is called the *declination* δ.

Stars seem to revolve around the pole once every day (Fig. 2.10). To define the second coordinate, we must again agree on a fixed direction, unaffected by the Earth's rotation. From a mathematical point of view, it does not matter which point on the equator is selected. However, for later purposes, it is more appropriate to employ a certain point with some valuable properties, which will be explained in the next section. This point is called the *vernal equinox*. Because it used to be in the constellation Aries (the Ram), it is also called the first point of Aries ant denoted by the sign of Aries, ♈. Now we can define the second coordinate as the angle from the vernal equinox measured along the equator. This angle is the *right ascension* α (or R.A.) of the object, measured counterclockwise from Υ.

Since declination and right ascension are independent of the position of the observer and the motions of the Earth, they can be used in star maps and catalogues. As will be explained later, in many telescopes one of the axes (the hour axis) is parallel to the rotation axis of the Earth. The other axis (declination axis) is perpendicular to the hour axis. Declinations can be read immediately on the declination dial of the telescope. But the zero point of the right ascension seems to move in the sky, due to the diurnal rotation of the Earth. So we cannot use the right ascension to find an object unless we know the direction of the vernal equinox.

Since the south meridian is a well-defined line in the sky, we use it to establish a local coordinate corresponding to the right ascension. The *hour angle* is measured clockwise from the meridian. The hour angle of an object is not a constant, but grows at a steady rate, due to the Earth's rotation. The hour angle of the vernal equinox is called the *sidereal time* Θ. Figure 2.11 shows that for any object,

$$\Theta = h + \alpha \,, \tag{2.11}$$

where h is the object's hour angle and α its right ascension.

Since hour angle and sidereal time change with time at a constant rate, it is practical to express them in units of time. Also the closely related right ascension is customarily given in time units. Thus 24 hours equals 360 degrees, 1 hour = 15 degrees, 1 minute of time = 15 minutes of arc, and so on. All these quantities are in the range [0 h, 24 h].

In practice, the sidereal time can be readily determined by pointing the telescope to an easily recognisable star and reading its hour angle on the hour angle dial of the telescope. The right ascension found in a catalogue is then added to the hour angle, giving the sidereal time at the moment of observation. For any other time, the sidereal time can be evaluated by adding the time elapsed since the observation. If we want to be accurate, we have to use a sidereal clock to measure time intervals. A sidereal clock runs 3 min 56.56 s fast a day as compared with an ordinary solar time clock:

$$\begin{aligned} &24\,\text{h solar time} \\ &\quad = 24\,\text{h } 3\,\text{min } 56.56\,\text{s sidereal time}\,. \end{aligned} \tag{2.12}$$

The reason for this is the orbital motion of the Earth: stars seem to move faster than the Sun across the sky; hence, a sidereal clock must run faster. (This is further discussed in Sect. 2.13.)

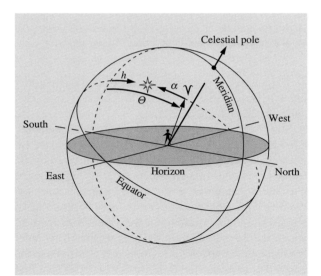

Fig. 2.11. The sidereal time Θ (the hour angle of the vernal equinox) equals the hour angle plus right ascension of any object

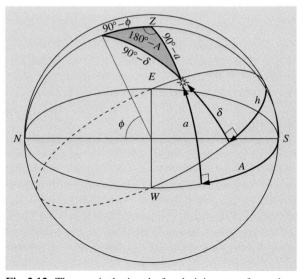

Fig. 2.12. The nautical triangle for deriving transformations between the horizontal and equatorial frames

Transformations between the horizontal and equatorial frames are easily obtained from spherical trigonometry. Comparing Figs. 2.6 and 2.12, we find that we must make the following substitutions into (2.5):

$$\psi = 90° - A , \quad \theta = a ,$$
$$\psi' = 90° - h , \quad \theta' = \delta , \quad \chi = 90° - \phi . \quad (2.13)$$

The angle ϕ in the last equation is the altitude of the celestial pole, or the latitude of the observer. Making the substitutions, we get

$$\sin h \cos \delta = \sin A \cos a ,$$
$$\cos h \cos \delta = \cos A \cos a \sin \phi + \sin a \cos \phi , \quad (2.14)$$
$$\sin \delta = - \cos A \cos a \cos \phi + \sin a \sin \phi .$$

The inverse transformation is obtained by substituting

$$\psi = 90° - h , \quad \theta = \delta ,$$
$$\psi' = 90° - A , \quad \theta' = a , \quad \chi = -(90° - \phi) , \quad (2.15)$$

whence

$$\sin A \cos a = \sin h \cos \delta ,$$
$$\cos A \cos a = \cos h \cos \delta \sin \phi - \sin \delta \cos \phi , \quad (2.16)$$
$$\sin a = \cos h \cos \delta \cos \phi + \sin \delta \sin \phi .$$

Since the altitude and declination are in the range $[-90°, +90°]$, it suffices to know the sine of one of these angles to determine the other angle unambiguously. Azimuth and right ascension, however, can have any value from $0°$ to $360°$ (or from 0 h to 24 h), and to solve for them, we have to know both the sine and cosine to choose the correct quadrant.

The altitude of an object is greatest when it is on the south meridian (the great circle arc between the celestial poles containing the zenith). At that moment (called *upper culmination*, or *transit*) its hour angle is 0 h. At the *lower culmination* the hour angle is $h = 12$ h. When $h = 0$ h, we get from the last equation in (2.16)

$$\sin a = \cos \delta \cos \phi + \sin \delta \sin \phi$$
$$= \cos(\phi - \delta) = \sin(90° - \phi + \delta) .$$

Thus the altitude at the upper culmination is

$$a_{max} = \begin{cases} 90° - \phi + \delta , & \text{if the object culminates} \\ & \text{south of zenith} , \\ 90° + \phi - \delta , & \text{if the object culminates} \\ & \text{north of zenith} . \end{cases} \quad (2.17)$$

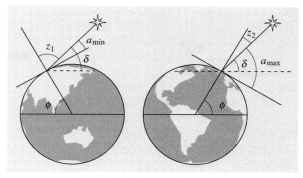

Fig. 2.13. The altitude of a circumpolar star at upper and lower culmination

The altitude is positive for objects with $\delta > \phi - 90°$. Objects with declinations less than $\phi - 90°$ can never be seen at the latitude ϕ. On the other hand, when $h = 12$ h we have

$$\sin a = - \cos \delta \cos \phi + \sin \delta \sin \phi$$
$$= - \cos(\delta + \phi) = \sin(\delta + \phi - 90°) ,$$

and the altitude at the lower culmination is

$$a_{min} = \delta + \phi - 90° . \quad (2.18)$$

Stars with $\delta > 90° - \phi$ will never set. For example, in Helsinki ($\phi \approx 60°$), all stars with a declination higher than $30°$ are such *circumpolar* stars. And stars with a declination less than $-30°$ can never be observed there.

We shall now study briefly how the (α, δ) frame can be established by observations. Suppose we observe a circumpolar star at its upper and lower culmination (Fig. 2.13). At the upper transit, its altitude is $a_{max} = 90° - \phi + \delta$ and at the lower transit, $a_{min} = \delta + \phi - 90°$. Eliminating the latitude, we get

$$\delta = \frac{1}{2}(a_{min} + a_{max}) . \quad (2.19)$$

Thus we get the same value for the declination, independent of the observer's location. Therefore we can use it as one of the absolute coordinates. From the same observations, we can also determine the direction of the celestial pole as well as the latitude of the observer. After these preparations, we can find the declination of any object by measuring its distance from the pole.

The equator can be now defined as the great circle all of whose points are at a distance of $90°$ from the pole. The zero point of the second coordinate (right ascension) can then be defined as the point where the Sun seems to cross the equator from south to north.

In practice the situation is more complicated, since the direction of Earth's rotation axis changes due to perturbations. Therefore the equatorial coordinate frame is nowadays defined using certain standard objects the positions of which are known very accurately. The best accuracy is achieved by using the most distant objects, *quasars* (Sect. 18.7), which remain in the same direction over very long intervals of time.

2.6 Rising and Setting Times

From the last equation (2.16), we find the hour angle h of an object at the moment its altitude is a:

$$\cos h = -\tan \delta \tan \phi + \frac{\sin a}{\cos \delta \cos \phi} . \qquad (2.20)$$

This equation can be used for computing rising and setting times. Then $a = 0$ and the hour angles corresponding to rising and setting times are obtained from

$$\cos h = -\tan \delta \tan \phi . \qquad (2.21)$$

If the right ascension α is known, we can use (2.11) to compute the sidereal time Θ. (Later, in Sect. 2.14, we shall study how to transform the sidereal time to ordinary time.)

If higher accuracy is needed, we have to correct for the refraction of light caused by the atmosphere of the Earth (see Sect. 2.9). In that case, we must use a small negative value for a in (2.20). This value, the *horizontal refraction*, is about $-34'$.

The rising and setting times of the Sun given in almanacs refer to the time when the upper edge of the Solar disk just touches the horizon. To compute these times, we must set $a = -50'$ ($= -34' - 16'$).

Also for the Moon almanacs give rising and setting times of the upper edge of the disk. Since the distance of the Moon varies considerably, we cannot use any constant value for the radius of the Moon, but it has to be calculated separately each time. The Moon is also so close that its direction with respect to the background stars varies due to the rotation of the Earth. Thus the rising and setting times of the Moon are defined as the

instants when the altitude of the Moon is $-34' - s + \pi$, where s is the apparent radius (15.5' on the average) and π the parallax (57' on the average). The latter quantity is explained in Sect. 2.9.

Finding the rising and setting times of the Sun, planets and especially the Moon is complicated by their motion with respect to the stars. We can use, for example, the coordinates for the noon to calculate estimates for the rising and setting times, which can then be used to interpolate more accurate coordinates for the rising and setting times. When these coordinates are used to compute new times a pretty good accuracy can be obtained. The iteration can be repeated if even higher precision is required.

2.7 The Ecliptic System

The orbital plane of the Earth, the *ecliptic*, is the reference plane of another important coordinate frame. The ecliptic can also be defined as the great circle on the celestial sphere described by the Sun in the course of one year. This frame is used mainly for planets and other bodies of the solar system. The orientation of the Earth's equatorial plane remains invariant, unaffected by annual motion. In spring, the Sun appears to move from the southern hemisphere to the northern one (Fig. 2.14). The time of this remarkable event as well as the direction to the Sun at that moment are called the *vernal equinox*. At the vernal equinox, the Sun's right ascension and declination are zero. The equatorial and ecliptic

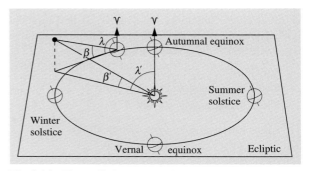

Fig. 2.14. The ecliptic geocentric (λ, β) and heliocentric (λ', β') coordinates are equal only if the object is very far away. The geocentric coordinates depend also on the Earth's position in its orbit

planes intersect along a straight line directed towards the vernal equinox. Thus we can use this direction as the zero point for both the equatorial and ecliptic coordinate frames. The point opposite the vernal equinox is the *autumnal equinox*, it is the point at which the Sun crosses the equator from north to south.

The *ecliptic latitude* β is the angular distance from the ecliptic; it is in the range $[-90°, +90°]$. The other coordinate is the *ecliptic longitude* λ, measured counterclockwise from the vernal equinox.

Transformation equations between the equatorial and ecliptic frames can be derived analogously to (2.14) and (2.16):

$$\sin\lambda\cos\beta = \sin\delta\sin\varepsilon + \cos\delta\cos\varepsilon\sin\alpha\;,$$
$$\cos\lambda\cos\beta = \cos\delta\cos\alpha\;, \qquad (2.22)$$
$$\sin\beta = \sin\delta\cos\varepsilon - \cos\delta\sin\varepsilon\sin\alpha\;,$$

$$\sin\alpha\cos\delta = -\sin\beta\sin\varepsilon + \cos\beta\cos\varepsilon\sin\lambda\;,$$
$$\cos\alpha\cos\delta = \cos\lambda\cos\beta\;, \qquad (2.23)$$
$$\sin\delta = \sin\beta\cos\varepsilon + \cos\beta\sin\varepsilon\sin\lambda\;.$$

The angle ε appearing in these equations is the *obliquity of the ecliptic*, or the angle between the equatorial and ecliptic planes. Its value is roughly $23°26'$ (a more accurate value is given in *Reduction of Coordinates, p. 36).

Depending on the problem to be solved, we may encounter *heliocentric* (origin at the Sun), *geocentric* (origin at the centre of the Earth) or *topocentric* (origin at the observer) coordinates. For very distant objects the differences are negligible, but not for bodies of the solar system. To transform heliocentric coordinates to geocentric coordinates or vice versa, we must also know the distance of the object. This transformation is most easily accomplished by computing the rectangular coordinates of the object and the new origin, then changing the origin and finally evaluating the new latitude and longitude from the rectangular coordinates (see Examples 2.4 and 2.5).

2.8 The Galactic Coordinates

For studies of the Milky Way Galaxy, the most natural reference plane is the plane of the Milky Way (Fig. 2.15). Since the Sun lies very close to that plane,

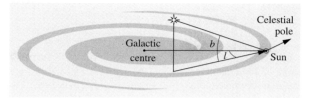

Fig. 2.15. The galactic coordinates l and b

we can put the origin at the Sun. The *galactic longitude l* is measured counterclockwise (like right ascension) from the direction of the centre of the Milky Way (in Sagittarius, $\alpha = 17$ h 45.7 min, $\delta = -29°00'$). The *galactic latitude b* is measured from the galactic plane, positive northwards and negative southwards. This definition was officially adopted only in 1959, when the direction of the galactic centre was determined from radio observations accurately enough. The old galactic coordinates l^{I} and b^{I} had the intersection of the equator and the galactic plane as their zero point.

The galactic coordinates can be obtained from the equatorial ones with the transformation equations

$$\sin(l_{\mathrm{N}} - l)\cos b = \cos\delta\sin(\alpha - \alpha_{\mathrm{P}})\;,$$
$$\cos(l_{\mathrm{N}} - l)\cos b = -\cos\delta\sin\delta_{\mathrm{P}}\cos(\alpha - \alpha_{\mathrm{P}})$$
$$+ \sin\delta\cos\delta_{\mathrm{P}}\;, \qquad (2.24)$$
$$\sin b = \cos\delta\cos\delta_{\mathrm{P}}\cos(\alpha - \alpha_{\mathrm{P}})$$
$$+ \sin\delta\sin\delta_{\mathrm{P}}\;,$$

where the direction of the Galactic north pole is $\alpha_{\mathrm{P}} = 12$ h 51.4 min, $\delta_{\mathrm{P}} = 27°08'$, and the galactic longitude of the celestial pole, $l_{\mathrm{N}} = 123.0°$.

2.9 Perturbations of Coordinates

Even if a star remains fixed with respect to the Sun, its coordinates can change, due to several disturbing effects. Naturally its altitude and azimuth change constantly because of the rotation of the Earth, but even its right ascension and declination are not quite free from perturbations.

Precession. Since most of the members of the solar system orbit close to the ecliptic, they tend to pull the equatorial bulge of the Earth towards it. Most of this "flattening" torque is caused by the Moon and the Sun.

Fig. 2.16. Due to precession the rotation axis of the Earth turns around the ecliptic north pole. Nutation is the small wobble disturbing the smooth precessional motion. In this figure the magnitude of the nutation is highly exaggerated

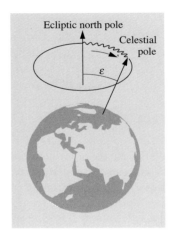

But the Earth is rotating and therefore the torque cannot change the inclination of the equator relative to the ecliptic. Instead, the rotation axis turns in a direction perpendicular to the axis and the torque, thus describing a cone once in roughly 26,000 years. This slow turning of the rotation axis is called *precession* (Fig. 2.16). Because of precession, the vernal equinox moves along the ecliptic clockwise about 50 seconds of arc every year, thus increasing the ecliptic longitudes of all objects at the same rate. At present the rotation axis points about one degree away from Polaris, but after 12,000 years, the celestial pole will be roughly in the direction of Vega. The changing ecliptic longitudes also affect the right ascension and declination. Thus we have to know the instant of time, or *epoch*, for which the coordinates are given.

Currently most maps and catalogues use the epoch J2000.0, which means the beginning of the year 2000, or, to be exact, the noon of January 1, 2000, or the Julian date 2,451,545.0 (see Sect. 2.15).

Let us now derive expressions for the changes in right ascension and declination. Taking the last transformation equation in (2.23),

$$\sin \delta = \cos \varepsilon \sin \beta + \sin \varepsilon \cos \beta \sin \lambda ,$$

and differentiating, we get

$$\cos \delta \, d\delta = \sin \varepsilon \cos \beta \cos \lambda \, d\lambda .$$

Applying the second equation in (2.22) to the right-hand side, we have, for the change in declination,

$$d\delta = d\lambda \sin \varepsilon \cos \alpha . \tag{2.25}$$

By differentiating the equation

$$\cos \alpha \cos \delta = \cos \beta \cos \lambda ,$$

we get

$$- \sin \alpha \cos \delta \, d\alpha - \cos \alpha \sin \delta \, d\delta = - \cos \beta \sin \lambda \, d\lambda ;$$

and, by substituting the previously obtained expression for $d\delta$ and applying the first equation (2.22), we have

$$\begin{aligned} \sin \alpha \cos \delta \, d\alpha &= d\lambda(\cos \beta \sin \lambda - \sin \varepsilon \cos^2 \alpha \sin \delta) \\ &= d\lambda(\sin \delta \sin \varepsilon + \cos \delta \cos \varepsilon \sin \alpha \\ &\quad - \sin \varepsilon \cos^2 \alpha \sin \delta) . \end{aligned}$$

Simplifying this, we get

$$d\alpha = d\lambda(\sin \alpha \sin \varepsilon \tan \delta + \cos \varepsilon) . \tag{2.26}$$

If $d\lambda$ is the annual increment of the ecliptic longitude (about $50''$), the precessional changes in right ascension and declination in one year are thus

$$\begin{aligned} d\delta &= d\lambda \sin \varepsilon \cos \alpha , \\ d\alpha &= d\lambda(\sin \varepsilon \sin \alpha \tan \delta + \cos \varepsilon) . \end{aligned} \tag{2.27}$$

These expressions are usually written in the form

$$\begin{aligned} d\delta &= n \cos \alpha , \\ d\alpha &= m + n \sin \alpha \tan \delta , \end{aligned} \tag{2.28}$$

where

$$\begin{aligned} m &= d\lambda \cos \varepsilon , \\ n &= d\lambda \sin \varepsilon \end{aligned} \tag{2.29}$$

are the *precession constants*. Since the obliquity of the ecliptic is not exactly a constant but changes with time, m and n also vary slowly with time. However, this variation is so slow that usually we can regard m and n as constants unless the time interval is very long. The values of these constants for some epochs are given in

Table 2.1. Precession constants m and n. Here, "a" means a tropical year

Epoch	m	n
1800	3.07048 s/a	1.33703 s/a $= 20.0554''$/a
1850	3.07141	1.33674	20.0511
1900	3.07234	1.33646	20.0468
1950	3.07327	1.33617	20.0426
2000	3.07419	1.33589	20.0383

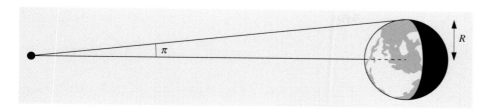

Fig. 2.17. The horizontal parallax π of an object is the angle subtended by the Earth's equatorial radius as seen from the object

Table 2.1. For intervals longer than a few decades a more rigorous method should be used. Its derivation exceeds the level of this book, but the necessary formulas are given in *Reduction of Coordinates (p. 36).

Nutation. The Moon's orbit is inclined with respect to the ecliptic, resulting in precession of its orbital plane. One revolution takes 18.6 years, producing perturbations with the same period in the precession of the Earth. This effect, *nutation*, changes ecliptic longitudes as well as the obliquity of the ecliptic (Fig. 2.16). Calculations are now much more complicated, but fortunately nutational perturbations are relatively small, only fractions of an arc minute.

Parallax. If we observe an object from different points, we see it in different directions. The difference of the observed directions is called the *parallax*. Since the amount of parallax depends on the distance of the observer from the object, we can utilize the parallax to measure distances. Human stereoscopic vision is based (at least to some extent) on this effect. For astronomical purposes we need much longer baselines than the distance between our eyes (about 7 cm). Appropriately large and convenient baselines are the radius of the Earth and the radius of its orbit.

Distances to the nearest stars can be determined from the *annual parallax*, which is the angle subtended by the radius of the Earth's orbit (called the *astronomical unit*, AU) as seen from the star. (We shall discuss this further in Sect. 2.10.)

By *diurnal parallax* we mean the change of direction due to the daily rotation of the Earth. In addition to the distance of the object, the diurnal parallax also depends on the latitude of the observer. If we talk about the parallax of a body in our solar system, we always mean the angle subtended by the Earth's equatorial radius (6378 km) as seen from the object (Fig. 2.17). This equals the apparent shift of the object with respect to

the background stars seen by an observer at the equator if (s)he observes the object moving from the horizon to the zenith. The parallax of the Moon, for example, is about $57'$, and that of the Sun $8.79''$.

In astronomy parallax may also refer to distance in general, even if it is not measured using the shift in the observed direction.

Aberration. Because of the finite speed of light, an observer in motion sees an object shifted in the direction of her/his motion (Figs. 2.18 and 2.19). This change of apparent direction is called *aberration*. To derive

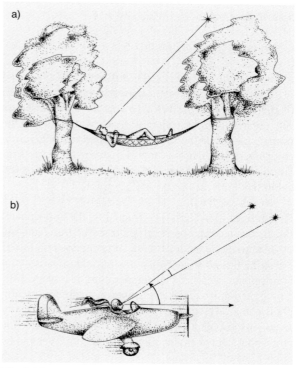

Fig. 2.18a,b. The effect of aberration on the apparent direction of an object. (**a**) Observer at rest. (**b**) Observer in motion

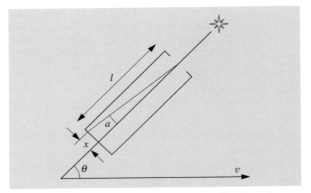

Fig. 2.19. A telescope is pointed in the true direction of a star. It takes a time $t = l/c$ for the light to travel the length of the telescope. The telescope is moving with velocity v, which has a component $v \sin \theta$, perpendicular to the direction of the light beam. The beam will hit the bottom of the telescope displaced from the optical axis by a distance $x = tv \sin \theta = l(v/c) \sin \theta$. Thus the change of direction in radians is $a = x/l = (v/c) \sin \theta$

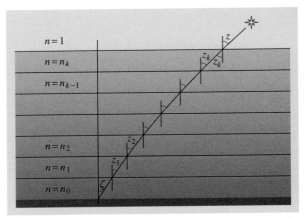

Fig. 2.20. Refraction of a light ray travelling through the atmosphere

the exact value we have to use the special theory of relativity, but for practical purposes it suffices to use the approximate value

$$a = \frac{v}{c} \sin \theta , \quad [a] = \text{rad} , \qquad (2.30)$$

where v is the velocity of the observer, c is the speed of light and θ is the angle between the true direction of the object and the velocity vector of the observer. The greatest possible value of the aberration due to the orbital motion of the Earth, v/c, called the *aberration constant*, is $21''$. The maximal shift due to the Earth's rotation, the diurnal aberration constant, is much smaller, about $0.3''$.

Refraction. Since light is refracted by the atmosphere, the direction of an object differs from the true direction by an amount depending on the atmospheric conditions along the line of sight. Since this refraction varies with atmospheric pressure and temperature, it is very difficult to predict it accurately. However, an approximation good enough for most practical purposes is easily derived. If the object is not too far from the zenith, the atmosphere between the object and the observer can be approximated by a stack of parallel planar layers, each of which has a certain index of refraction n_i (Fig. 2.20). Outside the atmosphere, we have $n = 1$.

Let the true zenith distance be z and the apparent one, ζ. Using the notations of Fig. 2.20, we obtain the following equations for the boundaries of the successive layers:

$$\sin z = n_k \sin z_k ,$$
$$\vdots$$
$$n_2 \sin z_2 = n_1 \sin z_1 ,$$
$$n_1 \sin z_1 = n_0 \sin \zeta ,$$

or

$$\sin z = n_0 \sin \zeta . \qquad (2.31)$$

When the *refraction angle* $R = z - \zeta$ is small and is expressed in radians, we have

$$n_0 \sin \zeta = \sin z = \sin(R + \zeta)$$
$$= \sin R \cos \zeta + \cos R \sin \zeta$$
$$\approx R \cos \zeta + \sin \zeta .$$

Thus we get

$$R = (n_0 - 1) \tan \zeta , \quad [R] = \text{rad} . \qquad (2.32)$$

The index of refraction depends on the density of the air, which further depends on the pressure and temperature. When the altitude is over $15°$, we can use an approximate formula

$$R = \frac{P}{273 + T} 0.00452° \tan(90° - a) , \qquad (2.33)$$

where a is the altitude in degrees, T temperature in degrees Celsius, and P the atmospheric pressure in hectopascals (or, equivalently, in millibars). At lower altitudes the curvature of the atmosphere must be taken into account. An approximate formula for the refraction is then

$$R = \frac{P}{273 + T} \frac{0.1594 + 0.0196a + 0.00002a^2}{1 + 0.505a + 0.0845a^2} \ . \tag{2.34}$$

These formulas are widely used, although they are against the rules of dimensional analysis. To get correct values, all quantities must be expressed in correct units. Figure 2.21 shows the refraction under different conditions evaluated from these formulas.

Altitude is always (except very close to zenith) increased by refraction. On the horizon the change is about 34′, which is slightly more than the diameter of the Sun. When the lower limb of the Sun just touches the horizon, the Sun has in reality already set.

Light coming from the zenith is not refracted at all if the boundaries between the layers are horizontal. Under some climatic conditions, a boundary (e. g. between cold and warm layers) can be slanted, and in this case, there can be a small zenith refraction, which is of the order of a few arc seconds.

Stellar positions given in star catalogues are *mean places*, from which the effects of parallax, aberration and nutation have been removed. The mean place of the date (i. e. at the observing time) is obtained by cor-

recting the mean place for the proper motion of the star (Sect. 2.10) and precession. The *apparent place* is obtained by correcting this place further for nutation, parallax and aberration. There is a catalogue published annually that gives the apparent places of certain references stars at intervals of a few days. These positions have been corrected for precession, nutation, parallax and annual aberration. The effects of diurnal aberration and refraction are not included because they depend on the location of the observer.

2.10 Positional Astronomy

The position of a star can be measured either with respect to some reference stars (relative astrometry) or with respect to a fixed coordinate frame (absolute astrometry).

Fig. 2.22. Astronomers discussing observations with the transit circle of Helsinki Observatory in 1904

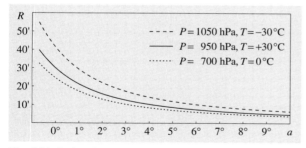

Fig. 2.21. Refraction at different altitudes. The refraction angle R tells how much higher the object seems to be compared with its true altitude a. Refraction depends on the density and thus on the pressure and temperature of the air. The *upper curves* give the refraction at sea level during rather extreme weather conditions. At the altitude of 2.5 kilometers the average pressure is only 700 hPa, and thus the effect of refraction smaller (*lowest curve*)

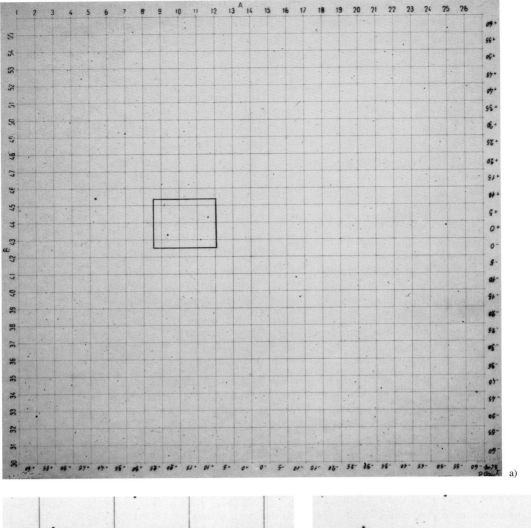

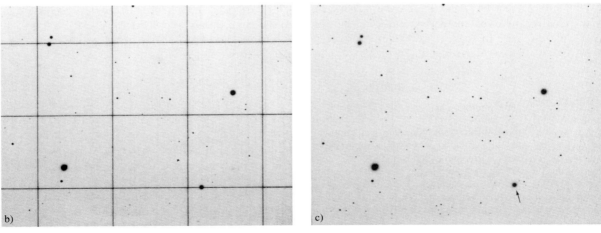

Absolute coordinates are usually determined using a *meridian circle*, which is a telescope that can be turned only in the meridional plane (Fig. 2.22). It has only one axis, which is aligned exactly in the east-west direction. Since all stars cross the meridian in the course of a day, they all come to the field of the meridian circle at some time or other. When a star culminates, its altitude and the time of the transit are recorded. If the time is determined with a sidereal clock, the sidereal time immediately gives the right ascension of the star, since the hour angle is $h = 0$ h. The other coordinate, the declination δ, is obtained from the altitude:

$$\delta = a - (90° - \phi) \,,$$

where a is the observed altitude and ϕ is the geographic latitude of the observatory.

Relative coordinates are measured on photographic plates (Fig. 2.23) or CCD images containing some known reference stars. The scale of the plate as well as the orientation of the coordinate frame can be determined from the reference stars. After this has been done, the right ascension and declination of any object in the image can be calculated if its coordinates in the image are measured.

All stars in a small field are almost equally affected by the dominant perturbations, precession, nutation, and aberration. The much smaller effect of parallax, on the other hand, changes the relative positions of the stars.

The shift in the direction of a star with respect to distant background stars due to the annual motion of the Earth is called the *trigonometric parallax* of the star. It gives the distance of the star: the smaller the parallax, the farther away the star is. Trigonometric parallax is, in fact, the only direct method we currently have of measuring distances to stars. Later we shall be introduced to some other, indirect methods, which require

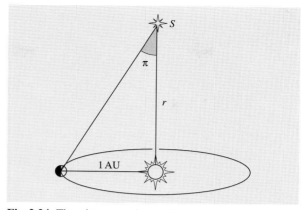

Fig. 2.24. The trigonometric parallax π of a star S is the angle subtended by the radius of the orbit of the Earth, or one astronomical unit, as seen from the star

certain assumptions on the motions or structure of stars. The same method of triangulation is employed to measure distances of earthly objects. To measure distances to stars, we have to use the longest baseline available, the diameter of the orbit of the Earth.

During the course of one year, a star will appear to describe a circle if it is at the pole of the ecliptic, a segment of line if it is in the ecliptic, or an ellipse otherwise. The semimajor axis of this ellipse is called the parallax of the star. It is usually denoted by π. It equals the angle subtended by the radius of the Earth's orbit as seen from the star (Fig. 2.24).

The unit of distance used in astronomy is *parsec* (pc). At a distance of one parsec, one astronomical unit subtends an angle of one arc second. Since one radian is about $206,265''$, 1 pc equals $206,265$ AU. Furthermore, because 1 AU $= 1.496 \times 10^{11}$ m, 1 pc $\approx 3.086 \times 10^{16}$ m. If the parallax is given in arc seconds, the distance is simply

$$r = 1/\pi \,, \quad [r] = \text{pc} \,, \quad [\pi] = '' \,. \tag{2.35}$$

In popular astronomical texts, distances are usually given in *light-years*, one light-year being the distance light travels in one year, or 9.5×10^{15} m. Thus one parsec is about 3.26 light-years.

The first parallax measurement was accomplished by *Friedrich Wilhelm Bessel* (1784–1846) in 1838. He found the parallax of 61 Cygni to be $0.3''$. The nearest star Proxima Centauri has a parallax of $0.762''$ and thus a distance of 1.31 pc.

◀ **Fig. 2.23.** (**a**) A plate photographed for the Carte du Ciel project in Helsinki on November 21, 1902. The centre of the field is at $\alpha = 18$ h 40 min, $\delta = 46°$, and the area is $2° \times 2°$. Distance between coordinate lines (exposed separately on the plate) is 5 minutes of arc. (**b**) The framed region on the same plate. (**c**) The same area on a plate taken on November 7, 1948. The bright star in the lower right corner (SAO 47747) has moved about 12 seconds of arc. The brighter, slightly drop-shaped star to the left is a binary star (SAO 47767); the separation between its components is $8''$

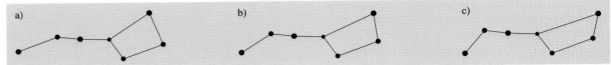

Fig. 2.25a–c. Proper motions of stars slowly change the appearance of constellations. (**a**) The Big Dipper during the last ice age 30,000 years ago, (**b**) nowadays, and (**c**) after 30,000 years

In addition to the motion due to the annual parallax, many stars seem to move slowly in a direction that does not change with time. This effect is caused by the relative motion of the Sun and the stars through space; it is called the *proper motion*. The appearance of the sky and the shapes of the constellations are constantly, although extremely slowly, changed by the proper motions of the stars (Fig. 2.25).

The velocity of a star with respect to the Sun can be divided into two components (Fig. 2.26), one of which is directed along the line of sight (the radial component or the *radial velocity*), and the other perpendicular to it (the tangential component). The tangential velocity results in the proper motion, which can be measured by taking plates at intervals of several years or decades. The proper motion μ has two components, one giving the change in declination μ_δ and the other, in right ascension, $\mu_\alpha \cos \delta$. The coefficient $\cos \delta$ is used to correct the scale of right ascension: hour circles (the great circles with $\alpha = $ constant) approach each other towards the poles, so the coordinate difference must be multiplied by $\cos \delta$ to obtain the true angular separation. The total proper motion is

$$\mu = \sqrt{\mu_\alpha^2 \cos^2 \delta + \mu_\delta^2} \; . \tag{2.36}$$

The greatest known proper motion belongs to Barnard's Star, which moves across the sky at the enormous speed of 10.3 arc seconds per year. It needs less than 200 years to travel the diameter of a full moon.

In order to measure proper motions, we must observe stars for decades. The radial component, on the other hand, is readily obtained from a single observation, thanks to the *Doppler effect*. By the Doppler effect we mean the change in frequency and wavelength of radiation due to the radial velocity of the radiation source. The same effect can be observed, for example, in the sound of an ambulance, the pitch being higher when the ambulance is approaching and lower when it is receding.

The formula for the Doppler effect for small velocities can be derived as in Fig. 2.27. The source of radiation transmits electromagnetic waves, the period of one cycle being T. In time T, the radiation approaches the observer by a distance $s = cT$, where c is the speed of propagation. During the same time, the source moves with respect to the observer a distance $s' = vT$, where v is the speed of the source, positive for a receding source and negative for an approaching one. We find that the length of one cycle, the wavelength λ, equals

$$\lambda = s + s' = cT + vT \; .$$

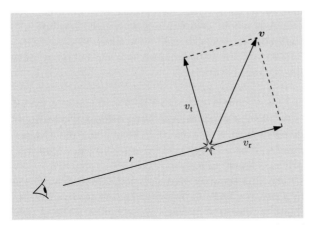

Fig. 2.26. The radial and tangential components, v_r and v_t of the velocity v of a star. The latter component is observed as proper motion

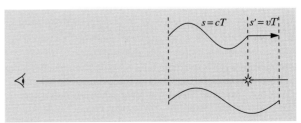

Fig. 2.27. The wavelength of radiation increases if the source is receding

If the source were at rest, the wavelength of its radiation would be $\lambda_0 = cT$. The motion of the source changes the wavelength by an amount

$$\Delta\lambda = \lambda - \lambda_0 = cT + vT - cT = vT ,$$

and the relative change $\Delta\lambda$ of the wavelength is

$$\frac{\Delta\lambda}{\lambda_0} = \frac{v}{c} . \tag{2.37}$$

This is valid only when $v \ll c$. For very high velocities, we must use the relativistic formula

$$\frac{\Delta\lambda}{\lambda_0} = \sqrt{\frac{1+v/c}{1-v/c}} - 1 . \tag{2.38}$$

In astronomy the Doppler effect can be seen in stellar spectra, in which the spectral lines are often displaced towards the blue (shorter wavelengths) or red (longer wavelengths) end of the spectrum. A *blueshift* means that the star is approaching, while a *redshift* indicates that it is receding.

The displacements due to the Doppler effect are usually very small. In order to measure them, a reference spectrum is exposed on the plate next to the stellar spectrum. Now that CCD-cameras have replaced photographic plates, separate calibration exposures of reference spectra are taken to determine the wavelength scale. The lines in the reference spectrum are produced by a light source at rest in the laboratory. If the reference spectrum contains some lines found also in the stellar spectrum, the displacements can be measured.

Displacements of spectral lines give the radial velocity v_r of the star, and the proper motion μ can be measured from photographic plates or CCD images. To find the tangential velocity v_t, we have to know the distance r, obtainable from e.g. parallax measurements. Tangential velocity and proper motion are related by

$$v_t = \mu r . \tag{2.39}$$

If μ is given in arc seconds per year and r in parsecs we have to make the following unit transformations to get v_t in km/s:

$$1 \text{ rad} = 206{,}265'' , \quad 1 \text{ year} = 3.156 \times 10^7 \text{ s} ,$$
$$1 \text{ pc} = 3.086 \times 10^{13} \text{ km} .$$

Hence

$$v_t = 4.74\, \mu r , \quad [v_t] = \text{km/s} ,$$
$$[\mu] = ''/\text{a} , \quad [r] = \text{pc} . \tag{2.40}$$

The total velocity v of the star is

$$v = \sqrt{v_r^2 + v_t^2} .$$

2.11 Constellations

At any one time, about 1000–1500 stars can be seen in the sky (above the horizon). Under ideal conditions, the number of stars visible to the naked eye can be as high as 3000 on a hemisphere, or 6000 altogether. Some stars seem to form figures vaguely resembling something; they have been ascribed to various mythological and other animals. This grouping of stars into constellations is a product of human imagination without any physical basis. Different cultures have different constellations, depending on their mythology, history and environment.

About half of the shapes and names of the constellations we are familiar with date back to Mediterranean antiquity. But the names and boundaries were far from unambiguous as late as the 19th century. Therefore the International Astronomical Union (IAU) confirmed fixed boundaries at its 1928 meeting.

The official boundaries of the constellations were established along lines of constant right ascension and declination for the epoch 1875. During the time elapsed since then, precession has noticeably turned the equatorial frame. However, the boundaries remain fixed with respect to the stars. So a star belonging to a constellation will belong to it forever (unless it is moved across the boundary by its proper motion).

The names of the 88 constellations confirmed by the IAU are given in Table C.21 at the end of the book. The table also gives the abbreviation of the Latin name, its genitive (needed for names of stars) and the English name.

In his star atlas *Uranometria* (1603) *Johannes Bayer* started the current practice to denote the brightest stars of each constellation by Greek letters. The brightest star is usually α (alpha), e. g. Deneb in the constellation Cygnus is α Cygni, which is abbreviated as α Cyg. The second brightest star is β (beta), the next one γ (gamma) and so on. There are, however, several exceptions to this rule; for example, the stars of the Big Dipper are named in the order they appear in the constellation. After the Greek alphabet has been exhausted, Latin letters can be employed. Another method is to use mbers, which are assigned in the order of increasing right ascension; e. g. 30 Tau is a bright binary star in the constellation Taurus. Moreover, variable stars have their special identifiers (Sect. 13.1). About two hundred bright stars have a proper name; e. g. the bright α Aur is called also Capella.

As telescopes evolved, more and more stars were seen and catalogued. It soon became impractical to continue this method of naming. Thus most of the stars are known only by their catalogue index numbers. One star may have many different numbers; e. g. the abovementioned Capella (α Aur) is number BD+45° 1077 in the Bonner Durchmusterung and HD 34029 in the Henry Draper catalogue.

2.12 Star Catalogues and Maps

The first actual star catalogue was published by *Ptolemy* in the second century; this catalogue appeared in the book to be known later as *Almagest* (which is a Latin corruption of the name of the Arabic translation, *Almijisti*). It had 1025 entries; the positions of these bright stars had been measured by *Hipparchos* 250 years earlier. Ptolemy's catalogue was the only widely used one prior to the 17th century.

The first catalogues still being used by astronomers were prepared under the direction of *Friedrich Wilhelm August Argelander* (1799–1875). Argelander worked in Turku and later served as professor of astronomy in Helsinki, but he made his major contributions in Bonn. Using a 72 mm telescope, he and his assistants measured the positions and estimated the magnitudes of 320,000 stars. The catalogue, *Bonner Durchmusterung*, contains nearly all stars brighter than magnitude 9.5 between the north pole and declination $-2°$. (Magnitudes are further discussed in Chap. 4.) Argelander's work was later used as a model for two other large catalogues covering the whole sky. The total number of stars in these catalogues is close to one million.

The purpose of these *Durchmusterungen* or general catalogues was to systematically list a great number of stars. In the *zone catalogues*, the main goal is to give the positions of stars as exactly as possible. A typical zone catalogue is the German *Katalog der Astronomischen Gesellschaft* (AGK). Twelve observatories, each measuring a certain region in the sky, contributed to this catalogue. The work was begun in the 1870's and completed at the turn of the century.

General and zone catalogues were based on visual observations with a telescope. The evolution of photography made this kind of work unnecessary at the end of the 19th century. Photographic plates could be stored for future purposes, and measuring the positions of stars became easier and faster, making it possible to measure many more stars.

A great international program was started at the end of the 19th century in order to photograph the entire sky. Eighteen observatories participated in this *Carte du Ciel* project, all using similar instruments and plates. The positions of stars were first measured with respect to a rectangular grid exposed on each plate (Fig. 2.23a). These coordinates could then be converted into declination and right ascension.

Positions of stars in catalogues are measured with respect to certain comparison stars, the coordinates of which are known with high accuracy. The coordinates of these reference stars are published in fundamental catalogues. The first such catalogue was needed for the AGK catalogue; it was published in Germany in 1879. This *Fundamental Katalog* (FK 1) gives the positions of over 500 stars.

The fundamental catalogues are usually revised every few decades. The latest edition, the FK 5, appeared in 1984. At the same time, a new system of astronomical constants was adopted. The catalogue contains 1535 fundamental and 3117 additional stars. The apparent places of the fundamental stars in the FK catalogues at 10-day intervals are published annually in the *Apparent Places of Fundamental Stars*.

A widely used catalogue is the SAO catalogue, published by the Smithsonian Astrophysical Observatory in the 1960's. It contains the exact positions, magnitudes, proper motions, spectral classifications, etc. of 258,997 stars brighter than magnitude 9. The catalogue is accompanied by a star map containing all the stars in the catalogue.

In the 1990's a large astrometric catalogue, *PPM* (Positions and Proper Motions), was published to replace the AGK and SAO catalogues. It contains all stars brighter than 7.5 magnitudes, and is almost complete to magnitude 8.5. Also many fainter stars are included. Altogether, the four volumes of the catalogue contain information on 378,910 stars.

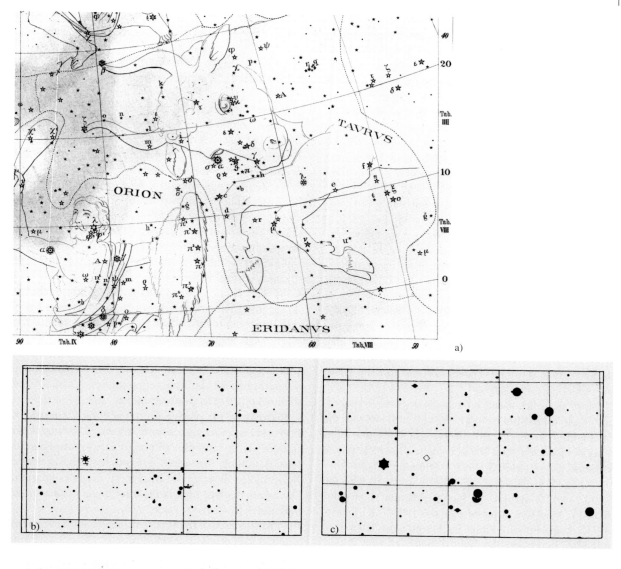

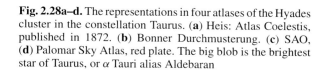

Fig. 2.28a–d. The representations in four atlases of the Hyades cluster in the constellation Taurus. (**a**) Heis: Atlas Coelestis, published in 1872. (**b**) Bonner Durchmusterung. (**c**) SAO, (**d**) Palomar Sky Atlas, red plate. The big blob is the brightest star of Taurus, or α Tauri alias Aldebaran

The European Space Agency (ESA) launched the first astrometric satellite Hipparcos in 1989. Although Hipparcos didn't reach the planned geosynchronous orbit, it gave exact positions of over a hundred thousand stars. The *Hipparcos catalogue*, based on the measurements of the satellite, contains astrometric and photometric data of 118,000 stars. The coordinates are precise to a couple of milliarcseconds. A less precise *Tycho catalogue* contains the data of about one million stars. The Gaia satellite, planned to be launched around 2010, is expected to improve the accuracy even to 10^{-5} seconds of arc.

Star maps have been published since ancient times, but the earliest maps were globes showing the celestial sphere as seen from the outside. At the beginning of the 17th century, a German, Johannes Bayer, published the first map showing the stars as seen from inside the celestial sphere, as we see them in the sky. Constellations were usually decorated with drawings of mythological figures. The *Uranometria Nova* (1843) by Argelander represents a transition towards modern maps: mythological figures are beginning to fade away. The map accompanying the *Bonner Durchmusterung* carried this evolution to its extreme. The sheets contain nothing but stars and coordinate lines.

Most maps are based on star catalogues. Photography made it possible to produce star maps without the cataloguing stage. The most important of such maps is a photographic atlas the full name of which is *The National Geographic Society – Palomar Observatory Sky Atlas*. The plates for this atlas were taken with the 1.2 m Schmidt camera on Mount Palomar. The Palomar Sky Atlas was completed in the 1950's. It consists of 935 pairs of photographs: each region has been photographed in red and blue light. The size of each plate is about 35 cm × 35 cm, covering an area of 6.6° × 6.6°. The prints are negatives (black stars on a light background), because in this way, fainter objects are visible. The limiting magnitude is about 19 in blue and 20 in red.

The Palomar atlas covers the sky down to −30°. Work to map the rest of the sky was carried out later at two observatories in the southern hemisphere, at Siding Spring Observatory in Australia, and at the European Southern Observatory (ESO) in Chile. The instruments and the scale on the plates are similar to those used earlier for the Palomar plates, but the atlas is distributed on film transparencies instead of paper prints.

For amateurs there are several star maps of various kinds. Some of them are mentioned in the references.

2.13 Sidereal and Solar Time

Time measurements can be based on the rotation of the Earth, orbital motion around the Sun, or on atomic clocks. The last-mentioned will be discussed in the next section. Here we consider the sidereal and solar times related to the rotation of the Earth.

We defined the sidereal time as the hour angle of the vernal equinox. A good basic unit is a *sidereal day*, which is the time between two successive upper culminations of the vernal equinox. After one sidereal day the celestial sphere with all its stars has returned to its original position with respect to the observer. The flow of sidereal time is as constant as the rotation of the Earth. The rotation rate is slowly decreasing, and thus the length of the sidereal day is increasing. In addition to the smooth slowing down irregular variations of the order of one millisecond have been observed.

Unfortunately, also the sidereal time comes in two varieties, apparent and mean. The *apparent sidereal time* is determined by the true vernal equinox, and so it is obtained directly from observations.

Because of the precession the ecliptic longitude of the vernal equinox increases by about 50″ a year. This motion is very smooth. Nutation causes more complicated wobbling. The *mean equinox* is the point where the vernal equinox would be if there were no nutation. The *mean sidereal time* is the hour angle of this mean equinox.

The difference of the apparent and mean sidereal time is called the *equation of equinoxes*:

$$\Theta_a - \Theta_M = \Delta\psi \cos\varepsilon , \qquad (2.42)$$

where ε is the obliquity of the ecliptic at the instant of the observation, and $\Delta\psi$, the nutation in longitude. This value is tabulated for each day e. g. in the *Astronomical Almanac*. It can also be computed from the formulae given in *Reduction of Coordinates. It is at most about one second, so it has to be taken into account only in the most precise calculations.

Figure 2.29 shows the Sun and the Earth at vernal equinox. When the Earth is at the point *A*, the Sun culminates and, at the same time, a new sidereal day

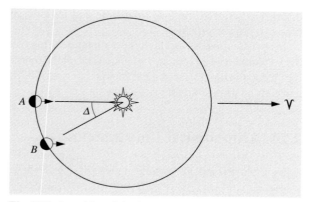

Fig. 2.29. One sidereal day is the time between two successive transits or upper culminations of the vernal equinox. By the time the Earth has moved from A to B, one sidereal day has elapsed. The angle Δ is greatly exaggerated; in reality, it is slightly less than one degree

begins in the city with the huge black arrow standing in its central square. After one sidereal day, the Earth has moved along its orbit almost one degree of arc to the point B. Therefore the Earth has to turn almost a degree further before the Sun will culminate. The *solar* or *synodic day* is therefore 3 min 56.56 s (sidereal time) longer than the sidereal day. This means that the beginning of the sidereal day will move around the clock during the course of one year. After one year, sidereal and solar time will again be in phase. The number of sidereal days in one year is one higher than the number of solar days.

When we talk about rotational periods of planets, we usually mean sidereal periods. The length of day, on the other hand, means the rotation period with respect to the Sun. If the orbital period around the Sun is P, sidereal rotation period τ_* and synodic day τ, we now know that the number of sidereal days in time P, P/τ_*, is one higher than the number of synodic days, P/τ:

$$\frac{P}{\tau_*} - \frac{P}{\tau} = 1 \,,$$

or

$$\frac{1}{\tau} = \frac{1}{\tau_*} - \frac{1}{P} \,. \tag{2.43}$$

This holds for a planet rotating in the direction of its orbital motion (counterclockwise). If the sense of rotation is opposite, or *retrograde*, the number of sidereal days in one orbital period is one less than the number

of synodic days, and the equation becomes

$$\frac{1}{\tau} = \frac{1}{\tau_*} + \frac{1}{P} \,. \tag{2.44}$$

For the Earth, we have $P = 365.2564$ d, and $\tau = 1$ d, whence (2.43) gives $\tau_* = 0.99727$ d $= 23$ h 56 min 4 s, solar time.

Since our everyday life follows the alternation of day and night, it is more convenient to base our timekeeping on the apparent motion of the Sun rather than that of the stars. Unfortunately, solar time does not flow at a constant rate. There are two reasons for this. First, the orbit of the Earth is not exactly circular, but an ellipse, which means that the velocity of the Earth along its orbit is not constant. Second, the Sun moves along the ecliptic, not the equator. Thus its right ascension does not increase at a constant rate. The change is fastest at the end of December (4 min 27 s per day) and slowest in mid-September (3 min 35 s per day). As a consequence, the hour angle of the Sun (which determines the solar time) also grows at an uneven rate.

To find a solar time flowing at a constant rate, we define a fictitious *mean sun*, which moves along the celestial equator with constant angular velocity, making a complete revolution in one year. By year we mean here the *tropical year*, which is the time it takes for the Sun to move from one vernal equinox to the next. In one tropical year, the right ascension of the Sun increases exactly 24 hours. The length of the tropical year is 365 d 5 h 48 min 46 s $= 365.2422$ d. Since the direction of the vernal equinox moves due to precession, the tropical year differs from the sidereal year, during which the Sun makes one revolution with respect to the background stars. One sidereal year is 365.2564 d.

Using our artificial mean sun, we now define an evenly flowing solar time, the *mean solar time* (or simply *mean time*) T_M, which is equal to the hour angle h_M of the centre of the mean sun plus 12 hours (so that the date will change at midnight, to annoy astronomers):

$$T_M = h_M + 12 \,\text{h} \,. \tag{2.45}$$

The difference between the true solar time T and the mean time T_M is called the *equation of time*:

$$\text{E.T.} = T - T_M \,. \tag{2.46}$$

(In spite of the identical abbreviation, this has nothing to do with a certain species of little green men.) The

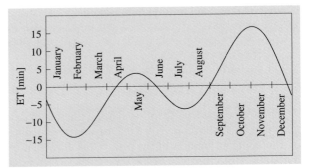

Fig. 2.30. Equation of time. A sundial always shows (if correctly installed) true local solar time. To find the local mean time the equation of time must be subtracted from the local solar time

greatest positive value of E.T. is about 16 minutes and the greatest negative value about −14 minutes (see Fig. 2.30). This is also the difference between the true noon (the meridian transit of the Sun) and the mean noon.

Both the true solar time and mean time are *local times*, depending on the hour angle of the Sun, real or artificial. If one observes the true solar time by direct measurement and computes the mean time from (2.46), a digital watch will probably be found to disagree with both of them. The reason for this is that we do not use local time in our everyday life; instead we use the *zonal time* of the nearest *time zone*.

In the past, each city had its own local time. When travelling became faster and more popular, the great variety of local times became an inconvenience. At the end of the 19th century, the Earth was divided into 24 zones, the time of each zone differing from the neighboring ones by one hour. On the surface of the Earth, one hour in time corresponds to 15° in longitude; the time of each zone is determined by the local mean time at one of the longitudes 0°, 15°, . . . , 345°.

The time of the zero meridian going through Greenwich is used as an international reference, Universal Time. In most European countries, time is one hour ahead of this (Fig. 2.31).

In summer, many countries switch to *daylight saving time*, during which time is one hour ahead of the ordinary time. The purpose of this is to make the time when people are awake coincide with daytime in order to save electricity, particularly in the evening, when people go to bed one hour earlier. During daylight saving time, the difference between the true solar time and the official time can grow even larger.

In the EU countries the daylight saving time begins on the last Sunday of March, at 1 o'clock UTC in the morning, when the clocks are moved forward to read 2 AM, and ends on the last Sunday of October at 1 o'clock.

2.14 Astronomical Time Systems

Time can be defined using several different phenomena:

1. The solar and sidereal times are based on the rotation of the Earth.
2. The standard unit of time in the current SI system, the second, is based on quantum mechanical atomary phenomena.
3. Equations of physics like the ones describing the motions of celestial bodies involve a time variable corresponding to an ideal time running at a constant pace. The ephemeris time and dynamical time discussed a little later are such times.

Observations give directly the apparent sidereal time as the hour angle of the true vernal equinox. From the apparent sidereal time the mean sidereal time can be calculated.

The *universal time* UT is defined by the equation

$$\begin{aligned} \text{GMST}(0\,\text{UT}) = {}& 24{,}110.54841\,\text{s} \\ & + T \times 8{,}640{,}184.812866\,\text{s} \\ & + T^2 \times 0.093104\,\text{s} \\ & - T^3 \times 0.0000062\,\text{s}\,, \end{aligned} \tag{2.47}$$

where GMST is the Greenwich mean sidereal time and T the Julian century. The latter is obtained from the Julian date J, which is a running number of the day (Sect. 2.15 and *Julian date, p. 38):

$$T = \frac{J - 2{,}451{,}545.0}{36{,}525}\,. \tag{2.48}$$

This gives the time elapsed since January 1, 2000, in Julian centuries.

Sidereal time and hence also UT are related to the rotation of the Earth, and thus contain perturbations due to the irregular variations, mainly slowing down, of the rotation.

In (2.47) the constant 8,640,184.812866 s tells how much sidereal time runs fast compared to the UT in

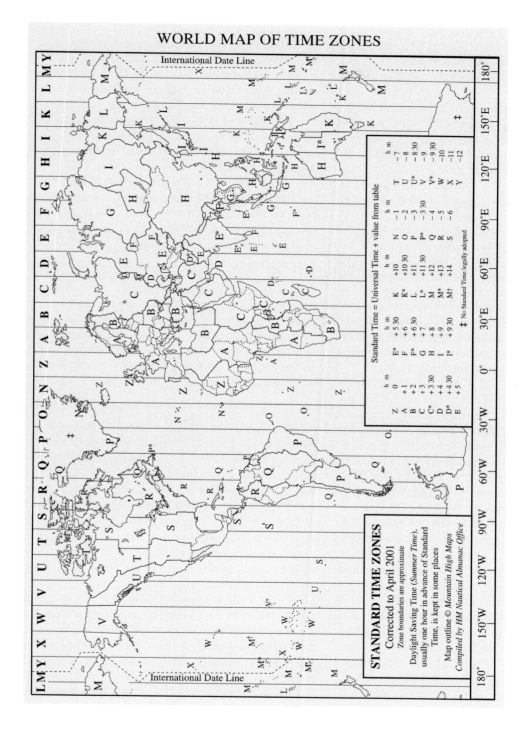

WORLD MAP OF TIME ZONES

Fig. 2.31. The time zones. The map gives the difference of the local zonal time from the Greenwich mean time (UT). During daylight saving time, one hour must be added to the given figures. When travelling across the date line westward, the date must be incremented by one day, and decremented if going eastward. For example, a traveller taking a flight from Honolulu to Tokyo on Monday morning will arrive on Tuesday, even though (s)he does not see a single night en route. (Drawing U.S. Nval Observatory)

a Julian century. As the rotation of the Earth is slowing down the solar day becomes longer. Since the Julian century T contains a fixed number of days, it will also become longer. This gives rise to the small correction terms in (2.47).

Strictly speaking this universal time is the time denoted by UT1. Observations give UT0, which contains a small perturbation due to the wandering of the geographical pole, or *polar variation*. The direction of the axis with respect to the solid surface varies by about $0.1''$ (a few metres on the surface) with a period of about 430 days (*Chandler period*). In addition to this, the polar motion contains a slow nonperiodic part.

The z axis of the astronomical coordinates is aligned with the angular momentum vector of the Earth, but the terrestrial coordinates refer to the axis at the epoch 1903.5. In the most accurate calculations this has to be taken into account.

Nowadays the SI unit of time, the second, is defined in a way that has nothing to do with celestial phenomena. Periods of quantum mechanical phenomena remain more stable than the motions of celestial bodies involving complicated perturbations.

In 1967, one second was defined as $9,192,631,770$ times the period of the light emitted by cesium 133 isotope in its ground state, transiting from hyperfine level $F = 4$ to $F = 3$. Later, this definition was revised to include small relativistic effects produced by gravitational fields. The relative accuracy of this atomic time is about 10^{-12}.

The *international atomic time*, TAI, was adopted as the basis of time signals in 1972. The time is maintained by the *Bureau International des Poids et Mesures* in Paris, and it is the average of several accurate atomic clocks.

Even before atomic clocks there was a need for an ideal time proceeding at a perfectly constant rate, corresponding to the time variable in the equations of Newtonian mechanics. The *ephemeris time* was such a time. It was used e. g. for tabulating ephemerides. The unit of ephemeris time was the *ephemeris second*, which is the length of the tropical year 1900 divided by $31,556,925.9747$. Ephemeris time was not known in advance. Only afterwards was it possible to determine the difference of ET and UT from observational data.

In 1984 ephemeris time was replaced by *dynamical time*. It comes in two varieties.

The *terrestrial dynamical time* (TDT) corresponds to the proper time of an observer moving with the Earth. The time scale is affected by the relativistic time dilation due to the orbital speed of the Earth. The rotation velocity depends on the latitude, and thus in TDT it is assumed that the observer is not rotating with the Earth. The zero point of TDT was chosen so that the old ET changed without a jump to TDT.

In 1991 a new standard time, the *terrestrial time* (TT), was adopted. Practically it is equivalent to TDT.

TT (or TDT) is the time currently used for tabulating ephemerides of planets and other celestial bodies. For example, the *Astronomical Almanac* gives the coordinates of the planets for each day at 0 TT.

The *Astronomical Almanac* also gives the difference

$$\Delta T = \mathrm{TDT} - \mathrm{UT} \tag{2.49}$$

for earlier years. For the present year and some future years a prediction extrapolated from the earlier years is given. Its accuracy is about 0.1 s. At the beginning of 1990 the difference was 56.7 s; it increases every year by an amount that is usually a little less than one second.

The terrestrial time differs from the atomic time by a constant offset

$$\mathrm{TT} = \mathrm{TAI} + 32.184 \text{ s} . \tag{2.50}$$

TT is well suited for ephemerides of phenomena as seen from the Earth. The equations of motion of the solar system, however, are solved in a frame the origin of which is the centre of mass or *barycentre* of the solar system. The coordinate time of this frame is called the *barycentric dynamical time*, TDB. The unit of TDB is defined so that, on the average, it runs at the same rate as TT, the difference containing only periodic terms depending on the orbital motion of the Earth. The difference can usually be neglected, since it is at most about 0.002 seconds.

Which of these many times should we use in our alarm-clocks? None of them. Yet another time is needed for that purpose. This official wall-clock time is called the *coordinated universal time*, UTC. The zonal time follows UTC but differs from it usually by an integral number of hours.

UTC is defined so that it proceeds at the same rate as TAI, but differs from it by an integral number of seconds. These *leap seconds* are used to adjust UTC so that the difference from UT1 never exceeds 0.9 seconds.

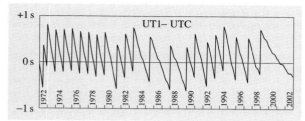

Fig. 2.32. The difference between the universal time UT1, based on the rotation of the Earth, and the coordinated universal time UTC during 1972–2002. Because the rotation of the Earth is slowing down, the UT1 will run slow of the UTC by about 0.8 seconds a year. Leap seconds are added to the UTC when necessary to keep the times approximately equal. In the graph these leap seconds are seen as one second jumps upward

A leap second is added either at the beginning of a year or the night between June and July.

The difference

$$\Delta AT = TAI - UTC \qquad (2.51)$$

is also tabulated in the *Astronomical Almanac*. According to the definition of UTC the difference in seconds is always an integer. The difference cannot be predicted very far to the future.

From (2.50) and (2.51) we get

$$TT = UTC + 32.184\,s + \Delta AT , \qquad (2.52)$$

which gives the terrestrial time TT corresponding to a given UTC. Table 2.2 gives this correction. The table is easy to extend to the future. When it is told in the news that a leap second will be added the difference will increase by one second. In case the number of leap seconds is not known, it can be approximated that a leap second will be added every 1.25 years.

The unit of the coordinated universal time UTC, atomic time TAI and terrestrial time TT is the same second of the SI system. Hence all these times proceed at the same rate, the only difference being in their zero points. The difference of the TAI and TT is always the same, but due to the leap seconds the UTC will fall behind in a slightly irregular way.

Culminations and rising and setting times of celestial bodies are related to the rotation of the Earth. Thus the sidereal time and hence the UT of such an event can be calculated precisely. The corresponding UTC cannot differ from the UT by more than 0.9 seconds, but the

Table 2.2. Differences of the atomic time and UTC (ΔAT) and the terrestrial time TT and UTC. The terrestrial time TT used in ephemerides is obtained by adding $\Delta AT + 32.184$ s to the ordinary time UTC

	ΔAT	TT – UTC
1.1.1972– 30.6.1972	10 s	42.184 s
1.7.1972–31.12.1972	11 s	43.184 s
1.1.1973–31.12.1973	12 s	44.184 s
1.1.1974–31.12.1974	13 s	45.184 s
1.1.1975–31.12.1975	14 s	46.184 s
1.1.1976–31.12.1976	15 s	47.184 s
1.1.1977–31.12.1977	16 s	48.184 s
1.1.1978–31.12.1978	17 s	49.184 s
1.1.1979–31.12.1979	18 s	50.184 s
1.1.1980– 30.6.1981	19 s	51.184 s
1.7.1981– 30.6.1982	20 s	52.184 s
1.7.1982– 30.6.1983	21 s	53.184 s
1.7.1983– 30.6.1985	22 s	54.184 s
1.7.1985–31.12.1987	23 s	55.184 s
1.1.1988–31.12.1989	24 s	56.184 s
1.1.1990–31.12.1990	25 s	57.184 s
1.1.1991– 30.6.1992	26 s	58.184 s
1.7.1992– 30.6.1993	27 s	59.184 s
1.7.1993– 30.6.1994	28 s	60.184 s
1.7.1994–31.12.1995	29 s	61.184 s
1.1.1996– 31.6.1997	31 s	62.184 s
1.7.1997–31.12.1998	32 s	63.184 s
1.1.1999–	33 s	64.184 s

exact value is not known in advance. The future coordinates of the Sun, Moon and planets can be calculated as functions of the TT, but the corresponding UTC can only be estimated.

2.15 Calendars

Our calendar is a result of long evolution. The main problem it must contend with is the incommensurability of the basic units, day, month and year: the numbers of days and months in a year are not integers. This makes it rather complicated to develop a calendar that takes correctly into account the alternation of seasons, day and night, and perhaps also the lunar phases.

Our calendar has its origin in the Roman calendar, which, in its earliest form, was based on the phases of the Moon. From around 700 B.C. on, the length of the year has followed the apparent motion of the Sun; thus originated the division of the year into twelve months. One month, however, still had a length roughly equal to the lunar cycle. Hence one year was only 354 days long. To keep the year synchronised with the seasons, a leap month had to be added to every other year.

Eventually the Roman calendar got mixed up. The mess was cleared by Julius Caesar in about 46 B.C., when the *Julian calendar* was developed upon his orders. The year had 365 days and a leap day was added to every fourth year.

In the Julian calendar, the average length of one year is 365 d 6 h, but the tropical year is 11 min 14 s shorter. After 128 years, the Julian year begins almost one day too late. The difference was already 10 days in 1582, when a calendar reform was carried out by Pope Gregory XIII. In the *Gregorian calendar*, every fourth year is a leap year, the years divisible by 100 being exceptions. Of these, only the years divisible by 400 are leap years. Thus 1900 was not a leap year, but 2000 was. The Gregorian calendar was adopted slowly, at different times in different countries. The transition period did not end before the 20th century.

Even the Gregorian calendar is not perfect. The differences from the tropical year will accumulate to one day in about 3300 years.

Since years and months of variable length make it difficult to compute time differences, especially astronomers have employed various methods to give each day a running number. The most widely used numbers are the *Julian dates*. In spite of their name, they are not related to the Julian calendar. The only connection is the length of a *Julian century* of 36,525 days, a quantity appearing in many formulas involving Julian dates. The Julian day number 0 dawned about 4700 B.C. The day number changes always at 12 : 00 UT. For example, the Julian day 2,451,545 began at noon in January 1, 2000. The Julian date can be computed using the formulas given in *Julian Date (p. 38).

Julian dates are uncomfortably big numbers, and therefore *modified Julian dates* are often used. The zero point can be e. g. January 1, 2000. Sometimes 0.5 is subtracted from the date to make it to coincide with the date corresponding to the UTC. When using such dates, the zero point should always be mentioned.

* Reduction of Coordinates

Star catalogues give coordinates for some standard epoch. In the following we give the formulas needed to reduce the coordinates to a given date and time.

The full reduction is rather laborious, but the following simplified version is sufficient for most practical purposes.

We assume that the coordinates are given for the epoch J2000.0.

1. First correct the place for proper motion unless it is negligible.
2. Precess the coordinates to the time of the observation. First we use the coordinates of the standard epoch (α_0, δ_0) to find a unit vector pointing in the direction of the star:

$$\boldsymbol{p}_0 = \begin{pmatrix} \cos \delta_0 \cos \alpha_0 \\ \cos \delta_0 \sin \alpha_0 \\ \sin \delta_0 \end{pmatrix} .$$

Precession changes the ecliptic longitude of the object. The effect on right ascension and declination can be calculated as three rotations, given by three rotation matrices. By multiplying these matrices we get the combined precession matrix that maps the previous unit vector to its precessed equivalent. A similar matrix can be derived for the nutation. The transformations and constants given here are based on the system standardized by the IAU in 1976.

The precession and nutation matrices contain several quantities depending on time. The time variables appearing in their expressions are

$$t = J - 2,451,545.0 ,$$
$$T = \frac{J - 2,451,545.0}{36,525} .$$

Here J is the Julian date of the observation, t the number of days since the epoch J2000.0 (i. e. noon of January 1, 2000), and T the same interval of time in Julian centuries.

The following three angles are needed for the precession matrix

$$\zeta = 2306.2181'' T + 0.30188'' T^2 + 0.017998'' T^3 ,$$
$$z = 2306.2181'' T + 1.09468'' T^2 + 0.018203'' T^3 ,$$
$$\theta = 2004.3109'' T - 0.42665'' T^2 - 0.041833'' T^3 .$$

The precession matrix is now

$$\boldsymbol{P} = \begin{pmatrix} P_{11} & P_{12} & P_{13} \\ P_{21} & P_{22} & P_{23} \\ P_{31} & P_{32} & P_{33} \end{pmatrix} .$$

The elements of this matrix in terms of the abovementioned angles are

$$P_{11} = \cos z \cos \theta \cos \zeta - \sin z \sin \zeta \,,$$
$$P_{12} = -\cos z \cos \theta \sin \zeta - \sin z \cos \zeta \,,$$
$$P_{13} = -\cos z \sin \theta \,,$$
$$P_{21} = \sin z \cos \theta \cos \zeta + \cos z \sin \zeta \,,$$
$$P_{22} = -\sin z \cos \theta \sin \zeta + \cos z \cos \zeta \,,$$
$$P_{23} = -\sin z \sin \theta \,,$$
$$P_{31} = \sin \theta \cos \zeta \,,$$
$$P_{32} = -\sin \theta \sin \zeta \,,$$
$$P_{33} = \cos \theta \,.$$

The new coordinates are now obtained by multiplying the coordinates of the standard epoch by the precession matrix:

$$\boldsymbol{p}_1 = \boldsymbol{P}\boldsymbol{p}_0 \,.$$

This is the mean place at the given time and date. If the standard epoch is not J2000.0, it is probably easiest to first transform the given coordinates to the epoch J2000.0. This can be done by computing the precession matrix for the given epoch and multiplying the coordinates by the inverse of this matrix. Inverting the precession matrix is easy: we just transpose it, i.e. interchange its rows and columns. Thus coordinates given for some epoch can be precessed to J2000.0 by multiplying them by

$$\boldsymbol{P}^{-1} = \begin{pmatrix} P_{11} & P_{21} & P_{31} \\ P_{12} & P_{22} & P_{32} \\ P_{13} & P_{23} & P_{33} \end{pmatrix} \,.$$

In case the required accuracy is higher than about one minute of arc, we have to do the following further corrections.

3. The full nutation correction is rather complicated. The nutation used in astronomical almanacs involves series expansions containing over a hundred terms. Very often, though, the following simple form is sufficient. We begin by finding the mean obliquity of the ecliptic at the observation time:

$$\varepsilon_0 = 23° \, 26' \, 21.448'' - 46.8150''T$$
$$- 0.00059''T^2 + 0.001813''T^3 \,.$$

The mean obliquity means that periodic perturbations have been omitted. The formula is valid a few centuries before and after the year 2000.

The true obliquity of the ecliptic, ε, is obtained by adding the nutation correction to the mean obliquity:

$$\varepsilon = \varepsilon_0 + \Delta\varepsilon \,.$$

The effect of the nutation on the ecliptic longitude (denoted usually by $\Delta\psi$) and the obliquity of the ecliptic can be found from

$$C_1 = 125° - 0.05295°t \,,$$
$$C_2 = 200.9° + 1.97129°t \,,$$
$$\Delta\psi = -0.0048° \sin C_1 - 0.0004° \sin C_2 \,,$$
$$\Delta\varepsilon = 0.0026° \cos C_1 + 0.0002° \cos C_2 \,.$$

Since $\Delta\psi$ and $\Delta\varepsilon$ are very small angles, we have, for example, $\sin \Delta\psi \approx \Delta\psi$ and $\cos \Delta\psi \approx 1$, when the angles are expressed in radians. Thus we get the nutation matrix

$$\boldsymbol{N} = \begin{pmatrix} 1 & -\Delta\psi \cos \varepsilon & -\Delta\psi \sin \varepsilon \\ \Delta\psi \cos \varepsilon & 1 & -\Delta\varepsilon \\ \Delta\psi \sin \varepsilon & \Delta\varepsilon & 1 \end{pmatrix} \,.$$

This is a linearized version of the full transformation. The angles here must be in radians. The place in the coordinate frame of the observing time is now

$$\boldsymbol{p}_2 = \boldsymbol{N}\boldsymbol{p}_1 \,.$$

4. The annual aberration can affect the place about as much as the nutation. Approximate corrections are obtained from

$$\Delta\alpha \cos \delta = -20.5'' \sin \alpha \sin \lambda$$
$$- 18.8'' \cos \alpha \cos \lambda \,,$$
$$\Delta\delta = 20.5'' \cos \alpha \sin \delta \sin \lambda$$
$$+ 18.8'' \sin \alpha \sin \delta \cos \lambda - 8.1'' \cos \delta \cos \lambda \,,$$

where λ is the ecliptic longitude of the Sun. Sufficiently accurate value for this purpose is given by

$$G = 357.528° + 0.985600°t \,,$$
$$\lambda = 280.460° + 0.985647°t$$
$$+ 1.915° \sin G + 0.020° \sin 2G \,.$$

These reductions give the apparent place of the date with an accuracy of a few seconds of arc. The effects of parallax and diurnal aberration are even smaller.

Example. The coordinates of Regulus (α Leo) for the epoch J2000.0 are

$$\alpha = 10\,\text{h}\ 8\,\text{min}\ 22.2\,\text{s} = 10.139500\,\text{h}\ ,$$
$$\delta = 11°\ 58'\ 02'' = 11.967222°\ .$$

Find the apparent place of Regulus on March 12, 1995.
We start by finding the unit vector corresponding to the catalogued place:

$$\boldsymbol{p}_0 = \begin{pmatrix} -0.86449829 \\ 0.45787318 \\ 0.20735204 \end{pmatrix}\ .$$

The Julian date is $J = 2{,}449{,}789.0$, and thus $t = -1756$ and $T = -0.04807666$. The angles of the precession matrix are $\zeta = -0.03079849°$, $z = -0.03079798°$ and $\theta = -0.02676709°$. The precession matrix is then

$$\boldsymbol{P} =$$
$$\begin{pmatrix} 0.99999931 & 0.00107506 & 0.00046717 \\ -0.00107506 & 0.99999942 & -0.00000025 \\ -0.00046717 & -0.00000025 & 0.99999989 \end{pmatrix}\ .$$

The precessed unit vector is

$$\boldsymbol{p}_1 = \begin{pmatrix} -0.86390858 \\ 0.45880225 \\ 0.20775577 \end{pmatrix}\ .$$

The angles needed for the nutation are $\Delta\psi = 0.00309516°$, $\Delta\varepsilon = -0.00186227°$, $\varepsilon = 23.43805403°$, which give the nutation matrix

$$\boldsymbol{N} =$$
$$\begin{pmatrix} 1 & -0.00004956 & -0.00002149 \\ 0.00004956 & 1 & 0.00003250 \\ 0.00002149 & -0.00003250 & 1 \end{pmatrix}\ .$$

The place in the frame of the date is

$$\boldsymbol{p}_2 = \begin{pmatrix} -0.86393578 \\ 0.45876618 \\ 0.20772230 \end{pmatrix}\ ,$$

whence

$$\alpha = 10.135390\,\text{h}\ ,$$
$$\delta = 11.988906°\ .$$

To correct for the aberration we first find the longitude of the Sun: $G = -1373.2° = 66.8°$, $\lambda = -8.6°$. The correction terms are then

$$\Delta\alpha = 18.25'' = 0.0050°$$
$$\Delta\delta = -5.46'' = -0.0015°\ .$$

Adding these to the previously obtained coordinates we get the apparent place of Regulus on March 12, 1995:

$$\alpha = 10.1357\,\text{h} = 10\,\text{h}\ 8\,\text{min}\ 8.5\,\text{s},$$
$$\delta = 11.9874° = 11°\ 59'\ 15''\ .$$

Comparison with the places given in the catalogue *Apparent Places of Fundamental Stars* shows that we are within about $3''$ of the correct place, which is a satisfactory result.

* Julian Date

There are several methods for finding the Julian date. The following one, developed by Fliegel and Van Flandern in 1968, is well adapted for computer programs. Let y be the year (with all four digits), m the month and d the day. The Julian date J at noon is then

$$J = 367y - \{7[y+(m+9)/12]\}/4$$
$$- \left(3\{[y+(m-9)/7]/100+1\}\right)/4$$
$$+ 275m/9 + d + 1721029\ .$$

The division here means an integer division, the decimal part being truncated: e.g. $7/3 = 2$ and $-7/3 = -2$.

Example. Find the Julian date on January 1, 1990. Now $y = 1990$, $m = 1$ and $d = 1$.

$$J = 367 \times 1990 - 7 \times [1990+(1+9)/12]/4$$
$$- 3 \times \{[1990+(1-9)/7]/100+1\}/4$$
$$+ 275 \times 1/9 + 1 + 1{,}721{,}029$$
$$= 730{,}330 - 3482 - 15 + 30 + 1 + 1{,}721{,}029$$
$$= 2{,}447{,}893\ .$$

Astronomical tables usually give the Julian date at 0 UT. In this case that would be $2{,}447{,}892.5$.
The inverse procedure is a little more complicated. In the following J is the Julian date at noon (so that it

will be an integer):

$a = J + 68,569$,

$b = (4a)/146,097$,

$c = a - (146,097b + 3)/4$,

$d = [4000(c + 1)]/1,461,001$,

$e = c - (1461d)/4 + 31$,

$f = (80e)/2447$,

$\mathrm{day} = e - (2447f)/80$,

$g = f/11$,

$\mathrm{month} = f + 2 - 12g$,

$\mathrm{year} = 100(b - 49) + d + g$.

Example. In the previous example we got $J = 2,447,893$. Let's check this by calculating the corresponding calendar date:

$a = 2,447,893 + 68,569 = 2,516,462$,

$b = (4 \times 2,516,462)/146,097 = 68$,

$c = 2,516,462 - (146,097 \times 68 + 3)/4 = 32,813$,

$d = [4000(32,813 + 1)]/1,461,001 = 89$,

$e = 32,813 - (1461 \times 89)/4 + 31 = 337$,

$f = (80 \times 337)/2447 = 11$,

$\mathrm{day} = 337 - (2447 \times 11)/80 = 1$,

$g = 11/11 = 1$,

$\mathrm{month} = 11 + 2 - 12 \times 1 = 1$,

$\mathrm{year} = 100(68 - 49) + 89 + 1 = 1990$.

Thus we arrived back to the original date.

Since the days of the week repeat in seven day cycles, the remainder of the division $J/7$ unambiguously determines the day of the week. If J is the Julian date at noon, the remainder of $J/7$ tells the day of the week in the following way:

$0 = \mathrm{Monday}$,

⋮

$5 = \mathrm{Saturday}$,

$6 = \mathrm{Sunday}$.

Example. The Julian date corresponding to January 1, 1990 was 2,447,893. Since $2,447,893 = 7 \times 349,699$, the remainder is zero, and the day was Monday.

2.16 Examples

Example 2.1 *Trigonometric Functions in a Rectangular Spherical Triangle*

Let the angle A be a right angle. When the figure is a plane triangle, the trigonometric functions of the angle B would be:

$$\sin B = b/a , \quad \cos B = c/a , \quad \tan B = b/c .$$

For the spherical triangle we have to use the equations in (2.7), which are now simply:

$\sin B \sin a = \sin b$,

$\cos B \sin a$
  $= \cos b \sin c$,

$\cos a = \cos b \cos c$.

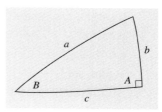

The first equation gives the sine of B:

$$\sin B = \sin b / \sin a .$$

Dividing the second equation by the third one, we get the cosine of B:

$$\cos B = \tan c / \tan a .$$

And the tangent is obtained by dividing the first equation by the second one:

$$\tan B = \tan b / \sin c .$$

The third equation is the equivalent of the Pythagorean theorem for rectangular triangles.

Example 2.2 *The Coordinates of New York City*

The geographic coordinates are 41° north and 74° west of Greenwich, or $\phi = +41°$, $\lambda = -74°$. In time units, the longitude would be $74/15\,\mathrm{h} = 4\,\mathrm{h}\ 56\,\mathrm{min}$ west of Greenwich. The geocentric latitude is obtained from

$$\tan \phi' = \frac{b^2}{a^2} \tan \phi = \left(\frac{6,356,752}{6,378,137} \right)^2 \tan 41°$$

$$= 0.86347 \quad \Rightarrow \quad \phi' = 40°\ 48'\ 34'' .$$

The geocentric latitude is $11'\ 26''$ less than the geographic latitude.

Example 2.3 The angular separation of two objects in the sky is quite different from their coordinate difference.

Suppose the coordinates of a star A are $\alpha_1 = 10$ h, $\delta_1 = 70°$ and those of another star B, $\alpha_2 = 11$ h, $\delta_2 = 80°$.

Using the Pythagorean theorem for plane triangles, we would get

$$d = \sqrt{(15°)^2 + (10°)^2} = 18° \ .$$

But if we use the third equation in (2.7), we get

$$
\begin{aligned}
\cos d &= \cos(\alpha_1 - \alpha_2) \\
&\quad \times \sin(90° - \delta_1)\sin(90° - \delta_2) \\
&\quad + \cos(90° - \delta_1)\cos(90° - \delta_2) \\
&= \cos(\alpha_1 - \alpha_2)\cos\delta_1\cos\delta_2 \\
&\quad + \sin\delta_1\sin\delta_2 \\
&= \cos 15°\cos 70°\cos 80° \\
&\quad + \sin 70°\sin 80° \\
&= 0.983 \ ,
\end{aligned}
$$

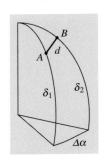

which yields $d = 10.6°$. The figure shows why the result obtained from the Pythagorean theorem is so far from being correct: hour circles (circles with $\alpha = $ constant) approach each other towards the poles and their angular separation becomes smaller, though the coordinate difference remains the same.

Example 2.4 Find the altitude and azimuth of the Moon in Helsinki at midnight at the beginning of 1996.

The right ascension is $\alpha = 2$ h 55 min 7 s $= 2.9186$ h and declination $\delta = 14°\ 42' = 14.70°$, the sidereal time is $\Theta = 6$ h 19 min 26 s $= 6.3239$ h and latitude $\phi = 60.16°$.

The hour angle is $h = \Theta - \alpha = 3.4053$ h $= 51.08°$. Next we apply the equations in (2.16):

$$
\begin{aligned}
\sin A\cos a &= \sin 51.08°\cos 14.70° = 0.7526 \ , \\
\cos A\cos a &= \cos 51.08°\cos 14.70°\sin 60.16° \\
&\quad - \sin 14.70°\cos 60.16° \\
&= 0.4008 \ , \\
\sin a &= \cos 51.08°\cos 14.70°\cos 60.16° \\
&\quad + \sin 14.70°\sin 60.16° \\
&= 0.5225.
\end{aligned}
$$

Thus the altitude is $a = 31.5°$. To find the azimuth we have to compute its sine and cosine:

$$\sin A = 0.8827 \ , \quad \cos A = 0.4701 \ .$$

Hence the azimuth is $A = 62.0°$. The Moon is in the southwest, 31.5 degrees above the horizon. Actually, this would be the direction if the Moon were infinitely distant.

Example 2.5 Find the topocentric place of the Moon in the case of the previous example.

The geocentric distance of the Moon at that time is $R = 62.58$ equatorial radii of the Earth. For simplicity, we can assume that the Earth is spherical.

We set up a rectangular coordinate frame in such a way that the z axis points towards the celestial pole and the observing site is in the xz plane. When the radius of the Earth is used as the unit of distance, the radius vector of the observing site is

$$
r_0 = \begin{pmatrix} \cos\phi \\ 0 \\ \sin\phi \end{pmatrix} = \begin{pmatrix} 0.4976 \\ 0 \\ 0.8674 \end{pmatrix} \ .
$$

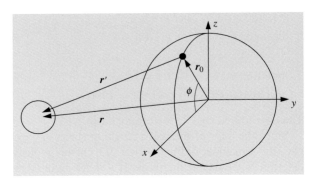

The radius vector of the Moon is

$$
r = R\begin{pmatrix} \cos\delta\cos h \\ -\cos\delta\sin h \\ \sin\delta \end{pmatrix} = 62.58\begin{pmatrix} 0.6077 \\ -0.7526 \\ 0.2538 \end{pmatrix} \ .
$$

The topocentric place of the Moon is

$$
r' = r - r_0 = \begin{pmatrix} 37.53 \\ -47.10 \\ 15.02 \end{pmatrix} \ .
$$

We divide this vector by its length 62.07 to get the unit vector e pointing to the direction of the Moon. This can be expressed in terms of the topocentric coordinates δ' and h':

$$e = \begin{pmatrix} 0.6047 \\ -0.7588 \\ 0.2420 \end{pmatrix} = \begin{pmatrix} \cos \delta' \cos h' \\ -\cos \delta' \sin h' \\ \sin \delta' \end{pmatrix},$$

which gives $\delta' = 14.00°$ and $h' = 51.45°$. Next we can calculate the altitude and azimuth as in the previous example, and we get $a = 30.7°$, $A = 61.9°$.

Another way to find the altitude is to take the scalar product of the vectors e and r_0, which gives the cosine of the zenith distance:

$$\cos z = e \cdot r_0 = 0.6047 \times 0.4976 + 0.2420 \times 0.8674$$
$$= 0.5108,$$

whence $z = 59.3°$ and $a = 90° - z = 30.7°$. We see that this is $0.8°$ less than the geocentric altitude; i.e. the difference is more than the apparent diameter of the Moon.

Example 2.6 The coordinates of Arcturus are $\alpha = 14\,h\,15.7\,min$, $\delta = 19° \, 1'1'$. Find the sidereal time at the moment Arcturus rises or sets in Boston ($\phi = 42° \, 19'$).

Neglecting refraction, we get

$$\cos h = -\tan 19° \, 11' \tan 42° \, 19'$$
$$= -0.348 \times 0.910 = -0.317.$$

Hence, $h = \pm 108.47° = 7\,h\,14\,min$. The more accurate result is

$$\cos h = -\tan 19° \, 11' \tan 42° \, 19'$$
$$- \frac{\sin 35'}{\cos 19° \, 11' \cos 42° 19'}$$
$$= -0.331,$$

whence $h = \pm 109.35° = 7\,h\,17\,min$. The plus and minus signs correspond to setting ant rising, respectively. When Arcturus rises, the sidereal time is

$$\Theta = \alpha + h = 14\,h\,16\,min - 7\,h\,17\,min$$
$$= 6\,h\,59\,min$$

and when it sets, the sidereal time is

$$\Theta = 14\,h\,16\,min + 7\,h\,17\,min$$
$$= 21\,h\,33\,min.$$

Note that the result is independent of the date: a star rises and sets at the same sidereal time every day.

Example 2.7 The proper motion of Aldebaran is $\mu = 0.20''/a$ and parallax $\pi = 0.048''$. The spectral line of iron at $\lambda = 440.5\,nm$ is displaced $0.079\,nm$ towards the red. What are the radial and tangential velocities and the total velocity?

The radial velocity is found from

$$\frac{\Delta\lambda}{\lambda} = \frac{v_r}{c}$$
$$\Rightarrow v_r = \frac{0.079}{440.5} \cdot 3 \times 10^8 \, m/s = 5.4 \times 10^4 \, m/s$$
$$= 54\,km/s.$$

The tangential velocity is now given by (2.40), since μ and π are in correct units:

$$v_t = 4.74\mu r = 4.74\mu/\pi = \frac{4.74 \times 0.20}{0.048} = 20\,km/s.$$

The total velocity is

$$v = \sqrt{v_r^2 + v_t^2} = \sqrt{54^2 + 20^2} \, km/s = 58\,km/s.$$

Example 2.8 Find the local time in Paris (longitude $\lambda = 2°$) at 12:00.

Local time coincides with the zonal time along the meridian 15° east of Greenwich. Longitude difference $15° - 2° = 13°$ equals $(13°/15°) \times 60\,min = 52$ minutes. The local time is 52 minutes less than the official time, or 11:08. This is mean solar time. To find the true solar time, we must add the equation of time. In early February, E.T. $= -14\,min$ and the true solar time is $11:08 - 14\,min = 10:54$. At the beginning of November, ET $= +16\,min$ and the solar time would be 11:24. Since $-14\,min$ and $+16\,min$ are the extreme values of E.T., the true solar time is in the range 10:54–11:24, the exact time depending on the day of the year. During daylight saving time, we must still subtract one hour from these times.

Example 2.9 *Estimating Sidereal Time*

Since the sidereal time is the hour angle of the vernal equinox V, it is $0\,h$ when V culminates or transits the south meridian. At the moment of the vernal equinox, the Sun is in the direction of ♈ and thus culminates at the same time as ♈. So the sidereal time at 12:00 local solar time is 0:00, and at the time of the vernal equinox, we have

$$\Theta = T + 12\,\text{h}\,,$$

where T is the local solar time. This is accurate within a couple of minutes. Since the sidereal time runs about 4 minutes fast a day, the sidereal time, n days after the vernal equinox, is

$$\Theta \approx T + 12\,\text{h} + n \times 4\,\text{min}\,.$$

At autumnal equinox ♈ culminates at 0:00 local time, and sidereal and solar times are equal.

Let us try to find the sidereal time in Paris on April 15 at 22:00, Central European standard time ($= 23:00$ daylight saving time). The vernal equinox occurs on the average on March 21; thus the time elapsed since the equinox is $10 + 15 = 25$ days. Neglecting the equation of time, the local time T is 52 minutes less than the zonal time. Hence

$$\begin{aligned}
\Theta &= T + 12\,\text{h} + n \times 4\,\text{min} \\
&= 21\,\text{h}\,8\,\text{min} + 12\,\text{h} + 25 \times 4\,\text{min} \\
&= 34\,\text{h}\,48\,\text{min} = 10\,\text{h}\,48\,\text{min}\,.
\end{aligned}$$

The time of the vernal equinox can vary about one day in either direction from the average. Therefore the accuracy of the result is roughly 5 min.

Example 2.10 Find the rising time of Arcturus in Boston on January 10.

In Example 2.6 we found the sidereal time of this event, $\Theta = 6\,\text{h}\,59\,\text{min}$. Since we do not know the year, we use the rough method of Example 2.9. The time between January 1 and vernal equinox (March 21) is about 70 days. Thus the sidereal time on January 1 is

$$\Theta \approx T + 12\,\text{h} - 70 \times 4\,\text{min} = T + 7\,\text{h}\,20\,\text{min}\,,$$

from which

$$\begin{aligned}
T &= \Theta - 7\,\text{h}\,20\,\text{min} = 6\,\text{h}\,59\,\text{min} - 7\,\text{h}\,20\,\text{min} \\
&= 30\,\text{h}\,59\,\text{min} - 7\,\text{h}\,20\,\text{min} = 23\,\text{h}\,39\,\text{min}\,.
\end{aligned}$$

The longitude of Boston is $71°$ W, and the Eastern standard time is $(4°/15°) \times 60\,\text{min} = 16$ minutes less, or 23:23.

Example 2.11 Find the sidereal time in Helsinki on April 15, 1982 at 20:00 UT.

The Julian date is $J = 2{,}445{,}074.5$ and

$$\begin{aligned}
T &= \frac{2{,}445{,}074.5 - 2{,}451{,}545.0}{36{,}525} \\
&= -0.1771526\,.
\end{aligned}$$

Next, we use (2.47) to find the sidereal time at 0 UT:

$$\begin{aligned}
\Theta_0 &= -1{,}506{,}521.0\,\text{s} = -418\,\text{h}\,28\,\text{min}\,41\,\text{s} \\
&= 13\,\text{h}\,31\,\text{min}\,19\,\text{s}\,.
\end{aligned}$$

Since the sidereal time runs 3 min 57 s fast a day as compared to the solar time, the difference in 20 hours will be

$$\frac{20}{24} \times 3\,\text{min}\,57\,\text{s} = 3\,\text{min}\,17\,\text{s}\,,$$

and the sidereal time at 20 UT will be 13 h 31 min 19 s + 20 h 3 min 17 s = 33 h 34 min 36 s = 9 h 34 min 36 s.

At the same time (at 22:00 Finnish time, 23:00 daylight saving time) in Helsinki the sidereal time is ahead of this by the amount corresponding to the longitude of Helsinki, $25°$, i.e. 1 h 40 min 00 s. Thus the sidereal time is 11 h 14 min 36 s.

2.17 Exercises

Exercise 2.1 Find the distance between Helsinki and Seattle along the shortest route. Where is the northernmost point of the route, and what is its distance from the North Pole? The longitude of Helsinki is $25°$E and latitude $60°$; the longitude of Seattle is $122°$W and latitude $48°$. Assume that the radius of the Earth is 6370 km.

Exercise 2.2 A star crosses the south meridian at an altitude of $85°$, and the north meridian at $45°$. Find the declination of the star and the latitude of the observer.

Exercise 2.3 Where are the following statements true?

a) Castor (α Gem, declination $\delta = 31°53'$) is circumpolar.

b) Betelgeuze (α Ori, $\delta = 7° 24'$) culminates at zenith.

c) α Cen ($\delta = -60° 50'$) rises to an altitude of $30°$.

Exercise 2.4   In his *Old Man and the Sea* Hemingway wrote:

> It was dark now as it becomes dark quickly after the Sun sets in September. He lay against the worn wood of the bow and rested all that he could. The first stars were out. He did not know the name of Rigel but he saw it and knew soon they would all be out and he would have all his distant friends.

How was Hemingway's astronomy?

Exercise 2.5   The right ascension of the Sun on June 1, 1983, was 4 h 35 min and declination $22° 00'$. Find the ecliptic longitude and latitude of the Sun and the Earth.

Exercise 2.6   Show that on the Arctic Circle the Sun

a) rises at the same sidereal time Θ_0 between December 22 and June 22,

b) sets at the same sidereal time Θ_0 between June 22 and December 22.

What is Θ_0?

Exercise 2.7   Derive the equations (2.24), which give the galactic coordinates as functions of the ecliptic coordinates.

Exercise 2.8   The coordinates of Sirius for the epoch 1900.0 were $\alpha = 6$ h 40 min 45 s, $\delta = -16° 35'$, and the components of its proper motion were $\mu_\alpha = -0.037$ s/a, $\mu_\delta = -1.12''a^{-1}$. Find the coordinates of Sirius for 2000.0. The precession must also be taken into account.

Exercise 2.9   The parallax of Sirius is $0.375''$ and radial velocity -8 km/s.

a) What are the tangential and total velocities of Sirius? (See also the previous exercise.)

b) When will Sirius be closest to the Sun?

c) What will its proper motion and parallax be then?

3. Observations and Instruments

Up to the end of the Middle Ages, the most important means of observation in astronomy was the human eye. It was aided by various mechanical devices to measure the positions of celestial bodies in the sky. The telescope was invented in Holland at the beginning of the 17th century, and in 1609 Galileo Galilei made his first astronomical observations with this new instrument. Astronomical photography was introduced at the end of the 19th century, and during the last few decades many kinds of electronic detectors have been adopted for the study of electromagnetic radiation from space. The electromagnetic spectrum from the shortest gamma rays to long radio waves can now be used for astronomical observations.

3.1 Observing Through the Atmosphere

With satellites and spacecraft, observations can be made outside the atmosphere. Yet, the great majority of astronomical observations are carried out from the surface of the Earth. In the preceding chapter, we discussed refraction, which changes the apparent altitudes of objects. The atmosphere affects observations in many other ways as well. The air is never quite steady, and there are layers with different temperatures and densities; this causes convection and turbulence. When the light from a star passes through the unsteady air, rapid changes in refraction in different directions result. Thus, the amount of light reaching a detector, e. g. the human eye, constantly varies; the star is said to *scintillate* (Fig. 3.1). Planets shine more steadily, since they are not point sources like the stars.

A telescope collects light over a larger area, which evens out rapid changes and diminishes scintillation. Instead, differences in refraction along different paths of light through the atmosphere smear the image and point sources are seen in telescopes as vibrating speckles. This phenomenon is called *seeing*, and the size of the seeing disc may vary from less than an arc second to several tens of arc seconds. If the size of the seeing disc is small, we speak of good seeing. Seeing and scintillation both tend to blot out small details when one looks through a telescope, for example, at a planet.

Some wavelength regions in the electromagnetic spectrum are strongly absorbed by the atmosphere. The most important transparent interval is the *optical window* from about 300 to 800 nm. This interval coincides with the region of sensitivity of the human eye (about 400–700 nm).

At wavelengths under 300 nm absorption by atmospheric ozone prevents radiation from reaching the ground. The ozone is concentrated in a thin layer at a height of about 20–30 km, and this layer protects the Earth from harmful ultraviolet radiation. At still shorter wavelengths, the main absorbers are O_2, N_2 and free atoms. Nearly all of the radiation under 300 nm is absorbed by the upper parts of the atmosphere.

At wavelengths longer than visible light, in the near-infrared region, the atmosphere is fairly transparent up

Fig. 3.1. Scintillation of Sirius during four passes across the field of view. The star was very low on the horizon. (Photo by Pekka Parviainen)

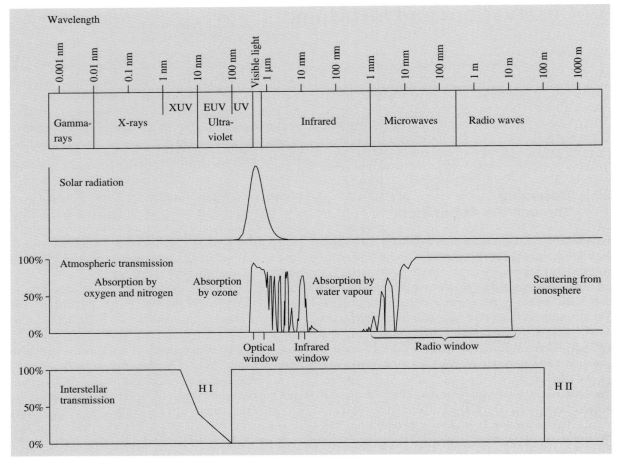

Fig. 3.2. The transparency of the atmosphere at different wavelengths. 100% transmission means that all radiation reaches the surface of the Earth. The radiation is also absorbed by interstellar gas, as shown in the lowermost very schematic figure. The interstellar absorption also varies very much depending on the direction (Chap. 15)

to 1.3 μm. There are some absorption belts caused by water and molecular oxygen, but the atmosphere gets more opaque only at wavelengths of longer than 1.3 μm. At these wavelengths, radiation reaches the lower parts of the atmosphere only in a few narrow windows. All wavelengths between 20 μm and 1 mm are totally absorbed. At wavelengths longer than 1 mm, there is the *radio window* extending up to about 20 m. At still longer wavelengths, the ionosphere in the upper parts of the atmosphere reflects all radiation (Fig. 3.2). The exact upper limit of the radio window depends on the strength of the ionosphere, which varies during the day. (The structure of the atmosphere is described in Chap. 7.)

At optical wavelengths (300–800 nm), light is scattered by the molecules and dust in the atmosphere, and the radiation is attenuated. Scattering and absorption together are called *extinction*. Extinction must be taken into account when one measures the brightness of celestial bodies (Chap. 4).

In the 19th century Lord Rayleigh succeeded in explaining why the sky is blue. Scattering caused by the molecules in the atmosphere is inversely proportional to the fourth power of the wavelength. Thus, blue light is scattered more than red light. The blue light we see all over the sky is scattered sunlight. The same phenomenon colours the setting sun red, because owing to the long, oblique path through the atmosphere, all the blue light has been scattered away.

In astronomy one often has to observe very faint objects. Thus, it is important that the background sky be as dark as possible, and the atmosphere as transparent as possible. That is why the large observatories have been built on mountain tops far from the cities. The air above an observatory site must be very dry, the number of cloudy nights few, and the seeing good.

Astronomers have looked all over the Earth for optimal conditions and have found some exceptional sites. In the 1970's, several new major observatories were founded at these sites. Among the best sites in the world are: the extinguished volcano Mauna Kea on Hawaii, rising more than 4000 m above the sea; the dry mountains in northern Chile; the Sonoran desert in the U.S., near the border of Mexico; and the mountains on La Palma, in the Canary Islands. Many older observatories are severely plagued by the lights of nearby cities (Fig. 3.3).

In radio astronomy atmospheric conditions are not very critical except when observing at the shortest wavelengths. Constructors of radio telescopes have much greater freedom in choosing their sites than optical astronomers. Still, radio telescopes are also often constructed in uninhabited places to isolate them from disturbing radio and television broadcasts.

3.2 Optical Telescopes

The telescope fulfills three major tasks in astronomical observations:

1. It collects light from a large area, making it possible to study very faint sources.
2. It increases the apparent angular diameter of the object and thus improves resolution.
3. It is used to measure the positions of objects.

The light-collecting surface in a telescope is either a lens or a mirror. Thus, optical telescopes are divided into two types, lens telescopes or *refractors* and mirror telescopes or *reflectors* (Fig. 3.4).

Geometrical Optics. Refractors have two lenses, the *objective* which collects the incoming light and forms an image in the focal plane, and the *eyepiece* which is a small magnifying glass for looking at the image (Fig. 3.5). The lenses are at the opposite ends of a tube which can be directed towards any desired point. The distance between the eyepiece and the focal plane can be adjusted to get the image into focus. The image formed

Fig. 3.3. Night views from the top of Mount Wilson. The upper photo was taken in 1908, the lower one in 1988. The lights of Los Angeles, Pasadena, Hollywood and more than 40 other towns are reflected in the sky, causing considerable disturbance to astronomical observations. (Photos by Ferdinand Ellerman and International Dark-Sky Association)

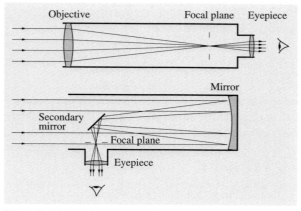

Fig. 3.4. A lens telescope or refractor and a mirror telescope or reflector

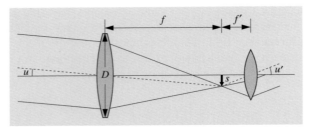

Fig. 3.5. The scale and magnification of a refractor. The object subtends an angle u. The objective forms an image of the object in the focal plane. When the image is viewed through the eyepiece, it is seen at an angle u'

by the objective lens can also be registered, e.g. on a photographic film, as in an ordinary camera.

The diameter of the objective, D, is called the *aperture* of the telescope. The ratio of the aperture D to the focal length f, $F = D/f$, is called the *aperture ratio*. This quantity is used to characterize the light-gathering power of the telescope. If the aperture ratio is large, near unity, one has a powerful, "fast" telescope; this means that one can take photographs using short exposures, since the image is bright. A small aperture ratio (the focal length much greater than the aperture) means a "slow" telescope.

In astronomy, as in photography, the aperture ratio is often denoted by f/n (e.g. $f/8$), where n is the focal length divided by the aperture. For fast telescopes this ratio can be $f/1 \ldots f/3$, but usually it is smaller, $f/8 \ldots f/15$.

The *scale* of the image formed in the focal plane of a refractor can be geometrically determined from Fig. 3.5. When the object is seen at the angle u, it forms an image of height s,

$$s = f \tan u \approx fu , \tag{3.1}$$

since u is a very small angle. If the telescope has a focal length of, for instance, 343 cm, one arc minute corresponds to

$$s = 343 \, \text{cm} \times 1'$$
$$= 343 \, \text{cm} \times (1/60) \times (\pi/180)$$
$$= 1 \, \text{mm} .$$

The *magnification* ω is (from Fig. 3.5)

$$\omega = u'/u \approx f/f' , \tag{3.2}$$

where we have used the equation $s = fu$. Here, f is the focal length of the objective and f' that of the eyepiece. For example, if $f = 100$ cm and we use an eyepiece with $f' = 2$ cm, the magnification is 50-fold. The magnification is not an essential feature of a telescope, since it can be changed simply by changing the eyepiece.

A more important characteristic, which depends on the aperture of the telescope, is the *resolving power*, which determines, for example, the minimum angular separation of the components of a binary star that can be seen as two separate stars. The theoretical limit for the resolution is set by the diffraction of light: The telescope does not form a point image of a star, but rather a small disc, since light "bends around the corner" like all radiation (Fig. 3.6).

The theoretical resolution of a telescope is often given in the form introduced by Rayleigh (see *Diffraction by a Circular Aperture, p. 78)

$$\sin \theta \approx \theta = 1.22 \, \lambda/D , \quad [\theta] = \text{rad} . \tag{3.3}$$

As a practical rule, we can say that two objects are seen as separate if the angular distance between them is

$$\theta \gtrsim \lambda/D , \quad [\theta] = \text{rad} . \tag{3.4}$$

This formula can be applied to optical as well as radio telescopes. For example, if one makes observations at a typical yellow wavelength ($\lambda = 550$ nm), the resolving power of a reflector with an aperture of 1 m is about $0.2''$. However, seeing spreads out the image to a diameter of typically one arc second. Thus, the theoretical diffraction limit cannot usually be reached on the surface of the Earth.

In photography the image is further spread in the photographic plate, decreasing the resolution as compared with visual observations. The grain size of photographic emulsions is about 0.01–0.03 mm, which is also the minimum size of the image. For a focal length of 1 m, the scale is 1 mm $= 206''$, and thus 0.01 mm corresponds to about 2 arc seconds. This is similar to the theoretical resolution of a telescope with an aperture of 7 cm in visual observations.

In practice, the resolution of visual observations is determined by the ability of the eye to see details.

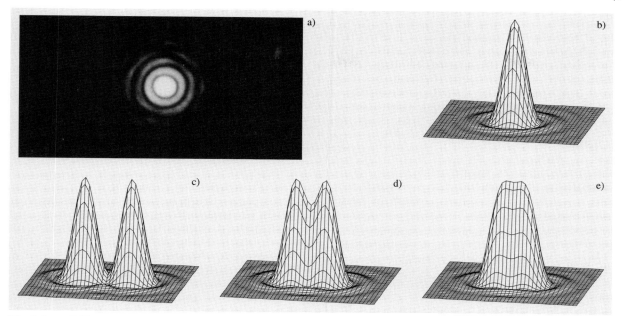

Fig. 3.6a–e. Diffraction and resolving power. The image of a single star (**a**) consists of concentric diffraction rings, which can be displayed as a mountain diagram (**b**). Wide pairs of stars can be easily resolved (**c**). For resolving close binaries, different criteria can be used. One is the Rayleigh limit $1.22\,\lambda/D$ (**d**). In practice, the resolution can be written λ/D, which is near the Dawes limit (**e**). (Photo (a) Sky and Telescope)

In night vision (when the eye is perfectly adapted to darkness) the resolving capability of the human eye is about $2'$.

The *maximum magnification* ω_{max} is the largest magnification that is worth using in telescopic observations. Its value is obtained from the ratio of the resolving capability of the eye, $e \approx 2' = 5.8 \times 10^{-4}$ rad, to the resolving power of the telescope, θ,

$$\omega_{\mathrm{max}} = e/\theta \approx eD/\lambda = \frac{5.8 \times 10^{-4}\,D}{5.5 \times 10^{-7}\,\mathrm{m}}$$

$$\approx D/1\,\mathrm{mm}\,. \tag{3.5}$$

If we use, for example, an objective with a diameter of 100 mm, the maximum magnification is about 100. The eye has no use for larger magnifications.

The *minimum magnification* ω_{min} is the smallest magnification that is useful in visual observations. Its value is obtained from the condition that the diameter of the *exit pupil L* of the telescope must be smaller than or equal to the pupil of the eye.

The exit pupil is the image of the objective lens, formed by the eyepiece, through which the light from the objective goes behind the eyepiece. From Fig. 3.7 we obtain

$$L = \frac{f'}{f}D = \frac{D}{\omega}\,. \tag{3.6}$$

Thus the condition $L \leq d$ means that

$$\omega \geq D/d\,. \tag{3.7}$$

In the night, the diameter of the pupil of the human eye is about 6 mm, and thus the minimum magnification of a 100 mm telescope is about 17.

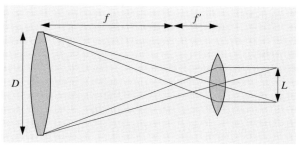

Fig. 3.7. The exit pupil L is the image of the objective lens formed by the eyepiece

Refractors. In the first refractors, which had a simple objective lens, the observations were hampered by the *chromatic aberration*. Since glass refracts different colours by different amounts, all colours do not meet at the same focal point (Fig. 3.8), but the focal length increases with increasing wavelength. To remove this aberration, *achromatic lenses* consisting of two parts were developed in the 18th century. The colour dependence of the focal length is much smaller than in single lenses, and at some wavelength, λ_0, the focal length has an extremum (usually a minimum). Near this point the change of focal length with wavelength is very small (Fig. 3.9). If the telescope is intended for visual observations, we choose $\lambda_0 = 550$ nm, corresponding to the maximum sensitivity of the eye. Objectives for photographic refractors are usually constructed with $\lambda_0 \approx 425$ nm, since normal photographic plates are most sensitive to the blue part of the spectrum.

By combining three or even more lenses of different glasses in the objective, the chromatic aberration can be corrected still better (as in apochromatic objectives). Also, special glasses have been developed where the wavelength dependences of the refractive index cancel out so well that two lenses already give a very good correction of the chromatic aberration. They have, however, hardly been used in astronomy so far.

The largest refractors in the world have an aperture of about one metre (102 cm in the Yerkes Observatory telescope (Fig. 3.10), finished in 1897, and 91 cm in the Lick Observatory telescope (1888)). The aperture ratio is typically $f/10 \ldots f/20$.

The use of refractors is limited by their small field of view and awkwardly long structure. Refractors are

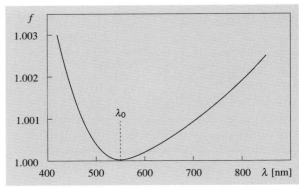

Fig. 3.9. The wavelength dependence of the focal length of a typical achromatic objective for visual observations. The focal length has a minimum near $\lambda = 550$ nm, where the eye is most sensitive. In bluer light ($\lambda = 450$ nm) or in redder light ($\lambda = 800$ nm), the focal length increases by a factor of about 1.002

used, e.g. for visual observations of binary stars and in various meridian telescopes for measuring the positions of stars. In photography they can be used for accurate position measurements, for example, to find parallaxes.

A wider field of view is obtained by using more complex lens systems, and telescopes of this kind are called *astrographs*. Astrographs have an objective made up of typically 3–5 lenses and an aperture of less than 60 cm. The aperture ratio is $f/5 \ldots f/7$ and the field of view about 5°. Astrographs are used to photograph large areas of the sky, e.g. for proper motion studies and for statistical brightness studies of the stars.

Reflectors. The most common telescope type in astrophysical research is the mirror telescope or reflector. As a light-collecting surface, it employs a mirror coated with a thin layer of aluminium. The form of the mirror is usually parabolic. A parabolic mirror reflects all light rays entering the telescope parallel to the main axis into the same focal point. The image formed at this point can be observed through an eyepiece or registered with a detector. One of the advantages of reflectors is the absence of chromatic aberration, since all wavelengths are reflected to the same point.

In the very largest telescopes, the observer can sit with his instruments in a special cage at the *primary*

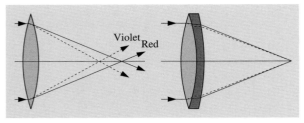

Fig. 3.8. Chromatic aberration. Light rays of different colours are refracted to different focal points (*left*). The aberration can be corrected with an achromatic lens consisting of two parts (*right*)

Fig. 3.10. The largest refractor in the world is at the Yerkes Observatory, University of Chicago. It has an objective lens with a diameter of 102 cm. (Photo by Yerkes Observatory)

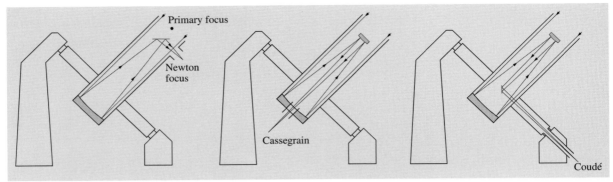

Fig. 3.11. Different locations of the focus in reflectors: primary focus, Newton focus, Cassegrain focus and coudé focus. The coudé system in this figure cannot be used for observations near the celestial pole. More complex coudé systems usually have three flat mirrors after the primary and secondary mirrors

focus (Fig. 3.11) without eclipsing too much of the incoming light. In smaller telescopes, this is not possible, and the image must be inspected from outside the telescope. In modern telescopes instruments are remotely controlled, and the observer must stay away from the telescope to reduce thermal turbulence.

In 1663 James Gregory (1638–1675) described a reflector. The first practical reflector, however, was built by Isaac Newton. He guided the light perpendicularly out from the telescope with a small flat mirror. Therefore the focus of the image in such a system is called the *Newton focus*. A typical aperture ratio of a Newtonian telescope is $f/3 \ldots f/10$. Another possibility is to bore a hole at the centre of the primary mirror and reflect the rays through it with a small hyperbolic secondary mirror in the front end of the telescope. In such a design, the rays meet in the *Cassegrain focus*. Cassegrain systems have aperture ratios of $f/8 \ldots f/15$.

The effective focal length (f_e) of a Cassegrain telescope is determined by the position and convexity of the secondary mirror. Using the notations of Fig. 3.12, we get

$$f_e = \frac{b}{a} f_p . \tag{3.8}$$

If we choose $a \ll b$, we have $f_e \gg f_p$. In this way one can construct short telescopes with long focal lengths. Cassegrain systems are especially well suited for spectrographic, photometric and other instruments, which can be mounted in the secondary focus, easily accessible to the observers.

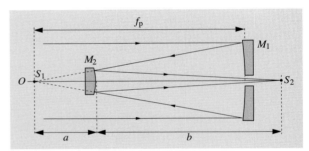

Fig. 3.12. The principle of a Cassegrain reflector. A concave (paraboloid) primary mirror M_1 reflects the light rays parallel to the main axis towards the primary focus S_1. A convex secondary mirror M_2 (hyperboloid) reflects the rays back through a small hole at the centre of the main mirror to the secondary focus S_2 outside the telescope

More complicated arrangements use several mirrors to guide the light through the declination axis of the telescope to a fixed *coudé focus* (from the French word *couder*, to bend), which can even be situated in a separate room near the telescope (Fig. 3.13). The focal length is thus very long and the aperture ratio $f/30 \ldots f/40$. The coudé focus is used mainly for accurate spectroscopy, since the large spectrographs can be stationary and their temperature can be held accurately constant. A drawback is that much light is lost in the reflections in the several mirrors of the coudé system. An aluminized mirror reflects about 80% of the light falling on it, and thus in a coudé system of, e.g. five mirrors (including the primary and secondary mirrors), only $0.8^5 \approx 30\%$ of the light reaches the detector.

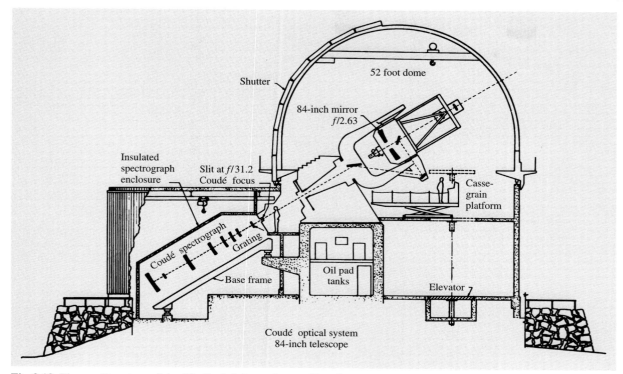

Fig. 3.13. The coudè system of the Kitt Peak 2.1 m reflector. (Drawing National Optical Astronomy Observatories, Kitt Peak National Observatory)

The reflector has its own aberration, *coma*. It affects images displaced from the optical axis. Light rays do not converge at one point, but form a figure like a comet. Due to the coma, the classical reflector with a paraboloid mirror has a very small correct field of view. The coma limits the diameter of the useful field to 2–20 minutes of arc, depending on the aperture ratio of the telescope. The 5 m Palomar telescope, for instance, has a useful field of view of about 4′, corresponding to about one-eighth of the diameter of the Moon. In practice, the small field of view can be enlarged by various correcting lenses.

If the primary mirror were spherical, there would be no coma. However, this kind of mirror has its own error, *spherical aberration*: light rays from the centre and edges converge at different points. To remove the spherical aberration, the Estonian astronomer *Bernhard Schmidt* developed a thin correcting lens that is placed in the way of the incoming light. Schmidt cameras (Figs. 3.14 and 3.15) have a very wide (about 7°), nearly faultless field of view, and the correcting lens is

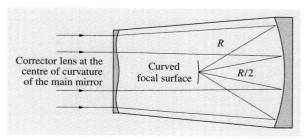

Fig. 3.14. The principle of the Schmidt camera. A correcting glass at the centre of curvature of a concave spherical mirror deviates parallel rays of light and compensates for the spherical aberration of the spherical mirror. (In the figure, the form of the correcting glass and the change of direction of the light rays have been greatly exaggerated.) Since the correcting glass lies at the centre of curvature, the image is practically independent of the incoming angle of the light rays. Thus there is no coma or astigmatism, and the images of stars are points on a spherical surface at a distance of $R/2$, where R is the radius of curvature of the spherical mirror. In photography, the plate must be bent into the form of the focal surface, or the field rectified with a corrector lens

so thin that it absorbs very little light. The images of the stars are very sharp.

In Schmidt telescopes the diaphragm with the correcting lens is positioned at the centre of the radius of curvature of the mirror (this radius equals twice the focal length). To collect all the light from the edges of the field of view, the diameter of the mirror must be larger than that of the correcting glass. The Palomar Schmidt camera, for example, has an aperture of 122 cm (correcting lens)/183 cm (mirror) and a focal length of 300 cm. The largest Schmidt telescope in the world is in Tautenburg, Germany, and its corresponding values are 134/203/400 cm.

A disadvantage of the Schmidt telescope is the curved focal plane, consisting of a part of a sphere. When the telescope is used for photography, the plate must be bent along the curved focal plane. Another possibility of correcting the curvature of the field of view is to use an extra correcting lens near the focal plane. Such a solution was developed by the Finnish astronomer Yrjö Väisälä in the 1930's, independently of Schmidt. Schmidt cameras have proved to be very effective in mapping the sky. They have been used to photograph the Palomar Sky Atlas mentioned in the previous chapter and its continuation, the *ESO/SRC Southern Sky Atlas*.

The Schmidt camera is an example of a *catadioptric telescope*, which has both lenses and mirrors. *Schmidt–Cassegrain telescopes* used by many amateurs are modifications of the Schmidt camera. They have a secondary mirror mounted at the centre of the correcting lens; the mirror reflects the image through a hole in the primary mirror. Thus the effective focal length can be rather long, although the telescope itself is very short. Another common catadioptric telescope is the *Maksutov* telescope. Both surfaces of the correcting lens as well as the primary mirror of a Maksutov telescope are concentric spheres.

Another way of removing the coma of the classical reflectors is to use more complicated mirror surfaces. The *Ritchey–Chrétien* system has hyperboloidal primary and secondary mirrors, providing a fairly wide useful field of view. Ritchey–Chrétien optics are used in many large telescopes.

Fig. 3.15. The large Schmidt telescope of the European Southern Observatory. The diameter of the mirror is 1.62 m and of the free aperture 1 m. (Photo ESO)

Mountings of Telescopes. A telescope has to be mounted on a steady support to prevent its shaking, and it must be smoothly rotated during observations. There are two principal types of mounting, *equatorial* and *azimuthal* (Fig. 3.16).

In the equatorial mounting, one of the axes is directed towards the celestial pole. It is called the *polar axis* or *hour axis*. The other one, the *declination axis*, is perpendicular to it. Since the hour axis is parallel to the axis of the Earth, the apparent rotation of the sky can be compensated for by turning the telescope around this axis at a constant rate.

The declination axis is the main technical problem of the equatorial mounting. When the telescope is pointing to the south its weight causes a force perpendicular to the axis. When the telescope is tracking an object and turns westward, the bearings must take an increasing load parallel with the declination axis.

In the azimuthal mounting, one of the axes is vertical, the other one horizontal. This mounting is easier to construct than the equatorial mounting and is more stable for very large telescopes. In order to follow the rotation of the sky, the telescope must be turned around both of the axes with changing velocities. The field of view will also rotate; this rotation must be compensated for when the telescope is used for photography.

If an object goes close to the zenith, its azimuth will change 180° in a very short time. Therefore, around the zenith there is a small region where observations with an azimuthal telescope are not possible.

The largest telescopes in the world were equatorially mounted until the development of computers made possible the more complicated guidance needed for azimuthal mountings. Most of the recently built large telescopes are already azimuthally mounted. Azimuthally mounted telescopes have two additional obvious places for foci, the *Nasmyth foci* at both ends of the horizontal axis.

The *Dobson mounting*, used in many amateur telescopes, is azimuthal. The magnification of the Newtonian telescope is usually small, and the telescope rests on pieces of teflon, which make it very easy to move. Thus the object can easily be tracked manually.

Another type of mounting is the *coelostat*, where rotating mirrors guide the light into a stationary telescope. This system is used especially in solar telescopes.

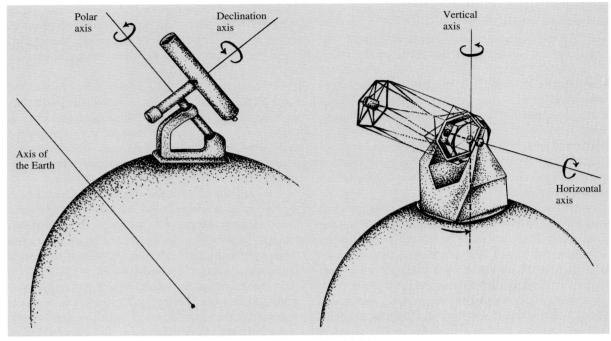

Fig. 3.16. The equatorial mounting (*left*) and the azimuthal mounting (*right*)

To measure absolute positions of stars and accurate time, telescopes aligned with the north–south direction are used. They can be rotated around one axis only, the east–west horizontal axis. *Meridian circles* or *transit instruments* with this kind of mounting were widely constructed for different observatories during the 19th century. A few are still used for astrometry, but they are now highly automatic like the meridian circle on La Palma funded by the Carlsberg foundation.

New Techniques. Detectors are already approaching the theoretical limit of efficiency, where all incident photons are registered. Ultimately, to detect even fainter objects the only solution is to increase the light gathering area, but also the mirrors are getting close to the practical maximum size. Thus, new technical solutions are needed.

One new feature is *active optics*, used e. g. in the ESO 3.5 metre NTT telescope (New Technology Telescope) at La Silla, Chile. The mirror is very thin, but its shape is kept exactly correct by a computer controlled support mechanism. The weight and production cost of such a mirror are much smaller compared with a conventional

thick mirror. Because of the smaller weight also the supporting structure can be made lighter.

Developing the support mechanism further leads to *adaptive optics*. A reference star (or an artificial beam) is monitored constantly in order to obtain the shape of the seeing disk. The shape of the main mirror or a smaller auxiliary mirror is adjusted up to hundreds of times a second to keep the image as concentrated as possible. Adaptive optics has been taken into use in the largest telescopes of the world from about the year 2000 on.

Fig. 3.17a–c. The largest telescopes in the world in 1947–2000. (**a**) For nearly 30 years, the 5.1 m Hale telescope on Mount Palomar, California, USA, was the largest telescope in the world. (**b**) The BTA, Big Azimuthal Telescope, is situated in the Caucasus in the southern Soviet Union. Its mirror has a diameter of 6 m. It was set in operation at the end of 1975. (**c**) The William M. Keck Telescope on the summit of Mauna Kea, Hawaii, was completed in 1992. The 10 m mirror consists of 36 hexagonal segments. (Photos Palomar Observatory, Spetsialnaya Astrofizitsheskaya Observatorya, and Roger Ressmeyer – Starlight for the California Association for Research in Astronomy)

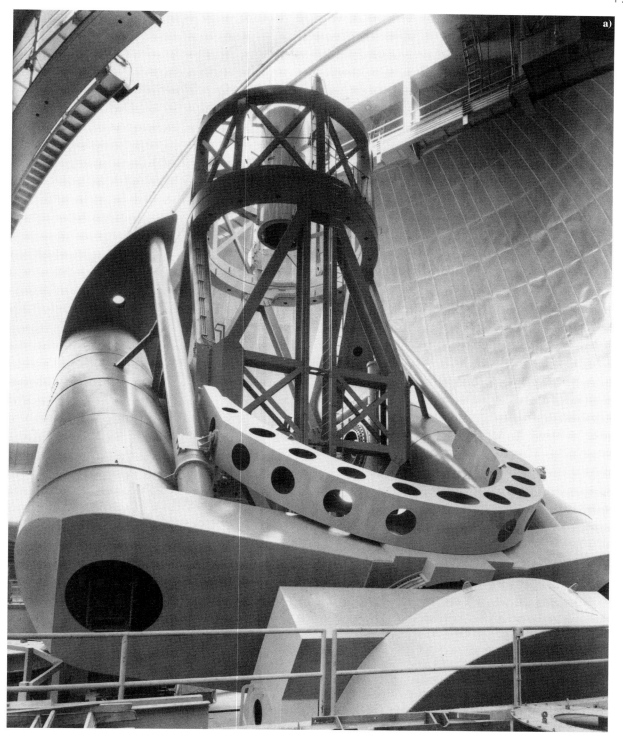

b)

c)

◄ **Fig. 3.18a–c.** Some new large telescopes. (**a**) The 8.1 m Gemini North telescope on Mauna Kea, Hawaii, was set in operation in 1999. Its twin, Gemini South, was dedicated in 2000. (**b**) The European Southern Observatory (ESO) was founded by Belgium, France, the Netherlands, Sweden and West Germany in 1962. Other European countries have joined them later. The VLT (Very Large Telescope) on Cerro Paranal in Northern Chile, was inaugurated in 1998–2000. (**c**) The first big Japanese telescope, the 8.3 m Subaru on Mauna Kea, Hawaii, started observations in 1999. (Photos National Optical Astronomy Observatories, European Southern Observatory and Subaru Observatory)

The mirrors of large telescopes need not be monolithic, but can be made of smaller pieces that are, e. g. hexagonal. These *mosaic mirrors* are very light and can be used to build up mirrors with diameters of several tens of metres (Fig. 3.19). Using active optics, the hexagons can be accurately focussed. The California Association for Research in Astronomy has constructed the William M. Keck telescope with a 10 m mosaic mirror. It is located on Mauna Kea, and the last segment was installed in 1992. A second, similar telescope Keck II was completed in 1996, the pair forming a huge binocular telescope.

The reflecting surface does not have to be continuous, but can consist of several separate mirrors. Such

Fig. 3.19. The mirror of a telescope can be made up of several smaller segments, which are much easier to manufacture, as in the Hobby–Eberle Telescope on Mount Fowlkes, Texas. The effective diameter of the mirror is 9.1 m. A similar telescope is being built in South Africa. (Photo MacDonald Observatory)

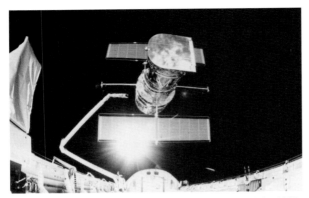

Fig. 3.20. The Hubble Space Telescope after launch in 1990. It is the most effective telescope in the world, since it operates above the atmospheric disturbances. (Photo Nasa)

a telescope was operating on Mount Hopkins, Arizona, in 1979–1999. It was the *Multiple-Mirror Telescope* (MMT) with six 1.8 m mirrors together corresponding to a single mirror having a diameter of 4.5 m. In 2000 the six mirrors were replaced by one 6.5 m mirror.

The European Southern Observatory has constructed its own multi-mirror telescope. ESO's Very Large Telescope (VLT) has four closely located mirrors (Fig. 3.18). The diameter of each mirror is eight metres, and the total area corresponds to one telescope with a 16 m mirror. The resolution is even better, since the "aperture", i.e. the maximum distance between the mirrors, is several tens of meters.

An important astronomical instruments of the 20th century is the *Hubble Space Telescope*, launched in 1990 (Fig. 3.20). It has a mirror with a diameter of 2.4 m. The resolution of the telescope (after the faulty optics was corrected) is near the theoretical diffraction limit, since there is no disturbing atmosphere. A second generation Space Telescope with a mirror of about 6 m is planned to be launched in about 2010.

The Space Telescope was the first large optical telescope in Earth orbit. In the future, satellites will continue to be mainly used for those wavelength regions where the radiation is absorbed by the atmosphere. Due to budgetary reasons, the majority of astronomical observations will still be carried out on the Earth, and great attention will be given to improving ground-based observatories and detectors.

3.3 Detectors and Instruments

Only a limited amount of information can be obtained by looking through a telescope with the unaided eye. Until the end of the 19th century this was the only way to make observations. The invention of photography in the middle of the 19th century brought a revolution in astronomy. The next important step forward in optical astronomy was the development of photoelectric photometry in the 1940's and 1950's. A new revolution, comparable to that caused by the invention of photography, took place in the middle of the 1970's with the introduction of different semiconductor detectors. The sensitivity of detectors has grown so much that today, a 60 cm telescope can be used for observations similar to those made with the Palomar 5 m telescope when it was set in operation in the 1940's.

The Photographic Plate. *Photography* has long been one of the most common methods of observation in astronomy. In astronomical photography glass plates were used, rather than film, since they keep their shape better, but nowadays they are no more manufactured, and CCD-cameras have largely replaced photography. The sensitive layer on the surface of the film or plate is made up of a silver halide, usually silver bromide, AgBr. A photon absorbed by the halide excites an electron that can move from one atom to another. A silver ion, Ag^+, can catch the electron, becoming a neutral atom. When the necessary amount of silver atoms have been accumulated at one place, they form a latent image. The latent image can be made into a permanent negative by treating the plate after exposure with various chemicals, which transform the silver bromide crystals enclosing the latent image into silver ("development"), and remove the unexposed crystals ("fixing").

The photographic plate has many advantages over the human eye. The plate can register up to millions of stars (picture elements) at one time, while the eye can observe at most one or two objects at a time. The image on a plate is practically permanent – the picture can be studied at any time. In addition, the photographic plate is cheap and easy to use, as compared to many other detectors. The most important feature of a plate is its capability to collect light over an extended time: the longer exposures are used, the more silver atoms are formed on the plate (the plate darkens). By increasing the exposure times, fainter objects can be photographed. The eye has no such capacity: if a faint object does not show through a telescope, it cannot been seen, no matter how long one stares.

One disadvantage of the photographic plate is its low sensitivity. Only one photon in a thousand causes a reaction leading to the formation of a silver grain. Thus the *quantum efficiency* of the plate is only 0.1%. Several chemical treatments can be used to sensitize the plate before exposure. This brings the quantum efficiency up to a few percent. Another disadvantage is the fact that a silver bromide crystal that has been exposed once does not register anything more, i. e. a saturation point is reached. On the other hand, a certain number of photons are needed to produce an image. Doubling the number of photons does not necessarily double the density (the 'blackness' of the image): the density of the plate depends nonlinearly on the amount of incoming light. The sensitivity of the plate is also strongly dependent on the wavelength of the light. For the reasons mentioned above the accuracy with which brightness can be measured on a photographic plate is usually worse than about 5%. Thus the photographic plate makes a poor photometer, but it can be excellently used, e. g. for measuring the positions of stars (positional astronomy) and for mapping the sky.

Photocathodes, Photomultipliers. A *photocathode* is a more effective detector than the photographic plate. It is based on the photoelectric effect. A light quantum, or photon, hits the photocathode and loosens an electron. The electron moves to the positive electrode, or anode, and gives rise to an electric current that can be measured. The quantum efficiency of a photocathode is about 10–20 times better than that of a photographic plate; optimally, an efficiency of 30% can be reached. A photocathode is also a linear detector: if the number of electrons is doubled, the outcoming current is also doubled.

The *photomultiplier* is one of the most important applications of the photocathode. In this device, the electrons leaving the photocathode hit a dynode. For each electron hitting the dynode, several others are released. When there are several dynodes in a row, the original weak current can be intensified a millionfold. The photomultiplier measures all the light entering it, but does not form an image. Photomultipliers are mostly

used in photometry, and an accuracy of 0.1–1% can be attained.

Photometers, Polarimeters. A detector measuring brightness, *a photometer*, is usually located behind the telescope in the Cassegrain focus. In the focal plane there is a small hole, the *diaphragm*, which lets through light from the object under observation. In this way, light from other stars in the field of view can be prevented from entering the photometer. A *field lens* behind the diaphragm refracts the light rays onto a photocathode. The outcoming current is intensified further in a preamplifier. The photomultiplier needs a voltage of 1000–1500 volts.

Observations are often made in a certain wavelength interval, instead of measuring all the radiation entering the detector. In this case a *filter* is used to prevent other wavelengths from reaching the photomultiplier. A photometer can also consist of several photomultipliers (Fig. 3.21), which measure simultaneously different

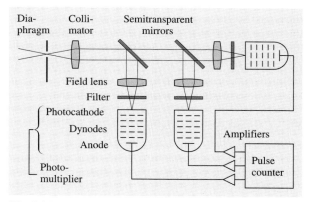

Fig. 3.21. The principle of a photoelectric multicolour photometer. Light collected by the telescope arrives from the left. The light enters the photometer through a small hole in the focal plane, the diaphragm. A lens collimates the light into a parallel beam. Semitransparent mirrors divide the beam to several photomultipliers. A field lens guides the light through a filter onto the photocathode of the photomultiplier. The quanta of light, photons, release electrons from the cathodes. The electrons are accelerated towards the dynodes with a voltage of about 1500 V. The electrons hitting the dynodes release still more electrons, and the current is greatly enhanced. Every electron emitted from the cathode gives rise to a pulse of up to 10^8 electrons at the anode; the pulse is amplified and registered by a pulse counter. In this way, the photons from the star are counted

wavelength bands. In such an instrument beam splitters or semitransparent mirrors split the light beam through fixed filters to the photomultipliers.

In a device called the *photopolarimeter*, a *polarizing filter* is used, either alone or in combination with other filters. The degree and direction of polarization can be found by measuring the intensity of the radiation with different orientations of the polarizers.

In practice, the diaphragm of a photometer will always also let through part of the background sky around the observed object. The measured brightness is in reality the combined brightness of the object and the sky. In order to find the brightness of the object, the background brightness must be measured separately and subtracted from the combined brightness. The accuracy of the measurements is decreased if long observation times are used and the background brightness undergoes fast changes. The problem can be solved by observing the brightness of the background sky and the object simultaneously.

Photometric observations are often relative. If one is observing, e. g. a variable star, a *reference star* close to the actual target is observed at regular intervals. Using the observations of this reference star it is possible to derive a model for the slow changes in the atmospheric extinction (see Chap. 4) and remove their effect. The instrument can be calibrated by observing some *standard stars*, whose brightness is known very accurately.

Image Intensifiers. Different *image intensifiers* based on the photocathode have been used since the 1960's. In the intensifier the information about the starting point of the electron on the photocathode is preserved and the intensified image is formed on a fluorescent screen. The image can then be registered, e. g. with a CCD camera. One of the advantages of the image intensifier is that even faint objects can be imaged using relatively short exposures, and observations can be made at wavelengths where the detector is insensitive.

Another common type of detector is based on the TV camera (*Vidicon camera*). The electrons released from the photocathode are accelerated with a voltage of a few kilovolts before they hit the electrode where they form an image in the form of an electric charge distribution. After exposure, the charge at different points of the electrode is read by scanning its surface with an electron beam row by row. This produces a video signal, which can be transformed into a visible image on a TV tube. The information can also be saved in digital form. In the most advanced systems, the scintillations caused by single electrons on the fluorescent screen of the image intensifier can be registered and stored in the memory of a computer. For each point in the image there is a memory location, called a picture element or *pixel*.

Since the middle of the 1970's, detectors using semiconductor techniques began to be used in increasing numbers. With semiconductor detectors a quantum efficiency of about 70–80% can be attained; thus, sensitivity cannot be improved much more. The wavelength regions suitable for these new detectors are much wider than in the case of the photographic plate. The detectors are also linear. Computers are used for collecting, saving and analyzing the output data available in digital form.

CCD Camera. The most important new detector is the *CCD camera* (Charge Coupled Device). The detector consists of a surface made up of light sensitive silicon diodes, arranged in a rectangular array of image elements or pixels. The largest cameras can have as many as 4096×4096 pixels, although most are considerably smaller.

A photon hitting the detector can release an electron, which will remain trapped inside a pixel. After the exposure varying potential differences are used to move the accumulated charges row by row to a readout buffer. In the buffer the charges are moved pixel by pixel to an analogy/digital converter, which transmits the digital value to a computer. Reading an image also clears the detector (Fig. 3.22). If the exposures are very short the readout times may take a substantial part of the observing time.

The CCD camera is nearly linear: the number of electrons is directly proportional to the number of photons. Calibration of the data is much easier than with photographic plates.

The quantum efficiency, i. e. the number of electrons per incident photon, is high, and the CCD camera is much more sensitive than a photographic plate. The sensitivity is highest in the red wavelength range, about 600–800 nm, where the quantum efficiency can be 80–90% or even higher.

The range of the camera extends far to the infrared. In the ultraviolet the sensitivity drops due to the absorption of the silicon very rapidly below about 500 nm. Two

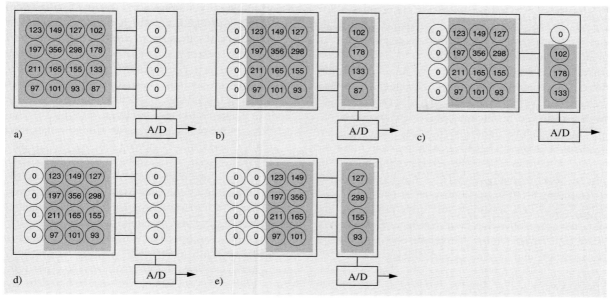

Fig. 3.22a–e. The principle of reading a CCD camera. (**a**) During an exposure electrons are trapped in potential wells corresponding to pixels of the camera. The number at each pixel shows the number of electrons. (**b**) After the exposure each horizontal line is moved one pixel to the right; the rightmost row moves to the readout buffer. (**c**) The contents of the buffer is moved down by one pixel. The lowermost charge moves to the A/D converter, which sends the number of electrons to the computer. (**d**) After moving the buffer down several times one vertical row has been read. (**e**) The image is again shifted right by one pixel. This procedure is repeated till the whole image is read

methods have been used to avoid this problem. One is to use a coating that absorbs the ultraviolet photons and emits light of longer wavelength. Another possibility is to turn the chip upside down and make it very thin to reduce the absorption.

The thermal noise of the camera generates *dark current* even if the camera is in total darkness. To reduce the noise the camera must be cooled. Astronomical CCD cameras are usually cooled with liquid nitrogen, which efficiently removes most of the dark current. However, the sensitivity is also reduced when the camera is cooled; so too cold is not good either. The temperature must be kept constant in order to obtain consistent data. For amateurs there are already moderately priced CCD cameras, which are electrically cooled. Many of them are good enough also for scientific work, if very high sensitivity is not required.

The dark current can easily be measured by taking exposures with the shutter closed. Subtracting this from the observed image gives the real number of electrons due to incident light.

The sensitivity of individual pixels may be slightly different. This can be corrected for by taking an image of an evenly illuminated field, like a twilight sky. This image is called a *flat-field*. When observations are divided by the flat-field, the error caused by different pixels is removed.

The CCD camera is very stable. Therefore it is not necessary to repeat the dark current and flat-field observations very frequently. Typically these calibration exposures are taken during evening and morning twilights, just before and after actual observations.

Cosmic rays are charged particles that can produce extraneous bright dots in CCD images. They are usually limited to one or two pixels, and are easily identified. Typically a short exposure of a few minutes contains a few traces of cosmic rays. Instead of a single long exposure it is usually better to take several short ones, clean the images from cosmic rays, and finally add the images on a computer.

A more serious problem is the *readout noise* of the electronics. In the first cameras it could be hundreds of electrons per pixel. In modern cameras it is a few electrons. This gives a limit to the faintest detectable signal: if the signal is weaker than the readout noise, it is indistinguishable from the noise.

Although the CCD camera is a very sensitive detector, even bright light cannot damage it. A photomultiplier, on the other hand, can be easily destroyed by letting in too much light. However, one pixel can only store a certain number of electrons, after which it becomes *saturated*. Excessive saturation can make the charge to overflow also to the neighboring pixels. If the camera becomes badly saturated it may have to be read several times to completely remove the charges.

The largest CCD cameras are quite expensive, and even they are still rather small compared with photographic plates and films. Therefore photography still has some use in recording extended objects.

Spectrographs. The simplest spectrograph is a prism that is placed in front of a telescope. This kind of device is called the *objective prism spectrograph*. The prism spreads out the different wavelengths of light into a spectrum which can be registered. During the exposure, the telescope is usually slightly moved perpendicularly to the spectrum, in order to increase the width of the spectrum. With an objective prism spectrograph, large numbers of spectra can be photographed, e. g. for spectral classification.

For more accurate information the *slit spectrograph* must be used (Fig. 3.23). It has a narrow slit in the focal plane of the telescope. The light is guided through the slit to a collimator that reflects or refracts all the light rays into a parallel beam. After this, the light is dispersed into a spectrum by a prism and focused with a camera onto a detector, which nowadays is usually a CCD camera. A comparison spectrum is exposed next to the stellar spectrum to determine the precise wavelengths. In modern spectrographs using CCD cameras, the comparison spectrum is usually exposed as a separate image. A big slit spectrograph is often placed at the coudé or Nasmyth focus of the telescope.

Instead of the prism a *diffraction grating* can be used to form the spectrum. A grating has narrow grooves, side by side, typically several hundred per millimetre. When light is reflected by the walls of the grooves, the adjoining rays interfere with each other and give rise to spectra of different orders. There are two kinds of gratings: *reflection* and *transmission gratings*. In a reflection grating no light is absorbed by the glass as in the prism or transmission grating. A grating usually has higher dispersion, or ability to spread the spectrum, than a prism. The dispersion can be increased by increasing the density of the grooves of the grating. In slit spectrographs the reflection grating is most commonly used.

Interferometers. The resolution of a big telescope is in practice limited by seeing, and thus increasing the aperture does not necessarily improve the resolution. To get nearer to the theoretical resolution limit set by diffraction (c. f. (3.3)), different *interferometers* can be used.

There are two types of optical interferometers. One kind uses an existing large telescope; the other a system of two or more separate telescopes. In both cases the light rays are allowed to interfere. By analyzing the outcoming interference pattern, the structures of close binaries can be studied, apparent angular diameters of the stars can be measured, etc.

One of the earliest interferometers was the *Michelson interferometer* that was built shortly before 1920 for the largest telescope of that time. In front of the telescope, at the ends of a six metre long beam, there were flat mirrors reflecting the light into the telescope. The form of the interference pattern changed when the separation of the mirrors was varied. In practice, the interference pattern was disturbed by seeing, and only a few positive results were obtained with this instrument.

The diameters of over 30 of the brightest stars have been measured using *intensity interferometers*. Such

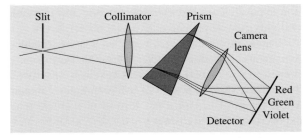

Fig. 3.23. The principle of the slit spectrograph. Light rays entering through a slit are collimated (made parallel to each other), dispersed into a spectrum by a prism and projected onto a photographic plate or a CCD

a device consists of two separate telescopes that can be moved in relation to each other. This method is suitable for the brightest objects only.

In 1970 the Frenchman *Antoine Labeyrie* introduced the principle of *speckle interferometry*. In traditional imaging the pictures from long exposures consist of a large number of instantaneous images, "speckles", that together form the seeing disc. In speckle interferometry very short exposures and large magnifications are used and hundreds of pictures are taken. When these pictures are combined and analyzed (usually in digital form), the actual resolution of the telescope can nearly be reached.

The accuracy of interferometric techniques was improved at the beginning of 00's. The first experiments to use the two 10 m Keck telescopes as one interferometer, were made in 2001. Similarly, the ESO VLT will be used as an interferometer.

3.4 Radio Telescopes

Radio astronomy represents a relatively new branch of astronomy. It covers a frequency range from a few megahertz (100 m) up to frequencies of about 300 GHz (1 mm), thereby extending the observable electromagnetic spectrum by many orders of magnitude. The low-frequency limit of the radio band is determined by the opacity of the ionosphere, while the high-frequency limit is due to the strong absorption from oxygen and water bands in the lower atmosphere. Neither of these limits is very strict, and under favourable conditions radio astronomers can work into the submillimetre region or through ionospheric holes during sunspot minima.

At the beginning of the 20th century attempts were made to observe radio emission from the Sun. These experiments, however, failed because of the low sensitivity of the antenna–receiver systems, and because of the opaqueness of the ionosphere at the low frequencies at which most of the experiments were carried out. The first observations of cosmic radio emission were later made by the American engineer Karl G. Jansky in 1932, while studying thunderstorm radio disturbances at a frequency of 20.5 MHz (14.6 m). He discovered radio emission of unknown origin, which varied within a 24 hour period. Somewhat later he identified the source of this radiation to be in the direction of the centre of our Galaxy.

The real birth of radio astronomy may perhaps be dated to the late 1930's, when Grote Reber started systematic observations with his homemade 9.5 m paraboloid antenna. Thereafter radio astronomy developed quite rapidly and has greatly improved our knowledge of the Universe.

Observations are made both in the continuum (broad band) and in spectral lines (radio spectroscopy). Much of our knowledge about the structure of our Milky Way comes from radio observations of the 21 cm line of neutral hydrogen and, more recently, from the 2.6 mm line of the carbon monoxide molecule. Radio astronomy has resulted in many important discoveries; e. g. both pulsars and quasars were first found by radio astronomical observations. The importance of the field can also be seen from the fact that the Nobel prize in physics has recently been awarded twice to radio astronomers.

A radio telescope collects radiation in an aperture or antenna, from which it is transformed to an electric signal by a receiver, called a radiometer. This signal is then amplified, detected and integrated, and the output is registered on some recording device, nowadays usually by a computer. Because the received signal is very weak, one has to use sensitive receivers. These are often cooled to minimize the noise, which could otherwise mask the signal from the source. Because radio waves are electromagnetic radiation, they are reflected and refracted like ordinary light waves. In radio astronomy, however, mostly reflecting telescopes are used.

At low frequencies the antennas are usually dipoles (similar to those used for radio or TV), but in order to increase the collecting area and improve the resolution, one uses dipole arrays, where all dipole elements are connected to each other.

The most common antenna type, however, is a parabolic reflector, which works exactly as an optical mirror telescope. At long wavelengths the reflecting surface does not need to be solid, because the long wavelength photons cannot see the holes in the reflector, and the antenna is therefore usually made in the form of a metal mesh. At high frequencies the surface has to be smooth, and in the millimetre-submillimetre range, radio astronomers even use large optical telescopes, which they equip with their own radiometers. To ensure a coherent amplification of the signal, the surface irregularities should be less than one-tenth of the wavelength used.

The main difference between a radio telescope and an optical telescope is in the recording of the signal. Radio telescopes are not imaging telescopes (except for synthesis telescopes, which will be described later); instead, a feed horn, which is located at the antenna focus, transfers the signal to a receiver. The wavelength and phase information is, however, preserved.

The resolving power of a radio telescope, θ, can be deduced from the same formula (3.4) as for optical telescopes, i.e. λ/D, where λ is the wavelength used and D is the diameter of the aperture. Since the wavelength ratio between radio and visible light is of the order of 10,000, radio antennas with diameters of several kilometres are needed in order to achieve the same resolution as for optical telescopes. In the early days of radio astronomy poor resolution was the biggest drawback for the development and recognition of radio astronomy. For example, the antenna used by Jansky had a fan beam with a resolution of about 30° in the narrower direction. Therefore radio observations could not be compared with optical observations. Neither was it possible to identify the radio sources with optical counterparts.

The world's biggest radio telescope is the Arecibo antenna in Puerto Rico, whose main reflector is fixed and built into a 305 m diameter, natural round valley covered by a metal mesh (Fig. 3.24). In the late 1970's the antenna surface and receivers were upgraded, enabling the antenna to be used down to wavelengths of 5 cm. The mirror of the Arecibo telescope is not parabolic but spherical, and the antenna is equipped with a movable feed system, which makes observations possible within a 20° radius around the zenith.

The biggest completely steerable radio telescope is the Green Bank telescope in Virginia, U.S.A., dedicated at the end of 2000. It is slightly asymmetric with a diameter of 100×110 m (Fig. 3.25). Before the Green Bank telescope, for over two decades the largest telescope was the Effelsberg telescope in Germany. This antenna has a parabolic main reflector with a diameter of 100 m. The inner 80 m of the dish is made of solid aluminium panels, while the outmost portion of the disk

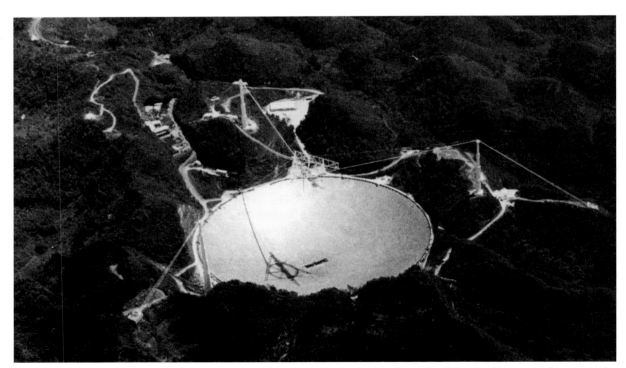

Fig. 3.24. The largest radio telescope in the world is the Arecibo dish in Puerto Rico. It has been constructed over a natural bowl and is 300 m in diameter. (Photo Arecibo Observatory)

Fig. 3.25. The largest fully steerable radio telescope is in Green Bank, Virginia. Its diameter is 100×110 m. (Photo NRAO)

is a metal mesh structure. By using only the inner portion of the telescope, it has been possible to observe down to wavelengths of 4 mm. The oldest and perhaps best-known big radio telescope is the 76 m antenna at Jodrell Bank in Britain, which was completed in the end of the 1950's.

The biggest telescopes are usually incapable of operating below wavelengths of 1 cm, because the surface cannot be made accurate enough. However, the millimetre range has become more and more important. In this wavelength range there are many transitions of interstellar molecules, and one can achieve quite high angular resolution even with a single dish telescope. At present, the typical size of a mirror of a millimetre telescope is about 15 m. The development of this field is rapid, and at present several big millimetre telescopes have started operations (Table C.24). Among them are the 40 m Nobeyama telescope in Japan, which can be used down to 3 mm, the 30 m IRAM telescope at Pico Veleta in Spain, which is usable down to 1 mm, and the 15 m Swedish-ESO Submillimetre Telescope (SEST) at La Silla, Chile, operating down to 0.6 mm (Fig. 3.26). The

largest project in the first decade of the 21st century is ALMA (Atacama Large Millimetre Array), which comprises of 64 telescopes with a diameter of 12 m (Fig. 3.27). It will be built as an international project by the United States, Europe and Japan.

As already mentioned, the resolving power of a radio telescope is far poorer than that of an optical telescope. The biggest radio telescopes can at present reach a resolution of 5 arc seconds, and that only at the very highest frequencies. To improve the resolution by increasing the size is difficult, because the present telescopes are already close to the practical upper limit. However, by combining radio telescopes and interferometers, it is possible to achieve even better resolution than with optical telescopes.

As early as 1891 Michelson used an interferometer for astronomical purposes. While the use of interferometers has proved to be quite difficult in the optical wavelength regime, interferometers are extremely useful in the radio region. To form an interferometer, one needs at least two antennas coupled together. The spacing between the antennas, D, is called the baseline. Let

Fig. 3.26. The 15 metre SEST submillimetre telescope at La Silla in Chile is located in a dry climate at an altitude of 2400 m. Observations can be made down to wavelengths of 0.6 mm. (Photo ESO)

us first assume that the baseline is perpendicular to the line of sight (Fig. 3.28). Then the radiation arrives at both antennas with the same phase, and the summed signal shows a maximum. However, due to the rotation of the Earth, the direction of the baseline changes, producing a phase difference between the two signals. The result is a sinusoidal interference pattern, in which minima occur when the phase difference is 180 degrees.

Fig. 3.27. The Atacama Large Millimetre Array (ALMA) will consist of 64 mirrors. It will be built in cooperation by Europe, U.S.A. and Japan. (Drawing ESO/NOAJ)

The distance between the peaks is given by

$$\theta D = \lambda \, ,$$

where θ is the angle the baseline has turned and λ is the wavelength of the received signal. The resolution of the interferometer is thus equal to that of an antenna with a linear size equal to D.

If the source is not a point source, the radiation emitted from different parts of the source will have phase differences when it enters the antennas. In this case the minima of the interference pattern will not be zero, but will have some positive value P_{min}. If we denote the maximum value of the interference pattern by P_{max}, the ratio

$$\frac{P_{max} - P_{min}}{P_{max} + P_{min}}$$

gives a measure of the source size (fringe visibility).

More accurate information about the source structure can be obtained by changing the spacing between the antennas, i.e. by moving the antennas with respect to each other. If this is done, interferometry is transformed into a technique called *aperture synthesis*.

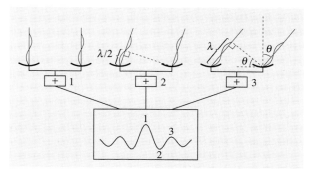

Fig. 3.28. The principle of an interferometer. If the radiation reaches the radio telescopes in the same phase, the waves amplify each other and a maximum is obtained in the combined radiation (cases *1* and *3*). If the incoming waves are in opposite phase, they cancel each other (case *2*)

The theory and techniques of aperture synthesis were developed by the British astronomer Sir Martin Ryle. In Fig. 3.29 the principle of aperture synthesis is illustrated. If the telescopes are located on an east–west track, the spacing between them, projected onto the sky, will describe a circle or an ellipse, depending on the position of the source as the the Earth rotates around its axis. If one varies the distance between the telescopes, one will get a series of circles or ellipses on the sky during a 12 hour interval. As we can see from Fig. 3.29, one does not have to cover all the spacings between the telescopes, because any antenna combination which has the same relative distance will describe the same path on the sky. In this way one can synthesize an antenna, a filled aperture, with a size equal to the maximum spacing between the telescopes. Interferometers working according to this principle are called *aperture synthesis telescopes*. If one covers all the spacings up to the maximum baseline, the result will be an accurate map of the source over the primary beam of an individual antenna element. Aperture synthesis telescopes therefore produce an image of the sky, i.e. a "*radio photograph*".

A typical aperture synthesis telescope consists of one fixed telescope and a number of movable telescopes, usually located on an east–west track, although T or Y configurations are also quite common. The number of telescopes used determines how fast one can synthesize a larger disk, because the number of possible antenna combinations increases as $n(n-1)$, where n is the number of telescopes. It is also possible to synthesize a large telescope with only one fixed and one movable telescope by changing the spacing between the telescopes every 12 hours, but then a full aperture synthesis can require several months of observing time. In order for this technique to work, the source must be constant, i.e. the signal cannot be time variable during the observing session.

The most efficient aperture synthesis telescope at present is the VLA (Very Large Array) in New Mexico, USA (Fig. 3.30). It consists of 27 paraboloid antennas, each with a diameter of 25 m, which are located on a Y-shaped track. The Y-formation was chosen because it provides a full aperture synthesis in 8 hours. Each antenna can be moved by a specially built carrier, and the locations of the telescopes are chosen to give optimal spacings for each configuration. In the largest

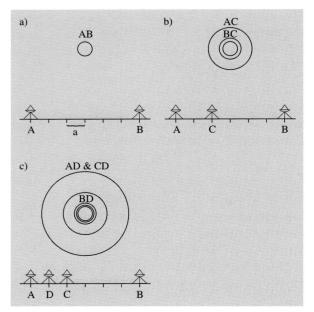

Fig. 3.29a–c. To illustrate the principle of aperture synthesis, let us consider an east–west oriented interferometer pointed towards the celestial north. Each antenna is identical, has a diameter D and operates at a wavelength λ. The minimum spacing between each antenna element is a, and the maximum spacing is $6a$. In (**a**) there are only two antennas, A and B, displaced by the maximum spacing $6a$. When the earth rotates, antennas A and B will, in the course of 12 hours, track a circle on the plane of the sky with a diameter $\lambda/(6a)$, the maximum resolution that can be achieved with this interferometer. In (**b**) the antenna C is added to the interferometer, thus providing two more baselines, which track the circles AC and BC with radii of $\lambda/(2a)$ and $\lambda/(4a)$, respectively. In (**c**) there is still another antenna D added to the interferometer. In this case two of the baselines are equal, AD and CD, and therefore only two new circles are covered on the plane of the sky. By adding more interferometer elements, one can fill in the missing parts within the primary beam, i.e. the beam of one single dish, and thus obtain a full coverage of the beam. It is also evident from (**c**), that not all of the antenna positions are needed to provide all the different spacings; some antenna spacings will in such a case be equal and therefore provide no additional information. Obtaining a full aperture synthesis with an east–west interferometer always takes 12 hours, if all spacings are available. Usually, however, several antenna elements are movable, in which case a full aperture synthesis can take a long time before all spacings are filled in

configuration each arm is about 21 km long, thereby resulting in an antenna with an effective diameter of 35 km. If the VLA is used in its largest configuration and

Fig. 3.30. The VLA at Socorro, New Mexico, is a synthesis telescope consisting of 27 movable antennas

at its highest frequency, 23 GHz (1.3 cm), the resolution achieved is 0.1 arc second, clearly superior to any optical telescope. Similar resolution can also be obtained with the British MERLIN telescope, where already existing telescopes have been coupled together by radio links. Other well-known synthesis telescopes are the Cambridge 5 km array in Britain and the Westerbork array in the Netherlands, both located on east–west tracks.

Even higher resolution can be obtained with an extension of the aperture synthesis technique, called *VLBI* (Very Long Baseline Interferometry). With the VLBI technique the spacing between the antennas is restricted only by the size of the Earth. VLBI uses existing antennas (often on different continents), which are all pointed towards the same source. In this case the signal is recorded together with accurate timing signals from atomic clocks. The data files are correlated against each other, resulting in maps similar to those obtained with a normal aperture synthesis telescope. With VLBI techniques it is possible to achieve resolutions of 0.0001″. Because interferometry is very sensitive to the distance between the telescopes, the VLBI technique also pro-

vides one of the most accurate methods to measure distances. Currently one can measure distances with an accuracy of a few centimetres on intercontinental baselines. This is utilized in geodetic VLBI experiments, which study continental drift and polar motion as a function of time.

In radio astronomy the maximum size of single antennas has also been reached. The trend is to build synthesis antennas, similar to the VLA in New Mexico. In the 1990's The United States built a chain of antennas extending across the whole continent, and the Australians have constructed a similar, but north–south antenna chain across their country.

More and more observations are being made in the submillimetre region. The disturbing effect of atmospheric water vapour becomes more serious at shorter wavelengths; thus, submillimetre telescopes must be located on mountain tops, like optical telescopes. All parts of the mirror are actively controlled in order to accurately maintain the proper form like in the new optical telescopes. Several new submillimetre telescopes are under construction.

3.5 Other Wavelength Regions

All wavelengths of the electromagnetic spectrum enter the Earth from the sky. However, as mentioned in Sect. 3.1, not all radiation reaches the ground. The wavelength regions absorbed by the atmosphere have been studied more extensively since the 1970's, using Earth-orbiting satellites. Besides the optical and radio regions, there are only some narrow wavelength ranges in the infrared that can be observed from high mountain tops.

The first observations in each new wavelength region were usually carried out from balloons, but not until rockets came into use could observations be made from outside the atmosphere. The first actual observations of an X-ray source, for instance, were made on a rocket flight in June 1962, when the detector rose above the atmosphere for about 6 minutes. Satellites have made it possible to map the whole sky in the wavelength regions invisible from the ground.

Gamma Radiation. Gamma ray astronomy studies radiation quanta with energies of 10^5–10^{14} eV. The boundary between gamma and X-ray astronomy, 10^5 eV, corresponds to a wavelength of 10^{-11} m. The boundary is not fixed; the regions of hard (= high-energy) X-rays and soft gamma rays partly overlap.

While ultraviolet, visible and infrared radiation are all produced by changes in the energy states of the electron envelopes of atoms, gamma and hard X-rays are produced by transitions in atomic nuclei or in mutual interactions of elementary particles. Thus observations of the shortest wavelengths give information on processes different from those giving rise to longer wavelengths.

The first observations of gamma sources were obtained at the end of the 1960's, when a device in the OSO 3 satellite (Orbiting Solar Observatory) detected gamma rays from the Milky Way. Later on, some satellites were especially designed for gamma astronomy, notably SAS 2, COS B, HEAO 1 and 3, and Granat. The most effective satellite so far, the Compton Gamma Ray Observatory, operated from 1991 to 2000.

The quanta of gamma radiation have energies a million times greater than those of visible light, but they cannot be observed with the same detectors. These observations are made with various *scintillation detectors*, usually composed of several layers of detector plates, where gamma radiation is transformed by the photoelectric effect into visible light, detectable by photomultipliers.

The energy of a gamma quantum can be determined from the depth to which it penetrates the detector. Analyzing the trails left by the quanta gives information on their approximate direction. The field of view is limited by the grating. The directional accuracy is low, and in gamma astronomy the resolution is far below that in other wavelength regions.

X-rays. The observational domain of X-ray astronomy includes the energies between 10^2 and 10^5 eV, or the wavelengths 10–0.01 nm. The regions 10–0.1 nm and 0.1–0.01 nm are called *soft* and *hard X-rays*, respectively. X-rays were discovered in the late 19th century. Systematic studies of the sky at X-ray wavelengths only became possible in the 1970's with the advent of satellite technology.

The first all-sky mapping was made in the early 1970's by SAS 1 (Small Astronomical Satellite), also called Uhuru. At the end of the 1970's, two High-Energy Astronomy Observatories, HEAO 1 and 2 (the latter called Einstein), mapped the sky with much higher sensitivity than Uhuru.

The Einstein Observatory was able to detect sources about a thousand times fainter than earlier X-ray telescopes. In optical astronomy, this would correspond to a jump from a 15 cm reflector to a 5 m telescope. Thus X-ray astronomy has developed in 20 years as much as optical astronomy in 300 years.

The latest X-ray satellites have been the American Chandra and the European XMM-Newton, both launched in 1999.

Besides satellites mapping the whole sky, there have been several satellites observing the X-ray radiation of the Sun. The first effective telescopes were installed in the Skylab space station, and they were used to study the Sun in 1973–74. In the 1990's, the European Soho started making regular X-ray observations of the Sun.

The first X-ray telescopes used detectors similar to those in gamma astronomy. Their directional accuracy was never better than a few arc minutes. The more precise X-ray telescopes utilize the principle of *grazing reflection* (Fig. 3.31). An X-ray hitting a surface perpendicularly is not reflected, but absorbed. If, however,

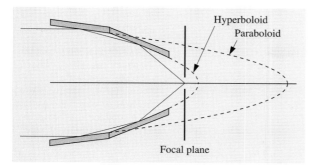

Fig. 3.31. X-rays are not reflected by an ordinary mirror, and the principle of grazing reflection must be used for collecting them. Radiation meets the paraboloid mirror at a very small angle, is reflected onto a hyperboloid mirror and further to a focal point. In practice, several mirrors are placed one inside another, collecting radiation in a common focus

X-rays meet the mirror nearly parallel to its surface, just grazing it, a high quality surface can reflect the ray.

The mirror of an X-ray reflector is on the inner surface of a slowly narrowing cone. The outer part of the surface is a paraboloid and the inner part a hyperboloid. The rays are reflected by both surfaces and meet at a focal plane. In practice, several tubes are installed one within another. For instance, the four cones of the Einstein Observatory had as much polished optical surface

as a normal telescope with a diameter of 2.5 m. The resolution in X-ray telescopes is of the order of a few arc seconds and the field of view about 1 deg.

The detectors in X-ray astronomy are usually *Geiger–Müller counters*, *proportional counters* or scintillation detectors. Geiger–Müller and proportional counters are boxes filled with gas. The walls form a cathode, and an anode wire runs through the middle of the box; in more accurate counters, there are several anode wires. An X-ray quantum entering the box ionizes the gas, and the potential difference between the anode and cathode gives rise to a current of electrons and positive ions.

Ultraviolet Radiation. Between X-rays and the optical region lies the domain of ultraviolet radiation, with wavelengths between 10 and 400 nm. Most ultraviolet observations have been carried out in the *soft UV* region, at wavelengths near those of optical light, since most of the UV radiation is absorbed by the atmosphere. The wavelengths below 300 nm are completely blocked out. The short wavelength region from 10 to 91.2 nm is called the *extreme ultraviolet* (EUV, XUV).

Extreme ultraviolet was one of the last regions of the electromagnetic radiation to be observed systematically. The reason for this is that the absorption of interstellar hydrogen makes the sky practically opaque

Fig. 3.32. The European X-ray satellite XMM-Newton was launched in 1999. (Drawing D. Ducros, XMM Team, ESA)

at these wavelengths. The visibility in most directions is limited to some hundred light years in the vicinity of the Sun. In some directions, however, the density of the interstellar gas is so low that even extragalactic objects can be seen. The first dedicated EUV satellite was the Extreme Ultraviolet Explorer (EUVE), operating in 1992–2000. It observed about a thousand EUV sources. In EUV grazing reflection telescopes similar to those used in X-ray astronomy are employed.

In nearly all branches of astronomy important information is obtained by observations of ultraviolet radiation. Many emission lines from stellar chromospheres or coronas, the Lyman lines of atomic hydrogen, and most of the radiation from hot stars are found in the UV domain. In the *near-ultraviolet*, telescopes can be made similar to optical telescopes and, equipped with a photometer or spectrometer, installed in a satellite orbiting the Earth.

The most effective satellites in the UV have been the European TD-1, the American Orbiting Astronomical Observatories OAO 2 and 3 (Copernicus), the International Ultraviolet Explorer IUE and the Soviet Astron. The instruments of the TD-1 satellite included both a photometer and a spectrometer. The satellite measured the magnitudes of over 30,000 stars in four different spectral regions between 135 and 274 nm, and registered UV spectra from over 1000 stars. The OAO satellites were also used to measure magnitudes and spectra, and OAO 3 worked for over eight years.

The IUE satellite, launched in 1978, was one of the most successful astronomical satellites. IUE had a 45 cm Ritchey-Chrétien telescope with an aperture ratio of $f/15$ and a field of view of 16 arc minutes. The satellite had two spectrographs to measure spectra of higher or lower resolution in wavelength intervals of 115–200 nm or 190–320 nm. For registration of the spectra, a Vidicon camera was used. IUE worked on the orbit for 20 years.

Infrared Radiation. Radiation with longer wavelengths than visible light is called infrared radiation. This region extends from about 1 micrometre to 1 millimetre, where the radio region begins. Sometimes the *near-infrared*, at wavelengths below 5 m, and the submillimetre domain, at wavelengths between 0.1 and 1 mm, are considered separate wavelength regions.

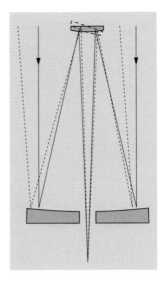

Fig. 3.33. Refractors are not suitable for infrared telescopes, because infrared radiation cannot penetrate glass. The Cassegrain reflectors intended especially for infrared observations have secondary mirrors nodding rapidly back and forth between the object and the background near the object. By subtracting the brightness of the background from the brightness of the object, the background can be eliminated

In infrared observations radiation is collected by a telescope, as in the optical region. The incoming radiation consists of radiation from the object, from the background and from the telescope itself. Both the source and the background must be continually measured, the difference giving the radiation from the object. The background measurements are usually made with a Cassegrain secondary mirror oscillating between the source and the background at a rate of, say, 100 oscillations per second, and thus the changing background can be eliminated. To register the measurements, semiconductor detectors are used. The detector must always be cooled to minimize its own thermal radiation. Sometimes the whole telescope is cooled.

Infrared observatories have been built on high mountain tops, where most of the atmospheric water vapour remains below. Some favourable sites are, e.g. Mauna Kea on Hawaii, Mount Lemon in Arizona and Pico del Teide on Tenerife. For observations in the far-infrared these mountains are not high enough; these observations are carried out, e.g. on aeroplanes. One of the best-equipped planes is the Kuiper Airborne Observatory, named after the well-known planetary scientist Gerard Kuiper.

Balloons and satellites are also used for infrared observations. The most successful infrared observatories so far have been the InfraRed Astronomy Satellite IRAS, built in cooperation by the U.S. and the Netherlands (Fig. 3.34), and the European Infrared Space

Fig. 3.34. The most productive infrared satellite so far has been the Dutch-American IRAS (Infrared Astronomical Satellite). It found over 200,000 new infrared objects while mapping the whole sky in 1983. (Photo Fokker)

Observatory ISO. A very succesful satellite was the 1989 launched COBE (Cosmic Background Explorer), which mapped the background radiation in submillimetre and infrared wavelengths. The Microwave Anisotropy Probe (MAP) has continued the work of COBE, starting in 2001.

3.6 Other Forms of Energy

Besides electromagnetic radiation, energy arrives from space in other forms: particles (*cosmic rays, neutrinos*) and *gravitational radiation*.

Cosmic Rays. Cosmic rays, consisting of electrons and totally ionized nuclei of atoms, are received in equal amounts from all directions. Their incoming directions do not reveal their origin, since cosmic rays are electri-

cally charged; thus their paths are continually changed when they move through the magnetic fields of the Milky Way. The high energies of cosmic rays mean that they have to be produced by high-energy phenomena like supernova explosions. The majority of cosmic rays are protons (nearly 90%) and helium nuclei (10%), but some are heavier nuclei; their energies lie between 10^8 and 10^{20} eV.

The most energetic cosmic rays give rise to *secondary radiation* when they hit molecules of the atmosphere. This secondary radiation can be observed from the ground, but primary cosmic rays can only be directly observed outside the atmosphere. The detectors used to observe cosmic rays are similar to those used in particle physics. Since Earth-based accelerators reach energies of only about 10^{12} eV, cosmic rays offer an excellent "natural" laboratory for particle physics. Many satellites and spacecraft have detectors for cosmic rays.

Neutrinos. Neutrinos are elementary particles with no electric charge and a mass equal to zero or, at any rate, less than $1/10,000$ of the mass of the electron. Most neutrinos are produced in nuclear reactions within stars; since they react very weakly with other matter, they escape directly from the stellar interior.

Neutrinos are very difficult to observe; the first method of detection was the radiochemical method. As a reactive agent, e. g. tetrachloroethene (C_2Cl_4) can be used. When a neutrino hits a chlorine atom, the chlorine is transformed into argon, and an electron is freed:

$$^{37}Cl + \nu \rightarrow {}^{37}Ar + e^- \ .$$

The argon atom is radioactive and can be observed. Instead of chlorine, lithium and gallium might be used to detect neutrinos. The first gallium detectors have been running in Italy and Russia from the end of the 1980's.

Another observation method is based on the Čerenkov radiation produced by neutrinos in extremely pure water. The flashes of light are registered with photomultipliers, and thus it is possible to find out the direction of the radiation. This method is used e. g. in the Japanese Kamiokande detector.

Neutrino detectors must be located deep under the ground to protect them from the secondary radiation caused by cosmic rays.

The detectors have observed neutrinos from the Sun, and the Supernova 1987A in the Large Magellanic Cloud was also observed in 1987.

Gravitational Radiation. Gravitational astronomy is as young as neutrino astronomy. The first attempts to measure gravitational waves were made in the 1960's. Gravitational radiation is emitted by accelerating masses, just as electromagnetic radiation is emitted by electric charges in accelerated motion. Detection of gravitational waves is very difficult, and they have yet to be directly observed.

The first type of gravitational wave antenna was the *Weber cylinder*. It is an aluminium cylinder which starts vibrating at its proper frequency when hit by a gravitational pulse. The distance between the ends of the cylinder changes by about 10^{-17} m, and the changes in the length are studied by strain sensors welded to the side of the cylinder.

Another type of modern gravity radiation detectors measures "spatial strain" induced by gravity waves and consists of two sets of mirrors in directions perpendicular to each other (Michelson interferometer), or one set of parallel mirrors (Fabry–Perot interferometer). The relative distances between the mirrors are monitored by laser interferometers. If a gravity pulse passes the detector, the distances change and the changes can be measured. The longest baseline between the mirrors is in the American LIGO (Laser Interferometer Gravitational-wave Observatory) system, about 25 km. LIGO is scheduled to start operation in 2002.

* Diffraction by a Circular Aperture

Consider a circular hole of radius R in the xy plane. Coherent light enters the hole from the direction of the negative z axis (see figure). We consider light rays leaving the hole parallel to the xz plane forming an angle θ with the z axis. The light waves interfere on a screen far away. The phase difference between a wave through a point (x, y) and a wave going through the centre of the hole can be calculated from the different path lengths $s = x \sin \theta$:

$$\delta = \frac{s}{\lambda} 2\pi = \frac{2\pi \sin \theta}{\lambda} x \equiv kx .$$

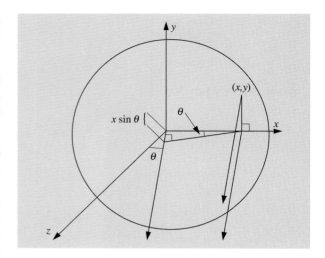

Thus, the phase difference δ depends on the x coordinate only. The sum of the amplitudes of the waves from a small surface element is proportional to the area of the element $dx\,dy$. Let the amplitude coming through the centre of the hole be $d\boldsymbol{a}_0 = dx\,dy\hat{\boldsymbol{i}}$. The amplitude coming from the point (x, y) is then

$$d\boldsymbol{a} = dx\,dy\,(\cos \delta\,\hat{\boldsymbol{i}} + \sin \delta\,\hat{\boldsymbol{j}}) .$$

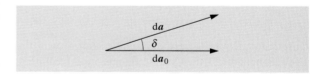

We sum up the amplitudes coming from different points of the hole:

$$\boldsymbol{a} = \int_{\text{Aperture}} d\boldsymbol{a}$$

$$= \int_{x=-R}^{R} \int_{y=-\sqrt{R^2-x^2}}^{\sqrt{R^2-x^2}} (\cos kx\,\hat{\boldsymbol{i}} + \sin kx\,\hat{\boldsymbol{j}})\,dy\,dx$$

$$= 2 \int_{-R}^{R} \sqrt{R^2 - x^2}(\cos kx\,\hat{\boldsymbol{i}} + \sin kx\,\hat{\boldsymbol{j}})\,dx .$$

Since sine is an odd function ($\sin(-kx) = -\sin(kx)$), we get zero when we integrate the second term. Cosine is an even function, and so

$$a \propto \int_0^R \sqrt{R^2 - x^2} \cos kx \, dx .$$

We substitute $x = Rt$ and define $p = kR = (2\pi r \sin\theta)/\lambda$, thus getting

$$a \propto \int_0^1 \sqrt{1 - t^2} \cos pt \, dt .$$

The zero points of the intensity observed on the screen are obtained from the zero points of the amplitude,

$$J(p) = \int_0^1 \sqrt{1 - t^2} \cos pt \, dt = 0 .$$

Inspecting the function $J(p)$, we see that the first zero is at $p = 3.8317$, or

$$\frac{2\pi R \sin\theta}{\lambda} = 3.8317 .$$

The radius of the diffraction disc in angular units can be estimated from the condition

$$\sin\theta = \frac{3.8317}{2\pi R} \approx 1.22 \frac{\lambda}{D} ,$$

where $D = 2R$ is the diameter of the hole.

In mirror telescopes diffraction is caused also by the support structure of the secondary mirror. If the aperture is more complex and only elementary mathematics is used calculations may become rather cumbersome. However, it can be shown that the diffraction pattern can be obtained as the Fourier transform of the aperture.

3.7 Examples

Example 3.1 The distance between the components of the binary star ζ Herculis is $1.38''$. What should the diameter of a telescope be to resolve the binary? If the focal length of the objective is 80 cm, what should the focal length of the eyepiece be to resolve the components, when the resolution of the eye is $2'$?

In the optical region, we can use the wavelength value of $\lambda \approx 550$ nm. The diameter of the objective is obtained from the equation for the resolution (3.4),

$$D \approx \frac{\lambda}{\theta} = \frac{550 \times 10^{-9}}{(1.38/3600) \times (\pi/180)} \text{ m}$$

$$= 0.08 \text{ m} = 8 \text{ cm} .$$

The required magnification is

$$\omega = \frac{2'}{1.38''} = 87 .$$

The magnification is given by

$$\omega = \frac{f}{f'} ,$$

and, thus, the focal length of the eyepiece should be

$$f' = \frac{f}{\omega} = \frac{80 \text{ cm}}{87} = 0.9 \text{ cm} .$$

Example 3.2 A telescope has an objective with a diameter of 90 mm and focal length of 1200 mm.

a) What is the focal length of an eyepiece, the exit pupil of which is 6 mm (about the size of the pupil of the eye)?

b) What is the magnification of such an eyepiece?

c) What is the angular diameter of the Moon seen through this telescope and eyepiece?

a) From Fig. 3.7 we get

$$L = \frac{f'}{f} D ,$$

whence

$$f' = f \frac{L}{D} = 1200 \text{ mm} \frac{6 \text{ mm}}{90 \text{ mm}}$$

$$= 80 \text{ mm} .$$

b) The magnification is $\omega = f/f' = 1200 \text{ mm}/80 \text{ mm} = 15$.

c) Assuming the angular diameter of the Moon is $\alpha = 31' = 0.52°$, its diameter through the telescope is $\omega\alpha = 7.8°$.

3.8 Exercises

Exercise 3.1 The Moon was photographed with a telescope, the objective of which had a diameter of 20 cm and focal length of 150 cm. The exposure time was 0.1 s.

a) What should the exposure time be, if the diameter of the objective were 15 cm and focal length 200 cm?

b) What is the size of the image of the Moon in both cases?

c) Both telescopes are used to look at the Moon with an eyepiece the focal length of which is 25 mm. What are the magnifications?

Exercise 3.2 The radio telescopes at Amherst, Massachusetts, and Onsala, Sweden, are used as an interferometer, the baseline being 2900 km.

a) What is the resolution at 22 GHz in the direction of the baseline?

b) What should be the size of an optical telescope with the same resolution?

4. Photometric Concepts and Magnitudes

Most astronomical observations utilize electromagnetic radiation in one way or another. We can obtain information on the physical nature of a radiation source by studying the energy distribution of its radiation. We shall now introduce some basic concepts that characterize electromagnetic radiation.

4.1 Intensity, Flux Density and Luminosity

Let us assume we have some radiation passing through a surface element dA (Fig. 4.1). Some of the radiation will leave dA within a solid angle $d\omega$; the angle between $d\omega$ and the normal to the surface is denoted by θ. The amount of energy with frequency in the range $[\nu, \nu+d\nu]$ entering this solid angle in time dt is

$$dE_\nu = I_\nu \cos\theta \, dA \, d\nu \, d\omega \, dt \, . \tag{4.1}$$

Here, the coefficient I_ν is the *specific intensity* of the radiation at the frequency ν in the direction of the solid angle $d\omega$. Its dimension is $\mathrm{W\,m^{-2}\,Hz^{-1}\,sterad^{-1}}$.

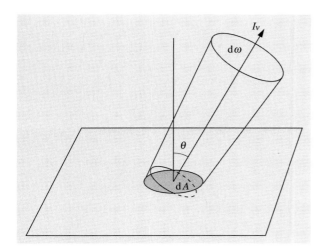

Fig. 4.1. The intensity I_ν of radiation is related to the energy passing through a surface element dA into a solid angle $d\omega$, in a direction θ

The projection of the surface element dA as seen from the direction θ is $dA_n = dA \cos\theta$, which explains the factor $\cos\theta$. If the intensity does not depend on direction, the energy dE_ν is directly proportional to the surface element perpendicular to the direction of the radiation.

The intensity including all possible frequencies is called the *total intensity I*, and is obtained by integrating I_ν over all frequencies:

$$I = \int_0^\infty I_\nu \, d\nu \, .$$

More important quantities from the observational point of view are the *energy flux* (L_ν, L) or, briefly, the *flux* and the *flux density* (F_ν, F). The flux density gives the power of radiation per unit area; hence its dimension is $\mathrm{W\,m^{-2}\,Hz^{-1}}$ or $\mathrm{W\,m^{-2}}$, depending on whether we are talking about the flux density at a certain frequency or about the total flux density.

Observed flux densities are usually rather small, and $\mathrm{W\,m^{-2}}$ would be an inconveniently large unit. Therefore, especially in radio astronomy, flux densities are often expressed in *Janskys*; one Jansky (Jy) equals $10^{-26}\,\mathrm{W\,m^{-2}\,Hz^{-1}}$.

When we are observing a radiation source, we in fact measure the energy collected by the detector during some period of time, which equals the flux density integrated over the radiation-collecting area of the instrument and the time interval.

The flux density F_ν at a frequency ν can be expressed in terms of the intensity as

$$F_\nu = \frac{1}{dA \, d\nu \, dt} \int_S dE_\nu$$
$$= \int_S I_\nu \cos\theta \, d\omega \, , \tag{4.2}$$

where the integration is extended over all possible directions. Analogously, the total flux density is

$$F = \int_S I \cos\theta \, d\omega \, .$$

For example, if the radiation is *isotropic*, i.e. if I is independent of the direction, we get

$$F = \int_S I \cos\theta \, d\omega = I \int_S \cos\theta \, d\omega \ . \qquad (4.3)$$

The solid angle element $d\omega$ is equal to a surface element on a unit sphere. In spherical coordinates it is (Fig. 4.2; also c.f. Appendix A.5):

$$d\omega = \sin\theta \, d\theta \, d\phi \ .$$

Substitution into (4.3) gives

$$F = I \int_{\theta=0}^{\pi} \int_{\phi=0}^{2\pi} \cos\theta \sin\theta \, d\theta \, d\phi = 0 \ ,$$

so there is no net flux of radiation. This means that there are equal amounts of radiation entering and leaving the surface. If we want to know the amount of radiation passing through the surface, we can find, for example, the radiation leaving the surface. For isotropic radiation this is

$$F_1 = I \int_{\theta=0}^{\pi/2} \int_{\phi=0}^{2\pi} \cos\theta \sin\theta \, d\theta \, d\phi = \pi I \ . \qquad (4.4)$$

In the astronomical literature, terms such as intensity and brightness are used rather vaguely. Flux density is hardly ever called flux density but intensity or (with luck) flux. Therefore the reader should always carefully check the meaning of these terms.

Flux means the power going through some surface, expressed in watts. The flux emitted by a star into a solid angle ω is $L = \omega r^2 F$, where F is the flux density observed at a distance r. *Total flux* is the flux passing through a closed surface encompassing the source. Astronomers usually call the total flux of a star the *luminosity* L. We can also talk about the luminosity L_ν at a frequency ν ($[L_\nu] = \text{W Hz}^{-1}$). (This must not be confused with the luminous flux used in physics; the latter takes into account the sensitivity of the eye.)

If the source (like a typical star) radiates isotropically, its radiation at a distance r is distributed evenly on a spherical surface whose area is $4\pi r^2$ (Fig. 4.3). If the flux density of the radiation passing through this surface is F, the total flux is

$$L = 4\pi r^2 F \ . \qquad (4.5)$$

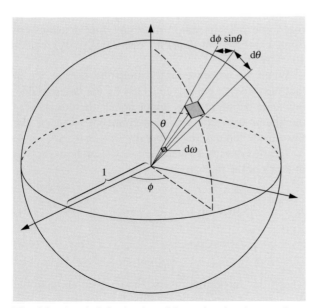

Fig. 4.2. An infinitesimal solid angle $d\omega$ is equal to the corresponding surface element on a unit sphere: $d\omega = \sin\theta \, d\theta \, d\phi$

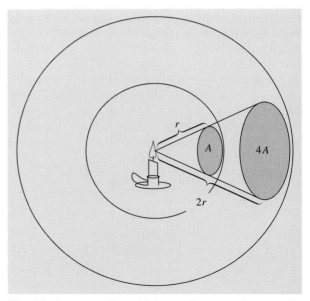

Fig. 4.3. An energy flux which at a distance r from a point source is distributed over an area A is spread over an area $4A$ at a distance $2r$. Thus the flux density decreases inversely proportional to the distance squared

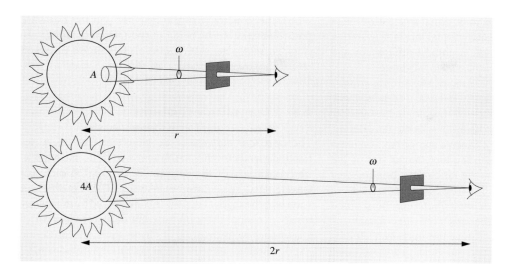

Fig. 4.4. An observer sees radiation coming from a constant solid angle ω. The area giving off radiation into this solid angle increases when the source moves further away ($A \propto r^2$). Therefore the surface brightness or the observed flux density per unit solid angle remains constant

If we are outside the source, where radiation is not created or destroyed, the luminosity does not depend on distance. The flux density, on the other hand, falls off proportional to $1/r^2$.

For extended objects (as opposed to objects such as stars visible only as points) we can define the *surface brightness* as the flux density per unit solid angle (Fig. 4.4). Now the observer is at the apex of the solid angle. The surface brightness is independent of distance, which can be understood in the following way. The flux density arriving from an area A is inversely proportional to the distance squared. But also the solid angle subtended by the area A is proportional to $1/r^2$ ($\omega = A/r^2$). Thus the surface brightness $B = F/\omega$ remains constant.

The *energy density* u of radiation is the amount of energy per unit volume (J m^{-3}):

$$u = \frac{1}{c} \int_S I \, d\omega \,. \tag{4.6}$$

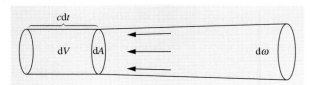

Fig. 4.5. In time dt, the radiation fills a volume $dV = c \, dt \, dA$, where dA is the surface element perpendicular to the propagation direction of the radiation

This can be seen as follows. Suppose we have radiation with intensity I arriving from a solid angle $d\omega$ perpendicular to the surface dA (Fig. 4.5). In the time dt, the radiation travels a distance $c \, dt$ and fills a volume $dV = c \, dt \, dA$. Thus the energy in the volume dV is (now $\cos\theta = 1$)

$$dE = I \, dA \, d\omega \, dt = \frac{1}{c} I \, d\omega \, dV \,.$$

Hence the energy density du of the radiation arriving from the solid angle $d\omega$ is

$$du = \frac{dE}{dV} = \frac{1}{c} I \, d\omega \,,$$

and the total energy density is obtained by integrating this over all directions. For isotropic radiation we get

$$u = \frac{4\pi}{c} I \,. \tag{4.7}$$

4.2 Apparent Magnitudes

As early as the second century B. C., Hipparchos divided the visible stars into six classes according to their apparent brightness. The first class contained the brightest stars and the sixth the faintest ones still visible to the naked eye.

The response of the human eye to the brightness of light is not linear. If the flux densities of three stars are in the proportion 1:10:100, the brightness difference of the first and second star seems to be equal to the difference of the second and third star. Equal brightness ratios correspond to equal apparent brightness differences: the human perception of brightness is logarithmic.

The rather vague classification of Hipparchos was replaced in 1856 by Norman R. Pogson. The new, more accurate classification followed the old one as closely as possible, resulting in another of those illogical definitions typical of astronomy. Since a star of the first class is about one hundred times brighter than a star of the sixth class, Pogson defined the ratio of the brightnesses of classes n and $n+1$ as $\sqrt[5]{100} = 2.512$.

The brightness class or *magnitude* can be defined accurately in terms of the observed flux density F ($[F] = \mathrm{W\,m^{-2}}$). We decide that the magnitude 0 corresponds to some preselected flux density F_0. All other magnitudes are then defined by the equation

$$m = -2.5 \lg \frac{F}{F_0} \,. \tag{4.8}$$

Note that the coefficient is exactly 2.5, not 2.512! Magnitudes are dimensionless quantities, but to remind us that a certain value is a magnitude, we can write it, for example, as 5 mag or 5^{m}.

It is easy to see that (4.8) is equivalent to Pogson's definition. If the magnitudes of two stars are m and $m+1$ and their flux densities F_m and F_{m+1}, respectively, we have

$$m - (m+1) = -2.5 \lg \frac{F_m}{F_0} + 2.5 \lg \frac{F_{m+1}}{F_0}$$
$$= -2.5 \lg \frac{F_m}{F_{m+1}} \,,$$

whence

$$\frac{F_m}{F_{m+1}} = \sqrt[5]{100} \,.$$

In the same way we can show that the magnitudes m_1 and m_2 of two stars and the corresponding flux densities F_1 and F_2 are related by

$$m_1 - m_2 = -2.5 \lg \frac{F_1}{F_2} \,. \tag{4.9}$$

Magnitudes extend both ways from the original six values. The magnitude of the brightest star, Sirius, is in fact negative -1.5. The magnitude of the Sun is -26.8 and that of a full moon -12.5. The magnitude of the faintest objects observed depends on the size of the telescope, the sensitivity of the detector and the exposure time. The limit keeps being pushed towards fainter objects; currently the magnitudes of the faintest observed objects are over 30.

4.3 Magnitude Systems

The *apparent magnitude m*, which we have just defined, depends on the instrument we use to measure it. The sensitivity of the detector is different at different wavelengths. Also, different instruments detect different wavelength ranges. Thus the flux measured by the instrument equals not the total flux, but only a fraction of it. Depending on the method of observation, we can define various magnitude systems. Different magnitudes have different zero points, i.e. they have different flux densities F_0 corresponding to the magnitude 0. The zero points are usually defined by a few selected standard stars.

In daylight the human eye is most sensitive to radiation with a wavelength of about 550 nm, the sensitivity decreasing towards red (longer wavelengths) and violet (shorter wavelengths). The magnitude corresponding to the sensitivity of the eye is called the *visual magnitude* m_{v}.

Photographic plates are usually most sensitive at blue and violet wavelengths, but they are also able to register radiation not visible to the human eye. Thus the *photographic magnitude* m_{pg} usually differs from the visual magnitude. The sensitivity of the eye can be simulated by using a yellow filter and plates sensitised to yellow and green light. Magnitudes thus observed are called *photovisual magnitudes* m_{pv}.

If, in ideal case, we were able to measure the radiation at all wavelengths, we would get the *bolometric magnitude* m_{bol}. In practice this is very difficult, since part of the radiation is absorbed by the atmosphere; also, different wavelengths require different detectors. (In fact there is a gadget called the bolometer, which, however, is not a real bolometer but an infrared detector.) The bolometric magnitude can be derived from the visual magnitude if we know the *bolometric correction* BC:

$$m_{\mathrm{bol}} = m_{\mathrm{v}} - \mathrm{BC} \,. \tag{4.10}$$

By definition, the bolometric correction is zero for radiation of solar type stars (or, more precisely, stars of the spectral class F5). Although the visual and bolometric magnitudes can be equal, the flux density corresponding to the bolometric magnitude must always be higher. The reason of this apparent contradiction is in the different values of F_0.

The more the radiation distribution differs from that of the Sun, the higher the bolometric correction is. The correction is positive for stars both cooler or hotter than the Sun. Sometimes the correction is defined as $m_{\mathrm{bol}} = m_v + BC$ in which case $BC \leq 0$ always. The chance for errors is, however, very small, since we must have $m_{\mathrm{bol}} \leq m_v$.

The most accurate magnitude measurements are made using photoelectric photometers. Usually filters are used to allow only a certain wavelength band to enter the detector. One of the multicolour magnitude systems used widely in photoeletric photometry is the UBV system developed in the early 1950's by *Harold L. Johnson* and *William W. Morgan*. Magnitudes are measured through three filters, U = ultraviolet, B = blue and V = visual. Figure 4.6 and Table 4.1 give the wavelength bands of these filters. The magnitudes observed through these filters are called U, B and V magnitudes, respectively.

The UBV system was later augmented by adding more bands. One commonly used system is the five colour UBVRI system, which includes R = red and I = infrared filters.

There are also other broad band systems, but they are not as well standardised as the UBV, which has been defined moderately well using a great number of

Table 4.1. Wavelength bands of the UBVRI and uvby filters and their effective ($\approx$ average) wavelengths

Magnitude		Band width [nm]	Effective wavelength [nm]
U	ultraviolet	66	367
B	blue	94	436
V	visual	88	545
R	red	138	638
I	infrared	149	797
u	ultraviolet	30	349
v	violet	19	411
b	blue	18	467
y	yellow	23	547

standard stars all over the sky. The magnitude of an object is obtained by comparing it to the magnitudes of standard stars.

In *Strömgren's* four-colour or *uvby system*, the bands passed by the filters are much narrower than in the UBV system. The uvby system is also well standardized, but it is not quite as common as the UBV. Other narrow band systems exist as well. By adding more filters, more information on the radiation distribution can be obtained.

In any multicolour system, we can define *colour indices*; a colour index is the difference of two magnitudes. By subtracting the B magnitude from U we get the colour index $U - B$, and so on. If the UBV system is used, it is common to give only the V magnitude and the colour indices $U - B$ and $B - V$.

The constants F_0 in (4.8) for U, B and V magnitudes have been selected in such a way that the colour indices $B - V$ and $U - B$ are zero for stars of spectral type A0 (for spectral types, see Chap. 8). The surface temperature of such a star is about $10,000\,\mathrm{K}$. For example, Vega (α Lyr, spectral class A0V) has $V = 0.03$, $B - V = U - B = 0.00$. The Sun has $V = -26.8$, $B - V = 0.62$ and $U - B = 0.10$.

Before the UBV system was developed, a colour index C.I., defined as

$$\mathrm{C.I.} = m_{\mathrm{pg}} - m_v \ ,$$

was used. Since m_{pg} gives the magnitude in blue and m_v in visual, this index is related to $B - V$. In fact,

$$\mathrm{C.I.} = (B - V) - 0.11 \ .$$

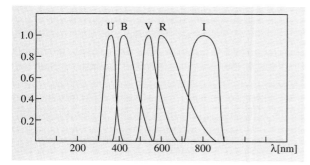

Fig. 4.6. Relative transmission profiles of filters used in the UBVRI magnitude system. The maxima of the bands are normalized to unity. The R and I bands are based on the system of Johnson, Cousins and Glass, which includes also infrared bands J, H, K, L and M. Previously used R and I bands differ considerably from these

4.4 Absolute Magnitudes

Thus far we have discussed only apparent magnitudes. They do not tell us anything about the true brightness of stars, since the distances differ. A quantity measuring the intrinsic brightness of a star is the *absolute magnitude*. It is defined as the apparent magnitude at a distance of 10 parsecs from the star (Fig. 4.7).

We shall now derive an equation which relates the apparent magnitude m, the absolute magnitude M and the distance r. Because the flux emanating from a star into a solid angle ω has, at a distance r, spread over an area ωr^2, the flux density is inversely proportional to the distance squared. Therefore the ratio of the flux density at a distance r, $F(r)$, to the flux density at a distance of 10 parsecs, $F(10)$, is

$$\frac{F(r)}{F(10)} = \left(\frac{10\,\mathrm{pc}}{r}\right)^2 .$$

Thus the difference of magnitudes at r and 10 pc, or the *distance modulus $m - M$*, is

$$m - M = -2.5 \lg \frac{F(r)}{F(10)} = -2.5 \lg \left(\frac{10\,\mathrm{pc}}{r}\right)^2$$

or

$$m - M = 5 \lg \frac{r}{10\,\mathrm{pc}} . \tag{4.11}$$

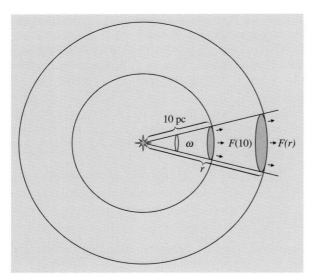

Fig. 4.7. The flux density at a distance of 10 parsecs from the star defines its absolute magnitude

For historical reasons, this equation is almost always written as

$$m - M = 5 \lg r - 5 , \tag{4.12}$$

which is valid *only* if the distance is expressed in parsecs. (The logarithm of a dimensional quantity is, in fact, physically absurd.) Sometimes the distance is given in kiloparsecs or megaparsecs, which require different constant terms in (4.12). To avoid confusion, we highly recommend the form (4.11).

Absolute magnitudes are usually denoted by capital letters. Note, however, that the U, B and V magnitudes are apparent magnitudes. The corresponding absolute magnitudes are M_U, M_B and M_V.

The absolute bolometric magnitude can be expressed in terms of the luminosity. Let the total flux density at a distance $r = 10\,\mathrm{pc}$ be F and let $F_\odot$ be the equivalent quantity for the Sun. Since the luminosity is $L = 4\pi r^2 F$, we get

$$M_{\mathrm{bol}} - M_{\mathrm{bol},\odot} = -2.5 \lg \frac{F}{F_\odot} = -2.5 \lg \frac{L/4\pi r^2}{L_\odot/4\pi r^2} ,$$

or

$$M_{\mathrm{bol}} - M_{\mathrm{bol},\odot} = -2.5 \lg \frac{L}{L_\odot} . \tag{4.13}$$

The absolute bolometric magnitude $M_{\mathrm{bol}} = 0$ corresponds to a luminosity $L_0 = 3.0 \times 10^{28}$ W.

4.5 Extinction and Optical Thickness

Equation (4.11) shows how the apparent magnitude increases (and brightness decreases!) with increasing distance. If the space between the radiation source and the observer is not completely empty, but contains some interstellar medium, (4.11) no longer holds, because part of the radiation is absorbed by the medium (and usually re-emitted at a different wavelength, which may be outside the band defining the magnitude), or scattered away from the line of sight. All these radiation losses are called the *extinction*.

Now we want to find out how the extinction depends on the distance. Assume we have a star radiating a flux L_0 into a solid angle ω in some wavelength range. Since the medium absorbs and scatters radiation, the flux L will now decrease with increasing distance r (Fig. 4.8). In a short distance interval $[r, r + dr]$, the extinction dL is proportional to the flux L and the distance travelled in the medium:

$$dL = -\alpha L \, dr \, . \tag{4.14}$$

The factor α tells how effectively the medium can obscure radiation. It is called the *opacity*. From (4.14) we see that its dimension is $[\alpha] = m^{-1}$. The opacity is zero for a perfect vacuum and approaches infinity when the substance becomes really murky. We can now define a dimensionless quantity, the *optical thickness* τ by

$$d\tau = \alpha \, dr \, . \tag{4.15}$$

Substituting this into (4.14) we get

$$dL = -L \, d\tau \, .$$

Next we integrate this from the source (where $L = L_0$ and $r = 0$) to the observer:

$$\int_{L_0}^{L} \frac{dL}{L} = -\int_{0}^{\tau} d\tau \, ,$$

which gives

$$L = L_0 e^{-\tau} \, . \tag{4.16}$$

Here, τ is the optical thickness of the material between the source and the observer and L, the observed flux. Now, the flux L falls off exponentially with increasing optical thickness. Empty space is perfectly transparent, i.e. its opacity is $\alpha = 0$; thus the optical thickness does not increase in empty space, and the flux remains constant.

Let F_0 be the flux density on the surface of a star and $F(r)$, the flux density at a distance r. We can express the fluxes as

$$L = \omega r^2 F(r) \, , \quad L_0 = \omega R^2 F_0 \, ,$$

where R is the radius of the star. Substitution into (4.16) gives

$$F(r) = F_0 \frac{R^2}{r^2} e^{-\tau} \, .$$

For the absolute magnitude we need the flux density at a distance of 10 parsecs, $F(10)$, which is still evaluated without extinction:

$$F(10) = F_0 \frac{R^2}{(10 \, \text{pc})^2} \, .$$

The distance modulus $m - M$ is now

$$\begin{aligned}
m - M &= -2.5 \lg \frac{F(r)}{F(10)} \\
&= 5 \lg \frac{r}{10 \, \text{pc}} - 2.5 \lg e^{-\tau} \\
&= 5 \lg \frac{r}{10 \, \text{pc}} + (2.5 \lg e)\tau
\end{aligned}$$

or

$$m - M = 5 \lg \frac{r}{10 \, \text{pc}} + A \, , \tag{4.17}$$

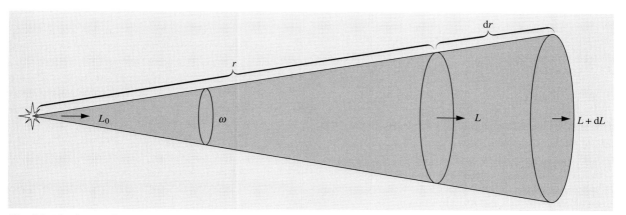

Fig. 4.8. The interstellar medium absorbs and scatters radiation; this usually reduces the energy flux L in the solid angle ω ($dL \leq 0$)

where $A \geq 0$ is the extinction in magnitudes due to the entire medium between the star and the observer. If the opacity is constant along the line of sight, we have

$$\tau = \alpha \int_0^r dr = \alpha r ,$$

and (4.17) becomes

$$m - M = 5 \lg \frac{r}{10 \, \text{pc}} + ar , \qquad (4.18)$$

where the constant $a = 2.5 \alpha \lg e$ gives the extinction in magnitudes per unit distance.

Colour Excess. Another effect caused by the interstellar medium is the *reddening of light*: blue light is scattered and absorbed more than red. Therefore the colour index $B - V$ increases. The visual magnitude of a star is, from (4.17),

$$V = M_V + 5 \lg \frac{r}{10 \, \text{pc}} + A_V , \qquad (4.19)$$

where M_V is the absolute visual magnitude and A_V is the extinction in the V passband. Similarly, we get for the blue magnitudes

$$B = M_B + 5 \lg \frac{r}{10 \, \text{pc}} + A_B .$$

The observed colour index is now

$$B - V = M_B - M_V + A_B - A_V ,$$

or

$$B - V = (B - V)_0 + E_{B-V} , \qquad (4.20)$$

where $(B - V)_0 = M_B - M_V$ is the *intrinsic colour* of the star and $E_{B-V} = (B - V) - (B - V)_0$ is the *colour excess*. Studies of the interstellar medium show that the ratio of the visual extinction A_V to the colour excess E_{B-V} is almost constant for all stars:

$$R = \frac{A_V}{E_{B-V}} \approx 3.0 .$$

This makes it possible to find the visual extinction if the colour excess is known:

$$A_V \approx 3.0 \, E_{B-V} . \qquad (4.21)$$

When A_V is obtained, the distance can be solved directly from (4.19), when V and M_V are known.

We shall study interstellar extinction in more detail in Sect. 15.1 ("Interstellar Dust").

Atmospheric Extinction. As we mentioned in Sect. 3.1, the Earth's atmosphere also causes extinction. The observed magnitude m depends on the location of the observer and the zenith distance of the object, since these factors determine the distance the light has to travel in the atmosphere. To compare different observations, we must first *reduce* them, i.e. remove the atmospheric effects somehow. The magnitude m_0 thus obtained can then be compared with other observations.

If the zenith distance z is not too large, we can approximate the atmosphere by a plane layer of constant thickness (Fig. 4.9). If the thickness of the atmosphere is used as a unit, the light must travel a distance

$$X = 1 / \cos z = \sec z \qquad (4.22)$$

in the atmosphere. The quantity X is the *air mass*. According to (4.18), the magnitude increases linearly with the distance X:

$$m = m_0 + kX , \qquad (4.23)$$

where k is the *extinction coefficient*.

The extinction coefficient can be determined by observing the same source several times during a night with as wide a zenith distance range as possible. The observed magnitudes are plotted in a diagram as a function of the air mass X. The points lie on a straight line the slope of which gives the extinction coefficient k. When

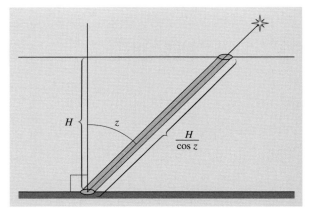

Fig. 4.9. If the zenith distance of a star is z, the light of the star travels a distance $H / \cos z$ in the atmosphere; H is the height of the atmosphere

this line is extrapolated to $X = 0$, we get the magnitude m_0, which is the apparent magnitude outside the atmosphere.

In practice, observations with zenith distances higher than 70° (or altitudes less than 20°) are not used to determine k and m_0, since at low altitudes the curvature of the atmosphere begins to complicate matters. The value of the extinction coefficient k depends on the observation site and time and also on the wavelength, since extinction increases strongly towards short wavelengths.

4.6 Examples

Example 4.1 Show that intensity is independent of distance.

Suppose we have some radiation leaving the surface element dA in the direction θ. The energy entering the solid angle $d\omega$ in time dt is

$$dE = I \cos \theta \, dA \, d\omega \, dt \,,$$

where I is the intensity. If we have another surface dA' at a distance r receiving this radiation from direction θ', we have

$$d\omega = dA' \cos \theta' / r^2 \,.$$

The definition of the intensity gives

$$dE = I' \cos \theta' \, dA' \, d\omega' \, dt \,,$$

where I' is the intensity at dA' and

$$d\omega' = dA \cos \theta / r^2 \,.$$

Substitution of $d\omega$ and $d\omega'$ into the expressions of dE gives

$$I \cos \theta \, d\theta \, dA \frac{dA' \cos \theta'}{r^2} \, dt$$
$$= I' \cos \theta' \, dA' \frac{dA \cos \theta}{r^2} \, dt \quad \Rightarrow \quad I' = I \,.$$

Thus the intensity remains constant in empty space.

Example 4.2 *Surface Brightness of the Sun*

Assume that the Sun radiates isotropically. Let R be the radius of the Sun, $F_\odot$ the flux density on the surface of the Sun and F the flux density at a distance r. Since the luminosity is constant,

$$L = 4\pi R^2 F_\odot = 4\pi r^2 F \,,$$

the flux density equals

$$F = F_\odot \frac{R^2}{r^2} \,.$$

At a distance $r \gg R$, the Sun subtends a solid angle

$$\omega = \frac{A}{r^2} = \frac{\pi R^2}{r^2} \,,$$

where $A = \pi R^2$ is the cross section of the Sun. The surface brightness B is

$$B = \frac{F}{\omega} = \frac{F_\odot}{\pi} \,.$$

Applying (4.4) we get

$$B = I_\odot \,.$$

Thus the surface brightness is independent of distance and equals the intensity. We have found a simple

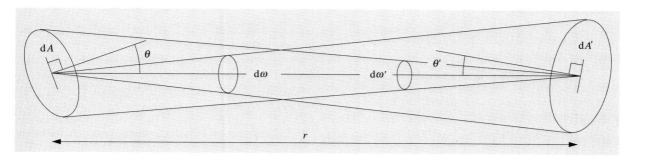

interpretation for the somewhat abstract concept of intensity.

The flux density of the Sun on the Earth, the *solar constant*, is $S_\odot \approx 1370$ W m^{-2}. The angular diameter of the Sun is $\alpha = 32'$, whence

$$\frac{R}{r} = \frac{\alpha}{2} = \frac{1}{2} \times \frac{32}{60} \times \frac{\pi}{180} = 0.00465 \text{ rad} .$$

The solid angle subtended by the Sun is

$$\omega = \pi \left(\frac{R}{r}\right)^2 = \pi \times 0.00465^2$$

$$= 6.81 \times 10^{-5} \text{ sterad} .$$

and the surface brightness

$$B = \frac{S_\odot}{\omega} = 2.01 \times 10^7 \text{ W m}^{-2} \text{ sterad}^{-1} .$$

Example 4.3 *Magnitude of a Binary Star*

Since magnitudes are logarithmic quantities, they can be a little awkward for some purposes. For example, we cannot add magnitudes like flux densities. If the magnitudes of the components of a binary star are 1 and 2, the total magnitude is certainly not 3. To find the total magnitude, we must first solve the flux densities from

$$1 = -2.5 \lg \frac{F_1}{F_0} , \quad 2 = -2.5 \lg \frac{F_2}{F_0} ,$$

which give

$$F_1 = F_0 \times 10^{-0.4} , \quad F_2 = F_0 \times 10^{-0.8} .$$

Thus the total flux density is

$$F = F_1 + F_2 = F_0(10^{-0.4} + 10^{-0.8})$$

and the total magnitude,

$$m = -2.5 \lg \frac{F_0(10^{-0.4} + 10^{-0.8})}{F_0}$$

$$= -2.5 \lg 0.5566 = 0.64 .$$

Example 4.4 The distance of a star is $r = 100$ pc and its apparent magnitude $m = 6$. What is its absolute magnitude?

Substitution into (4.11)

$$m - M = 5 \lg \frac{r}{10 \text{ pc}}$$

gives

$$M = 6 - 5 \lg \frac{100}{10} = 1 .$$

Example 4.5 The absolute magnitude of a star is $M = -2$ and the apparent magnitude $m = 8$. What is the distance of the star?

We can solve the distance r from (4.11):

$$r = 10 \text{ pc} \times 10^{(m-M)/5} = 10 \times 10^{10/5} \text{pc}$$

$$= 1000 \text{ pc} = 1 \text{ kpc} .$$

Example 4.6 Although the amount of interstellar extinction varies considerably from place to place, we can use an average value of 2 mag/kpc near the galactic plane. Find the distance of the star in Example 4.5, assuming such extinction.

Now the distance must be solved from (4.18):

$$8 - (-2) = 5 \lg \frac{r}{10} + 0.002\,r ,$$

where r is in parsecs. This equation cannot be solved analytically, but we can always use a numerical method. We try a simple iteration (Appendix A.7), rewriting the equation as

$$r = 10 \times 10^{2-0.0004\,r} .$$

The value $r = 1000$ pc found previously is a good initial guess:

$$r_0 = 1000$$

$$r_1 = 10 \times 10^{2-0.0004 \times 1000} = 398$$

$$r_2 = 693$$

$$\vdots$$

$$r_{12} = r_{13} = 584 .$$

The distance is $r \approx 580$ pc, which is much less than our earlier value 1000 pc. This should be quite obvious, since due to extinction, radiation is now reduced much faster than in empty space.

Example 4.7   What is the optical thickness of a layer of fog, if the Sun seen through the fog seems as bright as a full moon in a cloudless sky?

The apparent magnitudes of the Sun and the Moon are -26.8 and -12.5, respectively. Thus the total extinction in the cloud must be $A = 14.3$. Since

$$A = (2.5 \lg e)\tau ,$$

we get

$$\tau = A/(2.5 \lg e) = 14.3/1.086 = 13.2 .$$

The optical thickness of the fog is 13.2. In reality, a fraction of the light scatters several times, and a few of the multiply scattered photons leave the cloud along the line of sight, reducing the total extinction. Therefore the optical thickness must be slightly higher than our value.

Example 4.8   *Reduction of Observations*

The altitude and magnitude of a star were measured several times during a night. The results are given in the following table.

Altitude	Zenith distance	Air mass	Magnitude
50°	40°	1.31	0.90
35°	55°	1.74	0.98
25°	65°	2.37	1.07
20°	70°	2.92	1.17

By plotting the observations as in the following figure, we can determine the extinction coefficient k and the magnitude m_0 outside the atmosphere. This can be done graphically (as here) or using a least-squares fit.

Extrapolation to the air mass $X = 0$ gives $m_0 = 0.68$. The slope of the line gives $k = 0.17$.

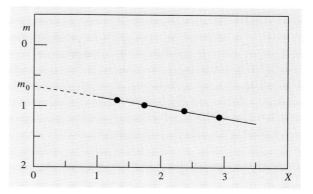

4.7   Exercises

Exercise 4.1   The total magnitude of a triple star is 0.0. Two of its components have magnitudes 1.0 and 2.0. What is the magnitude of the third component?

Exercise 4.2   The absolute magnitude of a star in the Andromeda galaxy (distance 690 kpc) is $M = 5$. It explodes as a supernova, becoming one billion (10^9) times brighter. What is its apparent magnitude?

Exercise 4.3   Assume that all stars have the same absolute magnitude and stars are evenly distributed in space. Let $N(m)$ be the number of stars brighter than m magnitudes. Find the ratio $N(m+1)/N(m)$.

Exercise 4.4   The V magnitude of a star is 15.1, $B - V = 1.6$, and absolute magnitude $M_V = 1.3$. The extinction in the direction of the star in the visual band is $a_V = 1$ mag kpc^{-1}. What is the intrinsic colour of the star?

Exercise 4.5   Stars are observed through a triple window. Each surface reflects away 15% of the incident light.

a) What is the magnitude of Regulus ($M_V = 1.36$) seen through the window?
b) What is the optical thickness of the window?

5. Radiation Mechanisms

In the previous chapters we have studied the physical properties and detection of electromagnetic radiation. Next we shall briefly discuss concepts related to emission and absorption of radiation. Since we can give here only a summary of some essential results without delving into quantum mechanical explanations, the reader interested in the details is advised to consult any good physics textbook.

5.1 Radiation of Atoms and Molecules

Electromagnetic radiation is emitted or absorbed when an atom or a molecule moves from one energy level to another. If the energy of the atom decreases by an amount ΔE, the atom emits or radiates a quantum of electromagnetic radiation, called a *photon*, whose frequency ν is given by the equation

$$\Delta E = h\nu , \tag{5.1}$$

where h is the *Planck constant*, $h = 6.6256 \times 10^{-34}$ J s. Similarly, if the atom receives or absorbs a photon of a frequency ν, its energy increases by $\Delta E = h\nu$.

The classical model describes an atom as a nucleus surrounded by a swarm of electrons. The nucleus consists of Z protons, each having a charge $+e$ and N electrically neutral neutrons; Z is the *charge number* of the atom and $A = Z + N$ is its *mass number*. A neutral atom has as many electrons (charge $-e$) as protons.

An energy level of an atom usually refers to an energy level of its electrons. The energy E of an electron cannot take arbitrary values; only certain energies are allowed: the energy levels are *quantized*. An atom can emit or absorb radiation only at certain frequencies ν_{if} corresponding to energy differences between some initial and final states i and f: $|E_i - E_f| = h\nu_{if}$. This gives rise to the *line spectrum*, specific for each element (Fig. 5.1). Hot gas under low pressure produces an *emission spectrum* consisting of such discrete lines. If the same gas is cooled down and observed against a source of white light (which has a continuous spectrum), the same lines are seen as dark *absorption lines*.

At low temperatures most atoms are in their lowest energy state, the *ground state*. Higher energy levels are *excitation states*; a transition from lower to higher state is called *excitation*. Usually the excited atom will return to the lower state very rapidly, radiating a photon (*spontaneous emission*); a typical lifetime of an excited state might be 10^{-8} seconds. The frequency of the emitted photon is given by (5.1). The atom may return to the lower state directly or through some intermediate states, emitting one photon in each transition.

Downward transitions can also be induced by radiation. Suppose our atom has swallowed a photon and become excited. Another photon, whose frequency ν corresponds to some possible downward transition from the excited state, can now irritate the atom, causing it to jump to a lower state, emitting a photon with the same frequency ν. This is called *induced* or *stimulated emission*. Photons emitted spontaneously leave the atom randomly in all directions with random phases: the radiation is isotropic and incoherent. Induced radiation, on the other hand, is *coherent*; it propagates in the same direction as and in phase with the inducing radiation.

The zero level of the energy states is usually chosen so that a bound electron has negative energy and a free electron positive energy (cf. the energy integral of planetary orbits, Chap. 6). If an electron with energy $E < 0$ receives more energy than $|E|$, it will leave the atom, which becomes an ion. In astrophysics ionization is often called a *bound-free* transition (Fig. 5.2). Unlike in excitation all values of energy ($E > 0$) are now possible. The extraneous part of the absorbed energy goes to the kinetic energy of the liberated electron. The inverse process, in which an atom captures a free electron, is the *recombination* or free–bound transition.

When an electron scatters from a nucleus or an ion without being captured, the electromagnetic interaction can change the kinetic energy of the electron producing *free–free* radiation. In a very hot gas ($T > 10^6$ K) hydrogen is fully ionized, and the free–free radiation is the most important source of emission. It is then usually called *thermal bremsstrahlung*. The latter part of the name derives from the fact that decelerating electrons hitting the anode of an X-ray tube emit similar

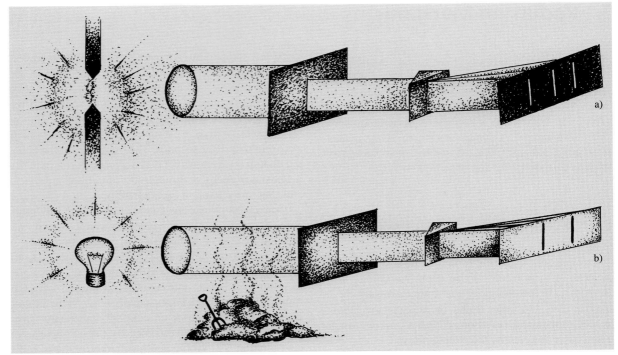

Fig. 5.1a,b. Origin of line spectra. (**a**) Emission spectrum. Atoms of glowing gas returning from excited states to lower states emit photons with frequencies corresponding to the energy difference of the states. Each element emits its own characteristic wavelengths, which can be measured by spread-ing the light into a spectrum with a prism or diffraction grating. (**b**) Absorption spectrum. When white light con-taining all wavelengths travels through gas, the wavelengths characteristic of the gas are absorbed

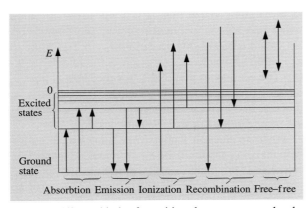

Fig. 5.2. Different kinds of transitions between energy levels. Absorption and emission occur between two bound states, whereas ionization and recombination occur between a bound and a free state. Interaction of an atom with an free electron can result in a free–free transition

radiation. In an analogous way the absorption process can be called a bound–bound transition.

Electromagnetic radiation is transverse wave motion; the electric and magnetic fields oscillate perpendicular to each other and also perpendicular to the direction of propagation. The light of an ordinary incandescent lamp has a random distribution of electric fields vi-brating in all directions. If the directions of electric fields in the plane perpendicular to the direction of propagation are not evenly distributed, the radiation is *polarized* (Fig. 5.3). The direction of polarization of *lin-early polarized* light means the plane determined by the electric vector and the direction of the light ray. If the electric vector describes a circle, the radiation is *circularly polarized*. If the amplitude of the elec-tric field varies at the same time, the polarization is *elliptic*.

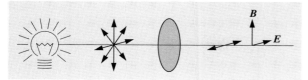

Fig. 5.3. Polarization of light. The light of an incandescent bulb contains all possible directions of vibration and is therefore unpolarized. Some crystals, for example, pass electric fields oscillating only in certain directions, and the transmitted part of the light becomes linearly polarized. E is the electric field and B the magnetic field

If polarized radiation travels through a magnetic field, the direction of the polarization will rotate. The amount of such *Faraday rotation* is proportional to the component of the magnetic field parallel to the line of sight, number of electrons along the line of sight, distance travelled, and square of the wavelength of the radiation.

Scattering is an absorption followed by an instantaneous emission at the same wavelength but usually in a new direction. On the macroscopic scale, radiation seems to be reflected by the medium. The light coming from the sky is sunlight scattered from atmospheric molecules. Scattered light is always polarized, the degree of polarization being highest in the direction perpendicular to the direction of the original radiation.

5.2 The Hydrogen Atom

The hydrogen atom is the simplest atom, consisting of a proton and an electron. According to the Bohr model the electron orbits the proton in a circular orbit. (In spite of the fact that this model has very little to do with reality, it can be successfully used to predict some properties of the hydrogen atom.) Bohr's first postulate says that the angular momentum of the electron must be a multiple of $\hbar$:

$$mvr = n\hbar , \tag{5.2}$$

where

$m =$ mass of the electron ,
$v =$ speed of the electron ,
$r =$ radius of the orbit ,

$n =$ the principal quantum number ,
$\qquad n = 1, 2, 3, \dots ,$
$\hbar = h/2\pi ,$
$h =$ the Planck constant .

The quantum mechanical interpretation of Bohr's first postulate is obvious: the electron is represented as a standing wave, and the "length of the orbit" must be a multiple of the de Broglie wavelength, $\lambda = \hbar/p = \hbar/mv$.

A charged particle in a circular orbit (and thus in accelerated motion) should emit electromagnetic radiation, losing energy, were it to obey the rules of classical electrodynamics. Therefore our electron should spiral down towards the nucleus. But obviously, Nature does not behave this way, and we have to accept Bohr's second postulate, which says that an electron moving in an allowed orbit around a nucleus does not radiate. Radiation is emitted only when the electron jumps from a higher energy state to a lower one. The emitted quantum has an energy $h\nu$, equal to the energy difference of these states:

$$h\nu = E_{n_2} - E_{n_1} . \tag{5.3}$$

We shall now try to find the energy of an electron in the state E_n. Coulomb's law gives the force pulling the electron towards the proton:

$$F = \frac{1}{4\pi\epsilon_0} \frac{e^2}{r_n^2} , \tag{5.4}$$

where

$\epsilon_0 =$ the vacuum permittivity
$\qquad = 8.85 \times 10^{-12} \, \text{N}^{-1} \, \text{m}^{-2} \, \text{C}^2 ,$
$e =$ the charge of the electron $= 1.6 \times 10^{-19} \, \text{C} ,$
$r_n =$ the distance between the electron
$\qquad$ and the proton .

The acceleration of a particle moving in a circular orbit of radius r_n is

$$a = \frac{v_n^2}{r_n} ,$$

and applying Newton's second law ($F = ma$), we get

$$\frac{mv_n^2}{r_n} = \frac{1}{4\pi\epsilon_0} \frac{e^2}{r_n^2}. \tag{5.5}$$

From (5.2) and (5.5) it follows that

$$v_n = \frac{e^2}{4\pi\epsilon_0 \hbar} \frac{1}{n} , \qquad r_n = \frac{4\pi\epsilon_0 \hbar^2}{me^2} n^2 .$$

The total energy of an electron in the orbit n is now

$$E_n = T + V = \frac{1}{2}mv_n^2 - \frac{1}{4\pi\epsilon_0}\frac{e^2}{r_n}$$

$$= -\frac{me^4}{32\pi^2\epsilon_0^2\hbar^2}\frac{1}{n^2} \equiv -C\frac{1}{n^2} ,$$

(5.6)

where C is a constant. For the ground state ($n = 1$), we get from (5.6)

$$E_1 = -2.18 \times 10^{-18}\,\text{J} = -13.6\,\text{eV} .$$

From (5.3) and (5.6) we get the energy of the quantum emitted in the transition $E_{n_2} \rightarrow E_{n_1}$:

$$h\nu = E_{n_2} - E_{n_1} = C\left(\frac{1}{n_1^2} - \frac{1}{n_2^2}\right) .$$

(5.7)

In terms of the wavelength λ this can be expressed as

$$\frac{1}{\lambda} = \frac{\nu}{c} = \frac{C}{hc}\left(\frac{1}{n_1^2} - \frac{1}{n_2^2}\right) \equiv R\left(\frac{1}{n_1^2} - \frac{1}{n_2^2}\right) ,$$

(5.8)

where R is the *Rydberg constant*, $R = 1.097 \times 10^7\,\text{m}^{-1}$.

Equation (5.8) was derived experimentally for $n_1 = 2$ by Johann Jakob Balmer as early as 1885. That is why we call the set of lines produced by transitions $E_n \rightarrow E_2$ the *Balmer series*. These lines are in the visible part of the spectrum. For historical reasons the Balmer lines are often denoted by symbols H_α, H_β, H_γ etc. If the electron returns to its ground state ($E_n \rightarrow E_1$), we get the *Lyman*

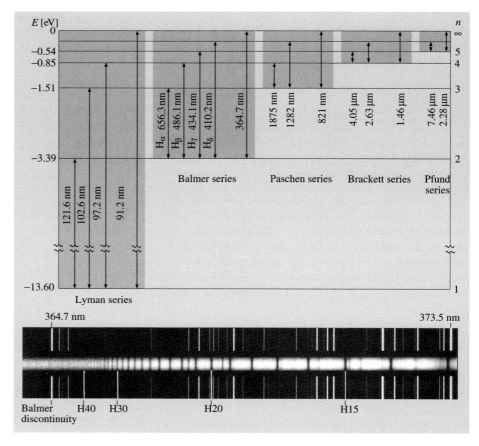

Fig. 5.4. Transitions of a hydrogen atom. The lower picture shows a part of the spectrum of the star HD193182. On both sides of the stellar spectrum we see an emission spectrum of iron. The wavelengths of the reference lines are known, and they can be used to calibrate the wavelengths of the observed stellar spectrum. The hydrogen Balmer lines are seen as dark absorption lines converging towards the Balmer ionization limit (also called the Balmer discontinuity) at $\lambda = 364.7$ nm to the left. The numbers (15, . . . , 40) refer to the quantum number n of the higher energy level. (Photo by Mt. Wilson Observatory)

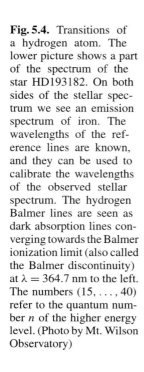

series, which is in the ultraviolet. The other series with specific names are the *Paschen series* ($n_1 = 3$), *Bracket series* ($n_1 = 4$) and *Pfund series* ($n_1 = 5$) (see Fig. 5.4).

5.3 Line Profiles

The previous discussion suggests that spectral lines would be infinitely narrow and sharp. In reality, however, they are somewhat broadened. We will now consider briefly the factors affecting the shape of a spectral line, called a *line profile*. An exact treatment would take us too deep into quantum mechanics, so we cannot go into the details here.

According to quantum mechanics everything cannot be measured accurately at the same time. For example, even in principle, there is no way to determine the x coordinate and the momentum p_x in the direction of the x axis with arbitrary precision simultaneously. These quantities have small uncertainties Δx and Δp_x, such that

$$\Delta x \Delta p_x \approx \hbar \,.$$

A similar relation holds for other directions, too. Time and energy are also connected by an uncertainty relation,

$$\Delta E \Delta t \approx \hbar \,.$$

The natural width of spectral lines is a consequence of this *Heisenberg uncertainty principle*.

If the average lifetime of an excitation state is T, the energy corresponding to the transition can only be determined with an accuracy of $\Delta E = \hbar/T = h/(2\pi T)$. From (5.1) it follows that $\Delta \nu = \Delta E/h$. In fact, the uncertainty of the energy depends on the lifetimes of both the initial and final states. The *natural width* of a line is defined as

$$\gamma = \frac{\Delta E_i + \Delta E_f}{\hbar} = \frac{1}{T_i} + \frac{1}{T_f} \,. \tag{5.9}$$

It can be shown that the corresponding line profile is

$$I_\nu = \frac{\gamma}{2\pi} \frac{I_0}{(\nu - \nu_0)^2 + \gamma^2/4} \,, \tag{5.10}$$

where ν_0 is the frequency at the centre of the line and I_0 the total intensity of the line. At the centre of the line the intensity per frequency unit is

$$I_{\nu_0} = \frac{2}{\pi\gamma} I_0 \,,$$

and at the frequency $\nu = \nu_0 + \gamma/2$,

$$I_{\nu_0 + \gamma/2} = \frac{1}{\pi\gamma} I_0 = \frac{1}{2} I_{\nu_0} \,.$$

Thus the width γ is the width of the line profile at a depth where the intensity is half of the maximum. This is called the *full width at half maximum* (FWHM).

Doppler Broadening. Atoms of a gas are moving the faster the higher the temperature of the gas. Thus spectral lines arising from individual atoms are shifted by the Doppler effect. The observed line consists of a collection of lines with different Doppler shifts, and the shape of the line depends on the number of atoms with different velocities.

Each Doppler shifted line has its characteristic natural width. The resulting line profile is obtained by giving each Doppler shifted line a weight proportional to the number of atoms given by the velocity distribution and integrating over all velocities. This gives rise to the *Voigt profile* (Fig. 5.5), which already describes most spectral lines quite well. The shapes of different profiles don't

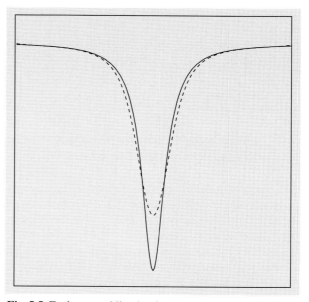

Fig. 5.5. Each spectral line has its characteristic natural width (*solid line*). Motions of particles broaden the line further due to the Doppler effect, resulting in the Voigt profile (*dashed line*). Both profiles have the same area

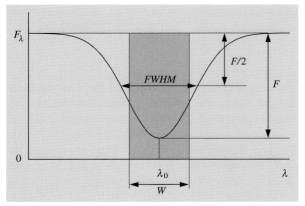

Fig. 5.6. The full width at half maximum (FWHM) of a spectral line is the width at the depth where the intensity is half of the maximum. The equivalent width W is defined so that the line and the shaded rectangle have the same area in the picture. The two measures are not generally the same, although they usually are close to each other

seem very different; the most obvious consequence of the broadening is that the maximum depth decreases.

One way to describe the width of a line is to give its full width at half maximum (Fig. 5.6). Due to Doppler broadening this is usually greater than the natural width. The *equivalent width* is another measure of a line strength. It is the area of a rectangular line that has the same area as the line profile and that emits no light at all. The equivalent width can be used to describe the energy corresponding to a line independently of the shape of the line profile.

5.4 Quantum Numbers, Selection Rules, Population Numbers

Quantum Numbers. The Bohr model needs only one quantum number, n, to describe all the energy levels of the electron. This can explain only the coarse features of an atom with a single electron.

Quantum mechanics describes the electron as a three dimensional wave, which only gives the probability of finding the electron in a certain place. Quantum mechanics has accurately predicted all the energy levels of hydrogen atoms. The energy levels of heavier atoms and molecules can also be computed; however, such calculations are very complicated. Also the existence of

quantum numbers can be understood from the quantum mechanical point of view.

The quantum mechanical description involves four quantum numbers, one of which is our n, the *principal quantum number*. The principal quantum number describes the quantized energy levels of the electron. The classical interpretation of discrete energy levels allows only certain orbits given by (5.6). The orbital angular momentum of the electron is also quantized. This is described by the *angular momentum quantum number l*. The angular momentum corresponding to a quantum number l is

$$L = \sqrt{l(l+1)}\hbar \, .$$

The classical analogy would be to allow some elliptic orbits. The quantum number l can take only the values

$$l = 0, 1, \ldots, n-1 \, .$$

For historical reasons, these are often denoted by the letters s, p, d, f, g, h, i, j.

Although l determines the magnitude of the angular momentum, it does not give its direction. In a magnetic field this direction is important, since the orbiting electron also generates a tiny magnetic field. In any experiment, only one component of the angular momentum can be measured at a time. In a given direction z (e. g. in the direction of the applied magnetic field), the projection of the angular momentum can have only the values

$$L_z = m_l \hbar \, ,$$

where m_l is the *magnetic quantum number*

$$m_l = 0, \pm 1, \pm 2, \ldots, \pm l \, .$$

The magnetic quantum number is responsible for the splitting of spectral lines in strong magnetic fields, known as the *Zeeman effect*. For example, if $l = 1$, m_l can have $2l + 1 = 3$ different values. Thus, the line arising from the transition $l = 1 \rightarrow l = 0$ will split into three components in a magnetic field (Fig. 5.7).

The fourth quantum number is the *spin* describing the intrinsic angular momentum of the electron. The spin of the electron is

$$S = \sqrt{s(s+1)}\hbar \, ,$$

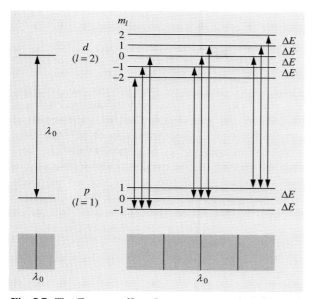

Fig. 5.7. The Zeeman effect. In strong magnetic fields each energy level of a hydrogen atom splits into $(2l + 1)$ separate levels, which correspond to different values of the magnetic quantum number $m_l = l, l - 1, \ldots, -l$. The energy differences of the successive levels have the same constant value ΔE. For example the p state $(l = 1)$ splits into three and the d state $(l = 2)$ into five sublevels. The selection rules require that in electric dipole transitions Δm_l equals 0 or ± 1, and only nine different transitions between p and d states are possible. Moreover, the transitions with the same Δm_l have the same energy difference. Thus the spectrum has only three separate lines

where the spin quantum number is $s = \frac{1}{2}$. In a given direction z, the spin is

$$S_z = m_s \hbar,$$

where m_s can have one of the two values:

$$m_s = \pm \frac{1}{2}.$$

All particles have a spin quantum number. Particles with an integral spin are called *bosons* (photon, mesons); particles with a half-integral spin are *fermions* (proton, neutron, electron, neutrino etc.).

Classically, spin can be interpreted as the rotation of a particle; this analogy, however, should not be taken too literally.

The total angular momentum J of an electron is the sum of its orbital and spin angular momentum:

$$J = L + S.$$

Depending on the mutual orientation of the vectors L and S the quantum number j of total angular momentum can have one of two possible values,

$$j = l \pm \frac{1}{2},$$

(except if $l = 0$, when $j = \frac{1}{2}$). The z component of the total angular momentum can have the values

$$m_j = 0, \pm 1, \pm 2, \ldots \pm j.$$

Spin also gives rise to the fine structure of spectral lines. Lines appear as close pairs or doublets.

Selection Rules. The state of an electron cannot change arbitrarily; transitions are restricted by selection rules, which follow from certain conservation laws. The selection rules express how the quantum numbers must change in a transition. Most probable are the *electric dipole transitions*, which make the atom behave like an oscillating dipole. The conservation laws require that in a transition we have

$$\Delta l = \pm 1,$$
$$\Delta m_l = 0, \pm 1.$$

In terms of the total angular momentum the selection rules are

$$\Delta l = \pm 1,$$
$$\Delta j = 0, \pm 1,$$
$$\Delta m_j = 0, \pm 1.$$

The probabilities of all other transitions are much smaller, and they are called *forbidden transitions*; examples are magnetic dipole transitions and all quadrupole and higher multipole transitions.

Spectral lines originating in forbidden transitions are called *forbidden lines*. The probability of such a transition is so low that under normal circumstances, the transition cannot take place before collisions force the electron to change state. Forbidden lines are possible only if the gas is extremely rarified (like in auroras and planetary nebulae).

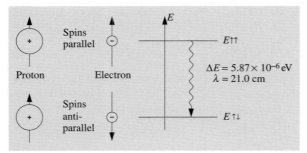

Fig. 5.8. The origin of the hydrogen 21 cm line. The spins of the electron and the proton may be either parallel or opposite. The energy of the former state is slightly larger. The wavelength of a photon corresponding to a transition between these states is 21 cm

The spins of an electron and nucleus of a hydrogen atom can be either parallel or antiparallel (Fig. 5.8). The energy of the former state is 0.0000059 eV higher. But the selection rules make an electric dipole transition between these states impossible. The transition, which is a magnetic dipole transition, has a very low probability, $A = 2.8 \times 10^{-15}$ s^{-1}. This means that the average lifetime of the higher state is $T = 1/A = 11 \times 10^6$ years. Usually collisions change the state of the electron well before this period of time has elapsed. But in interstellar space the density of hydrogen is so low and the total amount of hydrogen so great that a considerable number of these transitions can take place.

The wavelength of the radiation emitted by this transition is 21 cm, which is in the radio band of the spectrum. Extinction at radio wavelengths is very small, and we can observe more distant objects than by using optical wavelengths. The 21 cm radiation has been of crucial importance for surveys of interstellar hydrogen.

Population Numbers. The population number n_i of an energy state i means the number of atoms in that state

per unit volume. In thermal equilibrium, the population numbers obey the *Boltzmann distribution*:

$$\frac{n_i}{n_0} = \frac{g_i}{g_0} e^{-\Delta E/(kT)} \,, \tag{5.11}$$

where T is the temperature, k is the Boltzmann constant, $\Delta E = E_i - E_0 = h\nu$ is the energy difference between the excited and ground state, and g_i is the statistical weight of the level i (it is the number of different states with the same energy E_i). The subscript 0 always refers to the ground state. Often the population numbers differ from the values given by (5.11), but still we can define an *excitation temperature* T_{exc} in such a way that (5.11) gives correct population numbers, when T is replaced by T_{exc}. The excitation temperature may be different for different energy levels.

5.5 Molecular Spectra

The energy levels of an atom are determined by its electrons. In the case of a molecule, there are many more possibilities: atoms can vibrate around their equilibria and the molecule can rotate around some axis. Both vibrational and rotational states are quantized. Transitions between successive vibrational states typically involve photons in the infrared band, while transitions between rotational states involve photons in the microwave band. These combined with transitions of electrons produce a band spectrum, characteristic for molecules (Fig. 5.9). The spectrum has several narrow bands composed of a great number of lines.

5.6 Continuous Spectra

We have already mentioned some processes that produce continuous spectra. Continuous emission spectra

Fig. 5.9. Spectrum of carbon monoxide CO from 430 nm to 670 nm. The various bands correspond to different vibrational transitions. Each band is composed of numerous rotational lines. Near the right edge of each band the lines are so closely

packed that they overlap and at this resolution, the spectrum looks continuous. (R.W.B. Pearse, A.G. Gaydon: *The Identification of Molecular Spectra* (Chapman & Hall Ltd., London 1976) p. 394)

can originate in recombinations and free–free transitions. In recombination, an atom captures a free electron whose energy is not quantized; in free–free transitions, both initial and final states are unquantized. Thus the emission line can have any frequency whatsoever. Similarly, ionizations and free–free transitions can give rise to a continuous absorption spectrum.

Each spectrum contains a continuous component, or *continuum*, and spectral lines. Sometimes, however, the lines are so closely packed and so broad that they seem to form a nearly continuous spectrum.

When the pressure of hot gas is increased, the spectral lines begin to broaden. At high pressure, atoms bump into each other more frequently, and the close neighbors disturb the energy levels. When the pressure is high enough, the lines begin to overlap. Thus the spectrum of hot gas at high pressure is continuous. Electric fields also broaden spectral lines (the Stark effect).

In liquids and solids the atoms are more densely packed than in gaseous substances. Their mutual perturbations broaden the energy levels, producing a continuous spectrum.

5.7 Blackbody Radiation

A *blackbody* is defined as an object that does not reflect or scatter radiation shining upon it, but absorbs and re-emits the radiation completely. A blackbody is a kind of an ideal radiator, which cannot exist in the real world. Yet many objects behave very much as if they were blackbodies.

The radiation of a blackbody depends only on its temperature, being perfectly independent of its shape, material and internal constitution. The wavelength distribution of the radiation follows *Planck's law*, which is a function of temperature only. The intensity at a frequency ν of a blackbody at temperature T is

$$B_\nu(T) = B(\nu; T) = \frac{2h\nu^3}{c^2} \frac{1}{e^{h\nu/(kT)} - 1}, \qquad (5.12)$$

where

h = the Planck constant = 6.63×10^{-34} J s ,

c = the speed of light $\approx 3 \times 10^8$ m s^{-1} ,

k = the Boltzmann constant = 1.38×10^{-23} J K^{-1} .

By definition of the intensity, the dimension of B_ν is W m^{-2} Hz^{-1} sterad^{-1}.

Blackbody radiation can be produced in a closed cavity whose walls absorb all radiation incident upon them (and coming from inside the cavity). The walls and the radiation in the cavity are in equilibrium; both are at the same temperature, and the walls emit all the energy they receive. Since radiation energy is constantly transformed into thermal energy of the atoms of the walls and back to radiation, the blackbody radiation is also called *thermal radiation*.

The spectrum of a blackbody given by Planck's law (5.12) is continuous. This is true if the size of the radiator is very large compared with the dominant wavelengths. In the case of the cavity, this can be understood by considering the radiation as standing waves trapped in the cavity. The number of different wavelengths is larger, the shorter the wavelengths are compared with the size of the cavity. We already mentioned that spectra of solid bodies are continuous; very often such spectra can be quite well approximated by Planck's law.

We can also write Planck's law as a function of the wavelength. We require that $B_\nu \, d\nu = -B_\lambda \, d\lambda$. The wavelength decreases with increasing frequency; hence the minus sign. Since $\nu = c/\lambda$, we have

$$\frac{d\nu}{d\lambda} = -\frac{c}{\lambda^2}, \qquad (5.13)$$

whence

$$B_\lambda = -B_\nu \frac{d\nu}{d\lambda} = B_\nu \frac{c}{\lambda^2}, \qquad (5.14)$$

or

$$B_\lambda(T) = \frac{2hc^2}{\lambda^5} \frac{1}{e^{hc/(\lambda kT)} - 1}, \qquad (5.15)$$

$$[B_\lambda] = \text{W m}^{-2} \text{ m}^{-1} \text{ sterad}^{-1}.$$

The functions B_ν and B_λ are defined in such a way that the total intensity can be obtained in the same way using either of them:

$$B(T) = \int_0^\infty B_\nu \, d\nu = \int_0^\infty B_\lambda \, d\lambda.$$

Let us now try to find the total intensity using the first of these integrals:

$$B(T) = \int_0^\infty B_\nu(T) \, d\nu = \frac{2h}{c^2} \int_0^\infty \frac{\nu^3 \, d\nu}{e^{h\nu/(kT)} - 1}.$$

We now change the integration variable to $x = h\nu/(kT)$, whence $\mathrm{d}\nu = (kT/h)\mathrm{d}x$:

$$B(T) = \frac{2h}{c^2}\frac{k^4}{h^4}T^4 \int\limits_0^\infty \frac{x^3\,\mathrm{d}x}{e^x - 1} \ .$$

The definite integral in this expression is just a real number, independent of the temperature. Thus we find that

$$B(T) = AT^4 \ , \tag{5.16}$$

where the constant A has the value

$$A = \frac{2k^4}{c^2h^3}\frac{\pi^4}{15} \ . \tag{5.17}$$

(In order to get the value of A we have to evaluate the integral. There is no elementary way to do that. We can tell those who are familiar with all the exotic functions so beloved by theoretical physicists, that the integral can rather easily be expressed as $\Gamma(4)\zeta(4)$, where ζ is the Riemann zeta function and Γ is the gamma function. For integral values, $\Gamma(n)$ is simply the factorial $(n-1)!$. The difficult part is showing that $\zeta(4) = \pi^4/90$. This can be done by expanding $x^4 - x^2$ as a Fourier-series and evaluating the series at $x = \pi$.)

The flux density F for isotropic radiation of intensity B is (Sect. 4.1):

$$F = \pi B$$

or

$$F = \sigma T^4 \ . \tag{5.18}$$

This is the *Stefan-Boltzmann law*, and the constant σ $(= \pi A)$ is the *Stefan-Boltzmann constant*,

$$\sigma = 5.67 \times 10^{-8}\ \mathrm{W\,m^{-2}\,K^{-4}} \ .$$

From the Stefan-Boltzmann law we get a relation between the luminosity and temperature of a star. If the radius of the star is R, its surface area is $4\pi R^2$, and if the flux density on the surface is F, we have

$$L = 4\pi R^2 F \ .$$

If the star is assumed to radiate like a blackbody, we have $F = \sigma T^4$, which gives

$$L = 4\pi\sigma R^2 T^4 \ . \tag{5.19}$$

In fact this defines the *effective temperature* of the star, discussed in more detail in the next section.

The luminosity, radius and temperature of a star are interdependent quantities, as we can see from (5.19). They are also related to the absolute bolometric magnitude of the star. Equation (4.13) gives the difference of the absolute bolometric magnitude of the star and the Sun:

$$M_{\mathrm{bol}} - M_{\mathrm{bol},\odot} = -2.5\lg\frac{L}{L_\odot} \ . \tag{5.20}$$

But we can now use (5.19) to express the luminosities in terms of the radii and temperatures:

$$M_{\mathrm{bol}} - M_{\mathrm{bol},\odot} = -5\lg\frac{R}{R_\odot} - 10\lg\frac{T}{T_\odot} \ . \tag{5.21}$$

As we can see in Fig. 5.10, the wavelength of the maximum intensity decreases with increasing total intensity (equal to the area below the curve). We can find the wavelength $\lambda_{\max}$ corresponding to the maximum intensity by differentiating Planck's function $B_\lambda(T)$ with

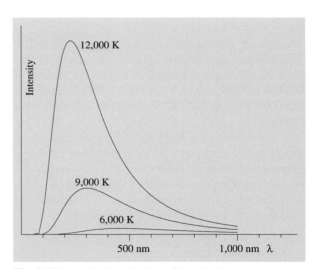

Fig. 5.10. Intensity distributions of blackbodies at temperature 12,000 K, 9000 K and 6000 K. Since the ratios of the temperatures are 4:3:2, the wavelengths of intensity maxima given by the Wien displacement law are in the proportions 1:4, 1:3 and 1:2, or 3, 4 and 6. The actual wavelengths of the maxima are 241.5 nm, 322 nm and 483 nm. The total intensities or the areas below the curves are proportional to 4^4, 3^4 and 2^4

respect to λ and finding zero of the derivative. The result is the *Wien displacement law*:

$$\lambda_{max} T = b = \text{const} , \qquad (5.22)$$

where the *Wien displacement constant b* is

$$b = 0.0028978 \text{ K m} .$$

We can use the same procedure to find the maximum of B_ν. But the frequency ν_{max} thus obtained is different from $\nu_{max} = c/\lambda_{max}$ given by (5.22). The reason for this is the fact that the intensities are given per unit frequency or unit wavelength, and the dependence of frequency on wavelength is nonlinear.

When the wavelength is near the maximum or much longer than λ_{max} Planck's function can be approximated by simpler expressions. When $\lambda \approx \lambda_{max}$ (or $hc/(\lambda kT) \gg 1$), we have

$$e^{hc/(\lambda kT)} \gg 1 .$$

In this case we get the *Wien approximation*

$$B_\lambda(T) \approx \frac{2hc^2}{\lambda^5} e^{-hc/(\lambda kT)} . \qquad (5.23)$$

When $hc/(\lambda kT) \ll 1$ ($\lambda \gg \lambda_{max}$), we have

$$e^{hc/\lambda kT} \approx 1 + hc/(\lambda kT) ,$$

which gives the *Rayleigh–Jeans approximation*

$$B_\lambda(T) \approx \frac{2hc^2}{\lambda^5} \frac{\lambda kT}{hc} = \frac{2ckT}{\lambda^4} . \qquad (5.24)$$

This is particularly useful in radio astronomy.

Classical physics predicted only the Rayleigh–Jeans approximation. Were (5.24) true for all wavelengths, the intensity would grow beyond all limits when the wavelength approaches zero, contrary to observations. This contradiction was known as the ultraviolet catastrophe.

5.8 Temperatures

Temperatures of astronomical objects range from almost absolute zero to millions of degrees. Temperature can be defined in a variety of ways, and its numerical value depends on the specific definition used. All these different temperatures are needed to describe different physical phenomena, and often there is no unique 'true' temperature.

Often the temperature is determined by comparing the object, a star for instance, with a blackbody. Although real stars do not radiate exactly like blackbodies, their spectra can usually be approximated by blackbody spectra after the effect of spectral lines has been eliminated. The resulting temperature depends on the exact criterion used to fit Planck's function to observations.

The most important quantity describing the surface temperature of a star is the *effective temperature* T_e. It is defined as the temperature of a blackbody which radiates with the same total flux density as the star. Since the effective temperature depends only on the total radiation power integrated over all frequencies, it is well defined for all energy distributions even if they deviate far from Planck's law.

In the previous section we derived the Stefan-Boltzmann law, which gives the total flux density as a function of the temperature. If we now find a value T_e of the temperature such that the Stefan-Boltzmann law gives the correct flux density F on the surface of the star, we have found the effective temperature. The flux density on the surface is

$$F = \sigma T_e^4 . \qquad (5.25)$$

The total flux is $L = 4\pi R^2 F$, where R is the radius of the star, and the flux density at a distance r is

$$F' = \frac{L}{4\pi r^2} = \frac{R^2}{r^2} F = \left(\frac{\alpha}{2}\right)^2 \sigma T_e^4 , \qquad (5.26)$$

where $\alpha = 2R/r$ is the observed angular diameter of the star. For direct determination of the effective temperature, we have to measure the total flux density and the angular diameter of the star. This is possible only in the few cases in which the diameter has been found by interferometry.

If we assume that at some wavelength λ the flux density F_λ on the surface of the star is obtained from Planck's law, we get the *brightness temperature* T_b. In the isotropic case we have then $F_\lambda = \pi B_\lambda(T_b)$. If the radius of the star is R and distance from the Earth r, the observed flux density is

$$F'_\lambda = \frac{R^2}{r^2} F_\lambda .$$

Again F_λ can be determined only if the angular diameter α is known. The brightness temperature T_b can then be solved from

$$F'_\lambda = \left(\frac{\alpha}{2}\right)^2 \pi B_\lambda(T_b) . \qquad (5.27)$$

Since the star does not radiate like a blackbody, its brightness temperature depends on the particular wavelength used in (5.27).

In radio astronomy, brightness temperature is used to express the intensity (or surface brightness) of the source. If the intensity at frequency ν is I_ν, the brightness temperature is obtained from

$$I_\nu = B_\nu(T_b) .$$

T_b gives the temperature of a blackbody with the same surface brightness as the observed source.

Since radio wavelengths are very long, the condition $h\nu \ll kT$ of the Rayleigh–Jeans approximation is usually satisfied (except for millimetre and submillimetre bands), and we can write Planck's law as

$$\begin{aligned}
B_\nu(T_b) &= \frac{2h\nu^3}{c^2} \frac{1}{e^{h\nu/(kT_b)} - 1} \\
&= \frac{2h\nu^3}{c^2} \frac{1}{1 + h\nu/(kT_b) + \ldots - 1} \\
&\approx \frac{2k\nu^2}{c^2} T_b .
\end{aligned}$$

Thus we get the following expression for the radio astronomical brightness temperature:

$$T_b = \frac{c^2}{2k\nu^2} I_\nu = \frac{\lambda^2}{2k} I_\nu . \qquad (5.28)$$

A measure of the signal registered by a radio telescope is the *antenna temperature* T_A. After the antenna temperature is measured, we get the brightness temperature from

$$T_A = \eta T_b , \qquad (5.29)$$

where η is the *beam efficiency* of the antenna (typically $0.4 \lesssim \eta \lesssim 0.8$). Equation (5.29) holds if the source is wide enough to cover the whole beam, i.e. the solid angle Ω_A from which the antenna receives radiation. If the solid angle subtended by the source, Ω_S, is smaller than Ω_A, the observed antenna temperature is

$$T_A = \eta \frac{\Omega_S}{\Omega_A} T_b , \qquad (\Omega_S < \Omega_A) . \qquad (5.30)$$

The *colour temperature* T_c can be determined even if the angular diameter of the source is unknown (Fig. 5.11). We only have to know the relative energy distribution in some wavelength range $[\lambda_1, \lambda_2]$; the absolute value of the flux is not needed. The observed flux density as a function of wavelength is compared with Planck's function at different temperatures. The temperature giving the best fit is the colour temperature in the interval $[\lambda_1, \lambda_2]$. The colour temperature is usually different for different wavelength intervals, since the shape of the observed energy distribution may be quite different from the blackbody spectrum.

A simple method for finding a colour temperature is the following. We measure the flux density F'_λ at two wavelengths λ_1 and λ_2. If we assume that the intensity distribution follows Planck's law, the ratio of these flux densities must be the same as the ratio obtained from Planck's law:

$$\frac{F'_{\lambda_1}(T)}{F'_{\lambda_2}(T)} = \frac{B_{\lambda_1}(T)}{B_{\lambda_2}(T)} = \frac{\lambda_2^5}{\lambda_1^5} \frac{e^{hc/(\lambda_2 kT)} - 1}{e^{hc/(\lambda_1 kT)} - 1} . \qquad (5.31)$$

The temperature T solved from this equation is a colour temperature.

The observed flux densities correspond to certain magnitudes m_{λ_1} and m_{λ_2}. The definition of magnitudes gives

$$m_{\lambda_1} - m_{\lambda_2} = -2.5 \lg \frac{F'_{\lambda_1}}{F'_{\lambda_2}} + \text{const} ,$$

where the constant term is a consequence of the different zero points of the magnitude scales. If the temperature

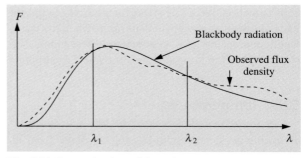

Fig. 5.11. Determination of the colour temperature. The ratio of the flux densities at wavelengths λ_1 and λ_2 gives the temperature of a blackbody with the same ratio. In general the result depends on the wavelengths chosen

is not too high, we can use the Wien approximation in the optical part of the spectrum:

$$m_{\lambda_1} - m_{\lambda_2} = -2.5 \lg \frac{B_{\lambda_1}}{B_{\lambda_2}} + \text{const}$$

$$= -2.5 \lg \left(\frac{\lambda_2}{\lambda_1}\right)^5$$

$$+ 2.5 \frac{hc}{kT} \left(\frac{1}{\lambda_1} - \frac{1}{\lambda_2}\right) \lg e + \text{const} .$$

This can be written as

$$m_{\lambda_1} - m_{\lambda_2} = a + b/T_c , \tag{5.32}$$

where a and b are constants. This shows that there is a simple relationship between the difference of two magnitudes and the colour temperature.

Strictly speaking, the magnitudes in (5.32) are monochromatic, but the same relation can be also used with broadband magnitudes like B and V. In that case, the two wavelengths are essentially the effective wavelengths of the B and V bands. The constant is chosen so that $B - V = 0$ for stars of the spectral type A0 (see Chap. 8). Thus the colour index $B - V$ also gives a colour temperature.

The *kinetic temperature* T_k, is related to the average speed of gas molecules. The kinetic energy of an ideal gas molecule as a function of temperature follows from the kinetic gas theory:

$$\text{Kinetic energy} = \frac{1}{2}mv^2 = \frac{3}{2}kT_k .$$

Solving for T_k we get

$$T_k = \frac{mv^2}{3k} , \tag{5.33}$$

where m is the mass of the molecule, v its average velocity (or rather its r.m.s velocity, which means that v^2 is the average of the squared velocities), and k, the Boltzmann constant. For ideal gases the pressure is directly proportional to the kinetic temperature (c.f. *Gas Pressure and Radiation Pressure, p. 230):

$$P = nkT_k , \tag{5.34}$$

where n is the number density of the molecules (molecules per unit volume). We previously defined the excitation temperature T_{exc} as a temperature which, if substituted into the Boltzmann distribution (5.11), gives the observed population numbers. If the distribution of atoms in different levels is a result of mutual collisions of the atoms only, the excitation temperature equals the kinetic temperature, $T_{\text{exc}} = T_k$.

The *ionization temperature* T_i is found by comparing the number of atoms in different states of ionization. Since stars are not exactly blackbodies, the values of excitation and ionization temperatures usually vary, depending on the element whose spectral lines were used for temperature determination.

In *thermodynamic equilibrium* all these various temperatures are equal.

5.9 Other Radiation Mechanisms

The radiation of a gas in thermodynamic equilibrium depends on the temperature and density only. In astrophysical objects deviations from thermodynamic equilibrium are, however, quite common. Some examples of *non-thermal radiation* arising under such conditions are mentioned in the following.

Maser and Laser (Fig. 5.12). The Boltzmann distribution (5.11) shows that usually there are fewer atoms in excited states than in the ground state. There are, however, means to produce a *population inversion*, an excited state containing more atoms than the ground state. This inversion is essential for both the *maser* and the *laser* (Microwave/Light Amplification by Stimulated Emission of Radiation). If the excited atoms are now illuminated with photons having energies equal to the excitation energy, the radiation will induce downward transitions. The number of photons emitted greatly exceeds the number of absorbed photons, and radiation is amplified. Typically the excited state is a *metastable state*, a state with a very long average lifetime, which means that the contribution of spontaneous emission is negligible. Therefore the resulting radiation is coherent and monochromatic. Several maser sources have been found in interstellar molecular clouds and dust envelopes around stars.

Synchrotron Radiation. A free charge in accelerated motion will emit electromagnetic radiation. Charged particles moving in a magnetic field follow helices around the field lines. As seen from the direction of

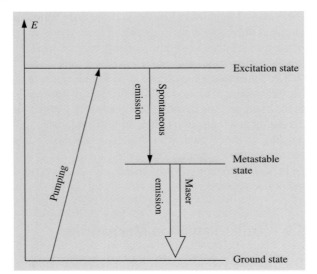

Fig. 5.12. The operational principle of the maser and the laser. A metastable state (a state with a relatively long average lifetime) stores atoms where they accumulate; there are more atoms in the metastable state than in the ground state. This population inversion is maintained by radiatively exciting atoms to a higher excitation state ("pumping"), from which they spontaneously jump down to the metastable state. When the atoms are illuminated by photons with energies equal to the excitation energy of the metastable state, the photons will induce more radiation of the same wavelength, and the radiation is amplified in geometric progression

the field, the motion is circular and therefore accelerated. The moving charge will radiate in the direction of its velocity vector. Such radiation is called *synchrotron radiation*. It will be further discussed in Chap. 15.

5.10 Radiative Transfer

Propagation of radiation in a medium, also called radiative transfer, is one of the basic problems of astrophysics. The subject is too complicated to be discussed here in any detail. The fundamental equation of radiative transfer is, however, easily derived.

Assume we have a small cylinder, the bottom of which has an area dA and the length of which is dr. Let I_ν be the intensity of radiation perpendicular to the bottom surface going into a solid angle $d\omega$ ($[I_\nu] = \mathrm{W\,m^{-2}\,Hz^{-1}\,sterad^{-1}}$). If the intensity changes by an amount dI_ν in the distance dr, the energy changes

by

$$dE = dI_\nu\, dA\, d\nu\, d\omega\, dt$$

in the cylinder in time dt. This equals the emission minus absorption in the cylinder. The absorbed energy is (c. f. (4.14))

$$dE_{\mathrm{abs}} = \alpha_\nu I_\nu\, dr\, dA\, d\nu\, d\omega\, dt\,, \tag{5.35}$$

where α_ν is the opacity of the medium at frequency ν. Let the amount of energy emitted per hertz at frequency ν into unit solid angle from unit volume and per unit time be j_ν ($[j_\nu] = \mathrm{W\,m^{-3}\,Hz^{-1}\,sterad^{-1}}$). This is called the *emission coefficient* of the medium. The energy emitted into solid angle $d\omega$ from the cylinder is then

$$dE_{\mathrm{em}} = j_\nu\, dr\, dA\, d\nu\, d\omega\, dt\,. \tag{5.36}$$

The equation

$$dE = -dE_{\mathrm{abs}} + dE_{\mathrm{em}}$$

gives then

$$dI_\nu = -\alpha_\nu I_\nu dr + j_\nu dr$$

or

$$\frac{dI_\nu}{\alpha_\nu\, dr} = -I_\nu + \frac{j_\nu}{\alpha_\nu}\,. \tag{5.37}$$

We shall denote the ratio of the emission coefficient j_ν to the absorption coefficient or opacity α_ν by S_ν:

$$S_\nu = \frac{j_\nu}{\alpha_\nu}\,. \tag{5.38}$$

S_ν is called the *source function*. Because $\alpha_\nu\, dr = d\tau_\nu$, where τ_ν is the optical thickness at frequency ν, (5.37) can be written as

$$\frac{dI_\nu}{d\tau_\nu} = -I_\nu + S_\nu\,. \tag{5.39}$$

Equation (5.39) is the basic equation of radiative transfer. Without solving the equation, we see that if $I_\nu < S_\nu$, then $dI_\nu/d\tau_\nu > 0$, and the intensity tends to increase in the direction of propagation. And, if $I_\nu > S_\nu$, then $dI_\nu/d\tau_\nu < 0$, and I_ν will decrease. In an equilibrium the emitted and absorbed energies are equal, in which case we find from (5.35) and (5.36)

$$I_\nu = j_\nu/\alpha_\nu = S_\nu\,. \tag{5.40}$$

Substituting this into (5.39), we see that $dI_\nu/d\tau_\nu = 0$. In thermodynamic equilibrium the radiation of the medium is blackbody radiation, and the source function is given by Planck's law:

$$S_\nu = B_\nu(T) = \frac{2h\nu^3}{c^2} \frac{1}{e^{h\nu/(kT)} - 1} .$$

Even if the system is not in thermodynamic equilibrium, it may be possible to find an excitation temperature T_{exc} such that $B_\nu(T_{exc}) = S_\nu$. This temperature may depend on frequency.

A formal solution of (5.39) is

$$I_\nu(\tau_\nu) = I_\nu(0)\, e^{-\tau_\nu} + \int_0^{\tau_\nu} e^{-(\tau_\nu - t)} S_\nu(t)\, dt . \quad (5.41)$$

Here $I_\nu(0)$ is the intensity of the background radiation, coming through the medium (e. g. an interstellar cloud) and decaying exponentially in the medium. The second term gives the emission in the medium. The solution is only formal, since in general, the source function S_ν is unknown and must be solved simultaneously with the intensity. If $S_\nu(\tau_\nu)$ is constant in the cloud and the background radiation is ignored, we get

$$I_\nu(\tau_\nu) = S_\nu \int_0^{\tau_\nu} e^{-(\tau_\nu - t)} dt = S_\nu(1 - e^{-\tau_\nu}) . \quad (5.42)$$

If the cloud is optically thick ($\tau_\nu \gg 1$), we have

$$I_\nu = S_\nu , \quad (5.43)$$

i.e. the intensity equals the source function, and the emission and absorption processes are in equilibrium.

An important field of application of the theory of radiative transfer is in the study of planetary and stellar atmospheres. In this case, to a good approximation, the properties of the medium only vary in one direction, say along the z axis. The intensity will then depend only on z and θ, where θ is the angle between the z axis and the direction of propagation of the radiation.

In applications to atmospheres it is customary to define the optical depth τ_ν in the vertical direction as

$$d\tau_\nu = -\alpha_\nu\, dz .$$

Conventionally z increases upwards and the optical depth inwards in the atmosphere. The vertical line el-

ement dz is related to that along the light ray, dr, according to

$$dz = dr \cos\theta .$$

With these notational conventions, (5.39) now yields

$$\cos\theta \frac{dI_\nu(z, \theta)}{d\tau_\nu} = I_\nu - S_\nu . \quad (5.44)$$

This is the form of the equation of radiative transfer usually encountered in the study of stellar and planetary atmospheres.

A formal expression for the intensity emerging from an atmosphere can be obtained by integrating (5.44) from $\tau_\nu = \infty$ (we assume that the bottom of the atmosphere is at infinite optical depth) to $\tau_\nu = 0$ (corresponding to the top of the atmosphere). This yields

$$I_\nu(0, \theta) = \int_0^\infty S_\nu\, e^{-\tau_\nu \sec\theta} \sec\theta\, d\tau_\nu . \quad (5.45)$$

This expression will be used later in Chap. 8 on the interpretation of stellar spectra.

5.11 Examples

Example 5.1 Find the wavelength of the photon emitted in the transition of a hydrogen atom from $n_2 = 110$ to $n_1 = 109$.

Equation (5.8) gives

$$\frac{1}{\lambda} = R\left(\frac{1}{n_1^2} - \frac{1}{n_2^2}\right)$$

$$= 1.097 \times 10^7\ \mathrm{m}^{-1}\left(\frac{1}{109^2} - \frac{1}{110^2}\right)$$

$$= 16.71\ \mathrm{m}^{-1} ,$$

whence

$$\lambda = 0.060\ \mathrm{m} .$$

This is in the radio band. Such radiation was observed for the first time in 1965 by an NRAO radio telescope.

Example 5.2 The effective temperature of a star is 12,000 K and the absolute bolometric magnitude 0.0. Find the radius of the star, when the effective temperature of the Sun is 5000 K and the absolute bolometric magnitude 4.7.

We can apply (5.21):

$$M_{bol} - M_{bol,\odot} = -5\lg\frac{R}{R_\odot} - 10\lg\frac{T}{T_\odot}$$

$$\Rightarrow \frac{R}{R_\odot} = \left(\frac{T_{e\odot}}{T_e}\right)^2 10^{-0.2(-M_{bol,\odot})}$$

$$= \left(\frac{5800}{12000}\right)^2 10^{-0.2(0.0-4.7)}$$

$$= 2.0 .$$

Thus the radius is twice the Solar radius.

Example 5.3 Derive the Wien displacement laws.

Let us denote $x = hc/(\lambda kT)$. Planck's law then becomes

$$B_\lambda(T) = \frac{2k^5 T^5}{h^4 c^3}\frac{x^5}{e^x - 1} .$$

For a given temperature, the first factor is constant. Thus, it is sufficient to find the maximum of the function $f(x) = x^5/(e^x - 1)$.

First we must evaluate the derivative of f:

$$f'(x) = \frac{5x^4(e^x - 1) - x^5 e^x}{(e^x - 1)^2}$$

$$= \frac{x^4 e^x}{(e^x - 1)^2}(5 - 5e^{-x} - x) .$$

By definition, x is always strictly positive. Hence $f'(x)$ can be zero only if the factor $5 - 5e^{-x} - x$ is zero. This equation cannot be solved analytically. Instead we write the equation as $x = 5 - 5e^{-x}$ and solve it by iteration:

$$x_0 = 5 , \quad \text{(this is just a guess)}$$

$$x_1 = 5 - 5e^{-x_0} = 4.96631 ,$$

$$\vdots$$

$$x_5 = 4.96511 .$$

Thus the result is $x = 4.965$. The Wien displacement law is then

$$\lambda_{max} T = \frac{hc}{xk} = b = 2.898 \times 10^{-3}\,\text{K m} .$$

In terms of frequency Planck's law is

$$B_\nu(T) = \frac{2h\nu^3}{c^2}\frac{1}{e^{h\nu/(kT)} - 1} .$$

Substituting $x = h\nu/(kT)$ we get

$$B_\nu(T) = \frac{2k^3 T^3}{h^2 c^2}\frac{x^3}{e^x - 1} .$$

Now we study the function $f(x) = x^3/(e^x - 1)$:

$$f'(x) = \frac{3x^2(e^x - 1) - x^3 e^x}{(e^x - 1)^2}$$

$$= \frac{x^2 e^x}{(e^x - 1)^2}(3 - 3e^{-x} - x) .$$

This vanishes, when $3 - 3e^{-x} - x = 0$. The solution of this equation is $x = 2.821$. Hence

$$\frac{cT}{\nu_{max}} = \frac{hc}{kx} = b' = 5.100 \times 10^{-3}\,\text{K m}$$

or

$$\frac{T}{\nu_{max}} = 1.701 \times 10^{-11}\,\text{K s} .$$

Note that the wavelength corresponding to ν_{max} is different from λ_{max}. The reason is that we have used two different forms of Planck's function, one giving the intensity per unit wavelength, the other per unit frequency.

Example 5.4 a) Find the fraction of radiation that a blackbody emits in the range $[\lambda_1, \lambda_2]$, where λ_1 and $\lambda_2 \gg \lambda_{max}$. b) How much energy does a 100 W incandescent light bulb radiate in the radio wavelengths, $\lambda \geq 1$ cm? Assume the temperature is 2500 K.

Since the wavelengths are much longer than λ_{max} we can use the Rayleigh–Jeans approximation $B_\lambda(T) \approx 2ckT/\lambda^4$. Then

$$B' = \int_{\lambda_1}^{\lambda_2} B_\lambda(T)d\lambda \approx 2ckT\int_{\lambda_1}^{\lambda_2}\frac{d\lambda}{\lambda^4}$$

$$= \frac{2ckT}{3}\left(\frac{1}{\lambda_1^3} - \frac{1}{\lambda_2^3}\right) ,$$

and hence

$$\frac{B'}{B_{tot}} = \frac{5c^3 h^3}{k^3\pi^4}\frac{1}{T^3}\left(\frac{1}{\lambda_1^3} - \frac{1}{\lambda_2^3}\right) .$$

Now the temperature is $T = 2500$ K and the wavelength range $[0.01\,\text{m}, \infty)$, and so

$$B' = 100\,\text{W} \times 1.529 \times 10^{-7}\frac{1}{2500^3}\frac{1}{0.01^3}$$

$$= 9.8 \times 10^{-10}\,\text{W} .$$

It is quite difficult to listen to the radio emission of a light bulb with an ordinary radio receiver.

Example 5.5 *Determination of Effective Temperature*

The observed flux density of Arcturus is

$$F' = 4.5 \times 10^{-8} \text{ W m}^{-2} .$$

Interferometric measurements give an angular diameter of $\alpha = 0.020''$. Thus, $\alpha/2 = 4.85 \times 10^{-8}$ radians. From (5.26) we get

$$T_{\mathrm{e}} = \left(\frac{4.5 \times 10^{-8}}{(4.85 \times 10^{-8})^2 \times 5.669 \times 10^{-8}} \right)^{1/4} \text{ K}$$

$$= 4300 \text{ K} .$$

Example 5.6 Flux densities at the wavelengths 440 nm and 550 nm are 1.30 and 1.00 W m^{-2} m^{-1}, respectively. Find the colour temperature.

If the flux densities at the wavelengths λ_1 and λ_2 are F_1 and F_2, respectively, the colour temperature can be solved from the equation

$$\frac{F_1}{F_2} = \frac{B_{\lambda_1}(T_{\mathrm{c}})}{B_{\lambda_2}(T_{\mathrm{c}})} = \left(\frac{\lambda_2}{\lambda_1} \right)^5 \frac{e^{hc/(\lambda_2 k T_{\mathrm{c}})} - 1}{e^{hc/(\lambda_1 k T_{\mathrm{c}})} - 1} .$$

If we denote

$$A = \frac{F_1}{F_2} \left(\frac{\lambda_1}{\lambda_2} \right)^5 ,$$

$$B_1 = \frac{hc}{\lambda_1 k} ,$$

$$B_2 = \frac{hc}{\lambda_2 k} ,$$

we get the equation

$$A = \frac{e^{B_2/T_{\mathrm{c}}} - 1}{e^{B_1/T_{\mathrm{c}}} - 1}$$

for the colour temperature T_{c}. This equation must be solved numerically.

In our example the constants have the following values:

$$A = \frac{1.00}{1.30} \left(\frac{550}{440} \right)^5 = 2.348 ,$$

$$B_1 = 32,700 \text{ K} , \quad B_2 = 26,160 \text{ K} .$$

By substituting different values for T_{c}, we find that $T_{\mathrm{c}} = 7545$ K satisfies our equation.

5.12 Exercises

Exercise 5.1 Show that in the Wien approximation the relative error of B_λ is

$$\frac{\Delta B_\lambda}{B_\lambda} = -e^{-hc/(\lambda k T)} .$$

Exercise 5.2 If the transition of the hydrogen atom $n+1 \rightarrow n$ were to correspond to the wavelength 21.05 cm, what would the quantum number n be? The interstellar medium emits strong radiation at this wavelength. Can this radiation be due to such transitions?

Exercise 5.3 The space is filled with background radiation, remnant of the early age of the universe. Currently the distribution of this radiation is similar to the radiation of a blackbody at the temperature of 2.7 K. What is $\lambda_{\max}$ corresponding to this radiation? What is its total intensity? Compare the intensity of the background radiation to the intensity of the Sun at the visual wavelengths.

Exercise 5.4 The temperature of a red giant is $T = 2500$ K and radius 100 times the solar radius.

a) Find the total luminosity of the star, and the luminosity in the visual band 400 nm $\leq \lambda \leq$ 700 nm.
b) Compare the star with a 100 W lamp that radiates 5% of its energy in the visual band. What is the distance of the lamp if it looks as bright as the star?

Exercise 5.5 The effective temperature of Sirius is 10,000 K, apparent visual magnitude −1.5, distance 2.67 kpc and bolometric correction 0.5. What is the radius of Sirius?

Exercise 5.6 The observed flux density of the Sun at $\lambda = 300$ nm is 0.59 W m^{-2} nm^{-1}. Find the brightness temperature of the Sun at this wavelength.

Exercise 5.7 The colour temperature can be determined from two magnitudes corresponding to two different wavelengths. Show that

$$T_c = \frac{7000 \text{ K}}{(B - V) + 0.47} \, .$$

The wavelengths of the B and V bands are 440 nm and 548 nm, respectively, and we assume that $B = V$ for stars of the spectral class A0, the colour temperature of which is about 15,000 K.

Exercise 5.8 The kinetic temperature of the plasma in the solar corona can reach 10^6 K. Find the average speed of the electrons in such a plasma.

6. Celestial Mechanics

Celestial mechanics, the study of motions of celestial bodies, together with spherical astronomy, was the main branch of astronomy until the end of the 19th century, when astrophysics began to evolve rapidly. The primary task of classical celestial mechanics was to explain and predict the motions of planets and their satellites. Several empirical models, like epicycles and Kepler's laws, were employed to describe these motions. But none of these models explained why the planets moved the way they did. It was only in the 1680's that a simple explanation was found for all these motions – Newton's law of universal gravitation. In this chapter, we will derive some properties of orbital motion. The physics we need for this is simple indeed, just Newton's laws. (For a review, see *Newton's Laws, p. 124)

This chapter is mathematically slightly more involved than the rest of the book. We shall use some vector calculus to derive our results, which, however, can be easily understood with very elementary mathematics. A summary of the basic facts of vector calculus is given in Appendix A.4.

6.1 Equations of Motion

We shall concentrate on the systems of only two bodies. In fact, this is the most complicated case that allows a neat analytical solution. For simplicity, let us call the bodies the Sun and a planet, although they could quite as well be a planet and its moon, or the two components of a binary star.

Let the masses of the two bodies be m_1 and m_2 and the radius vectors in some fixed inertial coordinate frame r_1 and r_2 (Fig. 6.1). The position of the planet relative to the Sun is denoted by $r = r_2 - r_1$. According to Newton's law of gravitation the planet feels a gravitational pull proportional to the masses m_1 and m_2 and inversely proportional to the square of the distance r. Since the force is directed towards the Sun, it can be expressed as

$$F = \frac{Gm_1m_2}{r^2}\frac{-r}{r} = -Gm_1m_2\frac{r}{r^3} ,$$ (6.1)

where G is the *gravitational constant*. (More about this in Sect. 6.5.)

Newton's second law tells us that the acceleration $\ddot{r}_2$ of the planet is proportional to the applied force:

$$F = m_2\ddot{r}_2 .$$ (6.2)

Combining (6.1) and (6.2), we get the *equation of motion* of the planet

$$m_2\ddot{r}_2 = -Gm_1m_2\frac{r}{r^3} .$$ (6.3)

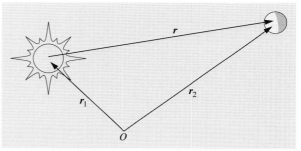

Fig. 6.1. The radius vectors of the Sun and a planet in an arbitrary inertial frame are r_1 and r_2, and $r = r_2 - r_1$ is the position of the planet relative to the Sun

Since the Sun feels the same gravitational pull, but in the opposite direction, we can immediately write the equation of motion of the Sun:

$$m_1\ddot{r}_1 = +Gm_1m_2\frac{r}{r^3} .$$ (6.4)

We are mainly interested in the relative motion of the planet with respect to the Sun. To find the equation of the relative orbit, we cancel the masses appearing on both sides of (6.3) and (6.4), and subtract (6.4) from (6.3) to get

$$\ddot{r} = -\mu\frac{r}{r^3} ,$$ (6.5)

where we have denoted

$$\mu = G(m_1 + m_2) .$$ (6.6)

The solution of (6.5) now gives the relative orbit of the planet. The equation involves the radius vector and its second time derivative. In principle, the solution should yield the radius vector as a function of time, $r = r(t)$. Unfortunately things are not this simple in practice; in fact, there is no way to express the radius vector as a function of time in a closed form (i.e. as a finite expression involving familiar elementary functions). Although there are several ways to solve the equation of motion, we must resort to mathematical manipulation in one form or another to figure out the essential properties of the orbit. Next we shall study one possible method.

6.2 Solution of the Equation of Motion

The equation of motion (6.5) is a second-order (i.e. contains second derivatives) vector valued differential equation. Therefore we need six integration constants or *integrals* for the complete solution. The solution is an infinite family of orbits with different sizes, shapes and orientations. A particular solution (e.g. the orbit of Jupiter) is selected by fixing the values of the six integrals. The fate of a planet is unambiguously determined by its position and velocity at any given moment; thus we could take the position and velocity vectors at some moment as our integrals. Although they do not tell us anything about the geometry of the orbit, they can be used as initial values when integrating the orbit numerically with a computer. Another set of integrals, the *orbital elements*, contains geometric quantities describing the orbit in a very clear and concrete way. We shall return to these later. A third possible set involves certain physical quantities, which we shall derive next.

We begin by showing that the angular momentum remains constant. The angular momentum of the planet in the heliocentric frame is

$$L = m_2 r \times \dot{r} . \tag{6.7}$$

Celestial mechanicians usually prefer to use the angular momentum divided by the planet's mass

$$k = r \times \dot{r} . \tag{6.8}$$

Let us find the time derivative of this:

$$\dot{k} = r \times \ddot{r} + \dot{r} \times \dot{r} .$$

The latter term vanishes as a vector product of two parallel vectors. The former term contains $\ddot{r}$, which is given by the equation of motion:

$$\dot{k} = r \times (-\mu r/r^3) = -(\mu/r^3) r \times r = 0 .$$

Thus k is a constant vector independent of time (as is L, of course).

Since the angular momentum vector is always perpendicular to the motion (this follows from (6.8)), the motion is at all times restricted to the invariable plane perpendicular to k (Fig. 6.2).

To find another constant vector, we compute the vector product $k \times \ddot{r}$:

$$k \times \ddot{r} = (r \times \dot{r}) \times (-\mu r/r^3)$$
$$= -\frac{\mu}{r^3} [(r \cdot r)\dot{r} - (r \cdot \dot{r})r] .$$

The time derivative of the distance r is equal to the projection of $\dot{r}$ in the direction of r (Fig. 6.3); thus, using the properties of the scalar product, we get $\dot{r} = r \cdot \dot{r}/r$, which gives

$$r \cdot \dot{r} = r\dot{r} . \tag{6.9}$$

Hence,

$$k \times \ddot{r} = -\mu(\dot{r}/r - r\dot{r}/r^2) = \frac{d}{dt}(-\mu r/r) .$$

The vector product can also be expressed as

$$k \times \ddot{r} = \frac{d}{dt}(k \times \dot{r}) ,$$

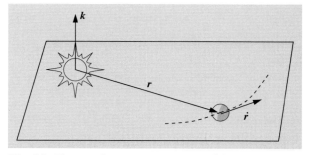

Fig. 6.2. The angular momentum vector k is perpendicular to the radius and velocity vectors of the planet. Since k is a constant vector, the motion of the planet is restricted to the plane perpendicular to k

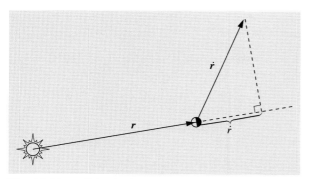

Fig. 6.3. The radial velocity $\dot{r}$ is the projection of the velocity vector $\dot{\boldsymbol{r}}$ in the direction of the radius vector $\boldsymbol{r}$

since $\boldsymbol{k}$ is a constant vector. Combining this with the previous equation, we have

$$\frac{\mathrm{d}}{\mathrm{d}t}(\boldsymbol{k} \times \dot{\boldsymbol{r}} + \mu \boldsymbol{r}/r) = 0$$

and

$$\boldsymbol{k} \times \dot{\boldsymbol{r}} + \mu \boldsymbol{r}/r = \text{const} = -\mu \boldsymbol{e} . \tag{6.10}$$

Since $\boldsymbol{k}$ is perpendicular to the orbital plane, $\boldsymbol{k} \times \dot{\boldsymbol{r}}$ must lie in that plane. Thus, $\boldsymbol{e}$ is a linear combination of two vectors in the orbital plane; so $\boldsymbol{e}$ itself must be in the orbital plane (Fig. 6.4). Later we shall see that it points to the direction where the planet is closest to the Sun in its orbit. This point is called the *perihelion*.

One more constant is found by computing $\dot{\boldsymbol{r}} \cdot \ddot{\boldsymbol{r}}$:

$$\dot{\boldsymbol{r}} \cdot \ddot{\boldsymbol{r}} = -\mu \dot{\boldsymbol{r}} \cdot \boldsymbol{r}/r^3 = -\mu r \dot{r}/r^3$$

$$= -\mu \dot{r}/r^2 = \frac{\mathrm{d}}{\mathrm{d}t}(\mu/r) .$$

Since we also have

$$\dot{\boldsymbol{r}} \cdot \ddot{\boldsymbol{r}} = \frac{\mathrm{d}}{\mathrm{d}t}\left(\frac{1}{2}\dot{\boldsymbol{r}} \cdot \dot{\boldsymbol{r}}\right) ,$$

we get

$$\frac{\mathrm{d}}{\mathrm{d}t}\left(\frac{1}{2}\dot{\boldsymbol{r}} \cdot \dot{\boldsymbol{r}} - \frac{\mu}{r}\right) = 0$$

or

$$\frac{1}{2}v^2 - \mu/r = \text{const} = h . \tag{6.11}$$

Here v is the speed of the planet relative to the Sun. The constant h is called the *energy integral*; the total energy of the planet is $m_2 h$. We must not forget that energy

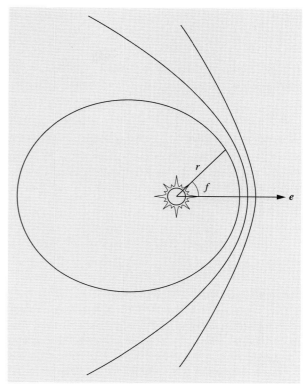

Fig. 6.4. The orbit of an object in the gravitational field of another object is a conic section: ellipse, parabola or hyperbola. Vector $\boldsymbol{e}$ points to the direction of the pericentre, where the orbiting object is closest to central body. If the central body is the Sun, this direction is called the perihelion; if some other star, periastron; if the Earth, perigee, etc. The true anomaly f is measured from the pericentre

and angular momentum depend on the coordinate frame used. Here we have used a heliocentric frame, which in fact is in accelerated motion.

So far, we have found two constant vectors and one constant scalar. It looks as though we already have seven integrals, i.e. one too many. But not all of these constants are independent; specifically, the following two relations hold:

$$\boldsymbol{k} \cdot \boldsymbol{e} = 0 , \tag{6.12}$$

$$\mu^2(e^2 - 1) = 2hk^2 , \tag{6.13}$$

where e and k are the lengths of $\boldsymbol{e}$ and $\boldsymbol{k}$. The first equation is obvious from the definitions of $\boldsymbol{e}$ and $\boldsymbol{k}$. To prove (6.13), we square both sides of (6.10) to get

$$\mu^2 e^2 = (\mathbf{k} \times \dot{\mathbf{r}}) \cdot (\mathbf{k} \times \dot{\mathbf{r}}) + \mu^2 \frac{\mathbf{r} \cdot \mathbf{r}}{r^2} + 2(\mathbf{k} \times \dot{\mathbf{r}}) \cdot \frac{\mu \mathbf{r}}{r} .$$

Since $\mathbf{k}$ is perpendicular to $\dot{\mathbf{r}}$, the length of $\mathbf{k} \times \dot{\mathbf{r}}$ is $|\mathbf{k}||\dot{\mathbf{r}}| = kv$ and $(\mathbf{k} \times \dot{\mathbf{r}}) \cdot (\mathbf{k} \times \dot{\mathbf{r}}) = k^2 v^2$. Thus, we have

$$\mu^2 e^2 = k^2 v^2 + \mu^2 + \frac{2\mu}{r}(\mathbf{k} \times \dot{\mathbf{r}} \cdot \mathbf{r}) .$$

The last term contains a scalar triple product, where we can exchange the dot and cross to get $\mathbf{k} \cdot \dot{\mathbf{r}} \times \mathbf{r}$. Next we reverse the order of the two last factors. Because the vector product is anticommutative, we have to change the sign of the product:

$$\mu^2 (e^2 - 1) = k^2 v^2 - \frac{2\mu}{r}(\mathbf{k} \cdot \mathbf{r} \times \dot{\mathbf{r}}) = k^2 v^2 - \frac{2\mu}{r} k^2$$

$$= 2k^2 \left(\frac{1}{2} v^2 - \frac{\mu}{r} \right) = 2k^2 h .$$

This completes the proof of (6.13).

The relations (6.12) and (6.13) reduce the number of independent integrals by two, so we still need one more. The constants we have describe the size, shape and orientation of the orbit completely, but we do not yet know where the planet is! To fix its position in the orbit, we have to determine where the planet is at some given instant of time $t = t_0$, or alternatively, at what time it is in some given direction. We use the latter method by specifying the time of perihelion passage, the *time of perihelion* τ.

6.3 Equation of the Orbit and Kepler's First Law

In order to find the geometric shape of the orbit, we now derive the equation of the orbit. Since $\mathbf{e}$ is a constant vector lying in the orbital plane, we choose it as the reference direction. We denote the angle between the radius vector $\mathbf{r}$ and $\mathbf{e}$ by f. The angle f is called the *true anomaly*. (There is nothing false or anomalous in this and other anomalies we shall meet later. Angles measured from the perihelion point are called anomalies to distinguish them from longitudes measured from some other reference point, usually the vernal equinox.) Using the properties of the scalar product we get

$$\mathbf{r} \cdot \mathbf{e} = re \cos f .$$

But the product $\mathbf{r} \cdot \mathbf{e}$ can also be evaluated using the definition of $\mathbf{e}$:

$$\mathbf{r} \cdot \mathbf{e} = -\frac{1}{\mu}(\mathbf{r} \cdot \mathbf{k} \times \dot{\mathbf{r}} + \mu \mathbf{r} \cdot \mathbf{r}/r)$$

$$= -\frac{1}{\mu}(\mathbf{k} \cdot \dot{\mathbf{r}} \times \mathbf{r} + \mu r) = -\frac{1}{\mu}(-k^2 + \mu r)$$

$$= \frac{k^2}{\mu} - r .$$

Equating the two expressions of $\mathbf{r} \cdot \mathbf{e}$ we get

$$r = \frac{k^2/\mu}{1 + e \cos f} . \tag{6.14}$$

This is the general equation of a *conic section* in polar coordinates (Fig. 6.4; see Appendix A.2 for a brief summary of conic sections). The magnitude of $\mathbf{e}$ gives the *eccentricity* of the conic:

$$e = 0 \qquad \text{circle} ,$$
$$0 < e < 1 \quad \text{ellipse} ,$$
$$e = 1 \qquad \text{parabola} ,$$
$$e > 1 \qquad \text{hyperbola} .$$

Inspecting (6.14), we find that r attains its minimum when $f = 0$, i.e. in the direction of the vector $\mathbf{e}$. Thus, $\mathbf{e}$ indeed points to the direction of the perihelion.

Starting with Newton's laws, we have thus managed to prove Kepler's first law:

> The orbit of a planet is an ellipse, one focus of which is in the Sun.

Without any extra effort, we have shown that also other conic sections, the parabola and hyperbola, are possible orbits.

6.4 Orbital Elements

We have derived a set of integrals convenient for studying the dynamics of orbital motion. We now turn to another collection of constants more appropriate for describing the geometry of the orbit. The following six quantities are called the *orbital elements* (Fig. 6.5):

– semimajor axis a,
– eccentricity e,
– inclination i (or ι),

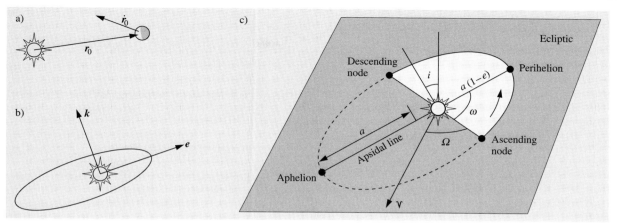

Fig. 6.5a–c. Six integration constants are needed to describe a planet's orbit. These constants can be chosen in various ways. (**a**) If the orbit is to be computed numerically, the simplest choice is to use the initial values of the radius and velocity vectors. (**b**) Another possibility is to use the angular momentum k, the direction of the perihelion e (the length of which gives the eccentricity), and the perihelion time τ. (**c**) The third method best describes the geometry of the orbit. The constants are the longitude of the ascending node Ω, the argument of perihelion ω, the inclination i, the semimajor axis a, the eccentricity e and the time of perihelion τ

– longitude of the ascending node Ω,
– argument of the perihelion ω,
– time of the perihelion τ.

The eccentricity is obtained readily as the length of the vector e. From the equation of the orbit (6.14), we see that the *parameter* (or semilatus rectum) of the orbit is $p = k^2/\mu$. But the parameter of a conic section is always $a|1 - e^2|$, which gives the semimajor axis, if e and k are known:

$$a = \frac{k^2/\mu}{|1 - e^2|} . \tag{6.15}$$

By applying (6.13), we get an important relation between the size of the orbit and the energy integral h:

$$a = \begin{cases} -\mu/2h , & \text{if the orbit is an ellipse ,} \\ \mu/2h , & \text{if the orbit is a hyperbola .} \end{cases} \tag{6.16}$$

For a bound system (elliptical orbit), the total energy and the energy integral are negative. For a hyperbolic orbit h is positive; the kinetic energy is so high that the particle can escape the system (or more correctly, recede without any limit). The parabola, with $h = 0$, is a limiting case between elliptical and hyperbolic orbits. In reality parabolic orbits do not exist, since hardly any object can have an energy integral exactly zero.

However, if the eccentricity is very close to one (as with many comets), the orbit is usually considered parabolic to simplify calculations.

The orientation of the orbit is determined by the directions of the two vectors k (perpendicular to the orbital plane) and e (pointing towards the perihelion). The three angles i, Ω and ω contain the same information.

The inclination i gives the obliquity of the orbital plane relative to some fixed reference plane. For bodies in the solar system, the reference plane is usually the ecliptic. For objects moving in the usual fashion, i.e. counterclockwise, the inclination is in the interval $[0°, 90°]$; for retrograde orbits (clockwise motion), the inclination is in the range $(90°, 180°]$. For example, the inclination of Halley's comet is $162°$, which means that the motion is retrograde and the angle between its orbital plane and the ecliptic is $180° - 162° = 18°$.

The longitude of the ascending node, Ω, indicates where the object crosses the ecliptic from south to north. It is measured counterclockwise from the vernal equinox. The orbital elements i and Ω together determine the orientation of the orbital plane, and they correspond to the direction of k, i.e. the ratios of its components.

The argument of the perihelion ω gives the direction of the perihelion, measured from the ascending node in the direction of motion. The same information is contained in the direction of e. Very often another angle, the *longitude of the perihelion* ϖ (pronounced as pi), is used instead of ω. It is defined as

$$\varpi = \Omega + \omega . \tag{6.17}$$

This is a rather peculiar angle, as it is measured partly along the ecliptic, partly along the orbital plane. However, it is often more practical than the argument of perihelion, since it is well defined even when the inclination is close to zero in which case the direction of the ascending node becomes indeterminate.

We have assumed up to this point that each planet forms a separate two-body system with the Sun. In reality planets interfere with each other by disturbing each other's orbits. Still their motions do not deviate very far from the shape of conic sections, and we can use orbital elements to describe the orbits. But the elements are no longer constant; they vary slowly with time. Moreover, their geometric interpretation is no longer quite as obvious as before. Such elements are *osculating elements* that would describe the orbit if all perturbations were to suddenly disappear. They can be used to find the positions and velocities of the planets exactly as if the elements were constants. The only difference is that we have to use different elements for each moment of time.

Table C.12 (at the end of the book) gives the mean orbital elements for the nine planets for the epoch J2000.0 as well as their first time derivatives. In addition to these secular variations the orbital elements suffer from periodic disturbances, which are not included in the table. Thus only approximate positions can be calculated with these elements. Instead of the time of perihelion the table gives the *mean longitude*

$$L = M + \omega + \Omega , \tag{6.18}$$

which gives directly the mean anomaly M (which will be defined in Sect. 6.7).

6.5 Kepler's Second and Third Law

The radius vector of a planet in polar coordinates is simply

$$\mathbf{r} = r\hat{\mathbf{e}}_r , \tag{6.19}$$

where $\hat{\mathbf{e}}_r$ is a unit vector parallel with $\mathbf{r}$ (Fig. 6.6). If the planet moves with angular velocity $\dot{f}$, the direction of this unit vector also changes at the same rate:

$$\dot{\hat{\mathbf{e}}}_r = \dot{f}\hat{\mathbf{e}}_f , \tag{6.20}$$

where $\hat{\mathbf{e}}_f$ is a unit vector perpendicular to $\hat{\mathbf{e}}_r$. The velocity of the planet is found by taking the time derivative of (6.19):

$$\dot{\mathbf{r}} = \dot{r}\hat{\mathbf{e}}_r + r\dot{\hat{\mathbf{e}}}_r = \dot{r}\hat{\mathbf{e}}_r + r\dot{f}\hat{\mathbf{e}}_f . \tag{6.21}$$

The angular momentum $\mathbf{k}$ can now be evaluated using (6.19) and (6.21):

$$\mathbf{k} = \mathbf{r} \times \dot{\mathbf{r}} = r^2 \dot{f}\hat{\mathbf{e}}_z , \tag{6.22}$$

where $\hat{\mathbf{e}}_z$ is a unit vector perpendicular to the orbital plane. The magnitude of $\mathbf{k}$ is

$$k = r^2 \dot{f} . \tag{6.23}$$

The *surface velocity* of a planet means the area swept by the radius vector per unit of time. This is obviously the time derivative of some area, so let us call it $\dot{A}$. In terms of the distance r and true anomaly f, the surface velocity is

$$\dot{A} = \frac{1}{2}r^2 \dot{f} . \tag{6.24}$$

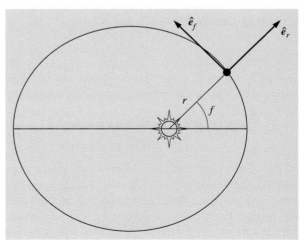

Fig. 6.6. Unit vectors $\hat{\mathbf{e}}_r$ and $\hat{\mathbf{e}}_f$ of the polar coordinate frame. The directions of these change while the planet moves along its orbit

By comparing this with the length of k (6.23), we find that

$$\dot{A} = \frac{1}{2}k .\qquad(6.25)$$

Since k is constant, so is the surface velocity. Hence we have Kepler's second law:

> The radius vector of a planet sweeps equal areas in equal amounts of time.

Since the Sun–planet distance varies, the orbital velocity must also vary (Fig. 6.7). From Kepler's second law it follows that a planet must move fastest when it is closest to the Sun (near perihelion). Motion is slowest when the planet is farthest from the Sun at *aphelion*.

We can write (6.25) in the form

$$dA = \frac{1}{2}k\,dt ,\qquad(6.26)$$

and integrate over one complete period:

$$\int\limits_{\text{orbital ellipse}} dA = \frac{1}{2}k \int\limits_0^P dt ,\qquad(6.27)$$

where P is the orbital period. Since the area of the ellipse is

$$\pi ab = \pi a^2 \sqrt{1 - e^2} ,\qquad(6.28)$$

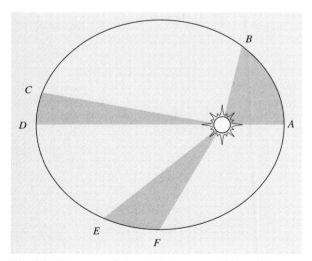

Fig. 6.7. The areas of the shaded sectors of the ellipse are equal. According to Kepler's second law, it takes equal times to travel distances AB, CD and EF

where a and b are the semimajor and semiminor axes and e the eccentricity, we get

$$\pi a^2 \sqrt{1 - e^2} = \frac{1}{2}kP .\qquad(6.29)$$

To find the length of k, we substitute the energy integral h as a function of semimajor axis (6.16) into (6.13) to get

$$k = \sqrt{G(m_1 + m_2)a\,(1 - e^2)} .\qquad(6.30)$$

When this is substituted into (6.29) we have

$$P^2 = \frac{4\pi^2}{G(m_1 + m_2)}\, a^3 .\qquad(6.31)$$

This is the exact form of Kepler's third law as derived from Newton's laws. The original version was

> The ratio of the cubes of the semimajor axes of the orbits of two planets is equal to the ratio of the squares of their orbital periods.

In this form the law is not exactly valid, even for planets of the solar system, since their own masses influence their periods. The errors due to ignoring this effect are very small, however.

Kepler's third law becomes remarkably simple if we express distances in astronomical units (AU), times in sidereal years (the abbreviation is unfortunately a, not to be confused with the semimajor axis, denoted by a somewhat similar symbol a) and masses in solar masses ($M_\odot$). Then $G = 4\pi^2$ and

$$a^3 = (m_1 + m_2)P^2 .\qquad(6.32)$$

The masses of objects orbiting around the Sun can safely be ignored (except for the largest planets), and we have the original law $P^2 = a^3$. This is very useful for determining distances of various objects whose periods have been observed. For absolute distances we have to measure at least one distance in metres to find the length of one AU. Earlier, triangulation was used to measure the parallax of the Sun or a minor planet, such as Eros, that comes very close to the Earth. Nowadays, radiotelescopes are used as radar to very accurately measure, for example, the distance to Venus. Since changes in the value of one AU also change all other distances, the International Astronomical Union decided in 1968 to adopt the value 1 AU $= 1.496000 \times 10^{11}$ m. The semimajor axis of Earth's orbit is then slightly over one AU.

But constants tend to change. And so, after 1984, the astronomical unit has a new value,

$$1\,\text{AU} = 1.49597870 \times 10^{11}\,\text{m}\,.$$

Another important application of Kepler's third law is the determination of masses. By observing the period of a natural or artificial satellite, the mass of the central body can be obtained immediately. The same method is used to determine masses of binary stars (more about this subject in Chap. 9).

Although the values of the AU and year are accurately known in SI-units, the gravitational constant is known only approximately. Astronomical observations give the product $G(m_1 + m_2)$, but there is no way to distinguish between the contributions of the gravitational constant and those of the masses. The gravitational constant must be measured in the laboratory; this is very difficult because of the weakness of gravitation. Therefore, if a precision higher than 2–3 significant digits is required, the SI-units cannot be used. Instead we have to use the solar mass as a unit of mass (or, for example, the Earth's mass after $Gm_\oplus$ has been determined from observations of satellite orbits).

6.6   Systems of Several Bodies

This far we have discussed systems consisting of only two bodies. In fact it is the most complex system for which a complete solution is known. The equations of motion are easily generalized, though. As in (6.5) we get the equation of motion for the body k, $k = 1, \ldots, n$:

$$\ddot{\boldsymbol{r}}_k = \sum_{i=1, i \neq k}^{i=n} Gm_i \frac{\boldsymbol{r}_i - \boldsymbol{r}_k}{|\boldsymbol{r}_i - \boldsymbol{r}_k|^3}\,, \tag{6.33}$$

where m_i is the mass of the ith body and $\boldsymbol{r}_i$ its radius vector. On the right hand side of the equation we now have the total gravitational force due to all other objects, instead of the force of just one body. If there are more than two bodies, these equations cannot be solved analytically in a closed form. The only integrals that can be easily derived in the general case are the total energy, total momentum, and total angular momentum.

If the radius and velocity vectors of all bodies are known for a certain instant of time, the positions at some other time can easily be calculated numerically from the equations of motion. For example, the planetary positions needed for astronomical yearbooks are computed by integrating the equations numerically.

Another method can be applied if the gravity of one body dominates like in the solar system. Planetary orbits can then be calculated as in a two-body system, and the effects of other planets taken into account as small perturbations. For these perturbations several series expansions have been derived.

The *restricted three-body problem* is an extensively studied special case. It consists of two massive bodies or *primaries*, moving on circular orbits around each other, and a third, massless body, moving in the same plane with the primaries. This small object does in no way disturb the motion of the primaries. Thus the orbits of the massive bodies are as simple as possible, and their positions are easily computed for all times. The problem is to find the orbit of the third body. It turns out that there is no finite expression for this orbit.

The Finnish astronomer *Karl Frithiof Sundman* (1873–1949) managed to show that a solution exists and derive a series expansion for the orbit. The series converges so slowly that it has no practical use, but as a mathematical result it was remarkable, since many mathematicians had for a long time tried to attack the problem without success.

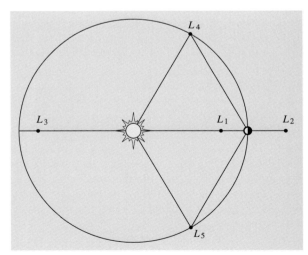

Fig. 6.8. The Lagrangian points of the restricted three-body problem. The points L_1, L_2 and L_3 are on the same line with the primaries, but the numbering may vary. The points L_4 and L_5 form equilateral triangles with the primaries

The three-body problem has some interesting special solutions. It can be shown that in certain points the third body can remain at rest with respect to the primaries. There are five such points, known as the *Lagrangian points* $L_1, \ldots, L_5$ (Fig. 6.8). Three of them are on the straight line determined by the primaries. These points are unstable: if a body in any of these points is disturbed, it will escape. The two other points, on the other hand, are stable. These points together with the primaries form equilateral triangles. For example, some *asteroids* have been found around the Lagrangian points L_4 and L_5 of Jupiter and Mars. The first of them were named after heroes of the Trojan war, and so they are called *Trojan asteroids*. They move around the Lagrangian points and can actually travel quite far from them, but they cannot escape. Fig. 7.38 shows two distinct condensations around the Lagrangian points of Jupiter.

6.7 Orbit Determination

Celestial mechanics has two very practical tasks: to determine orbital elements from observations and to predict positions of celestial bodies with known elements. Planetary orbits are already known very accurately, but new comets and minor planets are found frequently, requiring orbit determination.

The first practical methods for orbit determination were developed by *Johann Karl Friedrich Gauss* (1777–1855) at the beginning of the 19th century. By that time the first minor planets had been discovered, and thanks to Gauss's orbit determinations, they could be found and observed at any time.

At least three observations are needed for computing the orbital elements. The directions are usually measured from pictures taken a few nights apart. Using these directions, it is possible to find the corresponding absolute positions (the rectangular components of the radius vector). To be able to do this, we need some additional constraints on the orbit; we must assume that the object moves along a conic section lying in a plane that passes through the Sun. When the three radius vectors are known, the ellipse (or some other conic section) going through these three points can be determined. In practice, more observations are used. The elements determined are more accurate if there are more observations and if they cover the orbit more completely.

Although the calculations for orbit determination are not too involved mathematically, they are relatively long and laborious. Several methods can be found in textbooks of celestial mechanics.

6.8 Position in the Orbit

Although we already know everything about the geometry of the orbit, we still cannot find the planet at a given time, since we do not know the radius vector r as a function of time. The variable in the equation of the orbit is an angle, the true anomaly f, measured from the perihelion. From Kepler's second law it follows that f cannot increase at a constant rate with time. Therefore we need some preparations before we can find the radius vector at a given instant.

The radius vector can be expressed as

$$r = a(\cos E - e)\hat{\boldsymbol{i}} + b \sin E \hat{\boldsymbol{j}} , \qquad (6.34)$$

where $\hat{\boldsymbol{i}}$ and $\hat{\boldsymbol{j}}$ are unit vectors parallel with the major and minor axes, respectively. The angle E is the *eccentric anomaly*; its slightly eccentric definition is shown in Fig. 6.9. Many formulas of elliptical motion become very simple if either time or true anomaly is replaced

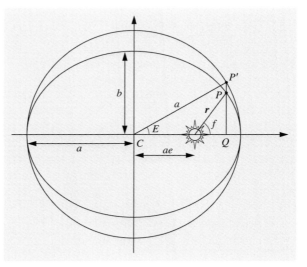

Fig. 6.9. Definition of the eccentric anomaly E. The planet is at P, and r is its radius vector

by the eccentric anomaly. As an example, we take the square of (6.34) to find the distance from the Sun:

$$r^2 = \boldsymbol{r} \cdot \boldsymbol{r}$$
$$= a^2(\cos E - e)^2 + b^2 \sin^2 E$$
$$= a^2[(\cos E - e)^2 + (1 - e^2)(1 - \cos^2 E)]$$
$$= a^2[1 - 2e \cos E + e^2 \cos^2 E] \, ,$$

whence

$$r = a \, (1 - e \cos E) \, . \tag{6.35}$$

Our next problem is to find how to calculate E for a given moment of time. According to Kepler's second law, the surface velocity is constant. Thus the area of the shaded sector in Fig. 6.10 is

$$A = \pi a b \frac{t - \tau}{P} \, , \tag{6.36}$$

where $t - \tau$ is the time elapsed since the perihelion, and P is the orbital period. But the area of a part of an ellipse is obtained by reducing the area of the corresponding part of the circumscribed circle by the axial ratio b/a. (As the mathematicians say, an ellipse is an

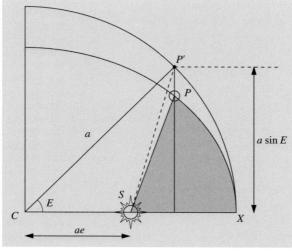

Fig. 6.10. The area of the shaded sector equals b/a times the area $SP'X$. $S =$ the Sun, $P =$ the planet, $X =$ the perihelion

affine transformation of a circle.) Hence the area of SPX is

$$A = \frac{b}{a}(\text{area of } SP'X)$$
$$= \frac{b}{a}(\text{area of the sector } CP'X$$
$$\quad - \text{area of the triangle } CP'S)$$
$$= \frac{b}{a}\left(\frac{1}{2}a \cdot aE - \frac{1}{2}ae \cdot a \sin E\right)$$
$$= \frac{1}{2}ab(E - e \sin E) \, .$$

By equating these two expressions for the area A, we get the famous *Kepler's equation*,

$$E - e \sin e = M \, , \tag{6.37}$$

where

$$M = \frac{2\pi}{P}(t - \tau) \tag{6.38}$$

is the *mean anomaly* of the planet at time t. The mean anomaly increases at a constant rate with time. It indicates where the planet would be if it moved in a circular orbit of radius a. For circular orbits all three anomalies f, E, and M are always equal.

If we know the period and the time elapsed after the perihelion, we can use (6.38) to find the mean anomaly. Next we must solve for the eccentric anomaly from Kepler's equation (6.37). Finally the radius vector is given by (6.35). Since the components of $\boldsymbol{r}$ expressed in terms of the true anomaly are $r \cos f$ and $r \sin f$, we find

$$\cos f = \frac{a(\cos E - e)}{r} = \frac{\cos E - e}{1 - e \cos E} \, ,$$
$$\sin f = \frac{b \sin E}{r} = \sqrt{1 - e^2} \frac{\sin E}{1 - e \cos E} \, . \tag{6.39}$$

These determine the true anomaly, should it be of interest.

Now we know the position in the orbital plane. This must usually be transformed to some other previously selected reference frame. For example, we may want to know the ecliptic longitude and latitude, which can later be used to find the right ascension and declination. These transformations belong to the realm of spherical astronomy and are briefly discussed in Examples 6.5–6.7.

6.9 Escape Velocity

If an object moves fast enough, it can escape from the gravitational field of the central body (to be precise: the field extends to infinity, so the object never really escapes, but is able to recede without any limit). If the escaping object has the minimum velocity allowing escape, it will have lost all its velocity at infinity (Fig. 6.11). There its kinetic energy is zero, since $v = 0$, and the potential energy is also zero, since the distance r is infinite. At infinite distance the total energy as well as the energy integral h are zero. The law of conservation of energy gives, then:

$$\frac{1}{2}v^2 - \frac{\mu}{R} = 0 \,, \qquad (6.40)$$

where R is the initial distance at which the object is moving with velocity v. From this we can solve the *escape velocity*:

$$v_e = \sqrt{\frac{2\mu}{R}} = \sqrt{\frac{2G(m_1 + m_2)}{R}} \,. \qquad (6.41)$$

For example on the surface of the Earth, v_e is about 11 km/s (if $m_2 \ll m_\oplus$).

The escape velocity can also be expressed using the orbital velocity of a circular orbit. The orbital period P as a function of the radius R of the orbit and the orbital velocity v_c is

$$P = \frac{2\pi R}{v_c} \,.$$

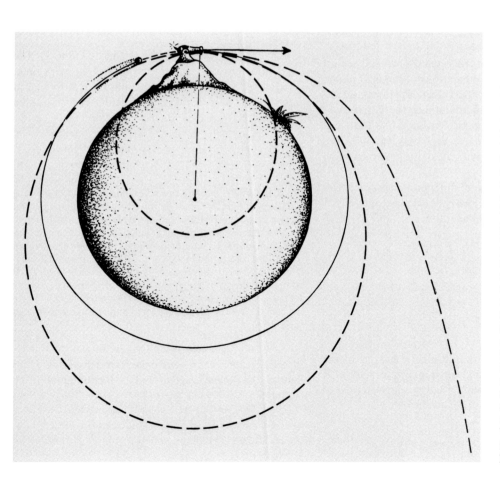

Fig. 6.11. A projectile is shot horizontally from a mountain on an atmosphereless planet. If the initial velocity is small, the orbit is an ellipse whose pericentre is inside the planet, and the projectile will hit the surface of the planet. When the velocity is increased, the pericentre moves outside the planet. When the initial velocity is v_c, the orbit is circular. If the velocity is increased further, the eccentricity of the orbit grows again and the pericentre is at the height of the cannon. The apocentre moves further away until the orbit becomes parabolic when the initial velocity is v_e. With even higher velocities, the orbit becomes hyperbolic

Substitution into Kepler's third law yields

$$\frac{4\pi^2 R^2}{v_c^2} = \frac{4\pi^2 R^3}{G(m_1 + m_2)} .$$

From this we can solve the velocity v_c in a circular orbit of radius R:

$$v_c = \sqrt{\frac{G(m_1 + m_2)}{R}} . \tag{6.42}$$

Comparing this with the expression (6.41) of the escape velocity, we see that

$$v_e = \sqrt{2} v_c . \tag{6.43}$$

6.10 Virial Theorem

If a system consists of more than two objects, the equations of motion cannot in general be solved analytically (Fig. 6.12). Given some initial values, the orbits can, of course, be found by numerical integration, but this does not tell us anything about the general properties of all possible orbits. The only integration constants available for an arbitrary system are the total momentum, angular momentum and energy. In addition to these, it is possible to derive certain statistical results, like the virial theorem. It concerns time averages only, but does not say anything about the actual state of the system at some specified moment.

Suppose we have a system of n point masses m_i with radius vectors r_i and velocities $\dot{r}_i$. We define a quantity A (the "virial" of the system) as follows:

$$A = \sum_{i=1}^{n} m_i \dot{r}_i \cdot r_i . \tag{6.44}$$

The time derivative of this is

$$\dot{A} = \sum_{i=1}^{n} (m_i \dot{r}_i \cdot \dot{r}_i + m_i \ddot{r}_i \cdot r_i) . \tag{6.45}$$

The first term equals twice the kinetic energy of the ith particle, and the second term contains a factor $m_i \ddot{r}_i$ which, according to Newton's laws, equals the force applied to the ith particle. Thus we have

$$\dot{A} = 2T + \sum_{i=1}^{n} F_i \cdot r_i , \tag{6.46}$$

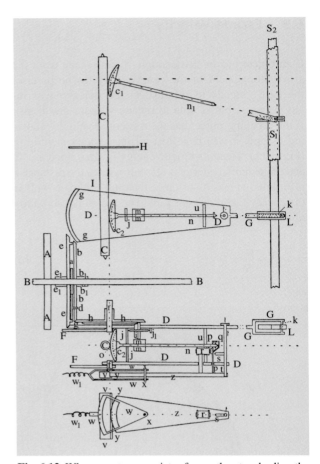

Fig. 6.12. When a system consists of more than two bodies, the equations of motion cannot be solved analytically. In the solar system the mutual disturbances of the planets are usually small and can be taken into account as small perturbations in the orbital elements. K.F. Sundman designed a machine to carry out the tedious integration of the perturbation equations. This machine, called the perturbograph, is one of the earliest analogue computers; unfortunately it was never built. Shown is a design for one component that evaluates a certain integral occurring in the equations. (The picture appeared in K.F. Sundman's paper in *Festskrift tillegnad Anders Donner* in 1915.)

where T is the total kinetic energy of the system. If $\langle x \rangle$ denotes the time average of x in the time interval $[0, \tau]$, we have

$$\langle \dot{A} \rangle = \frac{1}{\tau} \int_0^\tau \dot{A} \, dt = \langle 2T \rangle + \left\langle \sum_{i=1}^{n} F_i \cdot r_i \right\rangle . \tag{6.47}$$

If the system remains bounded, i. e. none of the particles escapes, all r_i's as well as all velocities will remain bounded. In such a case, A does not grow without limit, and the integral of the previous equation remains finite. When the time interval becomes longer ($\tau \to \infty$), $\langle \dot{A} \rangle$ approaches zero, and we get

$$\langle 2T \rangle + \left\langle \sum_{i=1}^{n} \boldsymbol{F}_i \cdot \boldsymbol{r}_i \right\rangle = 0 \,. \tag{6.48}$$

This is the general form of the virial theorem. If the forces are due to mutual gravitation only, they have the expressions

$$\boldsymbol{F}_i = -G m_i \sum_{j=1, \, j \neq i}^{n} m_j \frac{\boldsymbol{r}_i - \boldsymbol{r}_j}{r_{ij}^3} \,, \tag{6.49}$$

where $r_{ij} = |\boldsymbol{r}_i - \boldsymbol{r}_j|$. The latter term in the virial theorem is now

$$\sum_{i=1}^{n} \boldsymbol{F}_i \cdot \boldsymbol{r}_i = -G \sum_{i=1}^{n} \sum_{j=1, \, j \neq i}^{n} m_i m_j \frac{\boldsymbol{r}_i - \boldsymbol{r}_j}{r_{ij}^3} \cdot \boldsymbol{r}_i$$

$$= -G \sum_{i=1}^{n} \sum_{j=i+1}^{n} m_i m_j \frac{\boldsymbol{r}_i - \boldsymbol{r}_j}{r_{ij}^3} \cdot (\boldsymbol{r}_i - \boldsymbol{r}_j) \,,$$

where the latter form is obtained by rearranging the double sum, combining the terms

$$m_i m_j \frac{\boldsymbol{r}_i - \boldsymbol{r}_j}{r_{ij}^3} \cdot \boldsymbol{r}_i$$

and

$$m_j m_i \frac{\boldsymbol{r}_j - \boldsymbol{r}_i}{r_{ji}^3} \cdot \boldsymbol{r}_j = m_i m_j \frac{\boldsymbol{r}_i - \boldsymbol{r}_j}{r_{ij}^3} \cdot (-\boldsymbol{r}_j) \,.$$

Since $(\boldsymbol{r}_i - \boldsymbol{r}_j) \cdot (\boldsymbol{r}_i - \boldsymbol{r}_j) = r_{ij}^2$ the sum reduces to

$$-G \sum_{i=1}^{n} \sum_{j=i+1}^{n} \frac{m_i m_j}{r_{ij}} = U \,,$$

where U is the potential energy of the system. Thus, the virial theorem becomes simply

$$\langle T \rangle = -2 \langle U \rangle \,. \tag{6.50}$$

6.11 The Jeans Limit

We shall later study the birth of stars and galaxies. The initial stage is, roughly speaking, a gas cloud that begins to collapse due to its own gravitation. If the mass of the cloud is high enough, its potential energy exceeds the kinetic energy and the cloud collapses. From the virial theorem we can deduce that the potential energy must be at least twice the kinetic energy. This provides a criterion for the critical mass necessary for the cloud of collapse. This criterion was first suggested by *Sir James Jeans* in 1902.

The critical mass will obviously depend on the pressure P and density ρ. Since gravitation is the compressing force, the gravitational constant G will probably also enter our expression. Thus the critical mass is of the form

$$M = C P^a G^b \rho^c \,, \tag{6.51}$$

where C is a dimensionless constant, and the constants a, b and c are determined so that the right-hand side has the dimension of mass. The dimension of pressure is $\mathrm{kg}\,\mathrm{m}^{-1}\,\mathrm{s}^{-2}$, of gravitational constant $\mathrm{kg}^{-1}\,\mathrm{m}^3\,\mathrm{s}^{-2}$ and of density $\mathrm{kg}\,\mathrm{m}^{-3}$. Thus the dimension of the right-hand side is

$$\mathrm{kg}^{(a-b+c)}\,\mathrm{m}^{(-a+3b-3c)}\,\mathrm{s}^{(-2a-2b)} \,.$$

Since this must be kilograms ultimately, we get the following set of equations:

$$a - b + c = 1 \,, \quad -a + 3b - 3c = 0$$
$$-2a - 2b = 0 \,.$$

The solution of this is $a = 3/2$, $b = -3/2$ and $c = -2$. Hence the critical mass is

$$M_\mathrm{J} = C \frac{P^{3/2}}{G^{3/2} \rho^2} \,. \tag{6.52}$$

This is called the *Jeans mass*. In order to determine the constant C, we naturally must calculate both kinetic and potential energy. Another method based on the propagation of waves determines the diameter of the cloud, the *Jeans length* λ_J, by requiring that a disturbance of size λ_J grow unbounded. The value of the constant C depends on the exact form of the perturbation, but its typical values are in the range $[1/\pi, 2\pi]$. We can take $C = 1$ as well, in which case (6.52) gives a correct order of magnitude for the critical mass. If the mass of a cloud is much higher than M_J, it will collapse by its own gravitation.

In (6.52) the pressure can be replaced by the kinetic temperature T_k of the gas (see Sect. 5.8 for a definition). According to the kinetic gas theory, the pressure is

$$P = nkT_k , \qquad (6.53)$$

where n is the number density (particles per unit volume) and k is Boltzmann's constant. The number density is obtained by dividing the density of the gas ρ by the average molecular weight μ:

$$n = \rho/\mu ,$$

whence

$$P = \rho k T_k/\mu .$$

By substituting this into (6.52) we get

$$M_J = C \left(\frac{kT_k}{\mu G}\right)^{3/2} \frac{1}{\sqrt{\rho}} . \qquad (6.54)$$

* Newton's Laws

1. In the absence of external forces, a particle will remain at rest or move along a straight line with constant speed.
2. The rate of change of the momentum of a particle is equal to the applied force F:

$$\dot{p} = \frac{d}{dt}(mv) = F .$$

3. If particle A exerts a force F on another particle B, B will exert an equal but opposite force $-F$ on A.

If several forces $F_1, F_2, \ldots$ are applied on a particle, the effect is equal to that caused by one force F which is the vector sum of the individual forces ($F = F_1 + F_2 + \ldots$).

Law of gravitation: If the masses of particles A and B are m_A and m_B and their mutual distance r, the force exerted on A by B is directed towards B and has the magnitude $Gm_A\, m_B/r^2$, where G is a constant depending on the units chosen.

Newton denoted the derivative of a function f by $\dot{f}$ and the integral function by f'. The corresponding notations used by Leibniz were df/dt and $\int f\, dx$. Of Newton's notations, only the dot is still used, always signifying the time derivative: $\dot{f} \equiv d f/dt$. For example, the velocity $\dot{r}$ is the time derivative of r, the acceleration $\ddot{r}$ its second derivative, etc.

6.12 Examples

Example 6.1 Find the orbital elements of Jupiter on August 23, 1996.

The Julian date is $2{,}450{,}319$, hence from (6.17), $T = -0.0336$. By substituting this into the expressions of Table C.12, we get

$$a = 5.2033 ,$$
$$e = 0.0484 ,$$
$$i = 1.3053° ,$$
$$\Omega = 100.5448° ,$$
$$\varpi = 14.7460° ,$$
$$L = -67.460° = 292.540° .$$

From these we can compute the argument of perihelion and mean anomaly:

$$\omega = \varpi - \Omega = -85.7988° = 274.201° ,$$
$$M = L - \varpi = -82.2060° = 277.794° .$$

Example 6.2 *Orbital Velocity*

a) Comet Austin (1982g) moves in a parabolic orbit. Find its velocity on October 8, 1982, when the distance from the Sun was 1.10 AU.

The energy integral for a parabola is $h = 0$. Thus (6.11) gives the velocity v:

$$v = \sqrt{\frac{2\mu}{r}} = \sqrt{\frac{2GM_\odot}{r}}$$

$$= \sqrt{\frac{2 \times 4\pi^2 \times 1}{1.10}} = 8.47722 \text{ AU/a}$$

$$= \frac{8.47722 \times 1.496 \times 10^{11} \text{ m}}{365.2564 \times 24 \times 3600 \text{ s}} \approx 40 \text{ km/s} .$$

b) The semimajor axis of the minor planet 1982 RA is 1.568 AU and the distance from the Sun on October 8, 1982, was 1.17 AU. Find its velocity.

The energy integral (6.16) is now

$$h = -\mu/2a .$$

Hence

$$\frac{1}{2}v^2 - \frac{\mu}{r} = -\frac{\mu}{2a} ,$$

which gives

$$v = \sqrt{\mu\left(\frac{2}{r} - \frac{1}{a}\right)}$$

$$= \sqrt{4\pi^2\left(\frac{2}{1.17} - \frac{1}{1.568}\right)}$$

$$= 6.5044 \text{ AU/a} \approx 31 \text{ km/s} .$$

Example 6.3 In an otherwise empty universe, two rocks of 5 kg each orbit each other at a distance of 1 m. What is the orbital period?

The period is obtained from Kepler's third law:

$$P^2 = \frac{4\pi^2 a^3}{G(m_1 + m_2)}$$

$$= \frac{4\pi^2 1}{6.67 \times 10^{-11}(5+5)} \text{ s}^2$$

$$= 5.9 \times 10^{10} \text{ s}^2 ,$$

whence

$$P = 243,000 \text{ s} = 2.8 \text{ d} .$$

Example 6.4 The period of the Martian moon Phobos is 0.3189 d and the radius of the orbit 9370 km. What is the mass of Mars?

First we change to more appropriate units:

$$P = 0.3189 \text{ d} = 0.0008731 \text{ sidereal years} ,$$

$$a = 9370 \text{ km} = 6.2634 \times 10^{-5} \text{ AU} .$$

Equation (6.32) gives (it is safe to assume that $m_{\text{Phobos}} \ll m_{\text{Mars}}$)

$$m_{\text{Mars}} = a^3/P^2 = 0.000000322 \, M_\odot$$
$$(\approx 0.107 \, M_\oplus) .$$

Example 6.5 Derive formulas for a planet's heliocentric longitude and latitude, given its orbital elements and true anomaly.

We apply the sine formula to the spherical triangle of the figure:

$$\frac{\sin \beta}{\sin i} = \frac{\sin(\omega + f)}{\sin(\pi/2)}$$

or

$$\sin \beta = \sin i \sin(\omega + f) .$$

The sine-cosine formula gives

$$\cos(\pi/2) \sin \beta$$
$$= -\cos i \sin(\omega + f) \cos(\lambda - \Omega)$$
$$+ \cos(\omega + f) \sin(\lambda - \Omega) ,$$

whence

$$\tan(\lambda - \Omega) = \cos i \tan(\omega + f) .$$

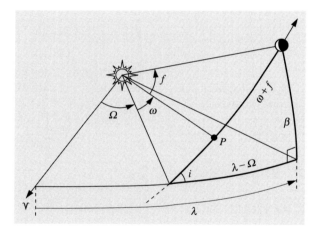

Example 6.6 Find the radius vector and heliocentric longitude and latitude of Jupiter on August 23, 1996.

The orbital elements were computed in Example 6.1:

$$a = 5.2033 \text{ AU} ,$$
$$e = 0.0484 ,$$
$$i = 1.3053° ,$$
$$\Omega = 100.5448° ,$$
$$\omega = 274.2012° ,$$
$$M = 277.7940° = 4.8484 \text{ rad} .$$

Since the mean anomaly was obtained directly, we need not compute the time elapsed since perihelion.

Now we have to solve Kepler's equation. It cannot be solved analytically, and we are obliged to take the brute force approach (also called numerical analysis) in the form of iteration. For iteration, we write the equation as

$$E_{n+1} = M + e \sin E_n ,$$

where E_n is the value found in the nth iteration. The mean anomaly is a reasonable initial guess, E_0. (N.B.: Here, all angles must be in radians; otherwise, nonsense results!) The iteration proceeds as follows:

$$E_0 = M = 4.8484 ,$$

$$E_1 = M + e \sin E_0 = 4.8004 ,$$

$$E_2 = M + e \sin E_1 = 4.8002 ,$$

$$E_3 = M + e \sin E_2 = 4.8002 ,$$

after which successive approximations no longer change, which means that the solution, accurate to four decimal places, is

$$E = 4.8002 = 275.0° .$$

The radius vector is

$$\boldsymbol{r} = a(\cos E - e)\,\hat{\boldsymbol{i}} + a\sqrt{1-e^2}\,\sin E\,\hat{\boldsymbol{j}}$$

$$= 0.2045\,\hat{\boldsymbol{i}} - 5.1772\,\hat{\boldsymbol{j}}$$

and the distance from the Sun,

$$r = a(1 - e \cos E) = 5.1813 \text{ AU} .$$

The signs of the components of the radius vector show that the planet is in the fourth quadrant. The true anomaly is

$$f = \arctan \frac{-5.1772}{0.2045} = 272.3° .$$

Applying the results of the previous example, we find the latitude and longitude:

$$\sin \beta = \sin i \, \sin(\omega + f)$$

$$= \sin 1.3° \sin(274.2° + 272.3°)$$

$$= -0.0026$$

$$\Rightarrow \beta = -0.15° ,$$

$$\tan(\lambda - \Omega) = \cos i \, \tan(\omega + f)$$

$$= \cos 1.3° \tan(274.2° + 272.3°)$$

$$= 0.1139$$

$$\Rightarrow \lambda = \Omega + 186.5°$$

$$= 100.5° + 186.5°$$

$$= 287.0° .$$

(We must be careful here; the equation for $\tan(\lambda - \Omega)$ allows two solutions. If necessary, a figure can be drawn to decide which is the correct one.)

Example 6.7 Find Jupiter's right ascension and declination on August 23, 1996.

In Example 6.6, we found the longitude and latitude, $\lambda = 287.0°$, $\beta = -0.15°$. The corresponding rectangular (heliocentric) coordinates are:

$$x = r \cos \lambda \cos \beta = 1.5154 \text{ AU} ,$$

$$y = r \sin \lambda \cos \beta = -4.9547 \text{ AU} ,$$

$$z = r \sin \beta = -0.0133 \text{ AU} .$$

Jupiter's ecliptic coordinates must be transformed to equatorial ones by rotating them around the x-axis by an angle ε, the obliquity of the ecliptic (see *Reduction of Coordinates, p. 36):

$$X_{\mathrm{J}} = x = 1.5154 \text{ AU} ,$$

$$Y_{\mathrm{J}} = y \cos \varepsilon - z \sin \varepsilon = -4.5405 \text{ AU} ,$$

$$Z_{\mathrm{J}} = y \sin \varepsilon + z \cos \varepsilon = -1.9831 \text{ AU} .$$

To find the direction relative to the Earth, we have to find where the Earth is. In principle, we could repeat the previous procedure with the orbital elements of the Earth. Or, if we are lazy, we could pick up the nearest *Astronomical Almanac*, which lists the equatorial coordinates of the Earth:

$$X_\oplus = 0.8815 \text{ AU} ,$$

$$Y_\oplus = -0.4543 \text{ AU} ,$$

$$Z_\oplus = -0.1970 \text{ AU} .$$

Then the position relative to the Earth is

$$X_0 = X_J - X_\oplus = 0.6339 \text{ AU} ,$$

$$Y_0 = Y_J - Y_\oplus = -4.0862 \text{ AU} ,$$

$$Z_0 = Z_J - Z_\oplus = -1.7861 \text{ AU} .$$

And finally, the right ascension and declination are

$$\alpha = \arctan(Y_0/X_0) = 278.82° = 18 \text{ h } 35 \text{ min} ,$$

$$\delta = \arctan \frac{Z_0}{\sqrt{X_0^2 + Y_0^2}} = -23.4° .$$

If the values given by the *Astronomical Almanac* are rounded to the same accuracy, the same result is obtained. We should not expect a very precise position since we have neglected all short-period perturbations in Jupiter's orbital elements.

Example 6.8 Which is easier, to send a probe to the Sun or away from the Solar system?

The orbital velocity of the Earth is about 30 km/s. Thus the escape velocity from the Solar system is $\sqrt{2} \times 30 \approx 42$ km/s. A probe that is sent from the Earth already has a velocity equal to the orbital velocity of the Earth. Hence an extra velocity of only 12 km/s is needed. In addition, the probe has to escape from the Earth, which requires 11 km/s. Thus the total velocity changes are about 23 km/s.

If the probe has to fall to the Sun it has to get rid of the orbital velocity of the Earth 30 km/s. In this case, too, the probe has first to be lifted from the Earth. Thus the total velocity change needed is 41 km/s. This is nearly impossible with current technology. Therefore a probe to be sent to the Sun is first directed close to some planet, and the gravitational field of the planet is used to accelerate the probe towards its final destination.

Example 6.9 An interstellar hydrogen cloud contains 10 atoms per cm^3. How big must the cloud be to collapse

due to its own gravitation? The temperature of the cloud is 100 K.

The mass of one hydrogen atom is 1.67×10^{-27} kg, which gives a density

$$\rho = n\mu = 10^7 \text{ m}^{-3} \times 1.67 \times 10^{-27} \text{ kg}$$

$$= 1.67 \times 10^{-20} \text{ kg/m}^3 .$$

The critical mass is

$$M_J = \left(\frac{1.38 \times 10^{-23} \text{ J/K} \times 100 \text{ K}}{1.67 \times 10^{-27} \text{ kg} \times 6.67 \times 10^{-11} \text{ N m}^2 \text{ kg}^{-2}} \right)^{3/2}$$

$$\times \frac{1}{\sqrt{1.67 \times 10^{-20} \text{ kg/m}^3}}$$

$$\approx 1 \times 10^{34} \text{ kg} \approx 5000 \, M_\odot .$$

The radius of the cloud is

$$R = \sqrt[3]{\frac{3}{4\pi} \frac{M}{\rho}} \approx 5 \times 10^{17} \text{ m} \approx 20 \text{ pc} .$$

6.13 Exercises

Exercise 6.1 Find the ratio of the orbital velocities at aphelion and perihelion v_a/v_p. What is this ratio for the Earth?

Exercise 6.2 The perihelion and aphelion of the orbit of Eros are 1.1084 and 1.8078 astronomical units from the Sun. What is the velocity of Eros when its distance from the Sun equals the mean distance of Mars?

Exercise 6.3 Find the radius of the orbit of a geostationary satellite; such a satellite remains always over the same point of the equator of the Earth. Are there areas on the surface of the Earth that cannot be seen from any geostationary satellite? If so, what fraction of the total surface area?

Exercise 6.4 From the angular diameter of the Sun and the length of the year, derive the mean density of the Sun.

Exercise 6.5   Find the mean, eccentric and true anomalies of the Earth one quarter of a year after the perihelion.

Exercise 6.6   The velocity of a comet is 5 m/s, when it is very far from the Sun. If it moved along a straight line, it would pass the Sun at a distance of 1 AU. Find the eccentricity, semimajor axis and perihelion distance of the orbit. What will happen to the comet?

Exercise 6.7   a) Find the ecliptic geocentric radius vector of the Sun on May 1, 1997 ($J = 2450570$).
b) What are the declination and right ascension of the Sun then?

7. The Solar System

The solar system consists of a central star, called *the Sun*, nine *planets*, dozens of *moons*, thousands of *minor planets*, and myriads of *comets* and *meteoroids*. The planets in order from the Sun are: *Mercury, Venus, Earth, Mars, Jupiter, Saturn, Uranus, Neptune*, and *Pluto*.

The planets from Mercury to Saturn are bright and well visible with a naked eye. Uranus and Neptune can be seen with a pair of binoculars, but discovering Pluto requires a moderate-sized telescope. In addition to the bright planets, only the brightest comets are visible with a naked eye.

Distances in the solar system are often measured in *astronomical units* (AU), the mean distance of the Sun and Earth. The semimajor axis of the orbit of Mercury is 0.39 AU, and the distance of Pluto is 100 times greater, 39.8 AU. The distance to the nearest star, *Proxima Centauri* is over 270,000 AU.

Gravitation controls the motion of the solar system bodies. The planetary orbits around the Sun (Fig. 7.1) are almost coplanar ellipses which deviate only slightly from circles. The innermost and the outermost planets, Mercury and Pluto, have the largest deviations. The orbital planes of *asteroids*, minor planets that circle the Sun mainly between the orbits of Mars and Jupiter, are often more tilted than the planes of the planetary orbits. All minor planets revolve in the same direction as the major planets; comets, however, may move in the opposite direction. Cometary orbits can be very elongated, even hyperbolic. Most of the moons circle their parent planets in the same direction as the planet moves around the Sun. Only the motions of the smallest particles, gas and dust are affected by the *solar wind, radiation pressure* and *magnetic fields*.

The planets can be divided into two physically different groups. Mercury, Venus, Earth, and Mars are called *terrestrial* (Earth-like) planets; they have a solid surface, are of almost equal size (diameters from 5000 to 12,000 km), and have quite a high mean density (4000–5000 kg m^{-3}; the density of water is 1000 kg m^{-3}). The planets from Jupiter to Neptune are called *Jovian* (Jupiter-like) or *giant planets*. The densities of the giant planets are about 1000–2000 kg m^{-3}, and most of their volume is liquid. Diameters are ten times greater than those of the terrestrial planets. Pluto is a unique object falling outside this classification. Very likely, Pluto belongs to the family of icy

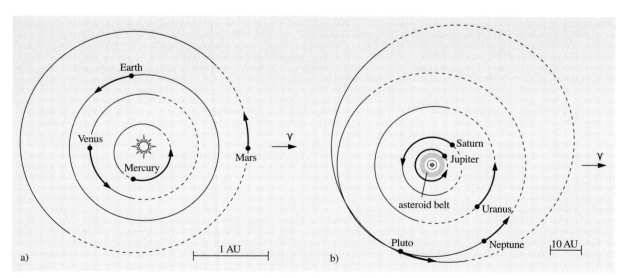

Fig. 7.1. (**a**) Planetary orbits from Mercury to Mars. The dashed line represents the part of the orbit below the ecliptic; the arrows show the distances travelled by the planets during one month (January 2000). (**b**) Planets from Jupiter to Pluto. The arrows indicate the distances travelled by the planets during the 10 year interval 2000–2010

bodies orbiting the Sun at the outer edges of the solar system.

Since the 1960's the planets have been studied with *spacecraft*. During the first half of 1900's the planetary science had almost disappeared under the extensive growth of other branches of astronomy. Most of the data we have today are collected by spacecraft, and many methods used in geosciences are nowadays applied in planetary studies. Pluto is the only planet which has not been studied by spacecraft, and there are no firm plans for a Pluto mission.

7.1 Planetary Configurations

The *apparent motions* of the planets are quite complicated, partly because they reflect the motion of the Earth around the Sun (Figs. 7.2 and 7.3). Normally the planets move eastward (*direct motion*, counterclockwise as seen from the Northern hemisphere) when compared with the stars. Sometimes the motion reverses to the opposite or *retrograde* direction. After a few weeks of retrograde motion, the direction is changed again, and the planet continues in the original direction. It is quite understandable that the ancient astronomers had great difficulties in explaining and modelling such complicated turns and loops.

Figure 7.4 explains some basic planetary configurations. A *superior planet* (planet outside the orbit of the Earth) is said to be in *opposition* when it is exactly opposite the Sun, i. e. when the Earth is between the planet and the Sun. When the planet is behind the Sun, it is in *conjunction*. In practise, the planet may not be exactly opposite or behind the Sun because the orbits of the planet and the Earth are not in the same plane. In astronomical almanacs oppositions and conjunctions are defined in terms of ecliptic longitudes. The longitudes of a body and the Sun differ by 180° at the moment of opposition; in conjunction the longitudes are equal. However, the right ascension is used if the other body is not the Sun. Those points at which the apparent motion of a planet turns toward the opposite direction are called *stationary points*. Opposition occurs in the middle of the retrograde loop.

Inferior planets (Mercury and Venus) are never in opposition. The configuration occurring when either of these planets is between the Earth and the Sun is called *inferior conjunction*. The conjunction corresponding to that of a superior planet is called *upper conjunction* or *superior conjunction*. The maximum (eastern or western) *elongation*, i. e. the angular distance of the planet

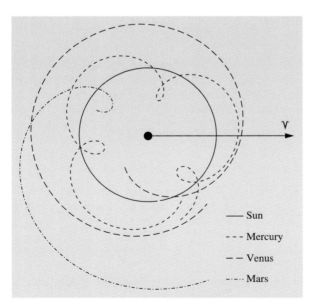

Fig. 7.2. Planetary motions are complex when viewed from the Earth. The orbits of the Sun, Mercury, Venus and Mars are shown here during the year 1995 in a reference frame fixed to the Earth and viewed from the north celestial pole

— Sun

--- Mercury

-- Venus

-·-· Mars

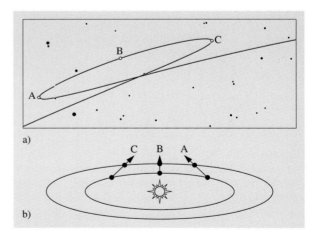

Fig. 7.3. (**a**) Apparent motion of Mars during the 1995 opposition. (**b**) Relative positions of the Earth and Mars. The projection of the Earth–Mars direction on the infinitely distant celestial sphere results in (**a**)

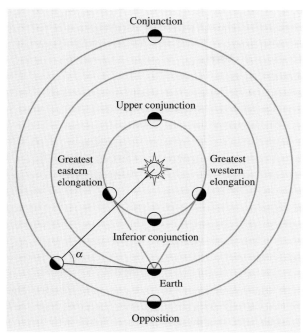

Fig. 7.4. Planetary configurations. The angle α (Sun–object–Earth) is the phase angle and ε (Sun–Earth–object) is the elongation

from the Sun is 28° for Mercury and 47° for Venus. Elongations are called eastern or western, depending on which side of the Sun the planet is seen. The planet is an "evening star" and sets after the Sun when it is in eastern elongation; in western elongation the planet is seen in the morning sky as a "morning star".

The *synodic period* is the time interval between two successive events (e. g. oppositions). The period which we used in the previous chapters is the *sidereal period*, the true time of revolution around the Sun, unique for each object. The synodic period depends on the difference of the sidereal periods of two bodies.

Let the sidereal periods of two planets be P_1 and P_2 (assume that $P_1 < P_2$). Their mean angular velocities (mean motions) are $2\pi/P_1$ and $2\pi/P_2$. After one synodic period $P_{1,2}$, the inner planet has made one full revolution more than the outer planet:

$$P_{1,2}\frac{2\pi}{P_1} = 2\pi + P_{1,2}\frac{2\pi}{P_2} ,$$

or

$$\frac{1}{P_{1,2}} = \frac{1}{P_1} - \frac{1}{P_2} . \tag{7.1}$$

The angle Sun–planet–Earth is called the *phase angle*, often denoted by the Greek letter α. The phase angle is between 0° and 180° in the case of Mercury and Venus. This means that we can see "full Venus", "half Venus", and so on, exactly as in the phases of the Moon. The phase angle range for the superior planets is more limited. For Mars the maximum phase is 41°, for Jupiter 11°, and for Pluto only 2°.

7.2 Orbit of the Earth

The *sidereal year* is the real orbital period of the Earth around the Sun. After one sidereal year, the Sun is seen at the same position relative to the stars. The length of the sidereal year is 365.2564 mean solar days (in fact, days of 86,400 SI seconds, but the difference is very small, and we may neglect it here). We noted earlier that, owing to precession, the direction of the vernal equinox moves along the ecliptic at about 50″ per year. This means that the Sun returns to the vernal equinox before one complete sidereal year has elapsed. This time interval, called the *tropical year*, is 365.2422 days. A third definition of the year is based on the perihelion passages of the Earth. Planetary perturbations cause a gradual change in the direction of the Earth's perihelion. The time interval between two perihelion passages is called the *anomalistic year*, the length of which is 365.2596 days, a little longer than the sidereal year.

The equator of the Earth is tilted about 23.5° with respect to the ecliptic. Owing to perturbations, this angle changes with time. If periodic terms are neglected, the *obliquity of the ecliptic* ε can be expressed as:

$$\varepsilon = 23°26'21.448'' - 46.8150''T$$
$$- 0.00059''T^2 + 0.001813''T^3 , \tag{7.2}$$

where T is the time elapsed since the epoch 2000.0 in Julian centuries (see Sect. 2.14). The expression is valid for a few centuries before and after the year 2000.

The declination of the Sun varies between $-\varepsilon$ and $+\varepsilon$ during the year. At any given time, the Sun is seen at zenith from one point on the surface of the Earth. The latitude of this point is the same as the declination of the Sun. At the latitudes $-\varepsilon$ (the *Tropic of Capricorn*) and $+\varepsilon$ (the *Tropic of Cancer*), the Sun is seen at zenith once every year, and between these latitudes twice a year. In the Northern hemisphere the Sun will not set if the latitude is greater than $90° - \delta$, where δ is the declination of the Sun.

The southernmost latitude where *the midnight Sun* can be seen is thus $90° - \varepsilon = 66.55°$. This is called the *Arctic Circle*. (The same holds true in the Southern hemisphere.) The Arctic Circle is the southernmost place where the Sun is (in theory) below the horizon during the whole day at the winter solstice. The sunless time lasts longer and longer when one goes north (south in the Southern hemisphere). At the poles, day and night last half a year each. In practise, refraction and location of the observing site will have a large influence on the visibility of the midnight Sun and the number of sunless days. Because refraction raises objects seen at the horizon, the midnight Sun can be seen a little further south than at the Arctic Circle. For the same reason the Sun can be seen simultaneously at both poles around the time of vernal and autumnal equinox.

The eccentricity of the Earth's orbit is about 0.0167. The distance from the Sun varies between 147–152 million km. The flux density of solar radiation varies somewhat at different parts of the Earth's orbit, but this has practically no effect on the seasons. In fact the Earth is at perihelion in the beginning of January, in the middle of the northern hemisphere's winter. The seasons are due to the obliquity of the ecliptic.

The energy received from the Sun depends on three factors. First the flux per unit area is proportional to $\sin a$, where a is the altitude of the Sun. In summer the altitude can have greater values than in winter, giving more energy per unit area. Another effect is due to the atmosphere: When the Sun is near the horizon, the radiation must penetrate thick atmospheric layers. This means large extinction and less radiation at the surface. The third factor is the length of the time the Sun is above the horizon. This is important at high latitudes, where the low altitude of the Sun is compensated by the long daylight time in summer. These effects are discussed in detail in Example 7.2.

7.3 The Orbit of the Moon

The Earth's satellite, *the Moon*, circles the Earth counterclockwise. One revolution, the *sidereal month*, takes about 27.322 days. In practise, a more important period is the *synodic month*, the duration of the Lunar phases (e. g. from full moon to full moon). In the course of one sidereal month the Earth has travelled almost 1/12 of its orbit around the Sun. The Moon still has about 1/12 of its orbit to go before the Earth–Moon–Sun configuration is again the same. This takes about 2 days, so the phases of the Moon are repeated every 29 days. More exactly, the length of the synodic month is 29.531 days.

The *new moon* is that instant when the Moon is in conjunction with the Sun. Almanacs define the phases of the Moon in terms of ecliptic longitudes; the longitudes of the new moon and the Sun are equal. Usually the new moon is slightly north or south of the Sun because the lunar orbit is tilted 5° with respect to the ecliptic.

About 2 days after the new moon, the waxing crescent moon can be seen in the western evening sky. About 1 week after the new moon, the *first quarter* follows, when the longitudes of the Moon and the Sun differ by 90°. The right half of the Moon is seen lit (left half when seen from the Southern hemisphere). The *full moon* appears a fortnight after the new moon, and 1 week after this the *last quarter*. Finally the waning crescent moon disappears in the glory of the morning sky.

The orbit of the Moon is approximately elliptic. The length of the semimajor axis is 384,400 km and the eccentricity 0.055. Owing to perturbations caused mainly by the Sun, the orbital elements vary with time. The minimum distance of the Moon from the centre of the Earth is 356,400 km, and the maximum distance 406,700 km. This range is larger than the one calculated from the semimajor axis and the eccentricity. The apparent angular diameter is in the range 29.4′–33.5′.

The rotation time of the Moon is equal to the sidereal month, so the same side of the Moon always faces the Earth. Such *synchronous rotation* is common among the satellites of the solar system: almost all large moons rotate synchronously.

The orbital speed of the Moon varies according to Kepler's second law. The rotation period, however, remains constant. This means that, at different phases of the lunar orbit, we can see slightly different parts of the surface. When the Moon is close to its perigee, its speed is greater than average (and thus greater than the mean rotation rate), and we can see more of the right-hand edge of the Moon's limb (as seen from the Northern hemisphere). Correspondingly, at the apogee we see "behind" the left edge. Owing to the *libration*, a total of 59% of the surface area can be seen from the Earth (Fig. 7.5). The libration is quite easy to see if one follows some detail at the edge of the lunar limb.

The orbital plane of the Moon is tilted only about 5° to the ecliptic. However, the orbital plane changes gradually with time, owing mainly to the perturbations caused by the Earth and the Sun. These perturbations cause the nodal line (the intersection of the plane of the ecliptic and the orbital plane of the Moon) to make one full revolution in 18.6 years. We have already encountered the same period in the nutation. When the ascending node of the lunar orbit is close to the vernal equinox, the Moon can be $23.5° + 5° = 28.5°$ north or south of the equator. When the descending node is close to the vernal equinox, the zone where the Moon can be found extends only $23.5° - 5° = 18.5°$ north or south of the equator.

The *nodical* or *draconic month* is the time in which the Moon moves from one ascending node back to the next one. Because the line of nodes is rotating, the nod-

ical month is 3 hours shorter than the sidereal month, i.e. 27.212 days. The orbital ellipse itself also precesses slowly. The orbital period from perigee to perigee, the *anomalistic month*, is 5.5 h longer than the sidereal month, or about 27.555 days.

Gravitational differences caused by the Moon and the Sun on different parts of the Earth's surface give rise to the *tides*. Gravitation is greatest at the sub-lunar point and smallest at the opposite side of the Earth. At these points, the surface of the seas is highest (high tide, *flood*). About 6 h after flood, the surface is lowest (low tide, *ebb*). The tide generated by the Sun is less than half of the lunar tide. When the Sun and the Moon are in the same direction with respect to the Earth (new moon) or opposite each other (full moon), the tidal effect reaches its maximum; this is called *spring tide*.

The sea level typically varies 1 m, but in some narrow straits, the difference can be as great as 15 m. Due to the irregular shape of the oceans, the true pattern of the

Fig. 7.5. Librations of the Moon can be seen in this pair of photographs taken when the Moon was close to the perigee and the apogee, respectively. (Helsinki University Observatory)

oceanic tide is very complicated. The solid surface of the Earth also suffers tidal effects, but the amplitude is much smaller, about 30 cm.

Tides generate friction, which dissipates the rotational and orbital kinetic energy of the Earth–Moon system. This energy loss induces some changes in the system. First, the rotation of the Earth slows down until the Earth also rotates synchronously, i.e. the same side of Earth will always face the Moon. Secondly, the semi-major axis of the orbit of the Moon increases, and the Moon drifts away about 3 cm per year.

* Tides

Let the tide generating body, the mass of which is M to be at point Q at a distance d from the centre of the Earth. The potential V at the point A caused by the body Q is

$$V(A) = \frac{GM}{s} , \tag{7.3}$$

where s is the distance of the point A from the body Q.

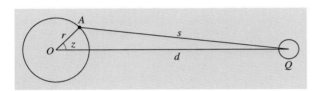

Applying the cosine law in the triangle OAQ, the distance s can be expressed in terms of the other sides and the angle $z = AOQ$

$$s^2 = d^2 + r^2 - 2dr \cos z ,$$

where r is the distance of the point A from the centre of the Earth. We can now rewrite (7.3)

$$V(A) = \frac{GM}{\sqrt{d^2 + r^2 - 2dr \cos z}} . \tag{7.4}$$

When the denominator is expanded into a Taylor series

$$(1+x)^{-\frac{1}{2}} \approx 1 - \frac{1}{2}x + \frac{3}{8}x^2 - \cdots$$

where

$$x = \frac{r^2}{d^2} - 2\frac{r}{d} \cos z$$

and ignoring all terms higher than or equal to $1/d^4$ one obtains

$$V(A) = \frac{GM}{d} + \frac{GM}{d^2} r \cos z + \frac{GMr^2}{d^3} \frac{1}{2}(3 \cos^2 z - 1) . \tag{7.5}$$

The gradient of the potential $V(A)$ gives a force vector per mass unit. The first term of (7.5) vanishes, and the second term is a constant and independent of r. It represents the central motion. The third term of the force vector, however, depends on r. It is the main term of the tidal force. As one can see, it depends inversely on the third power of the distance d. The tidal forces are diminished very rapidly when the distance of a body increases. Therefore the tidal force caused by the Sun is less than half of that of the Moon in spite of much greater mass of the Sun.

We may rewrite the third term of (7.5) as

$$V_2 = 2D \left(\cos^2 z - \frac{1}{3} \right) , \tag{7.6}$$

where

$$D = \frac{3}{4} GM \frac{r^2}{d^3}$$

is called *Doodson's tidal constant*. It's value for the Moon is 2.628 m^2 s^{-2} and for the Sun 1.208 m^2 s^{-2}. We can approximate that z is the zenith angle of the body. The zenith angle z can be expressed in terms of the hour angle h and declination δ of the body and the latitude ϕ of the observer

$$\cos z = \cos h \cos \delta \cos \phi + \sin \delta \sin \phi .$$

Inserting this into (7.6) we obtain after a lengthy algebraic operation

$$V_2 = D \Big(\cos^2 \phi \cos^2 \delta \cos 2h + \sin 2\phi \cos 2\delta \cos h + (3 \sin^2 \phi - 1) \left(\sin^2 \delta - \frac{1}{3} \right) \Big)$$
$$= D(S + T + Z) . \tag{7.7}$$

Equation (7.7) is the traditional basic equation of the tidal potential, the *Laplace's tidal equation*.

In (7.7) one can directly see several characteristics of tides. The term S causes the *semi-diurnal tide* because it depends on $\cos 2h$. It has two daily maxima and minima, separated by 12 hours, exactly as one can obtain in following the ebb and flood. It reaches its maximum at the equator and is zero at the poles ($\cos^2 \phi$).

The term T expresses the *diurnal tides* ($\cos h$). It has its maximum at the latitude $\pm 45°$ and is zero at the equator and at the poles ($\sin 2\phi$). The third term Z is independent of the rotation of the Earth. It causes the *long period tides*, the period of which is half the orbital period of the body (about 14 days in the case of the Moon and 6 months for the Sun). It is zero at the latitude $\pm 35.27°$ and has its maximum at the poles. Moreover, the time average of Z is non-zero, causing a permanent deformation of the Earth. This is called the *permanent tide*. It slightly increases the *flattening of the Earth* and it is inseparable from the flattening due to the rotation.

The total value of the tidal potential can be computed simply adding the potentials caused by the Moon and the Sun. Due to the tidal forces, the whole body of the Earth is deformed. The vertical motion Δr of the crust can be computed from

$$\Delta r = h \frac{V_2}{g} \approx 0.06 \, V_2 \, [\text{m}] , \qquad (7.8)$$

where g is the mean free fall acceleration, $g \approx 9.81 \, \text{m s}^{-2}$ and h is a dimensionless number, the *Love number*, $h \approx 0.6$, which describes the elasticity of the Earth. In the picture below, one can see the vertical motion of the crust in Helsinki, Finland ($\phi = 60°$, $\lambda = 25°$) in January 1995. The non-zero value of the temporal mean can already be seen in this picture.

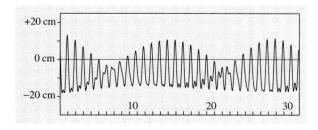

The tides have other consequences, too. Because the Earth rotates faster than the Moon orbits the Earth, the tidal bulge does not lie on the Moon–Earth line but is slightly ahead (in the direction of Earth's rotation), see below.

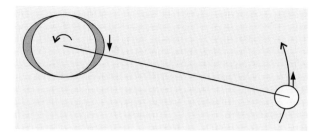

Due to the drag, the rotation of the Earth slows down by about 1–2 ms per century. The same reason has caused the Moon's period of rotation to slow down to its orbital period and the Moon faces the same side towards the Earth. The misaligned bulge pulls the Moon forward. The acceleration causes the increase in the semimajor axis of the Moon, about 3 cm per year.

7.4 Eclipses and Occultations

An *eclipse* is an event in which a body goes through the shadow of another body. The most frequently observed eclipses are the lunar eclipses and the eclipses of the large satellites of Jupiter. An *occultation* takes place when an occulting body goes in front of another object; typical examples are stellar occultations caused by the Moon. Generally, occultations can be seen only in a narrow strip; an eclipse is visible wherever the body is above the horizon.

Solar and lunar eclipses are the most spectacular events in the sky. A *solar eclipse* occurs when the Moon is between the Earth and the Sun (Fig. 7.6). (According to the definition, a solar eclipse is not an eclipse but an occultation!) If the whole disk of the Sun is behind the Moon, the eclipse is *total* (Fig. 7.7); otherwise, it is *partial*. If the Moon is close to its apogee, the apparent diameter of the Moon is smaller than that of the Sun, and the eclipse is *annular*.

A lunar eclipse is *total* if the Moon is entirely inside the umbral shadow of the Earth; otherwise the eclipse is *partial*. A partial eclipse is difficult to see with the unaided eye because the lunar magnitude remains almost unchanged. During the total phase the Moon is coloured

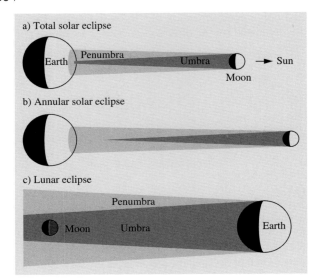

Fig. 7.6. (a) A total solar eclipse can be seen only inside a narrow strip; outside the zone of totality the eclipse is partial. (b) An eclipse is annular if the Moon is at apogee from where the shadow of the Moon does not reach the Earth. (c) A lunar eclipse is visible everywhere where the Moon is above the horizon

Fig. 7.7. The total eclipse of the Sun occurred in 1990 over Finland. (Photo Matti Martikainen)

deep red because some red light is refracted through the Earth's atmosphere.

If the orbital plane of the Moon coincided with the plane of the ecliptic, one solar and one lunar eclipse would occur every synodic month. However, the plane is tilted about 5°; therefore, at full moon, the Moon

must be close to the nodes for an eclipse to occur. The angular distance of the Moon from the node must be smaller than 4.6° for a total lunar eclipse, and 10.3° for a total solar eclipse.

Two to seven eclipses occur annually. Usually eclipses take place in a set of 1–3 eclipses, separated by an interval of 173 days. In one set there can be just one solar eclipse or a succession of solar, lunar and another solar eclipse. In one year, eclipses belonging to 2 or 3 such sets can take place.

The Sun and the (ascending or descending) node of the lunar orbit are in the same direction once every 346.62 days. Nineteen such periods ($= 6585.78$ days $= 18$ years 11 days) are very close to the length of 223 synodic months. This means that the Sun–Moon configuration and the eclipses are repeated in the same order after this period. This *Saros* period was already known to the ancient Babylonians.

During a solar eclipse the shadow of the Moon on Earth's surface is always less than 270 km wide. The shadow moves at least 34 km/min; thus the maximum duration of an eclipse is $7\frac{1}{2}$ minutes. The maximum duration of a lunar eclipse is 3.8 h, and the duration of the total phase is always shorter than 1.7 h.

Observations of the *stellar occultations* caused by the Moon formerly served as an accurate method for determining the lunar orbit. Because the Moon has no atmosphere, the star disappears abruptly in less than $1/50$ s. If a fast photometer is used for recording the event, the typical diffraction pattern can be seen. The shape of the diffraction is different for a binary star. In the first decades of radio astronomy the occultations of some radio sources were used for determining their exact positions.

The Moon moves eastwards, and stars are occulted by the dark edge of the Moon during the first quarter. Therefore occultation is easier to observe, and photometric measurements are possible; at the same time it is much more difficult to observe the appearance of an object. There are some bright stars and planets inside the 11° wide zone where the Moon moves, but the occultation of a bright, naked-eye object is quite rare.

Occultations are also caused by planets and asteroids. Accurate predictions are complicated because such an event is visible only in a very narrow path. The Uranian rings were found during an occultation in 1977, and the shapes of some asteroids have been studied during some favourable events, timed exactly by several observers located along the predicted path.

A *transit* is an event in which Mercury or Venus moves across the Solar disk as seen from the Earth. A transit can occur only when the planet is close to its orbital node at the time of inferior conjunction. Transits of Mercury occur about 13 times per century; transits of Venus only twice. The next transits of Mercury are: May 7, 2003; Nov 8, 2006; May 9, 2016 and Nov 11, 2019. The next three transits of Venus are: Jun 8, 2004; Jun 6, 2012 and Dec 11, 2117. In the 18th century the two transits of Venus were used for determining the value of the astronomical unit.

7.5 The Structure and Surfaces of Planets

Since 1960's a vast amount of data have been collected using spacecraft, either during a flyby, orbiting a body, or directly landing on the surface. This gives a great advantage compared to other astronomical observations. We may even speak of revolution: the solar system bodies have turned from astronomical objects to geophysical ones. Many methods traditionally used in various sibling branches of geophysics can now be applied to planetary studies.

The shape and irregularities of the gravitation field generated by a planet reflects its shape, internal structure and mass distribution. Also the surface gives certain indications on internal structure and processes.

The perturbations in the orbit of a satellite or spacecraft can be used in studying the internal structure of a planet. Any deviation from spherical symmetry is visible in the external gravitational field.

A rotating planet is always *flattened*. The amount of flattening depends on the rotation rate and the strength of the material; a liquid drop is more easily deformed than a rock. The shape of a rotating body in *hydrostatic equilibrium* can be derived from the equations of motion. If the rotation rate is moderate, the equilibrium shape of a liquid body is an ellipsoid of revolution. The shortest axis is the axis of rotation.

If R_e and R_p are the equatorial and polar radii, respectively, the shape of the planet can be expressed as

$$\frac{x^2}{R_e^2} + \frac{y^2}{R_e^2} + \frac{z^2}{R_p^2} = 1 \ .$$

The *dynamical flattening*, denoted by f is defined as

$$f = \frac{R_e - R_p}{R_e} \ . \tag{7.9}$$

Because $R_e > R_p$, the flattening f is always positive.

Asteroids are so small that they are not flattened by rotation. Bodies larger than about 1000 km in diameter are deformed into a symmetric shape due to gravity; the internal strength of the material keeps smaller bodies irregular in shape. The giant planets are in practise close to hydrostatic equilibrium, and their shape is determined by the rotation. The rotation period of Saturn is only 10.5 h, and its dynamical flattening is 1/10 which is easily visible.

The structure of the terrestrial planets (Fig. 7.8) can also be studied with *seismic waves*. The waves formed in an earthquake are reflected and refracted inside a planet like any other wave at the boundary of two different layers. The waves are longitudinal or transversal (*P* and *S waves*, respectively). Both can propagate in solid materials such as rock. However, only the longitudinal wave can penetrate liquids. One can determine whether a part of the interior material is in the liquid state and where the boundaries of the layers are by studying the recordings of seismometers placed on the surface of a planet. Naturally the Earth is the best-known body,

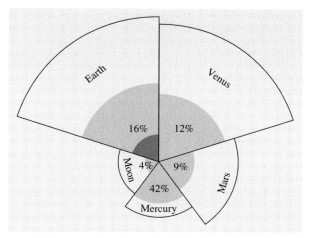

Fig. 7.8. Internal structure and relative sizes of the terrestrial planets. The percentage shows the volume of the core relative to the total volume of the planet. In the case of the Earth, the percentage includes both the outer and the inner core

but quakes of the Moon, Venus, and Mars have also been observed.

The terrestrial planets have an *iron-nickel core*. Mercury has the relatively largest core; Mars the smallest. The core of the Earth can be divided into an *inner* and an *outer core*. The outer core (2900–5150 km) is liquid but the inner core (from 5150 km to the centre) is solid.

Around the Fe–Ni core is a *mantle*, composed of *silicates* (compounds of silicon). The density of the outermost layers is about 3000 kg m^{-3}. The mean density of the terrestrial planets is 3500–5500 kg m^{-3}.

The internal structure of the giant planets (Fig. 7.9) cannot be observed with seismic waves since the planets do not have a solid surface. An alternative is to study the shape of the gravitational field by observing the orbit of a spacecraft when it passes (or orbits) the planet. This will give some information on the internal structure, but the details depend on the mathematical and physical models used for interpretation.

The mean densities of the giant planets are quite low; the density of Saturn, for example, is only 700 kg m^{-3}. (If Saturn were put in a gigantic bathtub, it would float on the water!) Most of the volume of a giant planet is a mixture of hydrogen and helium. In the centre, there is possibly a silicate core, the mass of which is a few Earth masses. The core is surrounded by a layer of *metallic hydrogen*. Due to the extreme pressure, hydrogen is not in its normal molecular form H_2, but dissociated into atoms. In this state, hydrogen is electrically conducting. The magnetic fields of the giant planets may originate in the layer of metallic hydrogen.

Closer to the surface, the pressure is lower and hydrogen is in molecular form. The relative thickness of the layers of metallic and molecular hydrogen vary from planet to planet. Uranus and Neptune may not have any layer of metallic hydrogen because their internal pressure is too low for dissociation of the hydrogen. Instead, a layer of "ices" surround the core. This is a layer of a water-dominant mixture of water, methane and ammonia. Under the high pressure and temperature the mixture is partly dissolved into its components and it behaves more like a molten salt and it is also electrically conductive like the metallic hydrogen.

On top of everything is a gaseous atmosphere, only a few hundred kilometres thick. The clouds at the top of the atmosphere form the visible "surface" of the giant planets.

The interior temperatures of the planets are considerably larger than the surface temperatures. For

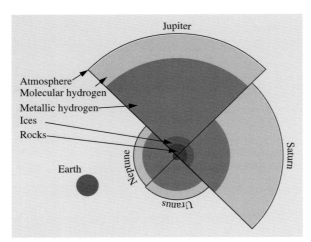

Fig. 7.9. Internal structure and relative sizes of the giant planets. Differences in size and distance from the Sun cause differences in the chemical composition and internal structure. Due to smaller size, Uranus and Neptune do not have any layer of metallic hydrogen. The Earth is shown in scale

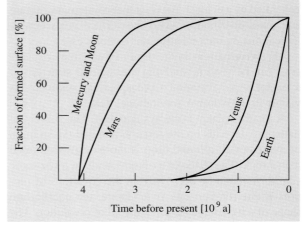

Fig. 7.10. Ages of the surfaces of Mercury, the Earth, the Moon and Mars. The curve represents the fraction of the surface which existed at a certain time. Most of the surface of the Moon, Mercury and Mars are more than 3500 million years old, whereas the surface of the Earth is mostly younger than 200 million years

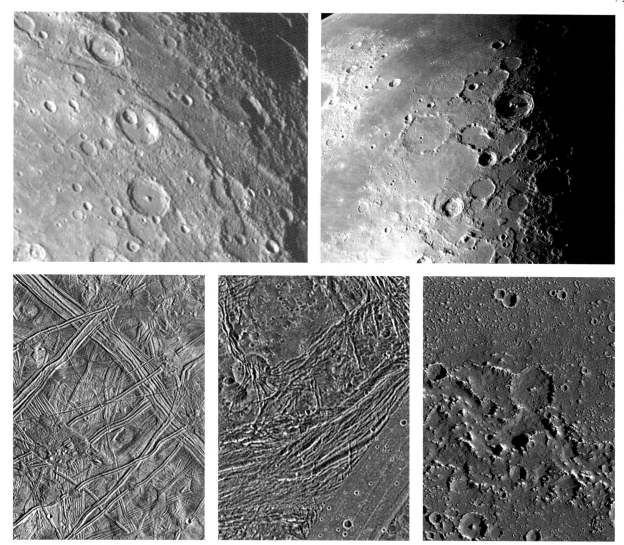

Fig. 7.11. The number of meteorite impact craters is an indicator of the age of the surface and the shapes of the craters give information on the strength of the material. The upper row shows Mercury (*left*) and the Moon, and the bottom row, the Jovian moons Europa (*left*), Ganymede (*centre*) and Callisto. The pictures of the Jovian moons were taken by the Galileo orbiter with a resolution of 150 metres/pixel. Europa has only a few craters, there are areas of different ages on the surface Ganymede and the surface of Callisto is the oldest. Note the grooves and ridges that indicate different geological processes. (NASA and DLR)

example, the temperature in the Earth's core is about 4500–5000 K, and in the core of Jupiter about 30,000 K.

A part of that heat is the remnant of the released potential energy from the gravitational contraction during the formation of planets. Decay of radioactive isotopes also releases heat. Soon after the formation of planets intense meteorite bombardment was an important source of heat. Together with heat from short-lived radioactive isotopes this caused *melting of terrestrial planets*. The planets were *differentiated*: the originally relatively homogeneous material became segregated into layers of different chemical composition. The heaviest elements sank into centre thus forming the Fe–Ni core.

The material of the giant planets is differentiated as well. In Saturn the differentiation may still be going on. Saturn is radiating about 2.8 times the heat it gets from the Sun, more than any other planet. This heat is suspected to originate from the separation of hydrogen and helium, where the heavier helium is gradually sinking toward the centre of the planet.

Planetary surfaces are modified by several geological processes. These include *continental drift, volcanism, meteorite impacts* and *climate*. The Earth is an example of a body whose surface has been renewed many times during past aeons. The age of the surface depends on the processes and thus implies the geological evolutionary history of the planet Figs. 7.10, 7.11).

Continental drift gives rise, for example, to mountain formation. The Earth is the only planet where *plate tectonics* is active today. On other terrestrial planets the process has either ceased long ago or has never occurred.

Volcanism is a minor factor on the Earth (at least now), but the surface of the Jovian moon Io is changing rapidly due to violent volcanic eruptions. Volcanoes have been observed on Mars and Venus, but not on the Moon.

Lunar craters are meteorite impact craters, common on almost every body with a solid surface. Meteorites are bombarding the planets continuously, but the rate has been diminishing since the beginnings of the solar system. The number of impact craters reflects the age of the surface (Fig. 7.11).

The Jovian moon Callisto is an example of a body with an ancient surface which is not fully inactive. Lack of small craters indicates some resurfacing process filling and degrading the minor surface features. The Earth is an example of a body, whose atmosphere both protects the surface and destroys the traces of impacts. All smaller meteorites are burned to ashes in the atmosphere (one need only note the number of shooting stars), and some larger bodies are bounced back to outer space. The traces on the surface are destroyed very quickly by erosion in less than a few million years. Venus is an even more extreme case where all small craters are missing due to a thick protective atmosphere.

Climate has the greatest influence on the Earth and Venus. Both planets have a thick atmosphere. On Mars, powerful dust storms deform the landscape, too, often covering the planet with yellowish dust clouds.

7.6 Atmospheres and Magnetospheres

Excluding Mercury, all major planets have an atmosphere. The composition, thickness, density and structure of the atmosphere vary from planet to planet, but some common features can be found (see, e.g., Figs. 7.12, 7.13).

Let us first study the dependence of the temperature T, pressure P, and density ρ on the height h. Let us consider a cylinder with a length dh. The change in the pressure dP from the height h to $h + dh$ is proportional to the mass of the gas in the cylinder:

$$dP = -g\rho\, dh \,, \tag{7.10}$$

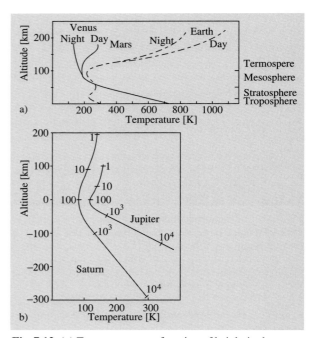

Fig. 7.12. (a) Temperature as a function of height in the atmospheres of Venus, Earth, and Mars. **(b)** Temperature profiles of the atmospheres of Jupiter and Saturn. The zero height is chosen to be the point where the pressure is 100 mbar. Numbers along the curves are pressures in millibars

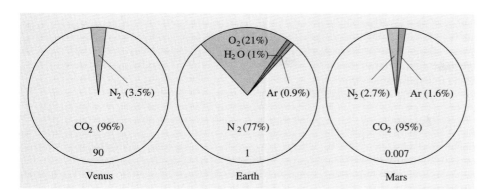

Fig. 7.13. Relative abundances of the most abundant gases in the atmospheres of Venus, Earth, and Mars. The number at the bottom of each circle denotes the surface pressure in atms

where g is the acceleration of gravity. Equation (7.10) is the *equation of hydrostatic equilibrium*. (It is discussed in detail in Chap. 10.)

As a first approximation, we may assume that g does not depend on height. In the case of the Earth, the error is only about 3% if g is considered constant from the surface to a height of 100 km.

The equation of state of the ideal gas

$$PV = NkT \tag{7.11}$$

gives the expression for the pressure P

$$P = \frac{\rho kT}{\mu}, \tag{7.12}$$

where N is the number of atoms or molecules, k is the Boltzmann constant, μ is the mass of one atom or molecule and

$$\rho = \frac{\mu N}{V}. \tag{7.13}$$

By using the equation of hydrostatic equilibrium (7.10) and the equation of state (7.12), we obtain

$$\frac{dP}{P} = -g\frac{\mu}{kT}dh.$$

Integration yields P as a function of height:

$$P = P_0 \exp\left(-\int_0^h \frac{\mu g}{kT}dh\right)$$

$$= P_0 \exp\left(-\int_0^h \frac{dh}{H}\right). \tag{7.14}$$

The variable H, which has the dimension of length, is called the *scale height*:

$$H = \frac{kT}{\mu g}. \tag{7.15}$$

The scale height defines the height at which the pressure has decreased by a factor e. H is a function of height, but here we may assume that it is constant. With this approximation, we obtain

$$-\frac{h}{H} = \ln \frac{P}{P_0}$$

or, using (7.12),

$$\frac{\rho T(h)}{\rho_0 T_0} = e^{-h/H}. \tag{7.16}$$

The scale height is an important parameter in many formulas describing the structure of the atmosphere (Table 7.1). For example, if the change of the pressure or the density is known as a function of height, the mean molecular weight of the atmosphere can be computed.

Table 7.1. Scale heights of some gases in the atmospheres of Venus, Earth, and Mars

Gas	Molecular weight [amu]	Earth H [km]	Venus H [km]	Mars H [km]
H_2	2	120	360	290
O_2	32	7	23	18
H_2O	18	13	40	32
CO_2	44	5	16	13
N_2	28	8	26	20
Temperature [K]		275	750	260
Acceleration of gravity [m/s²]		9.81	8.61	3.77

The scale height of the Jovian atmosphere was determined in 1952 when Jupiter occulted a star. With these observations, the scale height was calculated to be 8 km, and the mean molecular weight 3–5 amu (atomic mass unit, $1/12$ of the mass of ^{12}C). Thus the main components are hydrogen and helium, a result later confirmed by spacecraft data.

In terrestrial observations, infrared data are limited by water vapour and carbon dioxide. The scale height of CO_2 is 5 km, which means that the partial pressure is already halved at a height of 3.5 km. Thus infrared observations can be made on top of high mountains (like Mauna Kea in Hawaii). The scale height of water vapour is 13 km, but the relative humidity and hence the actual water content is very site- and time-dependent.

The scale height and the temperature of the atmosphere define the permanence of the atmosphere. If the speed of a molecule is greater than the escape velocity, the molecule will escape into space. The whole atmosphere could disappear in a relatively short time.

According to the kinetic gas theory, the mean velocity $\overline{v}$ of a molecule depends both on the kinetic temperature T_k of the gas and the mass m of the molecule:

$$\overline{v} = \sqrt{\frac{3kT_k}{m}} \ .$$

If the mass of a planet is M and its radius R, the escape velocity is

$$v_e = \sqrt{\frac{2GM}{R}} \ .$$

Even if the mean velocity is smaller than the escape velocity, the atmosphere can evaporate into space if there is enough time, since some molecules will always have velocities exceeding v_e. Assuming a velocity distribution, one can calculate the probability for $v > v_e$ Hence it is possible to estimate what fraction of the atmosphere will disappear in, say, 10^9 years. As a rule of thumb, it can be said that at least half of the atmosphere will remain over 1000 million years if the mean velocity $\overline{v} < 0.2v_e$.

The probability that a molecule close to the surface will escape is insignificantly small. The free mean path of a molecule is very small when the gas density is high (Fig. 7.14). Thus the escaping molecule is most probably leaving from the uppermost layers. The *critical layer* is defined as a height at which a molecule,

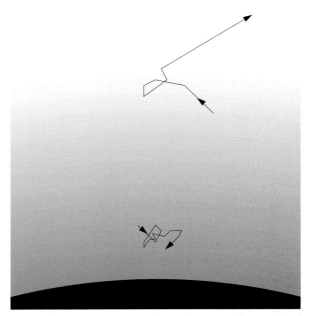

Fig. 7.14. Close to the surface, the mean free path of a molecule is smaller than higher in the atmosphere where the gas density is smaller. The escaping molecules originate close to the critical layer

moving upward, has a probability $1/e$ of hitting another molecule. The part of the atmosphere above the critical layer is called the *exosphere*. The exosphere of the Earth begins at a height of 500 km, where the kinetic temperature of the gas is 1500–2000 K and the pressure is lower than in the best terrestrial vacuums.

The *magnetosphere* is the "outer boundary" of a planet. Size and shape depend on the strength of the magnetic field of the planet and on the solar wind. The *solar wind* is a flux of charged particles, mostly electrons and protons, outflowing from the Sun. The speed of the wind at the distance of the Earth is about 400 km/s and the density 10 particles/cm^3 but both values can change considerably depending on the solar activity.

On the solar side there is a *bow shock* (Fig. 7.15), typically at a distance of a few tens of planetary radii (Table 7.2). At the bow shock, particles of the solar wind first hit the magnetosphere. The magnetosphere is limited by the *magnetopause*, flattened on the solar side and extended to a long tail on the opposite side. Charged particles inside the magnetopause are captured by the magnetic field and some particles are accelerated

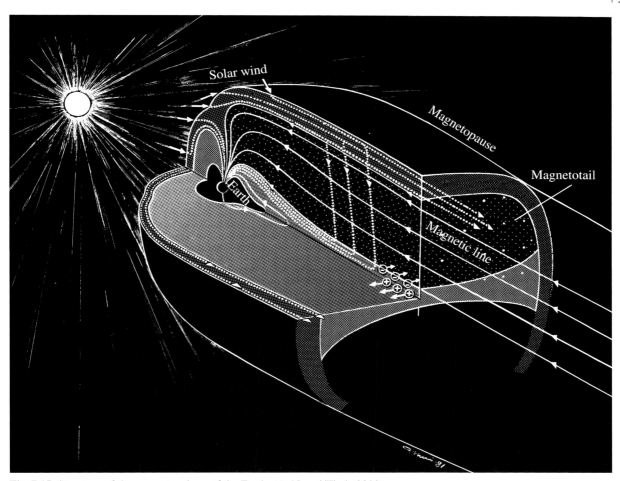

Fig. 7.15. Structure of the magnetosphere of the Earth. (A. Nurmi/Tiede 2000)

Table 7.2. Planetary magnetic fields

	Dipole moment (Earth = 1)	Field strength (gauss)[1]	Polarity[2]	Angle[3]	Magneto-pause[4]
Mercury	0.0007	0.003	⇑	14°	1.5
Venus	< 0.0004	< 0.00003	–	–	–
Earth	1.0	0.305	⇑	11°	10
Mars	< 0.0002	< 0.0003	–	–	–
Jupiter	20,000.	4.28	⇓	10°	80
Saturn	600.	0.22	⇓	< 1°	20
Uranus	50.	0.23	⇓	59°	20
Neptune	25.	0.14	⇓	47°	25

[1] at equator (1 gauss equals 10^{-4} T);
[2] ⇑ same as the Earth, ⇓ opposite;
[3] angle between magnetic and rotational axes;
[4] average magnetopause distance in the direction of the Sun in planetary radii

to great velocities. If the velocities are interpreted according to the kinetic gas theory, these velocities even correspond to millions of kelvins. However, the density, and thus the total energy, is very small. The "hottest" places found are around Jupiter and Saturn.

The region of space containing trapped charged particles, the radiation belts around the Earth, are named *van Allen's belts* (Fig. 7.16). These radiation zones were discovered by the first US satellite, Explorer 1, in 1958. The number of charged particles increases after strong solar bursts. Some of these particles "leak" to the atmosphere, resulting in *auroras*. Similar effects have also been detected in Jupiter, Saturn and Uranus.

The solar magnetic field arises from the turbulent motions of the electrically conductive matter. The energy driving the convection in the layer is coming from the nuclear fusion in the core. This, however, cannot explain planetary magnetism. Neither can the remanent primordial magnetic field explain it because the internal temperature of planets is well above the *Curie point* (about 850 K for magnetite). If the temperature is above the Curie point, ferromagnetic materials will lose their remanent magnetism.

The planetary *dynamo* generating the magnetic field requires that the planet is rotating and has a convective layer of electrically conductive material. Terrestrial planets have a liquid Fe–Ni core, or a liquid layer in the core, Jupiter and Saturn have a layer of liquid metallic hydrogen and Uranus and Neptune have a mixture of water, ammonia and methane. In all cases the temperature gradient between the bottom and top of the layer is large enough to cause the convection.

The strength of the magnetic field varies a lot from planet to planet. It can be characterised by the *dipole magnetic moment*. The magnetic moment of Jupiter is about 100 million times that of Mercury. The magnetic moment of the Earth is about 7.9×10^{25} gauss cm^3 that can be compared to the typical strong electromagnetic fields achieved in the laboratories, about $100,000$ gauss cm^3. Inducing such a strong field requires currents that are of the order of 10^9 Amperes. When divided by the cube of planetary radii, one gets an estimate of the field strength on the equator.

The alignment of the magnetic field with respect to the rotation axis of a planet differs from planet to planet (Fig. 7.17). Saturn's magnetic field is close to the ideal case where rotational axis and magnetic axis coincide. Also the Earth and Jupiter show reasonably good point dipole field with a tilt of about 10°. However, fields of Uranus and Neptune are both offset from the centre of the planet and tilted by about 50° from the rotation axis. This may indicate a different mechanism for the dynamo.

The magnetic fields of Mercury and the Earth have an opposite *polarity* than the fields of other planets. It is known that the polarity of the Earth's magnetic field has reversed several times over geologic time scales, previously about 750,000 years ago. There are some indications that the reversal of the polarity is beginning now because the field strength is declining about one percent per decade, magnetic poles are moving more rapidly and the field asymmetry is increasing. The whole process will take several thousand years during which the Earth's surface is more open to the cosmic rays.

The Galileo mission also revealed that the Jovian moon *Ganymede* has a magnetic field. The field is weak

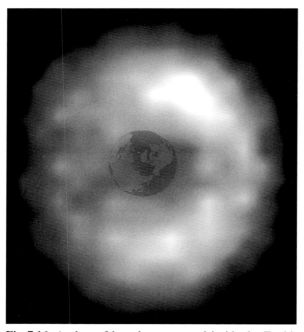

Fig. 7.16. A glow of hot plasma trapped inside the Earth's magnetosphere. The picture was taken by NASA's Imager for Magnetopause to Aurora Global Exploration (IMAGE) spacecraft on August 11, 2000 at 18:00 UT. The Sun is outside the picture area toward the top right corner. (NASA and the IMAGE science team)

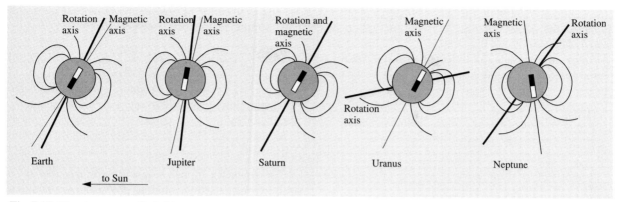

Fig. 7.17. Planetary magnetic fields

and too small to have a magnetotail or trapped particles around the moon. Callisto, which is of the same size, does not show any magnetosphere. Neither does our Moon have any global magnetic field.

7.7 Albedos

The planets and all other bodies of the solar system only reflect the radiation of the Sun (we may neglect here the thermal and radio wave radiation and concentrate mainly on the visual wavelengths). The brightness of a body depends on its distance from the Sun and the Earth, and on the albedo of its surface. The term *albedo* defines the ability of a body to reflect light.

If the luminosity of the Sun is $L_\odot$, the flux density at the distance r is (Fig. 7.18)

$$F = \frac{L_\odot}{4\pi r^2} . \tag{7.17}$$

If the radius of the planet is R, the area of its cross section is πR^2, and the total flux incident on the surface of the planet is

$$L_{\text{in}} = \pi R^2 \frac{L_\odot}{4\pi r^2} = \frac{L_\odot R^2}{4r^2} . \tag{7.18}$$

Only a part of the incident flux is reflected back. The other part is absorbed and converted into heat which is then emitted as a thermal emission from the planet. The *Bond albedo* A (or spherical albedo) is defined

as the ratio of the emergent flux to the incident flux ($0 \leq A \leq 1$). The flux reflected by the planet is thus

$$L_{\text{out}} = A L_{\text{in}} = \frac{A L_\odot R^2}{4r^2} . \tag{7.19}$$

The planet is observed at a distance Δ. If radiation is reflected isotropically, the observed flux density should be

$$F = \frac{L_{\text{out}}}{4\pi \Delta^2} . \tag{7.20}$$

In reality, however, radiation is reflected anisotropically. If we assume that the reflecting object is a homogeneous sphere, the distribution of the reflected radiation depends on the *phase angle α* only. Thus we can express the flux density observed at a distance Δ as

$$F = C\Phi(\alpha) \frac{L_{\text{out}}}{4\pi \Delta^2} . \tag{7.21}$$

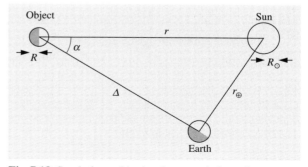

Fig. 7.18. Symbols used in the photometric formulas

The function Φ giving the phase angle dependence is called the *phase function*. It is normalised so that $\Phi(\alpha = 0°) = 1$.

Since all the radiation reflected from the planet is found somewhere on the surface of the sphere, we must have

$$\int_S C\Phi(\alpha) \frac{L_{out}}{4\pi\Delta^2} \, dS = L_{out} \qquad (7.22)$$

or

$$\frac{C}{4\pi\Delta^2} \int_S \Phi(\alpha) \, dS = 1 , \qquad (7.23)$$

where the integration is extended over the surface of the sphere of radius Δ. The surface element of such a sphere is $dS = \Delta^2 \, d\alpha \sin\alpha \, d\phi$, and we have

$$\int_S \Phi(\alpha) \, dS = \Delta^2 \int_{\alpha=0}^{\pi} \int_{\phi=0}^{2\pi} \Phi(\alpha) \sin\alpha \, d\alpha \, d\phi$$
$$= \Delta^2 2\pi \int_0^\pi \Phi(\alpha) \sin\alpha \, d\alpha . \qquad (7.24)$$

The normalisation constant C is

$$C = \frac{4\pi\Delta^2}{\int_S \Phi(\alpha) \, dS} = \frac{2}{\int_0^\pi \Phi(\alpha) \sin\alpha \, d\alpha} . \qquad (7.25)$$

The quantity

$$q = 2 \int_0^\pi \Phi(\alpha) \sin\alpha \, d\alpha \qquad (7.26)$$

is the *phase integral*. In terms of the phase integral the normalisation constant is

$$C = \frac{4}{q} . \qquad (7.27)$$

Remembering that $L_{out} = AL_{in}$, the equation (7.21) can be written in the form

$$F = \frac{CA}{4\pi} \Phi(\alpha) \frac{1}{\Delta^2} L_{in} . \qquad (7.28)$$

The first factor is intrinsic for each object, the second gives the phase angle dependence, the third the distance dependence and the fourth, the incident radiation power. The first factor is often denoted by

$$\Gamma = \frac{CA}{4\pi} . \qquad (7.29)$$

When we substitute here the expression of C (7.27), and solve for the Bond albedo, we get

$$A = \frac{4\pi\Gamma}{C} = \pi\Gamma\frac{4}{C} = \pi\Gamma q = pq . \qquad (7.30)$$

Here $p = \pi\Gamma$ is called the *geometric albedo* and q is the previously introduced phase integral. These quantities are related by

$$A = pq . \qquad (7.31)$$

The geometric albedo seems to have appeared as an arbitrary factor with no obvious physical interpretation. We'll now try to explain this quantity using a *Lambertian surface*. A Lambertian surface is defined as an absolutely white, diffuse surface which reflects all radiation, i. e. its Bond albedo is $A = 1$. Moreover, its surface brightness is the same for all viewing directions, which means that the phase function is

$$\Phi(\alpha) = \begin{cases} \cos\alpha , & \text{if } 0 \le \alpha \le \pi/2 , \\ 0 , & \text{otherwise} . \end{cases} \qquad (7.32)$$

In reality, no such surface exists but there are some materials which behave almost like a Lambertian surface. A wall with a mat white finish is a good approximation; although it doesn't reflect all incident light, the distribution of the reflected light is about right, and its brightness looks the same from all directions.

For a Lambertian surface the constant C is

$$C = \frac{2}{\int_0^\pi \Phi(\alpha) \sin\alpha \, d\alpha}$$
$$= \frac{2}{\int_0^{\pi/2} \cos\alpha \sin\alpha \, d\alpha} \qquad (7.33)$$
$$= \frac{2}{1/2} = 4 .$$

Thus the geometric albedo of a Lambertian surface is

$$p = \pi\Gamma = \frac{CA}{4} = \frac{4 \times 1}{4} = 1 . \qquad (7.34)$$

At the phase angle zero $\Phi(\alpha = 0°) = 1$ and the reflected flux density is

$$F = \frac{CA}{4\pi} \frac{1}{\Delta^2} L_{in} .$$

If we replace the object with a Lambertian surface of the same size, we get

$$F_L = \frac{4}{4\pi} \frac{1}{\Delta^2} L_{in} \ .$$

The ratio of these flux densities is

$$\frac{F}{F_L} = \frac{CA}{4} = \pi \Gamma = p \ . \tag{7.35}$$

Now we have found a physical interpretation for p: the geometric albedo is the ratio of the flux densities at phase angle $\alpha = 0°$ reflected by a planet and a Lambertian surface of the same cross section.

The geometric albedo depends on the reflectance of the surface but also on the phase function Φ. Many rough surfaces reflect most of the incident radiation directly backward. In such a case the geometric albedo p is greater than in the case of an isotropically reflecting surface. On some surfaces $p > 1$, and in the most extreme case, the specular reflection, $p = \infty$. The geometric albedo of solar system bodies vary between 0.03–1. The geometric albedo of the Moon is $p = 0.12$ and the greatest value, $p = 1.0$, has been measured for the Saturnian moon Enceladus.

It turns out that p can be derived from the observations, but the Bond albedo A can be determined only if the phase integral q is also known. That will be discussed in the next section.

7.8 Photometry, Polarimetry and Spectroscopy

Having defined the phase function and albedos we are ready to derive a formula for *planetary magnitudes*. The flux density of the reflected light is

$$F = \frac{CA}{4\pi} \Phi(\alpha) \frac{1}{\Delta^2} L_{in} \ .$$

We now substitute the incident flux

$$L_{in} = \frac{L_\odot R^2}{4r^2}$$

and the constant factor expressed in terms of the geometric albedo

$$\frac{CA}{4\pi} = \Gamma = \frac{p}{\pi} \ .$$

Thus we get

$$F = \frac{p}{\pi} \Phi(\alpha) \frac{1}{\Delta^2} \frac{L_\odot R^2}{4r^2} \ . \tag{7.36}$$

The observed solar flux density at a distance of $a = 1$ AU from the Sun is

$$F_\odot = \frac{L_\odot}{4\pi a^2} \ . \tag{7.37}$$

The ratio of these is

$$\frac{F}{F_\odot} = \frac{p\Phi(\alpha) R^2 a^2}{\Delta^2 r^2} \ . \tag{7.38}$$

If the apparent solar magnitude at a distance of 1 AU is $m_\odot$ and the apparent magnitude of the planet m we have

$$\begin{aligned}
m - m_\odot &= -2.5 \lg \frac{F}{F_\odot} \\
&= -2.5 \lg \frac{p\Phi(\alpha) R^2 a^2}{\Delta^2 r^2} \\
&= -2.5 \lg \frac{pR^2}{a^2} \frac{a^4}{\Delta^2 r^2} \Phi(\alpha) \\
&= -2.5 \lg p \frac{R^2}{a^2} - 2.5 \lg \frac{a^4}{\Delta^2 r^2} - 2.5 \lg \Phi(\alpha) \\
&= -2.5 \lg p \frac{R^2}{a^2} + 5 \lg \frac{\Delta r}{a^2} - 2.5 \lg \Phi(\alpha) \ .
\end{aligned} \tag{7.39}$$

If we denote

$$V(1, 0) \equiv m_\odot - 2.5 \lg \ p \frac{R^2}{a^2} \ , \tag{7.40}$$

then the magnitude of a planet can be expressed as

$$m = V(1, 0) + 5 \lg \frac{r\Delta}{a^2} - 2.5 \lg \Phi(\alpha) \ . \tag{7.41}$$

The first term $V(1, 0)$ depends only on the size of the planet and its reflection properties. So it is a quantity intrinsic to the planet, and it is called the *absolute magnitude* (not to be confused with the absolute magnitude in stellar astronomy!). The second term contains the distance dependence and the third one the dependence on the phase angle.

If the phase angle is zero, and we set $r = \Delta = a$, (7.41) becomes simply $m = V(1, 0)$. The absolute magnitude can be interpreted as the magnitude of a body if it is at a distance of 1 AU from the Earth and the Sun at a phase angle $\alpha = 0°$. As will be immediately noticed, this is physically impossible because the observer would be in the very centre of the Sun. Thus $V(1, 0)$ can never be observed.

The last term in (7.41) is the most problematic one. For many objects the phase function is not known very well. This means that from the observations, one can calculate only

$$V(1, \alpha) \equiv V(1, 0) - 2.5 \lg \Phi(\alpha) , \qquad (7.42)$$

which is the *absolute magnitude at phase angle* α. $V(1, \alpha)$, plotted as a function of the phase angle, is called the *phase curve* (Fig. 7.19). The phase curve extrapolated to $\alpha = 0°$ gives $V(1, 0)$.

By using (7.39) at $\alpha = 0°$, the geometric albedo can be solved for in terms of observed values:

$$p = \left(\frac{r\Delta}{aR}\right)^2 10^{-0.4(m_0 - m_\odot)} , \qquad (7.43)$$

where $m_0 = m(\alpha = 0°)$. As can easily be seen, p can be greater than unity but in the real world, it is normally well below that. A typical value for p is 0.1–0.5.

The Bond albedo can be determined only if the phase function Φ is known. Superior planets (and other bodies orbiting outside the orbit of the Earth) can be observed

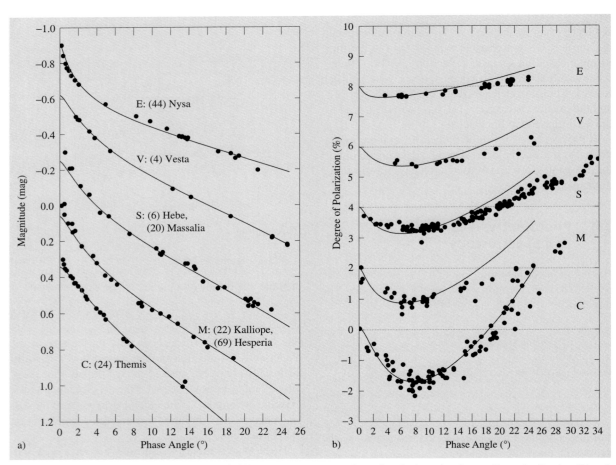

Fig. 7.19. The phase curves and polarization of different types of asteroids. The asteroid characteristics are discussed in more detail in Sect. 7.14. (From Muinonen *et al.*, *Asteroid photometric and polarimetric phase effects*, in Bottke, Binzel, Cellino, Paolizhi (Eds.) *Asteroids III*, University of Arizona Press, Tucson.)

only in a limited phase angle range, and therefore Φ is poorly known. The situation is somewhat better for the inferior planets. Especially in popular texts the Bond albedo is given instead of p (naturally without mentioning the exact names!). A good excuse for this is the obvious physical meaning of the former, and also the fact that the Bond albedo is normalised to [0, 1].

Opposition Effect. The brightness of an atmosphere-less body increases rapidly when the phase angle approaches zero. When the phase is larger than about 10°, the changes are smaller. This rapid brightening close to the opposition is called the *opposition effect*. The full explanation is still in dispute. A qualitative (but only partial) explanation is that close to the opposition, no shadows are visible. When the phase angle increases, the shadows become visible and the brightness drops. An atmosphere destroys the opposition effect.

The shape of the phase curve depends on the geometric albedo. It is possible to estimate the geometric albedo if the phase curve is known. This requires at least a few observations at different phase angles. Most critical is the range 0°–10°. A known phase curve can be used to determine the diameter of the body, e.g. the size of an asteroid. Apparent diameters of asteroids are so small that for ground based observations one has to use indirect methods, like polarimetric or radiometric (thermal radiation) observations. Beginning from the 1990's, imaging made during spacecraft fly-bys and with the Hubble Space Telescope have given also direct measures of the diameter and shape of asteroids.

Magnitudes of Asteroids. When the phase angle is greater than a few degrees, the magnitude of an asteroid depends almost linearly on the phase angle. Earlier this linear part was extrapolated to $\alpha = 0°$ to estimate the opposition magnitude of an asteroid. Due to the opposition effect the actual opposition magnitude can be considerably brighter.

In 1985 the IAU adopted the semi-empirical *HG system* where the magnitude of an asteroid is described by two constants H and G. Let

$$
\begin{aligned}
a_1 &= (1 - G) \times 10^{-0.4\,H} , \\
a_2 &= G \times 10^{-0.4\,H} .
\end{aligned}
\tag{7.44}
$$

The phase curve can be approximated by

$$
\begin{aligned}
V(1, \alpha) = -2.5 \\
\times \log\left[a_1 \exp\left(-3.33 \left(\tan\frac{\alpha}{2} \right)^{0.63} \right) \right. \\
\left. + a_2 \exp\left(-1.87 \left(\tan\frac{\alpha}{2} \right)^{1.22} \right) \right] .
\end{aligned}
\tag{7.45}
$$

When the phase angle is $\alpha = 0°$ (7.45) becomes

$$
\begin{aligned}
V(1, 0) &= -2.5 \log(a_1 + a_2) \\
&= -2.5 \log 10^{-0.4\,H} = H .
\end{aligned}
\tag{7.46}
$$

The constant H is thus the absolute magnitude and G describes the shape of the phase curve. If G is great, the phase curve is steeper and the brightness is decreasing rapidly with the phase angle. For very gentle slopes G can be negative. H and G can be determined with a least squares fit to the phase observations.

Polarimetric Observations. The light reflected by the bodies of the solar system is usually polarized. The amount of polarization depends on the reflecting material and also on the geometry: polarization is a function of the phase angle. The *degree of polarization P* is defined as

$$
P = \frac{F_\perp - F_\parallel}{F_\perp + F_\parallel} ,
\tag{7.47}
$$

where $F_\perp$ is the flux density of radiation, perpendicular to a fixed plane, and $F_\parallel$ is the flux density parallel to the plane. In solar system studies, polarization is usually referred to the plane defined by the Earth, the Sun, and the object. According to (7.47), P can be positive or negative; thus the terms "positive" and "negative" polarization are used.

The degree of polarization as a function of the phase angle depends on the surface structure and the atmosphere. The degree of polarization of the light reflected by the surface of an atmosphereless body is positive when the phase angle is greater than about 20°. Closer to opposition, polarization is negative. When light is reflected from an atmosphere, the degree of polarization as a function of the phase angle is more complicated. By combining observations with a theory of radiative transfer, one can compute atmosphere models. For example, the composition of Venus' atmosphere could be studied before any probes were sent to the planet.

Planetary Spectroscopy. The photometric and polarimetric observations discussed above were monochromatic. However, the studies of the atmosphere of Venus also used spectral information. Broadband UBV photometry or polarimetry is the simplest example of spectrophotometry (spectropolarimetry). The term spectrophotometry usually means observations made with several narrowband filters. Naturally, solar system objects are also observed by means of "classical" spectroscopy.

Spectrophotometry and polarimetry give information at discrete wavelengths only. In practise, the number of points of the spectrum (or the number of filters available)

is often limited to 20–30. This means that no details can be seen in the spectra. On the other hand, in ordinary spectroscopy, the limiting magnitude is smaller, although the situation is rapidly improving with the new generation detectors, such as the CCD camera.

The spectrum observed is the spectrum of the Sun. Generally, the planetary contribution is relatively small, and these differences can be seen when the solar spectrum is subtracted. The Uranian spectrum is a typical example (Fig. 7.20). There are strong absorption bands in the near-infrared. Laboratory measurements have shown that these are due to methane. A portion of the red light is also absorbed, causing the greenish colour of the planet. The general techniques of spectral observations are discussed in the context of stellar spectroscopy in Chap. 8.

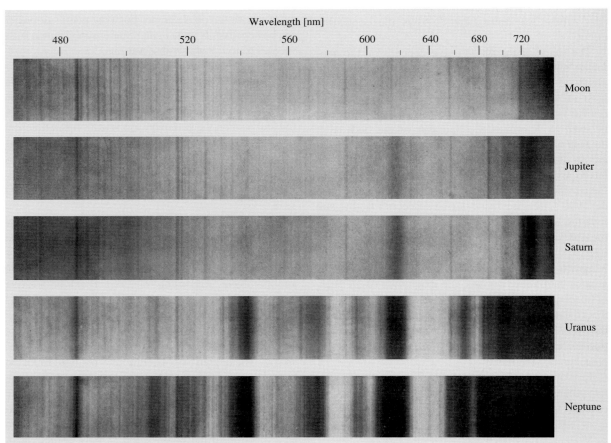

Fig. 7.20. Spectra of the Moon and the giant planets. Strong absorption bands can be seen in the spectra of Uranus and Neptune. (Lowell Observatory Bulletin **42** (1909))

7.9 Thermal Radiation of the Planets

Thermal radiation of the solar system bodies depends on the albedo and the distance from the Sun, i.e. on the amount of absorbed radiation. Internal heat is important in Jupiter and Saturn, but we may neglect it at this point.

By using the Stefan-Boltzmann law, the flux on the surface of the Sun can be expressed as

$$L = 4\pi R_\odot^2 \sigma T_\odot^4 .$$

If the Bond albedo of the body is A, the fraction of the radiation absorbed by the planet is $(1 - A)$. This is later emitted as heat. If the body is at a distance r from the Sun, the absorbed flux is

$$L_{abs} = \frac{R_\odot^2 \sigma T_\odot^4 \pi R^2}{r^2} (1 - A) . \qquad (7.48)$$

There are good reasons to assume that the body is in *thermal equilibrium*, i.e. the emitted and the absorbed fluxes are equal. If not, the body will warm up or cool down until equilibrium is reached.

Let us first assume that the body is rotating slowly. The dark side has had time to cool down, and the thermal radiation is emitted mainly from one hemisphere. The flux emitted is

$$L_{em} = 2\pi R^2 \sigma T^4 , \qquad (7.49)$$

where T is the temperature of the body and $2\pi R^2$ is the area of one hemisphere. In thermal equilibrium, (7.48) and (7.49) are equal:

$$\frac{R_\odot^2 T_\odot^4}{r^2} (1 - A) = 2T^4 ,$$

whence

$$T = T_\odot \left(\frac{1 - A}{2} \right)^{1/4} \left(\frac{R_\odot}{r} \right)^{1/2} . \qquad (7.50)$$

A body rotating quickly emits an approximately equal flux from all parts of its surface. The emitted flux is then

$$L_{em} = 4\pi R^2 \sigma T^4$$

and the temperature

$$T = T_\odot \left(\frac{1 - A}{4} \right)^{1/4} \left(\frac{R_\odot}{r} \right)^{1/2} . \qquad (7.51)$$

The theoretical temperatures obtained above are not valid for most of the major planets. The main "culprits" responsible here are the atmosphere and the internal heat. Measured and theoretical temperatures of some major planets are compared in Table 7.3. Venus is an extreme example of the disagreement between theoretical and actual figures. The reason is the *greenhouse effect*: radiation is allowed to enter, but not to exit. The same effect is at work in the Earth's atmosphere. Without the greenhouse effect, the mean temperature could be well below the freezing point and the whole Earth would be ice-covered.

7.10 Mercury

Mercury is the innermost planet of the solar system. Its diameter is 4800 km and its mean distance from the Sun 0.39 AU. The eccentricity of the orbit is 0.21, which means that the distance varies between 0.31 and 0.47 AU. Because of the high eccentricity, the surface temperature of the subsolar point varies substantially: at the perihelion, the temperature is about 700 K; at the aphelion, it is 100 K lower. Temperature variations on Mercury are the most extreme in the solar system because in the night side the temperature drops below 100 K.

Table 7.3. Theoretical and observed temperatures of some planets

	Albedo	Distance from the Sun [AU]	Theoretical temperature [K] (7.50)	(7.51)	Observed maximum temperature [K]
Mercury	0.06	0.39	525	440	700
Venus	0.76	0.72	270	230	750
Earth	0.36	1.00	290	250	310
Mars	0.16	1.52	260	215	290
Jupiter	0.73	5.20	110	90	130

The precession of the perihelion of Mercury is more than 0.15° per century. When the Newtonian perturbations are subtracted, there remains an excess of 43″. This is fully explained by the general theory of relativity. The explanation of the perihelion precession was one of the first tests of the general theory of relativity.

Mercury is always found in the vicinity of the Sun; its maximum elongation is only 28°. Observations are difficult because Mercury is always seen in a bright sky and close to the horizon. Moreover, when closest to the Earth in the inferior conjunction, the dark side of the planet is toward us.

The first maps of Mercury were drawn at the end of the 19th century but the reality of the details was not confirmed. As late as in the beginning of the 1960's, it was believed that Mercury always turns the same side toward the Sun. However, measurements of the thermal radio emission showed that the temperature of the night side is too high, about 100 K, instead of almost absolute zero. Finally, the rotation period was established by radar. One revolution around the Sun takes 88 days. The rotation period is two-thirds of this, 59 days. This means that every second time the planet is in, say, perihelion, the same hemisphere faces the Sun (Fig. 7.21). This

kind of spin–orbit coupling can result from tidal forces exerted by a central body on an object moving in a fairly eccentric orbit.

Re-examination of old observations revealed why Mercury had been presumed to rotate synchronously. Owing to its geometry, Mercury is easiest to observe in spring and autumn. In six months, Mercury orbits twice around the Sun, rotating exactly three times around its own axis. Consequently, during observations, the same side was always facing the Sun! The details visible on the surface are very obscure and the few exceptional observations were interpreted as observational errors.

The best (and thus far unique) data were received in 1974 and 1975, when the US space craft *Mariner 10* passed Mercury three times. The orbital period of Mariner 10 around the Sun was exactly twice the period of Mercury. The two-thirds-factor meant that the same side of the planet was illuminated during every fly-by! The other side is still unknown.

The Mariner 10 data revealed a moon-like landscape (Fig. 7.22). Mercury's surface is marked by craters and larger circular areas, caused by impacts of minor planets. The craters are 3000–4000 million years old, indicating that the surface is old and undisturbed by continental drift or volcanic eruptions. Most of Mercury's surface is covered by old and heavily cratered plains but there are some areas that are less saturated and the craters are less than 15 kilometres in diameter. These areas were probably formed as lava flows buried the older terrain.

The largest lava-filled circular area is the 1300 km wide *Caloris Basin*. The shock wave produced by the Caloris impact was focused to the antipodal point, breaking the crust into complex blocks in a large area, the diameter of which is about 100 km. There are also faults that were possibly produced by compression of the crust. The volume change probably was due to the cooling of the planet.

Mercury's relatively small size and proximity to the Sun, resulting in low gravity and high temperature, are the reasons for its lack of atmosphere. There is a layer made up of atoms blasted off the surface by the solar wind. The tenuous "atmosphere" is composed mainly of oxygen, sodium, and helium. The atoms quickly escape into space and are constantly replenished.

Due to the absence of an atmosphere, the temperature on Mercury drops very rapidly after sunset. The rota-

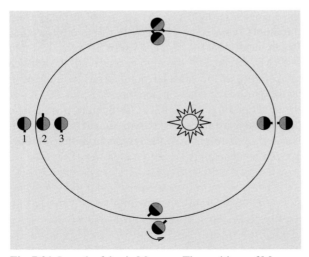

Fig. 7.21. Length of day in Mercury. The positions of Mercury during the first revolution are shown outside the ellipse. Upon returning to the aphelion, the planet has turned 540° (1½ revolutions). After two full cycles the planet has rotated three times around its axis and the same side points toward the Sun. The length of the day is 176 d, longer than on any other planet

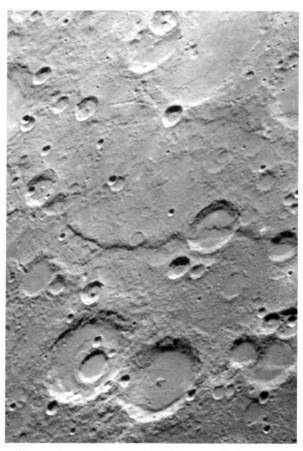

Fig. 7.22. (*Left*) A mosaic picture of Mercury. (NASA) (*Right*) Surface details on Mercury. One of the most prominent scarps photographed by Mariner 10 during it's first encounter with Mercury in 1974. The scarp is about 350 kilometres long and transects two craters 35 and 55 kilometres in diameter. It is up to 2 km high in some places and it appears to be a fault produced by compression of the crust. (NASA/JPL/Northwestern University)

tional axis is almost perpendicular to the orbital plane; therefore it is possible that, close to the poles, there are areas where the temperature is permanently below the freezing point. Radar echos from the surface of Mercury show several anomalously reflective and highly depolarized features at the north and south poles. Some of these areas can be addressed to the craters, the bottoms of which are permanently in shadow. One candidate of the radar-bright features is water ice that has survived in the permanent shadow.

The only relevant data concerning the interior of Mercury were obtained during the Mariner 10 fly-by when the gravity field was measured. Since Mercury has no satellites, the mass (and mass distribution) and density could not be determined before the force exerted by the gravitational field on a spacecraft was measured.

It has been said that Mercury looks like the Moon from the outside but is terrestrial from the inside. According to theoretical models, the internal structure is similar to that of the Earth but the core is substantially larger. The density of the planet is about the same as that of the Earth, indicating that the size of the Fe–Ni core is roughly about 75% of the planet's radius.

Due to the vicinity of the Sun, the temperature of the primeval nebula at the distance of Mercury was quite high during planetary formation. Thus the relative abundances of the volatile elements are smaller than on any other terrestrial planet.

Mercury has a weak magnetic field, about 1% as strong as that of the Earth. The presence of the magnetic field is unexpected because Mercury is much smaller than the Earth and it rotates slowly. According to the dynamo theory, a magnetic field is generated by flows in a liquid, electrically conducting core. The magnetic field cannot be a remnant from ancient times, since the internal temperature of the planet must have exceeded the critical Curie point. Therefore, it must be assumed that a part of the core is molten.

7.11 Venus

Venus is the brightest object in the sky, after the Sun and the Moon. Like Mercury, Venus can be seen only in the morning or in the evening sky. (It is sometimes possible to see Venus even if the Sun is above the horizon, if its exact position is known.) In antiquity, Venus was thought to be two different planets, *Hesperos* and *Phosphorus*, evening star and morning star.

The maximum elongation of Venus is about 47°. Venus is a remarkable object when shining in the dark sky at its brightest, 35 days before or after the inferior conjunction, when one-third of the surface is seen lit (Fig. 7.23). At the inferior conjunction, the Earth–Venus distance is only 42 million km. The diameter of Venus is about 12,000 km, which means that the apparent diameter can be as large as one arc minute. Under favourable conditions it is even possible to see the shape of the crescent Venus with binoculars. At the superior conjunction, the apparent diameter is only 10 arc seconds.

Venus is covered by clouds. Its surface is nowhere visible; only featureless yellowish cloud tops can be

Fig. 7.23. The phases of Venus were discovered by Galileo Galilei in 1610. This drawing illustrates how the apparent size of Venus changes with phase. The planet is far behind the Sun when the illuminated side faces the Earth

seen (Fig. 7.24). The rotation period was long unknown, and the measured 4-day period was the rotation time of the clouds. Finally, in 1962, radar measurements revealed that the rotation period is 243 days in a retrograde direction, i.e. opposite to the rotation of other planets. The axis of rotation is almost perpendicular to the orbital plane; the inclination is 177°.

The temperature at the cloud tops is about 250 K. Because the Bond albedo is as high as 75%, the surface temperature was believed to be moderate, even suitable for life. Opinions changed dramatically when thermal radio emission was measured at the end of the 1950's. This emission originates on the surface of the planet and can penetrate the clouds. The surface temperature turned out to be 750 K, well above the melting point of lead. The reason for this is the greenhouse effect. The outgoing infrared radiation is blocked by atmospheric carbon dioxide, the main component of the atmosphere.

The chemical composition of the Venusian atmosphere was known prior to the space age. Spectroscopic observations revealed CO_2, and some clues to the cloud composition were obtained from polarimetric observations. The famous French planetary astronomer *Bernard Lyot* made polarimetric observations in the 1920's, but not until decades later was it realised that his observations could be explained by assuming that light was scattered by liquid spherical particles whose index of refraction is 1.44. This is significantly higher than the index of refraction of water, 1.33. Moreover, water is not liquid at that temperature. A good candidate was sulphuric acid H_2SO_4. Later, spacecraft confirmed this interpretation.

Venus' atmosphere is very dry: the amount of water vapour present is only 1/1,000,000 of that in the Earth's atmosphere. One possible explanation is that, due to solar UV radiation, the water has dissociated to hydrogen and oxygen in the upper layers of the atmosphere, the former escaping into interplanetary space.

About 1% of the incident light reaches the surface of Venus; this light is deep red after travelling through clouds and the thick atmosphere. Most of the incident light, about 75%, is reflected back from the upper layers of clouds. The absorbed light is emitted back in infrared. The carbon dioxide atmosphere very effectively prevents the infrared radiation from escaping, and the temperature had not reached the equilibrium until at 750 K.

Fig. 7.24. *Left:* Venus in visible light imaged by the Galileo orbiter in February 1990. The cloud features are caused by winds that blow from east to west at about 100 m/s. *Right:* The northern hemisphere of Venus in a computer-generated picture of the radar observations. The north pole is at the centre of the image of the Magellan synthetic aperture radar mosaic. (NASA/JPL)

The pressure of the atmosphere at the surface is 90 atm. The visibility is several kilometres, and even in the clouds, a few hundred metres. The densest clouds are at a height of 50 km, but their thickness is only 2–3 km. Above this, there are haze-like layers which form the visible "surface" of the planet. The uppermost clouds move rapidly; they rotate around the planet in about 4 days, pushed by strong winds powered by the Sun. The sulphuric acid droplets do not rain on the Venusian surface but they evaporate in the lower atmosphere before reaching the surface.

Mariner 2 (1962) was the first spacecraft to encounter the planet. Five years later, the Soviet Venera 4 sent the first data from below the clouds, and the first pictures of the surface were sent by Venera 9 and 10 in 1975. The first radar map was completed in 1980, after 18 months of mapping by the US Pioneer Venus 1. The best and the most complete maps (about 98% of the planet's surface) were made using the synthetic aperture radar observations of the Magellan spacecraft in 1990–1994. The resolution of the maps is as high as 100 m and the elevation was measured with a resolution of 30 metres.

Radar mapping revealed canyons, mountains, craters, volcanoes and other volcanic formations (Fig. 7.25). The surface of Venus is covered by about 20% of lowland plains, 70% of gently rolling uplands and lava flows, and 10% of highlands. There are only two major highland areas. The largest continent, *Aphrodite Terra*, is close to the equator of Venus; its size is similar to South America. Another large continent at the latitude 70° N is called *Ishtar Terra*, where the highest mountain on Venus, the 12 km high *Maxwell Montes* is situated. (IAU has decided that the Venusian nomenclature has to be feminine. Maxwell Montes, after the famous physicist James Clerk Maxwell, is an exception.)

Unlike the Earth, volcanic features are quite evenly distributed all over the surface of Venus. There is no evidence of massive tectonic movement although local deformations can exist. Almost all volcanism on Venus seems to involve fluid lava flows without any explosive eruptions. Due to the high air pressure, Venusian lavas need a much higher gas content than the Earth lavas to erupt explosively. The main gas driving lava explosions on the Earth is water, which does not exist on Venus.

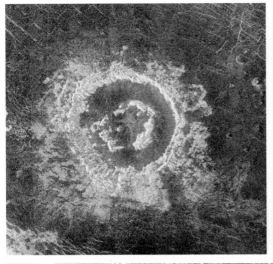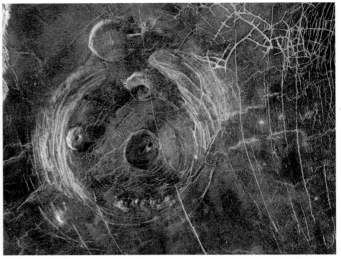

Fig. 7.25. Surface features of Venus. (*Top left*): A Magellan image of a 50 km peak-ring crater Barton at 27.4° N and 337.5° E. (*Top right*): A Magellan radar image of a region 300 km across, located in a vast plain to the south of Aphrodite Terra. The large circular structure near the centre of the image is a corona, approximately 200 km in diameter. North of the corona is a 35 km flat-topped volcanic construct known as a pancake dome. Complex fracture patterns like in the upper right of the image are often observed in association with coronas and various volcanic features. (NASA/JPL). (*Bottom*): The surface of Venus photographed by the Venera 14 lander in March 1982

Venus has more volcanoes than any other planet in the solar system. Over 1500 major volcanoes or volcanic features are known, and there may even be one million smaller ones. Most are shield volcanoes, but there are also many complex features. None are known to be active at present, although large variations of sulphur dioxide in the atmosphere may indicate that some volcanoes are active.

Flat-topped volcanic constructs known as *pancake domes* are probably formed by the eruption of an extremely viscous lava. A *corona* is a circular trench surrounding an elevated plain, the diameter of which can be as big as several hundreds of kilometres. They are possibly examples of local hot spots, mantle upwellings that have expanded and formed bulges. When the flow has stopped, the bulge has sunk and formed a set of ring mountains.

In other places fluid lava flows have produced long, sinuous channels extending for hundreds of kilometres.

Most of the Venusian *impact craters* are undeformed. This indicates that the Venusian surface must be young because erosion, volcanism and tectonic forces should affect the craters, too. Resurfacing processes may frequently cover the old craters, and all craters visible are therefore young, presumably less than 500 million years. There are no impact crates smaller than about 1.5–2 km because smaller meteoroids are burned in the thick atmosphere.

The Earth and Venus are almost equal in size, and their interiors are assumed to be similar. Venus has an iron core about 3000 km in radius and a molten rocky mantle covering the majority of the planet. Probably due to its slow rotation, however, Venus has no magnetic field. The analyses made by the Venera landers have shown that the surface material of Venus is similar to terrestrial granite and basalt (Fig. 7.25).

Venus has no satellites.

7.12 The Earth and the Moon

The third planet from the Sun, the *Earth*, and its satellite, the *Moon*, form almost a double planet. The relative size of the Moon is larger than that of any other satellite, excluding the moon of Pluto. Usually satellites are much smaller than their parent planets.

The Earth is a unique body, since a considerable amount of free water is found on its surface. This is possible only because the temperature is above the freezing point and below the boiling point of water and the atmosphere is thick enough. The Earth is also the only planet where life is known to exist. (Whether it is intelligent or not is yet to be resolved. . .). The moderate temperature and the water are essential for terrestrial life, although some life forms can be found in extreme conditions.

The diameter of the Earth is 12,000 km. At the centre, there is an iron–nickel core where the temperature is 5000 K, the pressure 3×10^{11} N m^{-2} and the density 12,000 kg m^{-3} (Fig. 7.26).

The core is divided into two layers, *inner* and *outer core*. The inner core, below 5150 km comprises only of 1.7% of the mass of the Earth. It is solid because of high pressure. The nonexistence of the seismic transverse S waves below a depth of 2890 km indicates that the outer core is molten. However, the speed of the longitudinal P waves change rapidly at a depth of 5150 km showing an obvious phase transition. It has been discovered that the solid inner core rotates with respect to the outer core and mantle.

The outer core comprises about 31% of the mass of the Earth. It is a hot, electrically conducting layer of liquid Fe–Ni where the convective motions take place. There are strong currents in the conductive layer that are responsible for the magnetic field.

Between the outer core and the lower mantle there is a 200 km thick transition layer. Although this D'' *layer* is often included as a part of the lower mantle, seismic

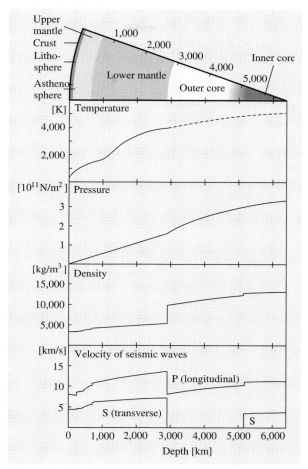

Fig. 7.26. Internal structure of the Earth. The speed of the seismic waves, density, pressure, and temperature are shown as a function of depth

discontinuities suggest that it might differ chemically from the lower mantle.

A silicate *mantle* extends from 2890 km upward up to a depth of few tens of kilometres. The part below 650 km is often identified as the *lower mantle*. It contains about 49% of the mass and is composed mainly of silicon, magnesium, and oxygen but some iron, calcium, and aluminium may also exist. The major minerals are olivine $(Mg, Fe)_2SiO_4$ and pyroxene $(Mg, Fe)SiO_3$. Under pressure the material behaves like a viscous liquid or an amorphous medium, resulting in slow vertical flows.

Between the lower and upper mantle there is a 250 km thick *transition region* or *mesosphere*. It is the source of basaltic magmas and is rich in calcium and aluminium. The upper mantle, between some tens of kilometres down to 400 km contains about 10% of the mass. Part of the upper mantle, called the *asthenosphere*, might be partially molten.

A thin *crust* floats on the mantle. The thickness of the crust is only 10–70 km; it is thickest below high mountain ranges such as the Himalayas and thinnest below the mid-ocean basins. The seismic discontinuity showing the border between the crust and mantle was discovered in 1909 by the Croatian scientist *Andrija Mohorovičić*, and it is now known as the *Moho discontinuity*.

The basaltic *oceanic crust* is very young, mostly less than 100 million years and nowhere more than 200 Ma. It is made through tectonic activity at the *mid-ocean ridges*. The *continental crust* is mainly composed of crystalline rocks that are dominated by quartz (SiO_2) and feldspars (metal-poor silicates). Because the continental crust is lighter than the oceanic crust (average densities are about 2700 kg m^{-3} and 3000 kg m^{-3}, respectively), the continents are floating on top of other layers, and currently they are neither created nor destroyed.

The *lithosphere* is the rigid outer part of the Earth (crust and the topmost part of the upper mantle). Below that is the partially molten asthenosphere where the damping of seismic waves is stronger than in the rigid lithosphere.

The lithosphere is not a single rigid and seamless layer; instead it is divided into more than 20 individual plates. The *plate tectonics* ("continental drift") is powered by the motion of the material in the mantle. New material is flowing up at the mid-ocean ridges, pushing the tectonic plates apart. New oceanic crust is generated at the rate of 17 km^3 per year. The Earth is the only planet that shows any large-scale tectonic activity. The history of the motion can be studied by using e. g. the paleomagnetic data of magnetic orientation of crystallised rocks.

At the end of the Precambrian era, about 700 million years ago, more than half of the continents were together forming the continent known as *Gondwana*, containing Africa, South America, Australia and Antarctica. About 350 million years ago Gondwana was on the South Pole but it moved toward the equator before the final breakup. Mutual collisions formed new mountains and finally in the beginning of the Mesozoic era, about 200 million years ago, all the continents were joined into one supercontinent, *Pangaea*.

Quite soon the flow pattern in the mantle changed and the Pangaea broke up. The Atlantic Ocean is still growing and new material is flowing up at the *mid-Atlantic ridge*. North America is drifting away from Europe at the rate of a few centimetres per year (your fingernails are growing at the same speed). At the same time, parts of the Pacific oceanic plate are disappearing below other plates. When an oceanic crust is pushed below a continental crust, a zone of active volcanoes is created. The earthquakes in the *subduction zones* can even originate 600 km below the surface. In the mid-ocean ridges, the depth is only tens of kilometres (Fig. 7.27).

Mountains are formed when two plates collide. The push of the African plate toward the Eurasian plate formed the Alps about 45 million years ago. The collision of the Indian plate created the Himalayas some 40 million years ago, and they are still growing.

Most of the surface is covered with water which condensed from the water vapour released in volcanic eruptions. The primordial atmosphere of the Earth was very different from the modern one; there was, for example, no oxygen. When organic chemical processes started in the oceans more than 2×10^9 years ago, the amount of oxygen rapidly increased (and was poison to the first forms of life!). The original carbon dioxide is now mainly concentrated in carbonate rocks, such as limestone, and the methane was dissociated by solar UV radiation.

The Earth's main atmospheric constituents are nitrogen (77% by volume) and oxygen (21%). Other gases, such as argon, carbon dioxide, and water vapour are present in minor amounts. The chemical composition is unchanged in the lower part of the atmosphere, called the *troposphere*. Most of the climatic phenomena occur in the troposphere, which reaches up to 8–10 km. The height of the layer is variable, being lowest at the poles, and highest at the equator, where it can extend up to 18 km.

The layer above the troposphere is the *stratosphere*, extending up to 60 km. The boundary between the troposphere and the stratosphere is called the *tropopause*. In the troposphere, the temperature decreases 5–7 K/km,

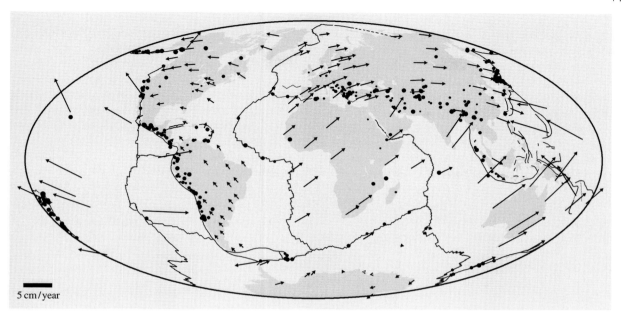

Fig. 7.27. The tectonic plates. The dots on the map indicate the location of earthquakes with magnitudes greater than 5 in the years 1980–1989. Arrows show the velocities observed with permanent GPS (Global Positioning System) tracking stations. The velocity scale is shown at lower left

5 cm/year

but in the stratosphere it begins to rise, due to the absorption of solar radiation by carbon dioxide, water vapour and ozone. The ozone layer, which shields the Earth from the solar UV radiation, is at a height of 20–25 km.

A total of 99% of air is in the troposphere and stratosphere. The stratopause at a height of 50–60 km separates the stratosphere from the *mesosphere*.

The mesosphere extends up to 85 km. In this layer, the temperature decreases again, reaching the minimum of about −90 °C at the height of 80–90 km in the *mesopause*. Chemicals in the mesosphere are mostly in an excited state, as they absorb energy from the Sun.

Above the mesopause is the *thermosphere* that extends up to 500 kilometres. The temperatures increases with altitude and can be above 1200 °C at the height of 500 km. The gas is in the form of a fully ionised plasma. Therefore, the layer above the mesopause is also called the *ionosphere*.

The density of air below a height of 150 km is high enough to cause colliding meteoroids to burn into ashes due to friction. It also plays an important role in radio communications, since radio waves are reflected by the ionosphere. Auroras are phenomena of the upper part of the ionosphere.

The thermosphere goes over into the *exosphere* at about 500 km. There the air pressure is much lower than in the best laboratory vacuums.

The magnetic field of the Earth is generated by flows in its core. The field is almost a dipole but there are considerable local and temporal variations. The mean field strength close to the equator is 3.1×10^{-5} Tesla (0.31 Gauss). The dipole is tilted 11° with respect to the Earth's axis, but the direction gradually changes with time. Moreover, the magnetic north and south poles have exchanged places several times during the past million years. More details are explained in Sect. 7.6 and in Figs. 7.15, 7.16, 7.17, and in Table 7.2.

The Moon. Our nearest neighbour in space is the *Moon*. Dark and light areas are visible even with the naked eye. For historical reasons, the former are called seas or *maria* (from Latin, *mare*, sea, pl. maria). The lighter areas are uplands but the maria have nothing in common with terrestrial seas, since there is no water on the Moon.

Fig. 7.28. Hurricane Elena in the Gulf of Mexico, with wind speeds in excess of 170 kilometres per hour was photographed from the space shuttle Discovery on September 1, 1985. Compare this to the Great Red Spot of Jupiter in Fig. 7.42. (NASA)

Numerous craters, all meteorite impacts, can be seen, even with binoculars or a small telescope (Fig. 7.29). The lack of atmosphere, volcanism, and tectonic activity help to preserve these formations.

The Moon is the best-known body after the Earth. The first man landed on the Moon in 1969 during the *Apollo 11* flight. A total of over 2000 samples, weighing 382 kg, were collected during the six Apollo flights

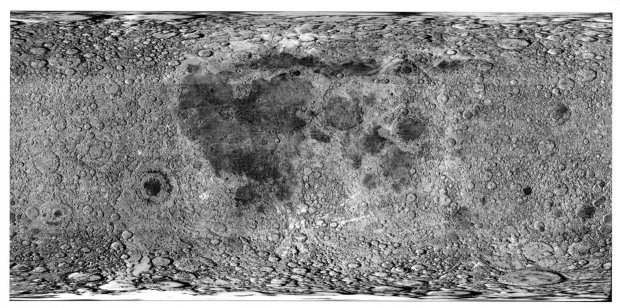

Fig. 7.29. A map of the Lunar surface, composed of images taken by the Clementine space probe in 1994. Note the large areas of maria in the Lunar near side, at the centre of the figure, as compared to the almost complete absence of the maria at the Lunar far side. (US Naval Observatory)

(Fig. 7.30). Moreover, the unmanned Soviet Luna spacecraft collected and returned about 310 grams of Lunar soil. Instruments placed on the Moon by the Apollo astronauts operated as long as eight years. These included seismometers, which detected moonquakes and meteorite impacts, and passive laser reflectors which made exact Earth–Moon distance measurements possible. The reflectors are still used for *Lunar laser ranging* (LLR) measurements.

Seismometric and gravimetric measurements have supplied basic information on the internal structure of the Moon. Moonquakes take place at a depth of 800–1000 km, considerably deeper than earthquakes, and they are also much weaker than on the Earth. Most of the quakes occur at the boundary of the solid mantle, the *lithosphere*, and the *asthenosphere* (Fig. 7.31). The transversal S waves cannot penetrate the asthenosphere, indicating that it is at least partially molten. Tidal forces may generate at least some of the moonquakes because most of them occur close to perigee or apogee.

Lunar orbiters have observed local mass concentrations, *mascons*, beneath the maria. These are large basaltic blocks, formed after the huge impacts which produced the maria. The craters were filled by lava flows during the next billion years or so in several phases. This can be seen, e.g. in the area of *Mare Imbrium*. Large maria were formed about 4×10^9 years ago when meteorite bombardment was much heavier than today. The last 3×10^9 years have been quite peaceful, without any major events.

The centre of mass is not at the geometric centre of the Moon but about 2.5 km away due to the 20–30 km thick basaltic plates below the large maria. Moreover, the thickness of the crust varies, being the thickest at the far side of the Moon, about 100 km. On the near side the thickness of the crust is about 60 km.

The mean density of the Moon is 3400 kg m^{-3}, which is comparable to that of basaltic lavas on the Earth. The Moon is covered with a layer of soil with scattered rocks, *regolith*. It consists of the debris blasted out by meteorite impacts. The original surface is nowhere visible. The thickness of the regolith is estimated to be at least tens of metres. A special type of rock, *breccia*, which is a fragment of different rocks compacted and welded together by meteor impacts, is found everywhere on the Moon.

The maria are mostly composed of dark basalts, which form from rapid cooling of massive lava flows. The highlands are largely composed of *anorthosite*, an

Fig. 7.30. Apollo 17 astronaut Harrison Schmitt on the Moon in 1972. (NASA)

igneous rock that forms when lava cools more slowly than in the case of basalts. This implies that the rocks of the maria and highlands cooled at different rates from the molten state and were formed under different conditions.

Data returned by the *Lunar Prospector* and *Clementine* spacecraft indicated that water ice is present at both the north and south lunar poles. Data indicates that there may be nearly pure water ice buried beneath the dry regolith. The ice is concentrated at the bottoms of deep valleys and craters that are in a permanent shadow where the temperature is below 100 K.

The Moon has no global magnetic field. Some of the rocks have remanent magnetism indicating a possible global magnetic field early in the Moon's history. Without the atmosphere and magnetic field, the solar wind can reach the Moon's surface directly. The ions from the solar wind have embedded in the regolith. Thus samples returned by the Apollo missions proved valuable in studies of the solar wind.

The origin of the Moon is still uncertain; it has, however, not been torn off from the Earth at the Pacific Ocean, as is sometimes believed. The Pacific is less than 200 million years old and formed as a result of

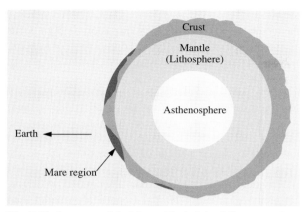

Fig. 7.31. Structure of the Moon. The height differences of the surface are strongly exaggerated

continental drift. Also, the chemical composition of the lunar soil is different from that of terrestrial material.

Recently it was suggested that the Moon was formed in the early stages of the formation of the Earth, when a lot of protoplanet embryos were orbiting the Sun. An off-axis collision of a Mars-size body resulted in ejection of a large amount of debris, a part of which then accreted to form the Moon. Differences in chemical compositions of the modern Earth and the Moon can be explained with the theory, as well as the orientation and

evolution of the Moon's orbit and the Earth's relatively fast spin rate.

* Atmospheric Phenomena

The best-known atmospheric phenomenon is the *rainbow*, which is due to the refraction of light from water droplets. The radius of the arc of the rainbow is about 41° and the width, 1.7°. The centre of the arc is opposite the Sun (or any other source of light). When the light is refracted inside a water droplet, it is divided into a spectrum, where the red colour is at the outer edge and blue, at the inner edge. Light can be reflected twice inside the droplet, resulting in a secondary rainbow outside the primary one. The colours of the secondary rainbow are in reversed order and its radius is 52°. A rainbow caused by the Moon is usually very weak and colourless, since the human eye is incapable of resolving colours of a dim object.

A *halo* results when the solar or lunar light is reflected from atmospheric ice crystals. The most common halo is a 22° arc or circle around the Sun or the Moon. Usually the halo is white, but occasionally even bright colours can be seen. Another common form is the side lobes which are at the same height as the Sun but at a distance of 22° from it. All other forms of halo are

Left: A typical halo; *Right*: Auroras (Photos P. Parviainen)

less common. The best "weather" for halos is when there are cirrostratus or cirrus clouds or an icy fog in the sky.

Noctilucent clouds are thin formations of cloud, at a height of approximately 80 km. The clouds contain particles, which are less than one micron in diameter, and become visible only when the Sun (which is below the horizon) illuminates the clouds. Most favourable conditions are at the northern latitudes during the summer nights when the Sun is only a few degrees below the horizon.

The night sky is never absolutely dark. One reason (in addition to light pollution) is the *airglow* or light emitted by excited atmospheric molecules. Most of the radiation is in the infrared domain, but e. g. the forbidden line of oxygen at 558 nm, has also been detected.

The same greenish oxygen line is clearly seen in *auroras*, which are formed at a height of 80–300 km. Auroras can be seen mainly from relatively high northern or southern latitudes because the Earth's magnetic field forces charged particles, coming from the Sun, close toward the magnetic poles. Alaska and northern Scandinavia are the best places to observe auroras. Occasionally, auroras can be seen as far south as 40°. They

are usually greenish or yellow-green, but red auroras have been observed, too. They most commonly appear as arcs, which are often dim and motionless, or as belts, which are more active and may contain rapidly varying vertical rays.

Meteors (also called shooting stars although they have nothing to do with stars) are small grains of sand, a few micrograms or grams in weight, which hit the Earth's atmosphere. Due to friction, the body heats up and starts to glow at a height of 100 km. Some 20–40 km lower, the whole grain has burnt to ashes. The duration of a typical meteor is less than a second. The brightest meteors are called *bolides* (magnitude smaller than about −2). Even larger particles may survive down to the Earth. Meteors are further discussed in Sect. 7.18.

7.13 Mars

Mars is the outermost of the terrestrial planets. Its diameter is only half of that of the Earth. Seen through a telescope, Mars seems to be a reddish disk with dark spots and white polar caps. The polar caps wax and wane with the Martian seasons, indicating that they are

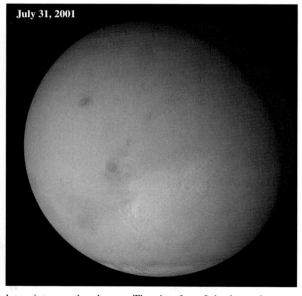

Fig. 7.32. Two pictures of Mars, taken by the Mars Global Surveyor in June and July 2001. The view from June (*left*) shows the Tharsis volcanic region, Valles Marineris and the late winter south polar cap. The view from July shows the same regions, but most of the details are hidden by dust storms and haze. (NASA/JPL/Malin Space Science Systems)

composed of ice. Darker areas were suspected to be vegetation. At the end of the 19th century, an Italian astronomer, *Giovanni Schiaparelli* claimed that there are canals on Mars.

In the United States, the famous planetary astronomer *Percival Lowell* studied the canals and even published books on the topic. Martians were also very popular in science fiction literature. Now the canals are known to be nonexistent, an optical illusion when the obscure details at the limit of visibility seem to form straight lines, canals. Finally, the first clear pictures by Mariner 4 in 1965 buried even the most optimistic hopes concerning life on Mars. Later spacecraft revealed more details of the planet.

Mars is a superior planet, which means that it is most easily observable when it is closest to the Earth, i.e. during opposition, when the planet is above the horizon all night long.

The rotation axis of Mars is tilted 25° to the ecliptic, about the same amount as the Earth's axis. A Martian day is only half an hour longer than a terrestrial day. Mars' orbit is significantly elliptical, resulting in temperature variations of about 30 °C at the subsolar point between the aphelion and perihelion. This has a major influence on the climate. Huge dust storms are occasionally seen on Mars (Fig. 7.32). Usually the storms begin when Mars is at the perihelion. Heating of the surface gives rise to large temperature differences that in turn cause strong winds. The wind-driven dust absorbs more heat and finally the whole planet is covered by a dust storm where the wind speeds exceed 100 m/s.

The atmosphere of Mars is mainly composed of carbon dioxide (95%). It contains only 2% nitrogen and 0.1–0.4% oxygen. The atmosphere is very dry: if all the moisture were condensed on the surface, the water layer would be thinner than 0.1 mm. Even the minor amount of water vapour is sufficient to occasionally form some thin clouds or haze.

The air pressure is only 5–8 mbar. A part of the atmosphere has escaped but it is probable that Mars never had a thick atmosphere. The primordial atmosphere of Mars was, however, somewhat similar to that of the Earth. Almost all of its carbon dioxide was used up to form carbonate rocks. Because there are no plate tectonics on Mars, the carbon dioxide was not recycled back into the atmosphere as on the Earth. Therefore, the

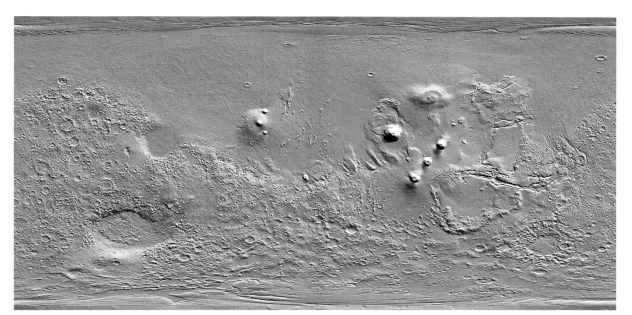

Fig. 7.33. A topographic shade map of Mars made from the Mars Global Surveyor data. The most prominent features are the large shield volcanoes in the northern hemisphere and the Valles Marineris canyon system that is more than 3000 km long and up to 8 km deep. (MOLA Science Team/NASA)

greenhouse effect on Mars is significantly smaller than on the Earth.

Craters were already found in the first pictures. The southern hemisphere is especially marked by craters, indicating that the original surface is still visible there. The largest impacts, *Hellas* and *Argyre* are about 2000 km in diameter. On the other hand, the northern hemisphere has an abundance of large lava basins and volcanoes (Fig. 7.33). The surface is younger than in the southern hemisphere. The largest volcano, *Olympus Mons* (Fig. 7.34), protrudes more than 20 km above the surrounding terrain. The diameter at the bottom is about 600 km.

There are no active volcanoes on Mars. The mare-like plains on Mars are of the same age as the Lunar maria, about 3×10^9 years old. Volcanism in the highland and mare-like plains stopped at that time, but the giant shield volcanoes are much younger, possibly $1-2 \times 10^9$ years. The youngest lava flows on Olympus Mons are possibly less than 100 million years old. Mars shows no sign of plate tectonics. It has no mountain chains, nor any global patterns of volcanism.

There are also several canyons, the largest of which is *Valles Marineris* (Fig. 7.33). Its length is 5000 km, width 200 km, and depth about 6 km. Compared with Valles Marineris, the Grand Canyon is merely a scratch on the surface.

Ancient riverbeds (Fig. 7.34), too small to be seen from the Earth, were also discovered by spacecraft. Rivers were probably formed soon after the formation of Mars itself, when there was a great deal of water and the atmospheric pressure and temperature were higher. At present, the temperature and air pressure on Mars are too low for free water to exist, although there have been speculations on warm weather cycles in the more recent history of the planet. The mean temperature is now below $-50\,°C$ and, on a warm summer day, the temperature can rise close to zero near the equator. Most

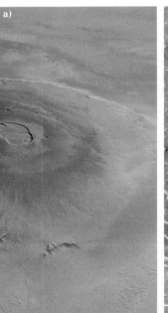
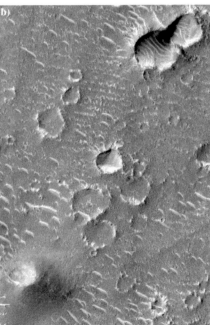
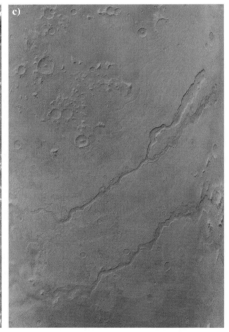

Fig. 7.34a–c. Volcanoes, impact craters and rivers. (**a**) Mars Global Surveyor wide-angle view of Olympus Mons in April 1998. (**b**) Small impact craters and sand dunes wuth a resolution of 1.5 m per pixel. The picture covers a 1.5 km wide portion of Isidis Planitia. (**c**) Three major valley systems east of the Hellas plains. These valleys have probably been formed by large outbursts of liquid water but the age of the valleys is unknown. The valleys are all roughly 1 km deep and 10–40 km wide. The picture covers an area approximately 800 km across. (Mars Global Surveyor, 2000) (NASA/JPL/Malin Space Science Systems)

Fig. 7.35. The 360 degree panorama was taken by the Mars Pathfinder Lander in 1997. The Sojourner rover is visible near the centre of the panorama, in front of the ramp. (NASA/JPL)

of the water is contained in kilometres deep permafrost below the surface and in the polar caps. The theory was confirmed in 2002, when the Mars Odyssey spacecraft detected a large supply of subsurface water ice of a wide area near the south pole. The ice is mixed into the soil a meter below the surface.

The *polar caps* are composed both of water and carbon dioxide ice. The northern cap is almost season-independent, extending down to latitude 70°. On the other hand, the southern cap, which reaches to the latitude $-60°$ in the southern winter, disappears almost totally during the summer. The southern cap consists mostly of CO_2 ice. The permanent parts are of ordinary water ice, since the temperature, $-73\,°C$, is too high for CO_2 ice. The water ice layers can be hundreds of metres thick.

The dark areas are not vegetation, but loose dust, moved around by strong winds. These winds raise the dust high into the atmosphere, colouring the Martian sky red. The Mars landers have revealed a reddish regolithic surface, scattered with boulders (Fig. 7.35). The red colour is caused mainly by iron oxide, rust; already

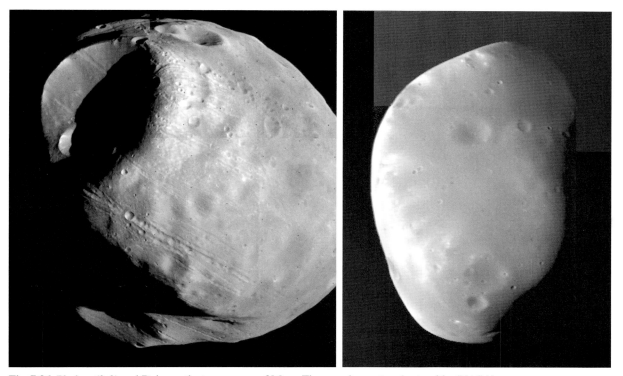

Fig. 7.36. Phobos (*left*) and Deimos, the two moons of Mars. They can be captured asteroids. (NASA)

in the 1950's, the existence of limonite ($2\,FeO_3\,3\,H_2O$) was deduced from polarization measurements. The on-site analysis showed that the soil consists of 13% iron and 21% silicon. The abundance of sulphur was found to be ten times that found on the Earth.

The interior of Mars is not well known. Most likely, Mars has a dense core about 1700 km in radius, a molten rocky mantle somewhat denser than the Earth's and a thin crust. Data from the Mars Global Surveyor indicates that the crust of Mars is about 80 km thick in the southern hemisphere but only about 35 km thick in the northern one. The relatively low density compared with other terrestrial planets indicates that its core probably contains a relatively large fraction of sulphur in addition to iron.

The Mars Global Surveyor confirmed in 1997 a weak magnetic field. It is probably a remnant of an earlier global field that has since disappeared. This has important implications for the structure of Mars' interior. There are no electric currents creating a magnetic field and therefore the core may be (at least partially) solid.

Three biological experiments of the Viking landers in 1976 searched for signs of life. No organic compounds were found – however, the biological tests did give some unexpected results. A closer look at the results indicated no life, but some uncommon chemical reactions.

Mars has two moons, *Phobos* and *Deimos* (Fig. 7.36). The size of Phobos is roughly 27 km × 21 km × 19 km, and the orbital period around Mars is only 7 h 39 min. In the Martian sky, Phobos rises in the west and sets in the east. Deimos is smaller. Its diameter is 15 km × 12 km × 11 km. There are craters on both moons. Polarimetric and photometric results show that they are composed of material resembling carbonaceous chondrite meteorites.

7.14 Asteroids

Asteroids, or minor planets, orbit the Sun mainly between Mars and Jupiter. Most of the asteroids are in the asteroid belt with distances of 2.2–3.3 AU from the

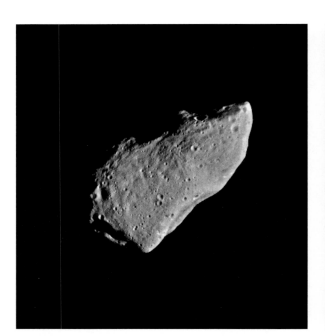

Fig. 7.37. *Left:* Asteroid (951) Gaspra was photographed by the Galileo spacecraft in October 1991. The illuminated part of the asteroid is about 16 × 12 km. The smallest craters in this view are about 300 m across. *Right:* A mosaic of asteroid

(433) Eros was taken by the NEAR spacecraft from a distance of 200 km. The crater on top is about 5 km in diameter. The NEAR spacecraft orbited Eros for one year and finally landed on it in 2001. (JPL/NASA)

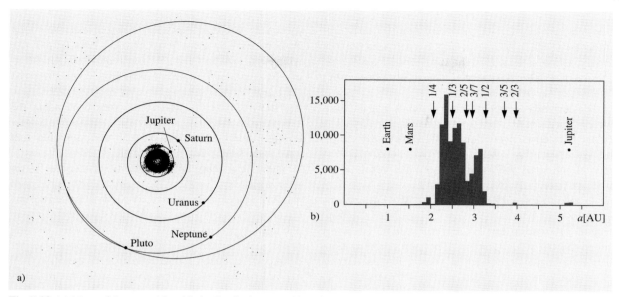

Fig. 7.38. (**a**) Most of the asteroids orbit the Sun in the asteroid belt between Mars and Jupiter. The figure shows the positions of about 96,000 catalogued asteroids on January 1, 2000, and the orbits and the positions of some major planets. The orbital elements of the asteroids are from the Lowell Observatory data base. (**b**) The total number of asteroids as a function of the distance from the Sun. Each bin corresponds to 0.1 AU. The empty areas, the Kirkwood gaps, are at those points, where the orbital period of an asteroid is in a simple ratio to the orbital period of Jupiter

Sun (Fig. 7.38). The most distant asteroids are beyond the orbit of Pluto, and there are a number of asteroids that come closer to the Sun than the Earth.

An asteroid observer needs a telescope, since even the brightest asteroids are too faint to be seen with the naked eye. Asteroids are points of light like a star, even if seen through a large telescope; only their slow motion against the stellar background reveals that they are members of the solar system. The rotation of an asteroid gives rise to a regular light variation. The amplitude of light variation is in most cases well below 1 magnitude and typical rotation periods range from 4 to 15 hours.

The first asteroid was discovered in 1801 by *Giuseppe Piazzi*, and at the end of year 2002 there were more than 40,000 numbered asteroids. The number of catalogued asteroids increases currently by thousands every year. It has been estimated that almost half a million asteroids larger than 1 km exist in the asteroid belt. Yet the total mass of the asteroids is less than 1/1000 of the mass of the Earth. The first and the biggest is 1 Ceres; its diameter is 1000 km (Fig. 7.39).

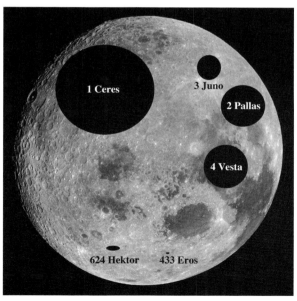

Fig. 7.39. Sizes of some asteroids compared with the Moon. (Moon image, NASA)

The centre of the asteroid belt is at a distance of approximately 2.8 AU, as predicted by the *Titius–Bode law* (Sect. 7.19). According to a formerly popular theory, asteroids were thought to be debris from the explosion of a planet. This theory, like catastrophe theories in general, has been abandoned.

The currently accepted theory assumes that asteroids were formed simultaneously with the major planets. The primeval asteroids were large chunks, most of them orbiting between the orbits of Mars and Jupiter. Due to mutual collisions and fragmentation, the present asteroids are debris of those primordial bodies which were never able to form a large planet. Some of the biggest asteroids may be those original bodies. The orbital elements of some asteroids are very similar. These are called the *Hirayama families*. They are probably remnants of a single, large body that was broken into a group of smaller asteroids. There are tens of identified Hirayama families, the largest ones including Hungarias, Floras, Eos, Themis, and Hildas (named after the main asteroid in the group).

The distribution of asteroids inside the asteroid belt is uneven (Fig. 7.38); they seem to avoid some areas known as the *Kirkwood gaps*. The most prominent void areas are at distances where the orbital period of an asteroid around the Sun (given by Kepler's third law) is in the ratio 1:3, 2:5, 3:7, or 1:2 to the orbital period of Jupiter. The motion of an asteroid orbiting in such a gap would be in resonance with Jupiter, and even small perturbations would tend to grow with time. The body would eventually be moved to another orbit. However, the resonance effects are not so simple: sometimes an orbit is "locked" to a resonance, e. g. the *Trojans* move along the same orbit as Jupiter (1:1 resonance), and the *Hilda group* is in the 2:3 resonance.

Some minor families of asteroids orbit the Sun outside the main belt. These include the above-mentioned Trojans, which orbit 60° behind and ahead of Jupiter. The Trojans, which are close to the special points L_4 and L_5 of the solution of the restricted three-body problem. At these Lagrangian points, a massless body can remain stationary with respect to the massive primaries (in this case, Jupiter and the Sun). In fact, the asteroids are oscillating around the stationary points, but the mean orbits can be shown to be stable against perturbations.

Another large family is the *Apollo-Amor asteroids*. The perihelia of Apollo and Amor are inside the Earth's orbit and between the orbits of the Earth and Mars, respectively. These asteroids are all small, less than 30 km in diameter. The most famous is 433 Eros (Fig. 7.37), which was used in the early 20th century for determining the length of the astronomical unit. When closest to the Earth, Eros is at a distance of only 20 million km and the distance can be directly measured using the trigonometric parallax. Some of the Apollo-Amor asteroids could be remnants of short-period comets that have lost all their volatile elements.

There is a marginal probability that some Earth-crossing asteroids will collide with the Earth. It has been estimated that, on the average, a collision of a large asteroid causing a global catastrophe may take place once in one million years. Collisions of smaller bodies, causing damage similar to a nuclear bomb may happen once per century. It has been estimated that there are 500–1000 *near-Earth asteroids* larger than one kilometre in diameter but possibly tens of thousands smaller objects. Programs have been started to detect and catalogue all near-Earth asteroids and to predict the probabilities of hazardous collisions.

Distant asteroids form the third large group outside the main asteroid belt. The first asteroid belonging to this group (2060) Chiron, was discovered in 1977. Chiron's aphelion is close to the orbit of Uranus and the perihelion is slightly inside the orbit of Saturn. Distant asteroids are very faint and thus difficult to find.

Already in the 1950's *Gerard Kuiper* suggested that comet-like debris from the formation of the solar system can exist beyond the orbit of Neptune as an additional source of comets to the more distant Oort cloud. Later, computer simulations of the solar system's formation showed that a disk of debris should form at the outer edge of the solar system. The disk is now known as the *Kuiper belt* (Fig. 7.40).

The first Trans-Neptunian asteroid (1992 QB1) was discovered in 1992, and in mid-2002 there were more than 500 known members. The total number of Kuiper belt objects larger than 100 km in diameter is estimated to be over 70,000. The Kuiper belt objects are remnants from the early accretion phases of the solar system. This may also be the origin of Pluto and Neptune's moon Triton. Several of the trans-Neptunian objects are in or near a 3:2 orbital period resonance with Neptune, the same resonance as Pluto.

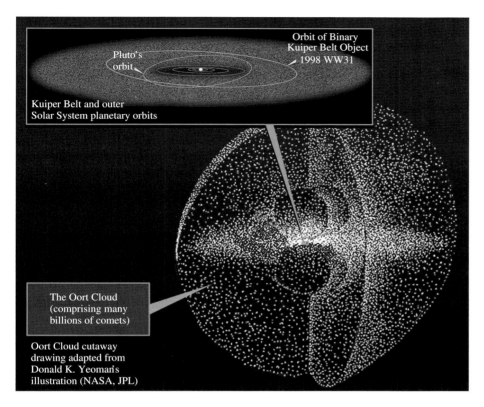

Orbit of Binary
Kuiper Belt Object
1998 WW31

Pluto's
orbit

Kuiper Belt and outer
Solar System planetary orbits

The Oort Cloud
(comprising many
billions of comets)

Oort Cloud cutaway
drawing adapted from
Donald K. Yeoman's
illustration (NASA, JPL)

Fig. 7.40. The Kuiper Belt is a disk-shaped cloud of distant icy bodies inside the halo of the Oort cloud. The short-period comets originate in the Kuiper belt, whereas a huge amount of icy bodies that form a source of long period comets resides in the Oort cloud (see Sect. 7.18). (JPL/NASA)

The Kuiper belt objects have a comet-like structure and composition. Therefore, they could be classified as comets as well, including (2060) Chiron.

The exact sizes of asteroids were long unknown. *Edward E. Barnard* of the Lick Observatory determined visually the diameters of (1) Ceres, (2) Vesta, (3) Juno, and (4) Pallas in the 1890's. Practically no other reliable results existed prior to the 1960's, when indirect methods applying photometry and spectroscopy were adopted. Moreover, several stellar occultations caused by asteroids have been observed since 1980's.

The first images of asteroids were obtained in the early 1990's. In 1991 the Galileo spacecraft passed asteroid (951) Gaspra, and in 1993 asteroid (243) Ida, on its long way to Jupiter (see Sect. 7.15). Finally, in 2001, the NEAR spacecraft landed on asteroid (433) Eros after orbiting it for one year.

The images of asteroids (Fig. 7.37) show irregular, crater-filled bodies with regolith and pulverised rock on their surface. Some asteroids may once have been two separate objects that merged into one. In 1992 asteroid (4179) Toutatis passed the Earth only by 4 million kilometres. Radar images revealed a two-body system, where the components were touching each other. Double asteroids may be quite common, and there exist light curves of some asteroids which have been interpreted as results of twin bodies. Another example of a twin asteroid is 243 Ida that has a "moon", a smaller body gravitationally bound to it.

The composition of main belt asteroids is similar to that of iron, stone and iron-stone meteorites. Most asteroids can be divided into three groups, according to their photometric and polarimetric properties. 95% of the classified asteroids belong to the types C and S types. Metal-rich M type asteroids are rarer.

About 75 percent of asteroids belong to the type C type. The *C asteroids* are dark due to radiation darkening (geometric albedo $p \approx 0.06$ or less), and they contain a considerable amount of carbon (mnemonic C for carbon). They resemble *stony meteorites*. The material is undifferentiated and thus they belong to the most primordial bodies of the solar system. The reflectivity of silicate-rich *S asteroids* is higher and their spectra are close to those of *stone-iron meteorites*. Their spectra show signs of silicates, such as olivine, e.g. fosterite Mg_2SiO_4 or fayalite Fe_2SiO_4. M type asteroids have more metals, mostly nickel and iron; they have undergone at least a partial differentiation.

The compositions of trans-Neptunian objects are different. They resemble comets, and most of the short period comets originate from this zone (see Sect. 7.18).

7.15 Jupiter

The realm of terrestrial planets ends at the asteroid belt. Outside this, the relative abundance of volatile elements is higher and the original composition of the solar nebula is still preserved in the giant planets. The first and largest is *Jupiter*. Its mass is 2.5 times the total mass of all other planets, almost $1/1000$ of the solar mass. The bulk of Jupiter is mainly hydrogen and helium. The relative abundance of these elements are approximately the same as in the Sun, and the density is of the same order of magnitude, namely $1330 \, kg \, m^{-3}$.

During oppositions, the angular diameter of Jupiter is as large as $50''$. The dark *belts* and lighter *zones* are visible even with a small telescope. These are cloud formations, parallel to the equator (Fig. 7.41). The most famous detail is the *Great Red Spot*, a huge cyclone, rotating counterclockwise once every six days. The spot was discovered by *Giovanni Cassini* in 1655; it has survived for centuries, but its true age is unknown (Fig. 7.42).

The rotation of Jupiter is rapid; one revolution takes 9 h 55 min 29.7 s. This is the period determined from the variation of the magnetic field, and it reflects the speed of Jupiter's interiors where the magnetic field is born. As might be expected, Jupiter does not behave like a rigid body. The rotation period of the clouds is about five minutes longer in the polar region than at the equator. Due to its rapid rotation, Jupiter is nonspherical; flattening is as large as $1/15$.

Fig. 7.41. A composed image of Jupiter taken by the Cassini spacecraft in December 2000. The resolution is 114 km/pixel. The dark dot is the shadow of the moon Europa. (NASA/JPL/University of Arizona)

Fig. 7.42. Jupiter's Great Red Spot and its surroundings with several smaller ovals as seen by Voyager 1 in 1979. Cloud details of 160 kilometres are visible. (NASA)

There is possibly an iron-nickel core in the centre of Jupiter. The mass of the core is probably equal to a few tens of Earth masses. The core is surrounded by a layer of metallic liquid hydrogen, where the temperature is over 10,000 K and the pressure, three million atm. Owing to this huge pressure, the hydrogen is dissociated into single atoms, a state unknown in ordinary laboratory environments. In this exotic state, hydrogen has many features typical of metals. This layer is electrically conductive, giving rise to a strong magnetic field. Closer to the surface where the pressure is lower, the hydrogen is present as normal molecular hydrogen, H_2. At the top there is a 1000 km thick atmosphere.

The atmospheric state and composition of Jupiter has been accurately measured by the spacecraft. *In situ* observations were obtained in 1995, when the probe of the Galileo spacecraft was dropped into Jupiter's atmosphere. It survived nearly an hour before crushing under the pressure, collecting the first direct measurements of Jupiter's atmosphere.

Belts and *zones* are stable cloud formations (Fig. 7.41). Their width and colour may vary with time, but the semi-regular pattern can be seen up to the latitude 50°. The colour of the polar areas is close to that of the belts. The belts are reddish or brownish, and the motion of the gas inside a belt is downward. The gas flows upward in the white zones. The clouds in the zones are slightly higher and have a lower temperature than those in the belts. Strong winds or jet streams blow along the zones and belts. The speed of the wind reaches 150 m/s at some places in the upper atmosphere. According to the measurements of the Galileo probe, the wind speeds in the lower cloud layers can reach up to 500 m/s. This indicates that the winds in deeper atmospheric layers are driven by the outflowing flux of the internal heat, not the solar heating.

The colour of the *Great Red Spot* (GRS) resembles the colour of the belts (Fig. 7.42). Sometimes it is almost colourless, but shows no signs of decrepitude. The GRS is 14,000 km wide and 30,000–40,000 km long. Some smaller red and white spots can also be observed on Jupiter, but their lifetime is generally much less than a few years.

The ratio of helium to hydrogen in the deep atmosphere is about the same as in the Sun. The results of the Galileo spacecraft gave considerably higher abundance than previous estimates. It means that there are no significant differentiation of helium, i. e. helium is not sinking to the interior of the planet as was expected according to the earlier results. Other compounds found in the atmosphere include methane, ethane and ammonia. The temperature in the cloud tops is about 130 K.

Jupiter radiates twice the amount of heat that it receives from the Sun. This heat is a remnant of the energy released in the gravitational contraction during the formation of the planet. Thus Jupiter is still gradually cooling. The internal heat is transferred outward

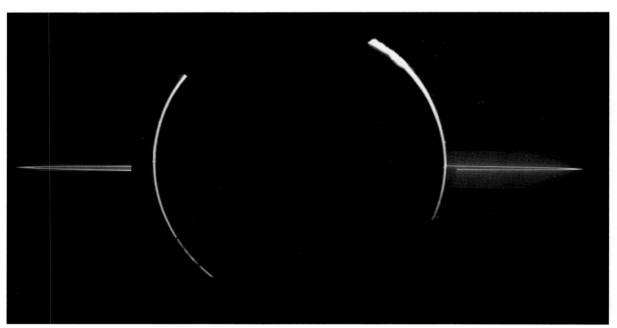

Fig. 7.43. Mosaic of Jupiter's ring system taken by the Galileo spacecraft when the spacecraft was in Jupiter's shadow looking back toward the Sun. Jupiter's ring system is composed of three parts: a thin outermost ring, a flat main ring, and an innermost doughnut-shaped halo. These rings are made up of dust-sized particles that originate from Io, or are blasted off from the nearby inner satellites by small impacts. (NASA/University of Arizona)

by convection; this gives rise to flows in the metallic hydrogen, causing the strong magnetic field (Fig. 7.44).

The ring of Jupiter (Fig. 7.43) was discovered in 1979. The innermost toroid-shaped halo is between 92,000–122,500 km from Jupiter's centre. It consists of dust falling from the main ring toward the planet. The main ring extends from the halo boundary out to about 128,940 km, just inside the orbit of the moon Adrastea. The ring particles are small, a few microns only, and they scatter light forward much more effectively than backward. Therefore, they were not discovered prior the Voyager flyby. A ring consisting of such small particles cannot be stable, and new material must enter the ring continuously. The most probable source is Io.

The two faint outermost rings are fairly uniform in nature. The inner of them extends from the orbit of Adrastea out to the orbit of Amalthea at 181,000 km. The fainter outermost ring extends out to Thebe's orbit at 221,000 km.

Jupiter's rings and moons exist within an intense radiation belt of Jupiter's magnetic field. The magnetosphere extends 3–7 million kilometres toward the Sun, depending on the strength of the solar wind. In the opposite direction it stretches to a distance of at least 750 million kilometres, behind Saturn's orbit.

Jupiter is an intense radio source. Its radio emission can be divided into three components, namely thermal millimetre and centimetre radiation, nonthermal decimetric radiation and burstal-decametric radiation. The nonthermal emission is most interesting; it is partly synchrotron radiation, generated by relativistic electrons in the Jovian magnetosphere. Its intensity varies in phase with Jupiter's rotation; thus the radio emission can be used for determining the exact rotation rate. The decametric bursts are related to the position of the innermost large moon, Io, and are possibly generated by the million Ampere electric current observed between Jupiter and the plasma torus at the orbit of Io.

In mid-2002 there were 39 known moons of Jupiter. The four largest, *Io*, *Europa*, *Ganymede* and *Callisto* are called the *Galilean satellites* (Fig. 7.45), in honour of Galileo Galilei, who discovered them in 1610. The Galilean satellites can already be seen with ordinary binoculars. They are the size of the Moon or even planet Mercury. The other moons are small, most of them only a few kilometres in diameter.

Owing to tidal forces, the orbits of Io, Europa and Ganymede have been locked into a resonance, so that their longitudes λ strictly satisfy the equation

$$\lambda_{\mathrm{Io}} - 3\lambda_{\mathrm{Europa}} + 2\lambda_{\mathrm{Ganymede}} = 180° . \tag{7.52}$$

Hence the moons can never be in the same direction when seen from Jupiter.

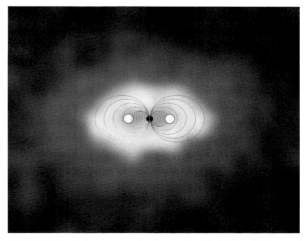

Fig. 7.44. *Left:* NASA Hubble Space Telescope close-up view of an aurora on Jupiter. The image shows the main oval of the aurora, centred over the magnetic north pole, and diffuse emissions inside the polar cap. (NASA, John Clarke/University of Michigan) *Right:* The image taken on January 2001 by NASA's Cassini spacecraft shows the bubble of charged particles trapped in the magnetosphere. The magnetic field and the torus of the ionised material from the volcanoes of Io are drawn over the image. (NASA/JPL/Johns Hopkins University)

Fig. 7.45. (*Top*) The Galilean satellites of Jupiter. From *left* to *right*: Io, Europa, Ganymede, and Callisto (NASA/DLR). (*Right page top*) Surface details of Io and Europa. (*Right page bottom*) Surface details of Ganymede and Callisto (NASA/Brown University, NASA/JPL)

Io is the innermost Galilean satellite. It is a little larger than the Moon. Its surface is spotted by numerous calderas, volcanoes without a mountain. The molten material is ejected up to a height of 250 km, and a part of the gas gets into Io's orbit. The volcanic activity on Io is much stronger than on the Earth. There is a huge bulk of the permanent tide raised by Jupiter. Due to the orbital perturbations caused by Europa and Ganymede the orbit of Io is slightly elliptical and therefore the orbital speed varies. The tidal bulk is forced to move with respect to the surface. This generates friction, which is transformed to heat. This heat keeps the sulphur compounds molten beneath the colourful surface of Io. No traces of impact craters are visible. The whole surface is new, being renewed continuously by eruptions. There is no water on Io.

Europa is the smallest of the Galilean satellites, a little smaller than the Moon. The surface is ice-covered and the geometric albedo is as high as 0.6. The surface is smooth with only a few features more than a hundred metres high. Most of the markings seem to be albedo features with very low relief. Only a few impact craters have been found indicating that the surface is young. The surface is renewed by fresh water, trickling from the internal ocean. Galileo spacecraft has found a very weak magnetic field. The field varies periodically as it passes through Jupiter's magnetic field. This shows that there is a conducting material beneath Europa's surface, most likely a salty ocean that could even be 100 km deep. At the centre, there is a solid silicate core.

Ganymede is the largest moon in the solar system. Its diameter is 5300 km; it is larger than the planet Mercury. The density of craters on the surface varies, indicating that there are areas of different ages. Ganymede's surface is partly very old, highly cratered dark regions, and somewhat younger but still ancient lighter regions marked with an extensive array of grooves and ridges. They have a tectonic origin, but the details of the formations are unknown. About 50% of the mass of the moon is water or ice, the other half being silicates (rocks). Contrary to Callisto, Ganymede is differentiated: a small iron or iron/sulphur core surrounded by a rocky silicate mantle with an icy (or liquid water) shell on top. Ganymede has a weak magnetic field.

Callisto is the outermost of the large moons. It is dark; its geometric albedo is less than 0.2. Callisto seems to be undifferentiated, with only a slight increase of rock toward the centre. About 40% of Callisto is ice and 60% rock/iron. The ancient surface is peppered by meteorite craters; no signs of tectonic activity are visible. However, there have been some later processes, because small craters have mostly been obliterated and ancient craters have collapsed.

The currently known moons can be divided into four groups: small inner regulars, Galilean satellites, prograde irregulars, and retrograde irregulars. The orbits of the inner group are inclined about 35° to the equator of Jupiter. Most of the outermost moons are in eccentric and/or retrograde orbits. It is possible that these are small asteroids captured by Jupiter.

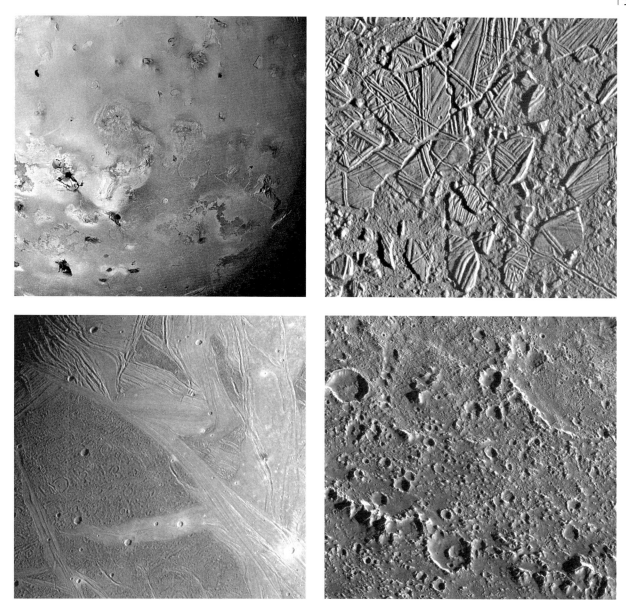

7.16 Saturn

Saturn is the second largest planet. Its diameter is about 120,000 km, ten times the diameter of the Earth, and the mass, 95 Earth masses. The density is only 700 kg m^{-3}, less than the density of water. The rotation axis is tilted about 27° with respect to the orbital plane, so every 15 years, the northern or the southern pole is well observable.

The rotation period is 10 h 39.4 min, determined from the periodic variation of the magnetic field. Due to the rapid rotation, Saturn is flattened; the flattening

Fig. 7.46. Saturn and its rings. Three satellites (Tethys, Dione, and Rhea) are seen to the left of Saturn, and the shadows of Mimas and Tethys are visible on Saturn's cloud tops. (NASA/JPL)

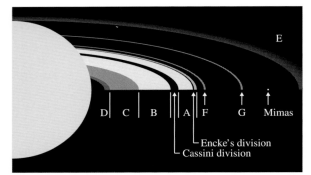

Fig. 7.47. A schematic drawing of the structure of the Saturnian rings

tures than those on Jupiter, because a haze, composed of hydrogen, ammonium and methane floats above the cloud tops. Furthermore, Saturn is farther from the Sun than Jupiter and thus has a different energy budget.

The temperature at the cloud tops is about 94 K. Close to the equator the wind speeds exceed 400 m/s and the zone in which the direction of the wind remains the same extends 40° from the equator. Such high speeds cannot be explained with external solar heat, but the reason for the winds is the internal flux of heat.

is 1/10, which can be easily seen even with a small telescope.

The internal structure of Saturn resembles that of Jupiter. Due to its smaller size, the metallic hydrogen layer is not so thick as on Jupiter. The thermal radiation of Saturn is 2.8 times that of the incoming solar flux. The heat excess originates from the differentiation of helium. The helium atoms are gradually sinking inward and the released potential energy is radiated out as a thermal radiation. The abundance of helium in Saturn's atmosphere is only about half of that on Jupiter.

The winds, or jet streams, are similar to those of Jupiter but Saturn's appearance is less colourful. Viewed from the Earth, Saturn is a yellowish disk without any conspicuous details. The clouds have fewer fea-

Fig. 7.48. At a close distance, the rings can be seen to be divided into thousands of narrow ringlets. (JPL/NASA)

Saturn's most remarkable feature is a thin *ring system* (Fig. 7.47, 7.48), lying in the planet's equatorial plane. The Saturnian rings can be seen even with a small telescope. The rings were discovered by *Galileo Galilei* in 1610; only 45 years later did *Christian Huygens* establish that the formation observed was actually a ring, and not two oddly behaving bulbs, as they appeared to Galileo. In 1857 *James Clerk Maxwell* showed theoretically that the rings cannot be solid but must be composed of small particles.

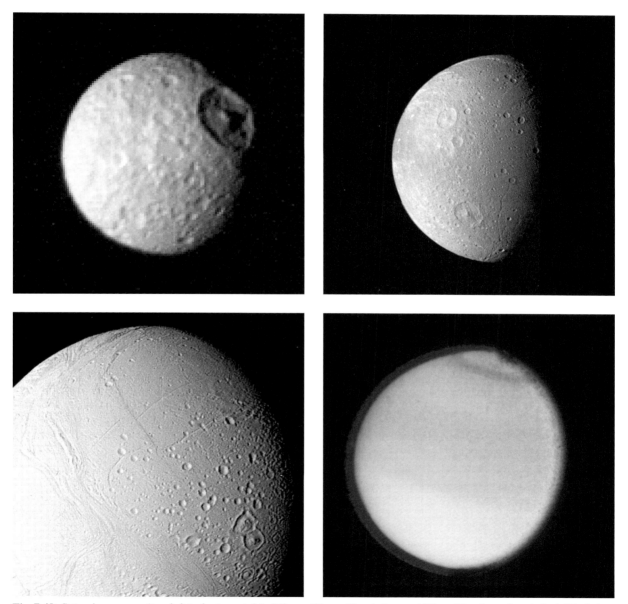

Fig. 7.49. Saturnian moons (*top left* to *bottom right*): Mimas, Dione, Enceladus, and Titan. Titan's surface is invisible below the thick atmosphere and clouds. (JPL/NASA)

The rings are made of normal water ice. The size of the ring particles ranges from microns to truck-size chunks. Most of the particles are in range of centimetres to metres. The width of the ring system is more than 60,000 km (about the radius of Saturn) and the thickness, at most 100 m, and possibly only a few metres.

According to Earth-based observations, the rings are divided into three parts, called simply A, B, and C. The innermost C ring is 17,000 km wide and consists of very thin material. There is some material even inside this (referred to as the D ring), and a haze of particles may extend down to the clouds of Saturn.

The B ring is the brightest ring. Its total width is 26,000 km, but the ring is divided into thousands of narrow ringlets, seen only by the spacecraft (Fig. 7.48). From the Earth, the ring seems more or less uniform. Between B and A, there is a 3000 km wide gap, the *Cassini division*. It is not totally void, as was previously believed; some material and even narrow ringlets have been found in the division by the Voyager space probes.

The A ring is not divided into narrow ringlets as clearly as the B ring. There is one narrow but obvious gap, *Encke's division*, close to the outer edge of the ring. The outer edge is very sharp, due to the "shepherd" moon, some 800 km outside the ring. The moon prevents the ring particles from spreading out to larger orbits. It is possible that the appearance of B is due to yet-undiscovered moonlets inside the ring.

The F ring, discovered in 1979, is about 3000 km outside A. The ring is only a few hundred kilometres wide. On both sides there is a small moon; these shepherds prevent the ring from spreading. An interior moon passing a ring particle causes the particle to move to a larger orbit. Similarly, at the outer edge of the ring, a second moon forces the particles inward. The net result is that the ring is kept narrow.

Outside the F ring, there are some zones of very sparse material, sometimes referred to as the G and E rings. These are merely collections of small particles.

The Saturnian rings were possibly formed together with Saturn and are not debris from some cosmic catastrophe, like remnants of a broken moon. The total mass of the rings is 10^{-7} of the mass of Saturn. If all ring particles were collected together, they would form an ice ball, 600 km in diameter.

A total of 30 moons of Saturn are known, 12 of them discovered in the year 2000. Most of the large Saturnian moons (Fig. 7.49) were observed by Pioneer 11 and Voyager 1 and 2. Large moons (excluding Titan) are composed mainly of ice. The temperature of the primeval nebula at the distance of Saturn was so low that bodies of pure ice could form and survive.

Some moons are dynamically interesting; some have an exotic geological past. Outside the F ring, there are two moonlets, *Epimetheus* and *Janus*, almost in the same orbit; the difference of the semimajor axes is about 50 km, less than the radii of the moons. The inner moon is gaining on the outer one. The moons will not collide, since the speed of the trailing moon increases and the moon moves outward. The speed of the leading moon decreases and it drops inward. The moons exchange their roles roughly every four years. There are also several shepherding satellites, like *Atlas*, *Prometheus* and *Pandora* that keep rings in their place. Their gravitational pull prevents ring particles from drifting away.

The innermost of the "old" moons is *Mimas*. There is a huge crater on Mimas' surface with a diameter of 100 km and a depth of 9 km (Fig. 7.49). Bigger craters exist in the solar system, but relative to the size of the parent body, this is almost the biggest crater there could be room for (otherwise the crater would be bigger than Mimas itself). On the opposite side, some grooves can be seen, possibly signifying that impact has almost torn the moon apart.

The surface of the next moon, *Enceladus*, consists of almost pure ice, and one side is nearly craterless. Craters and grooves can be found on the other hemisphere. Tidal forces result in volcanic activity where water (not lava or other "hot" material) is discharged to the surface.

Titan is the largest of the Saturnian moons. Its diameter is 5100 km, so it is only slightly smaller than Jupiter's moon Ganymede. Titan is the only moon with an atmosphere. The atmosphere is mainly nitrogen (99%), and the pressure at the surface is 1.5–2 bar; the temperature is about 90 K. Reddish clouds form the visible surface some 200 km above the solid body.

7.17 Uranus, Neptune and Pluto

The planets from Mercury to Saturn were already known in antiquity. Uranus, Neptune and Pluto can only be observed with a telescope. Uranus and Neptune are giants, similar to Jupiter and Saturn. Pluto is a small icy world, totally different from any other planet.

Uranus. The famous German-English amateur astronomer *William Herschel* discovered Uranus in 1781. Herschel himself first thought that the new object was a comet. However, the extremely slow motion revealed that the body was far beyond the orbit of Saturn. Based on the first observations, the Finnish astronomer *Anders Lexell* calculated a circular orbit. He was one of the first to propose that the newly discovered object was a planet. *Johann Bode* of the Berlin Observatory suggested the name Uranus but more than five decades passed before the name was unanimously accepted.

The mean distance of Uranus is 19 AU, and the orbital period 84 years. The inclination of the rotation axis is 98°, which is totally different from the other planets. Due to this uncommon geometry, the poles are either lit or in darkness for decades. The rotation period, confirmed by the Voyager 2 magnetometric measurements in 1986, is 17.3 hours; the exact period had been uncertain prior to the fly-by.

Uranus is greenish, as viewed through a telescope. Its colour is due to the strong methane absorption bands in the near-infrared. A part of the red light is also absorbed, leaving the green and blue part of the spectrum untouched. Uranus is almost featureless (Fig. 7.50) because its clouds are below a thick haze or smog.

The strong *limb darkening* makes the terrestrial determination of the Uranus' size difficult. Therefore, the radius was not accurately determined until 1977 during a stellar occultation caused by Uranus. The rings of Uranus were discovered at the same time.

The internal structure of Uranus is thought to be slightly different from that of other giant planets. Above the innermost rocky core, there is a layer of water, which, in turn, is surrounded by a mantle of hydrogen and helium. The mixture of water and ammonia and methane therein are dissociated to ions under the heavy pressure. This mixture behaves more like a molten salt than water. The convection flows in this electrically conductive "sea" give rise to the Uranian magnetic field. The strength of the magnetic field at the cloud tops is comparable to the terrestrial field. However, Uranus is much larger then the Earth, so the true strength of the field is 50 times greater than that of the Earth. The Uranian magnetic field is tilted 60° with respect to the rotation axis. No other planet has such a highly inclined magnetic field.

 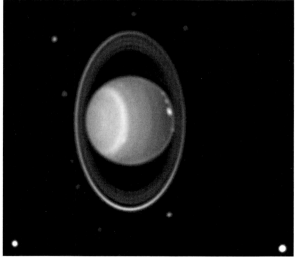

Fig. 7.50. Two views of Uranus. The left picture shows Uranus as it would appear to the naked eye. (NASA). At the right there is a Hubble Space Telescope view of Uranus surrounded by its rings. Also 10 of its 20 known satellites are visible in the original picture. (Seidelmann, U.S. Naval Observatory, and NASA)

Fig. 7.51. *Left:* The rings of Uranus are very narrow and composed of a dark material. Nine rings are visible in the picture of Voyager in 1986. *Right:* Rings seen in the light scattered forward when the Voyager spacecraft was in the shadow of the planet. (NASA)

The Uranian rings (Fig. 7.51) were discovered in 1977, during a stellar occultation. Secondary occultations were observed before and after the main event. A total of 11 rings are known. Nine were discovered by Earth-based observations and one during the Voyager 2 fly-by. The 11th ring is inside the other ten and is broad and diffuse. All other rings are dark and very narrow, only a few hundred metres or a few kilometres wide. The Voyager 2 results showed that the rings contain very little dust, unlike those of Jupiter and Saturn. The mean size of the ring particles is more than 1 metre. The ring particles are darker than practically any known material in the solar system; the cause of this dark colour is unknown.

There are 20 moons orbiting around Uranus, ten of which were discovered by Voyager 2. The geological history of some moons is puzzling, and many features reminiscent of an active past can be found.

The innermost of the large moons, *Miranda*, is one of the most peculiar objects discovered (Fig. 7.52). It has several geological formations also found elsewhere (but here they are all mixed together), in addition to the quite unique V-shaped formations. It is possible that Miranda's present appearance is the result of a vast collision that broke the moon apart; some pieces may have later settled down, inside out. Another peculiar object is Umbriel. It belongs to the ever increasing family of unusual dark bodies (such as the Uranian rings, one side of Iapetus and Halley's comet). The dark surface of Umbriel is covered by craters without any traces of geological activity.

Neptune. The orbit of Uranus was already well known in the beginning of the 19th century. However, some unknown perturbations displaced Uranus from its predicted orbit. Based on these perturbations, *John Couch Adams*, of Cambridge, and *Urbain Jean-Joseph Le Verrier*, of Paris, independently predicted the position of the unknown perturbing planet.

The new planet was discovered in 1846 by *Johann Gottfried Galle* at the Berlin Observatory; Le Verrier's prediction was found to be only 1° off. The discovery gave rise to a heated controversy as to who should be given the honour of the discovery, since Adams' calculations were not published outside the Cambridge Observatory. When the quarrel was settled years later, both men were equally honoured. The discovery of Neptune was also a great triumph of the Newtonian theory of gravitation.

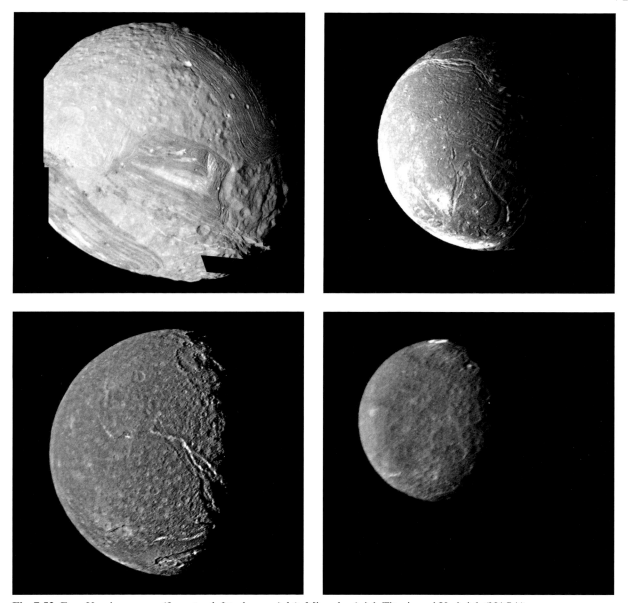

Fig. 7.52. Four Uranian moons (from *top left* to *lower right*): Miranda, Ariel, Titania and Umbriel. (NASA)

The semimajor axis of the orbit of Neptune is 30 AU and the orbital period around the Sun 165 years. The internal rotation period, confirmed by Voyager 2 in 1989, is 16 hours 7 minutes and the rotation period of the outer layers of the clouds is about 17 hours. The obliquity of the rotation axis is 29° but the magnetic field is tilted some 50° with respect to the rotation axis. The magnetic field is tilted like in Uranus, but the field strength is much smaller.

The density of Neptune is $1660 \, \text{kg m}^{-3}$, and the diameter 48,600 km. Thus the density of Neptune is higher than that of other giant planets. The internal

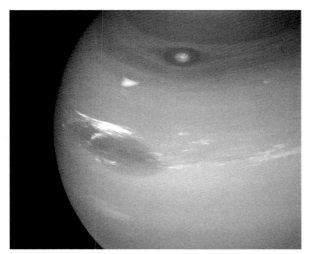

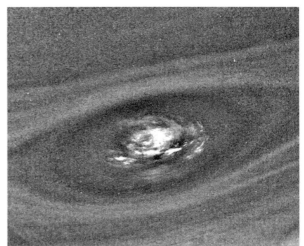

Fig. 7.53. (*Left*) Neptune shows more features than Uranus. In the picture of Voyager 2 the Great Dark Spot, accompanied by bright, white clouds is well visible. Their appearance is changing rapidly. To the south of the Great Dark Spot is a bright feature and still farther south is another dark spot. Each feature moves eastward at a different velocity. (*Right*) Details of the Southern Dark Spot. The V-shaped structure near the right edge of the bright area indicates that the spot rotates clockwise. Unlike the Great Red Spot on Jupiter, which rotates counterclockwise, the material in the Neptune's dark oval will be descending. (NASA/JPL)

structure is quite simple: The core, composed of silicates (rocks) is about 16000 km in diameter. This is surrounded by a layer of water and liquid methane and the outermost gaseous layer, atmosphere, is mainly composed of hydrogen and helium, methane and ethane being a minor components.

Cloud structures are more complicated than on Uranus, and some dark spots, like in Jupiter, were visible during the Voyager fly-by (Fig. 7.53). The speed of the winds are high, up to 400 m/s.

Like other giants, Neptune also has rings (Fig. 7.54). The rings were discovered by Voyager 2, although their existence was already expected prior the fly-by. Two relatively bright but very narrow rings are at a distance of 53,000 and 62,000 km from the centre of the planet. Moreover, there are some faint areas of fine dust.

There are 8 known moons, six of which were discovered by Voyager 2. The largest of the moons, *Triton,*

Fig. 7.54. The rings of Neptune. Ring particles are small and best visible in the forward scattered light. There are several brightenings in the outermost ring. One of the rings appears to have a twisted structure. Neptune at left is overexposed. (NASA/JPL)

Fig. 7.55. The southern hemisphere of Triton, Neptune's largest satellite in a picture taken in 1989 by Voyager 2. The dark spots may indicate eruptions of "icy volcanoes". Voyager 2 images showed active geyser-like eruptions spewing nitrogen gas and dark dust particles several kilometres into the atmosphere. (NASA)

is 2700 km in diameter, and it has a thin atmosphere, mainly composed of nitrogen. The albedo is high: Triton reflects 60–80% of the incident light. The surface is relatively young, without any considerable impact craters (Fig. 7.55). There are some active "geysers" of liquid nitrogen, which partly explains the high albedo and the lack of the craters. The low surface tempera-

ture of Triton, 37 K, means that the nitrogen is solid and covers the surface like snow. It is the lowest surface temperature known in the solar system.

Pluto. Pluto is the outermost planet of the solar system. It was discovered in 1930 at the Lowell Observatory, Arizona, after an extensive photographic search (Fig. 7.56). This search had already been initiated in the beginning of the century by *Percival Lowell*, on the basis of the perturbations observed in the orbits of Uranus and Neptune. Finally, *Clyde Tombaugh* discovered Pluto less than 6° off the predicted position. However, Pluto turned out to be far too small to cause any perturbations on Uranus or Neptune. Thus the discovery was purely accidental, and the perturbations observed were not real, but caused by minor errors of old observations.

Pluto has no visible disk as seen with terrestrial telescopes; instead, it resembles a point, like a star. This fact gave an upper limit for the diameter of Pluto, which turned out to be about 3000 km. The exact mass was unknown until the discovery of the Plutonian moon, *Charon*, in 1978. The mass of Pluto is only 0.2% of the mass of the Earth. The orbital period of Charon is 6.39 days, and this is also the period of rotation of both bodies. Pluto and Charon rotate synchronously, each turning the same side towards the other body. The rotation axis of Pluto is close to the orbital plane: the tilt is 122°.

Fig. 7.56. A small portion of the pair of pictures where Pluto was discovered in 1930. The planet is marked with an arrow. (Lowell Observatory)

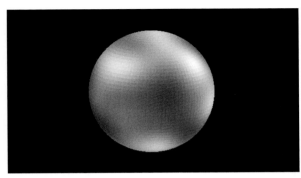

Fig. 7.57. Even with large terrestrial telescopes Pluto is seen only as a point of light. The best views have been obtained from the Hubble Space Telescope showing some albedo differences on the surface. (Alan Stern/Southwest Research Institute, Marc Buie/Lowell Observatory, NASA and ESA)

Mutual occultations of Pluto and Charon in 1985–1987 gave accurate diameters of each body: The diameter of Pluto is 2300 km and that of Charon, 1200 km. The density of Pluto turned out to be about 2100 kg m^{-3}. Thus Pluto is not a huge iceball but about 2/3 of its mass is composed of rocks. The relative small abundance of ices is possibly due to the low temperature during the planetary accretion when most of the free oxygen was combined with carbon forming carbon monoxide. The computed lower limit for water ice is about 30% which is fairly close to the value observed in Pluto.

Pluto has a thin methane atmosphere and there is possibly a thin haze over the surface. The surface pressure is 10^{-5}–10^{-6} atm. It has been speculated that when Pluto is far from perihelion, the whole atmosphere will become frozen and fall on the surface.

The orbit of Pluto is very different from other planetary orbits. The eccentricity is 0.25 and the inclination is 17°. During its 250 year orbit, Pluto is closer to the Sun than Neptune for 20 years; one such period lasted from 1979 to 1999. There is no danger of Pluto and Neptune colliding, since Pluto is high above the ecliptic when at the distance of Neptune. Pluto's orbital period is in a 3:2 resonance with Neptune.

A Trans-Plutonian planet, "*Planet-X*", was sought for decades, but in vain, and a lot of speculations on it's nature existed. Discovery of trans-Neptunian objects and thus the confirmation of the Kuiper belt in 1990's settled the question. The Kuiper belt objects resemble Pluto, and there are a lot of Pluto-like objects in the outer edge of the solar system (see Fig. 7.40). This far, Pluto has been the largest one discovered, but it is possible that there are more such distant bodies of the same size.

7.18 Minor Bodies of the Solar System

So far we have considered only planets, planetary satellites and asteroids. There are also other bodies in the solar system, like *comets*, *meteoroids* and *interplanetary dust*. However, there are no distinct borders between different types of objects. As we have already seen, some asteroids, or even Pluto, have same features or origin as the comets, and some near-Earth asteroids are possibly cometary remnants where all volatile elements have disappeared. Thus our classification is based more on the visual appearance and tradition than on real physical differences.

Fig. 7.58. *Top:* Comet Mrkos in 1957. (Palomar Observatory). *Lower left:* Nucleus of comet Borrelly shows a variety of surface structures on the eight kilometre long body. The image was taken 3400 km from the comet by Deep Space 1 in 1999. (JPL/NASA) *Lower right:* A composite image of the nucleus of comet P/Halley taken by ESA Giotto spacecraft in 1986. The size of the nucleus is approximately 13×7 km. Dust jets are originating from two regions on the nucleus. (ESA/Max Planck Institut für Aeronomie)

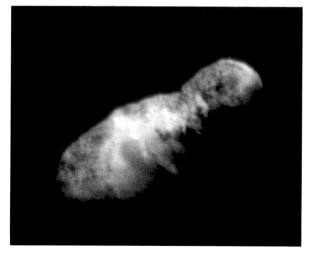

Comets. Comets are agglomerates of ice, snow, and dust; a typical diameter is of the order of 10 km or less. The nucleus contains icy chunks and frozen gases with embedded rock and dust. At its centre, there can be a small, rocky core.

A comet is invisible when far from the Sun; when it gets closer than about 2 AU, the heat of the Sun starts to melt the ice and snow. The outflowing gas and dust form an envelope, the *coma* around the nucleus. Radiation pressure and the solar wind push ionised gas and dust away from the Sun, resulting in the typical long-tailed shape of a comet (Fig. 7.58).

The tail is always pointing away from the Sun, a fact which was noticed in the 16th century. Usually, there are two tails, an *ion tail* (gas tail) and a *dust tail*. The partly ionised gas and very fine dust in the ion tail are driven by the solar wind. Some of the light is reflected solar light, but the brightness of the ion tail is mostly due to emission by the excited atoms. The dust tail is caused by the radiation pressure. Because the velocities of the particles of the dust tail are lower than the velocities in the ion tail, the dust tail is often more curved than the ion tail.

Fred Whipple introduced in 1950's a *"dirty snowball"* theory to describe the cometary structure. According to this model, cometary nuclei are composed of ice mixed with gravel and dust. The observations have revealed that the classical dirty snowball model is not quite accurate; at least the surface is more dirt than snow, also containing organic compounds. Several chemical compounds have been observed, including water ice, which probably makes up 75–80% of the volatile material. Other common compounds are carbon monoxide (CO), carbon dioxide (CO_2), methane (CH_4), ammonia (NH_3), and formaldehyde (H_2CO).

The most famous (and also best known) periodic comet is *Halley's comet*. Its orbital period is about 76 years; it was last in perihelion in 1986. During the last apparition, the comet was also observed by spacecraft, revealing the solid cometary body itself for the first time. Halley is a 13×7 km, peanut-shaped chunk whose surface is covered by an extremely black layer of a possibly tar-like organic or other similar material. Violent outbursts of gas and dust make an exact prediction of its brightness impossible, as often noticed when cometary magnitudes have been predicted. Near the perihelion, several tons of gas and dust burst out every second.

Cometary material is very loose. Ablation of gas and dust, large temperature variations and tidal forces sometimes cause the whole comet to break apart. *Comet Shoemaker–Levy 9* which impacted into Jupiter in 1994 was torn apart two years earlier when it passed Jupiter at a distance of 21,000 km (Fig. 7.60). The impact of Shoemaker–Levy 9 showed that there can be density variation (and perhaps variation in composition, too) inside the original cometary body.

Comets are rather ephemeral things, surviving only a few thousand revolutions around the Sun or less. The *short-period comets* are all newcomers and can survive only a short time here, in the central part of the solar system.

Since comets in the central solar system are rapidly destroyed, there has to be some source of new short-period comets. In 1950 *Jan Oort* discovered a strong peak for aphelia of long period comets at a distance of about 50,000 AU, and that there is no preferential direction from which comets come (Fig. 7.61). He proposed that there is a vast cloud of comets at the outer reaches of the solar system, now know as the *Oort cloud* (Fig. 7.40). The total mass of the Oort cloud is estimated to be tens of Earth masses, containing more than 10^{12} comets.

A year later *Gerard Kuiper* showed that there is a separate population of comets. Many of the short period comets, with periods less than 200 years, have the orbital inclination less than 40°, and they orbit the Sun in

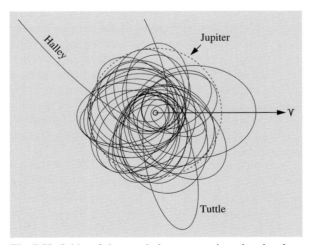

Fig. 7.59. Orbits of short period comets projected to the plane of the ecliptic

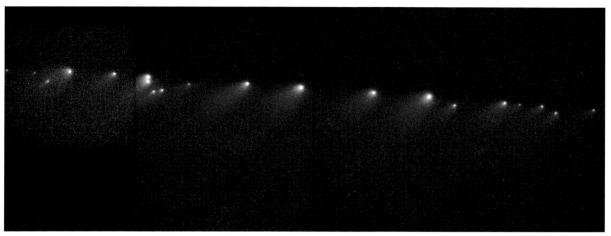

Fig. 7.60. Comet Shoemaker–Levy 9 five months before its collision to Jupiter as seen by the Hubble Space Telescope. (JPL/NASA)

the same direction as the Earth. The orbital inclination of long period comets are not peaked around the plane of the ecliptic but they are more random. Kuiper argued that the short period comets originate from a separate population of comets that resides in a disk-like cloud beyond the orbit of Neptune. The area is now known as the *Kuiper belt* (Fig. 7.40).

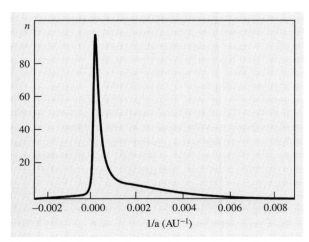

Fig. 7.61. A schematic diagram of the distribution of the semi-major axes of long-period comets. The abscissa is the inverse of the semimajor axis, $1/a$ [AU]$^{-1}$. The Oort cloud is visible as a strong peak at the very small positive values of $1/a$. The orbits shown here are the "original orbits", i.e. computed backward in time to remove all known perturbations

Occasionally perturbations from passing stars send some of the comets in the Oort cloud into orbits, which bring them into the central parts of the solar system, where they are seen as *long-period comets*. Around a dozen "new" comets are discovered each year. Most of these are visible only with a telescope, and only a couple of times per decade one can see a bright naked-eye comet.

Some of the long period comets are put into short period orbits by the perturbations of Jupiter and Saturn, whereas some others can be ejected from the solar system. However, there are no comets that have been proven to come from interstellar space, and the relative abundances of several isotopes in the cometary matter are the same as in other bodies of our solar system.

The origin of the Oort cloud and Kuiper belt is different. The Oort cloud objects were formed near the giant planets and have been ejected to the outer edge of the solar system by gravitational perturbations soon after the formation of the solar system. Small objects beyond the orbit of Neptune had no such interactions and they remained near the accretion disk.

Meteoroids. Solid bodies smaller than asteroids are called meteoroids. The boundary between asteroids and meteoroids, however, is diffuse; it is a matter of taste whether a ten metre body is called an asteroid or a meteoroid. We could say that it is an asteroid if it has been observed so often that its orbital elements are known.

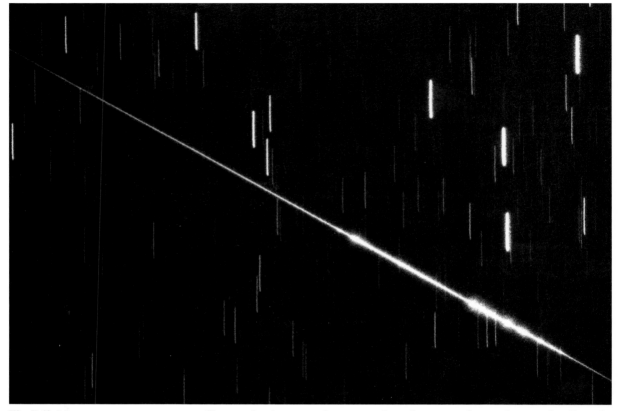

Fig. 7.62. Meteors are easy to capture on film: one just leaves a camera loaded with a sensitive film on a tripod with the shutter open for an hour or so. Stars make curved trails on the film. (L. Häkkinen)

When a meteoroid hits the atmosphere, an optical phenomenon, called a *meteor* ("shooting star") is seen (Fig. 7.62). The smallest bodies causing meteors have a mass of about 1 gram; the (micro)meteoroids smaller than this do not result in optical phenomena. However, even these can be observed with radar which is able to detect the column of ionised air. Micrometeoroids can also be studied with particle detectors installed in satellites and space crafts. Bright meteors are called *bolides*.

The number of meteoroids increases rapidly as their size diminishes. It has been estimated that at least 10^5 kg of meteoritic material falls on the Earth each day. Most of this material is micrometeoroids and causes no visible phenomena.

Due to perspective, all meteors coming from the same direction seem to radiate from the same point. Such *meteor streams* (meteor showers) are, e. g. the *Perseids* in August and the *Geminides* in December; the names are given according to the constellation in which the radiation point seems to be. On the average, one can see a few sporadic meteors per hour. During a strong meteor shower one can see even tens of meteors per minute, although a normal rate is some tens per hour.

Most of the meteoroids are small and burn to ashes at a height of 100 km. However, larger bodies may come through and fall to the Earth. These are called *meteorites*. The relative speed of a typical meteoroid varies in the range 10–70 km/s. The speed of the largest bodies does not diminish in the atmosphere; thus, they hit the Earth at their cosmic speeds, resulting in large impact craters. Smaller bodies slow down and drop like stones. The impacts of large bodies are discussed in Sect. 7.14.

Thousands of meteorites have been found. One of the best places to find meteorites is the Antarctic, where the pieces are carried by ice to the edge of the continent.

Iron meteorites or irons, composed of almost pure nickel-iron, comprise about one quarter of all meteorites. Actually the irons are in a minority among meteoroids, but they survive their violent voyage through the atmosphere more easily than weaker bodies. Three-quarters are *stony meteorites*, or *stone-iron meteorites*.

Meteoroids themselves can be divided into three groups of roughly equal size. One-third is ordinary stones, *chondrites*. The second class contains weaker *carbonaceous chondrites* and the third class includes cometary material, loose bodies of ice and snow which are unable to survive down to the Earth.

Many meteor streams are in the same orbit as a known comet, so at least some meteoroids are of cometary origin. Near a perihelion passage, every second several tons of gravel is left on the orbit of a comet. There are several examples of meteorites that have their origin in the Moon or Mars. Debris of large impacts may have been ejected into space and finally ended up on the Earth. Some meteoroids are debris of asteroids.

Interplanetary Dust. Two faint light phenomena, namely *zodiacal light* and *gegenschein* (counterglow) make it possible to observe interplanetary dust, small dust particles reflecting the light of the Sun (Fig. 7.63). This weak glow can be seen above the rising or setting Sun (zodiacal light) or exactly opposite the Sun (gegenschein). The interplanetary dust is concentrated near the plane of the ecliptic. The typical sizes of the particles are in the range of $10-100\,\mu\mathrm{m}$.

Solar Wind. Elementary particles hitting the Earth originate both in the Sun and outside the solar system. Charged particles, mainly protons, electrons and alpha particles (helium nuclei) flow continuously out of the Sun. At the distance of the Earth, the speed of this solar wind is $300-500\,\mathrm{km/s}$. The particles interact with the solar magnetic field. The strength of the solar magnetic field at the Earth's distance is about $1/1000$ of that of the Earth. Particles coming from outside the solar system are called cosmic rays (Sect. 15.8).

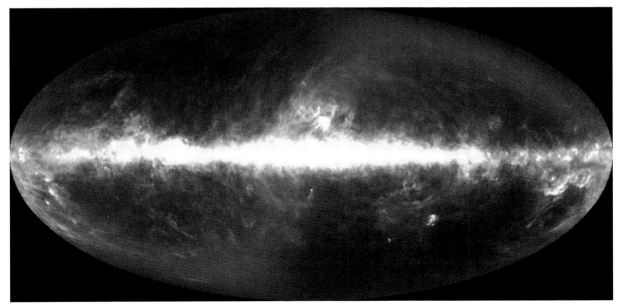

Fig. 7.63. A projection of the entire infrared sky created from observations of the COBE satellite. The bright horizontal band is the Milky Way. The dust of the solar system, visible on the Earth as zodiacal light is an S-shaped glow across the image. (G. Greaney and NASA)

7.19 Origin of the Solar System

Cosmogony is a branch of astronomy which studies the origin of the solar system. The first steps of the planetary formation processes are closely connected to star formation.

The solar system has some very distinct features which have to be explained by any serious cosmogonical theory. These include:

– planetary orbits are almost coplanar and also parallel to the solar equator;
– orbits are almost circular;
– planets orbit the Sun counterclockwise, which is also the direction of solar rotation;
– planets also rotate around their axes counterclockwise (excluding Venus, Uranus and Pluto);
– planetary distances approximately obey the empirical Titius-Bode law, i. e.

$$a = 0.4 + 0.3 \times 2^n , \\ n = -\infty, 0, 1, 2, \cdots \tag{7.53}$$

where the semimajor axis a is expressed in AU;
– planets have 98% of the angular momentum of the solar system but only 0.15% of the total mass;
– terrestrial and giant planets exhibit physical and chemical differences;
– the structure of planetary satellite systems resembles miniature solar systems.

The first modern cosmogonical theories were introduced in the 18th century. One of the first cosmogonists was *Immanuel Kant*, who in 1755 presented his *nebular hypothesis*. According to this theory, the solar system condensed from a large rotating nebula. Kant's nebular hypothesis is surprisingly close to the basic ideas of modern cosmogonical models. In a similar vein, *Pierre Simon de Laplace* suggested in 1796 that the planets have formed from gas rings ejected from the equator of the collapsing Sun.

The main difficulty of the nebular hypothesis was its inability to explain the distribution of angular momentum in the solar system. Although the planets represent less than 1% of the total mass, they possess 98% of the angular momentum. There appeared to be no way of achieving such an unequal distribution. A second objection to the nebular hypothesis was that it provided no mechanism to form planets from the postulated gas rings.

Already in 1745, *Georges Louis Leclerc de Buffon* had proposed that the planets were formed from a vast outflow of solar material, ejected upon the impact of a large comet. Various *catastrophe theories* were popular in the 19th century and in the first decades of the 20th century when the cometary impact was replaced by a close encounter with another star. The theory was developed, e. g. by *Forest R. Moulton* (1905) and *James Jeans* (1917).

Strong tidal forces during the closest approach would tear some gas out of the Sun; this material would later accrete into planets. Such a close encounter would be an extremely rare event. Assuming a typical star density of 0.15 stars per cubic parsec and an average relative velocity of 20 km/s, only a few encounters would have taken place in the whole Galaxy during the last 5×10^9 years. The solar system could be a unique specimen.

The main objection to the collision theory is that most of the hot material torn off the Sun would be captured by the passing star, rather than remaining in orbit around the Sun. There also was no obvious way how the material could form a planetary system.

In the face of the dynamical and statistical difficulties of the collision theory, the nebular hypothesis was revised and modified in the 1940's. In particular, it became clear that magnetic forces and gas outflow could efficiently transfer angular momentum from the Sun to the planetary nebula. The main principles of planetary formation are now thought to be reasonably well understood.

The oldest rocks found on the Earth are about 3.7×10^9 years old; some lunar and meteorite samples are somewhat older. When all the facts are put together, it can be estimated that the Earth and other planets were formed about 4.6×10^9 years ago. On the other hand, the age of the Galaxy is at least twice as high, so the overall conditions have not changed significantly during the lifetime of the solar system. Moreover, there is even direct evidence nowadays, such as other planetary systems and protoplanetary disks, *proplyds* (Fig. 7.64).

The Sun and practically the whole solar system simultaneously condensed from a rotating collapsing cloud of dust and gas, the density of which was some 10,000 atoms or molecules per cm^3 and the temperature 10–50 K (Fig. 7.65). The elements heavier than

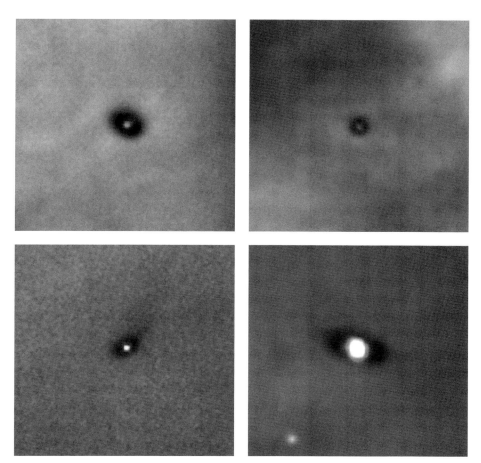

Fig. 7.64. Hubble Space Telescope images of four protoplanetary disks, "*proplyds*", around young stars in the Orion nebula. The disk diameters are two to eight times the diameter of our solar system. There is a T Tauri star in the centre of each disk. (Mark McCaughrean/Max-Planck-Institute for Astronomy, C. Robert O'Dell/Rice University, and NASA)

helium were formed in the interiors of stars of preceding generations, as will be explained in Sect. 11.8. The collapse of the cloud was initiated perhaps by a shock wave emanating from a nearby supernova explosion.

When the originally spherical cloud collapsed, particles inside the cloud collided with each other. Rotation of the cloud allowed the particles to sink toward the same plane, perpendicular to the rotation axis of the cloud, but prevented them from moving toward the axis. This explains why the planetary orbits are in the same plane.

The mass of the proto-Sun was larger than the mass of the modern Sun. The flat disk in the plane of the ecliptic contained perhaps 1/10 of the total mass. Moreover, far outside, the remnants of the outer edges of the original cloud were still moving toward the centre. The Sun was losing its angular momentum to the surrounding gas by means of the magnetic field. When nuclear reactions were ignited, a strong solar wind carried away more angular momentum from the Sun. The final result was the modern, slowly rotating Sun.

The small particles in the disk were accreting to larger clumps by means of continuous mutual collisions, resulting finally in asteroid-size bodies, *planetesimals*. The gravitation of the clumps pulled them together, forming ever growing seeds of planets. When these protoplanets were large enough, they started to accrete gas and dust from the surrounding cloud. Some minor clumps were orbiting planets; these became moons. Mutual perturbations may have prevented planetesimals in the current asteroid belt from ever being able to become "grown-up" planets. Moreover, reso-

Fig. 7.65a–g. A schematic plot on the formation of the solar system. (**a**) A large rotating cloud, the mass of which was 3–4 solar masses, began to condense. (**b**) The innermost part condensed most rapidly and a disk of gas and dust formed around the proto-sun. (**c**) Dust particles in the disk collided with each other forming larger particles and sinking rapidly to a single plane. (**d**) Particles clumped together into planetesimals which were of the size of present asteroids. (**e**) These clumps drifted together, forming planet-size bodies which began (**f**) to collect gas and dust from the surrounding cloud. (**g**) The strong solar wind "blew" away extra gas and dust; the planet formation was finished

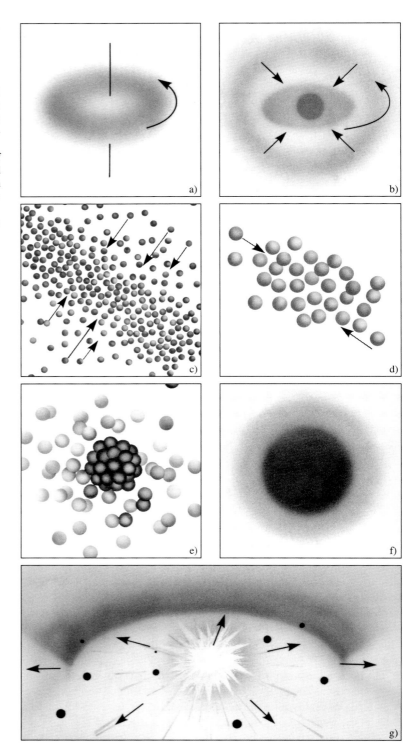

Table 7.4. True distances of the planets from the Sun and distances according to the Titius–Bode law (7.53)

Planet	n	Calculated distance [AU]	True distance [AU]
Mercury	$-\infty$	0.4	0.4
Venus	0	0.7	0.7
Earth	1	1.0	1.0
Mars	2	1.6	1.5
Ceres	3	2.8	2.8
Jupiter	4	5.2	5.2
Saturn	5	10.0	9.2
Uranus	6	19.6	19.2
Neptune	7	38.8	30.1
Pluto	8	77.2	39.5

nances could explain the Titius-Bode law: the planets were able to accrete in very limited zones only (Table 7.4).

The temperature distribution of the primordial cloud explains the differences of the chemical composition of the planets (Fig. 7.66). The volatile elements (such as hydrogen and helium, and ices) are almost totally absent in the innermost planets. The planets from Mercury to Mars are composed of "rocks", relatively heavy material which condenses above 500 K or so. The relative abundance of this material in the primeval nebula was only 0.4%. Thus the masses of terrestrial planets are relatively small. More than 99% of the material was left over.

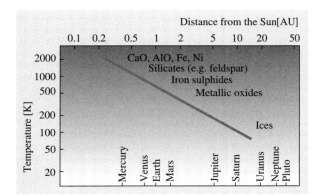

Fig. 7.66. Temperature distribution in the solar system during planet formation. The present chemical composition of the planets reflects this temperature distribution. The approximate condensing temperatures of some compounds have been indicated

At the distance of Mercury, the temperature was about 1400 K. At this temperature, iron and nickel compounds begin to condense. The relative abundance of these compounds is greatest on Mercury and smallest on Mars, where the temperature was only 450 K. Thus the amount of iron(II)oxide, FeO, is relatively high on Mars, whereas there is practically no FeO on Mercury.

At the distance of Saturn, the temperature was so low that bodies of ice could form; e. g. some moons of Saturn are almost pure water ice. Because 98.2% of the primordial material was hydrogen and helium, the abundance of hydrogen and helium in Jupiter and Saturn is close to those values. However, the relative importance of ices became more prominent at the distance of Uranus and Neptune. A considerable amount of the mass of these planets can be water.

Meteorite bombardment, contraction and radioactive decay produced a great deal of heat after the planetary formation. This gave rise to the partial melting of some terrestrial planets, resulting in *differentiation* of material: the heavy elements sank to the centre and the light dross was left to float on the surface.

The left over material wandered among the planets. Planetary perturbations caused bodies in unstable orbits to collide with planets or to be slung outer edges of the solar system, as happened to the bodies now in the Oort cloud. Asteroids remained in their stable orbits. On the outskirts of the solar system, bodies of ice and dust, such as the Kuiper belt objects, could also survive.

The beginning of the solar nuclear reactions meant the end of planetary formation. The Sun was in the T Tauri phase losing mass to a strong solar wind. The mass loss was as high as $10^{-7}\,M_\odot/a$. However, this phase was relatively short and the total mass loss did not exceed $0.1\,M_\odot$. The solar wind "blew" away the interplanetary gas and dust, and the gas accretion to the planets was over.

The solar wind or radiation pressure has no effect on millimetre- and centimetre-sized particles. However, they will drift into the Sun because of the *Poynting–Robertson effect*, first introduced by John P. Poynting in 1903. Later H.P. Robertson derived the effect by using the theory of relativity. When a small body absorbs and emits radiation, it loses its orbital angular momentum and the body spirals to the Sun. At the distance of the asteroid belt, this process takes only a million years or so.

7.20 Other Solar Systems

Planetary systems are common around stars. However, such systems are difficult to observe, since the brightness of a planet is much less than the brightness of the nearby star.

Theoretical models predict that planetesimals are accreting to planets much like in our solar system. Most calculations have dealt with single stars around which stable planetary orbits are easier to find. One of the best examples is the accretion disk around β Pictoris (Fig. 7.67). There are also several other examples of disks, see for example Fig. 7.64. However, stable orbits can also be found around binary stars either close to one of the components, or far from both.

There is a way to detect possible planets by observing the spectrum of a star. A planet orbiting a star causes small periodic perturbations which can be seen as Doppler shifts in the spectral lines. The first candi-

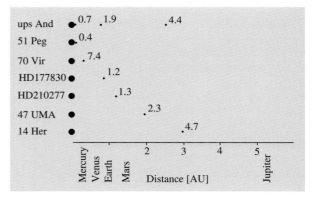

Fig. 7.68. Orbits and masses of some extrasolar planetary systems. The mass of a planet shown beside each planet is given in units of Jupiter mass

date was discovered in 1992 orbiting a pulsar. The first planet orbiting a normal star, *51 Pegasi*, was discovered in 1995, and since then an ever increasing number of candidates have been detected.

Most of the planetary systems have a planet or planets that are more massive than Jupiter and their distance from the central star is small, 0.1–1 AU (Fig. 7.68). Such bodies are easier to discover than smaller and more distant planets but it is only a question of observing accuracy. With the aid of new giant telescopes and adaptive optics, the observing accuracy will improve so that it will also become possible to detect smaller planets. However, we are still far from direct imaging of planets outside our solar system.

Our picture on other solar systems has changed rapidly in the beginning of the new millennium. Speculations have changed to discoveries and we now have several examples of planetary systems.

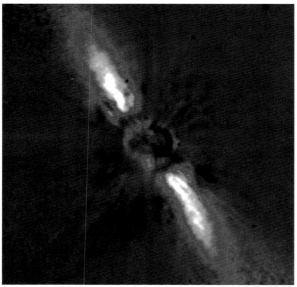

Fig. 7.67. A detailed image of the circumstellar disk around β Pictoris was obtained with the ESO ADONIS adaptive optics system at the 3.6 m telescope at La Silla, Chile, and the Observatoire de Grenoble coronagraph at the wavelength of 1.25 microns in 1996. The full extent of the disk is about 1500 AU. The area around the star masked by the coronagraph is only 24 AU (corresponding to a distance from the Sun to halfway between Uranus and Neptune). No planets are directly seen but their gravitational effects can be detected as a bending of the main plane in the inner part of the disk. (ESO)

7.21 Examples

Example 7.1 *Sidereal and Synodic Period*

The time interval between two successive oppositions of Mars is 779.9 d. Calculate the semimajor axis of Mars' orbit.

The synodic period is 779.9 d = 2.14 years. We obtain from (7.1)

$$\frac{1}{P_2} = \frac{1}{1} - \frac{1}{2.14} = 0.53 \quad \Rightarrow \quad P_2 = 1.88 \text{ a} .$$

By using Kepler's third law, ($m \ll M_\odot$), the semimajor axis is found to be

$$a = P^{2/3} = 1.88^{2/3} = 1.52 \text{ AU} .$$

Example 7.2 *Solar Energy Flux on the Earth*

Calculate the diurnal solar energy flux per unit area at the distance of the Earth.

The solar flux density outside the Earth's atmosphere (the solar constant) is $S_0 = 1370 \text{ W/m}^2$. Consider a situation at latitude ϕ, when the solar declination is δ. If the atmospheric extinction is neglected, the flux density on the surface is

$$S = S_0 \sin a ,$$

where a is the elevation of the Sun. We can write $\sin a$ as a function of latitude, declination, and hour angle h:

$$\sin a = \sin \delta \sin \phi + \cos \delta \cos \phi \cos h .$$

On a cloudless day, the energy is received between sunrise and sunset. The corresponding hour angles can be obtained from the equation above, when $a = 0$:

$$\cos h_0 = - \tan \delta \tan \phi .$$

In the course of one day, the energy received on a unit area is

$$W = \int_{-h_0}^{h_0} S \, dt .$$

The hour angle h is expressed in radians, so the time t is

$$t = \frac{h}{2\pi} P ,$$

where $P = 1 \text{ d} = 24 \text{ h}$. The total energy is thus

$$W = \int_{-h_0}^{h_0} S_0 (\sin \delta \sin \phi + \cos \delta \cos \phi \cos h) \frac{P}{2\pi} \, dh$$

$$= \frac{S_0 P}{\pi} (h_0 \sin \delta \sin \phi + \cos \delta \cos \phi \sin h_0) ,$$

where

$$h_0 = \arccos(- \tan \delta \tan \phi) .$$

For example near the equator ($\phi = 0°$) $\cos h_0 = 0$ and

$$W(\phi = 0°) = \frac{S_0 P}{\pi} \cos \delta .$$

At those latitudes where the Sun will not set, $h_0 = \pi$ and

$$W_{\text{circ}} = S_0 P \sin \delta \sin \phi .$$

Near the poles, the Sun is always circumpolar when above the horizon, and so

$$W(\phi = 90°) = S_0 P \sin \delta .$$

Interestingly enough, during the summer when the declination of the Sun is large, the polar areas receive more energy than the areas close to the equator. This is true when

$$W(\phi = 90°) > W(\phi = 0°)$$

$$\Leftrightarrow S_0 P \sin \delta > S_0 P \cos \delta / \pi$$

$$\Leftrightarrow \tan \delta > 1/\pi$$

$$\Leftrightarrow \delta > 17.7° .$$

The declination of the Sun is greater than this about two months every summer.

However, atmospheric extinction diminishes these values, and the loss is at its greatest at the poles, where the elevation of the Sun is always relatively small. Radiation must penetrate thick layers of the atmosphere and the path length is comparable to $1/\sin a$. If it is assumed that the fraction k of the flux density reaches the surface when the Sun is at zenith, the flux density when the Sun is at the elevation a is

$$S' = S_0 \sin a \, k^{1/\sin a} .$$

The total energy received during one day is thus

$$W = \int_{-h_0}^{h_0} S' \, dt = \int_{-h_0}^{h_0} S_0 \sin a \, k^{1/\sin a} \, dt .$$

This cannot be solved in a closed form and numerical methods must be used.

The figure on next page shows the daily received energy W [kWh/m^2] during a year at latitudes $\phi = 0°$, $60°$, and $90°$ without extinction, and when $k = 0.8$, which is close to the real value.

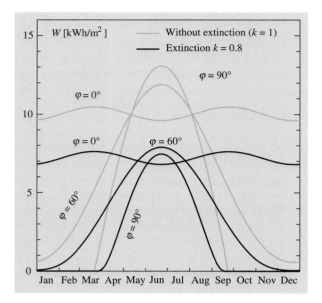

Example 7.3 *Magnitude of a Planet*

The apparent magnitude of Mars during the 1975 opposition was $m_1 = -1.6$ and the distance to the Sun, $r_1 = 1.55$ AU. During the 1982 opposition, the distance was $r_2 = 1.64$ AU. Calculate the apparent magnitude in the 1982 opposition.

At opposition, the distance of Mars from the Earth is $\Delta = r - 1$. The observed flux density depends on the distances to the Earth and the Sun,

$$F \propto \frac{1}{r^2 \Delta^2} \ .$$

Using the magnitude formula (4.9) we obtain

$$m_1 - m_2 = -2.5 \lg \frac{r_2^2 (r_2 - 1)^2}{r_1^2 (r_1 - 1)^2}$$

$$\Rightarrow m_2 = m_1 + 5 \lg \frac{r_2 (r_2 - 1)}{r_1 (r_1 - 1)}$$

$$= -1.6 + 5 \lg \frac{1.64 \times 0.64}{1.55 \times 0.55} \approx -1.1 \ .$$

The same result is obtained if (7.41) is separately written for both oppositions.

Example 7.4 *The Brightness of Venus*

Find the instant when Venus is brightest if the brightness is proportional to the projected size of the illuminated surface. The orbits are assumed to be circular.

The size of the illuminated surface is the area of the semicircle $ACE\pm$ half the area of the ellipse $ABCD$. The semiaxes of the ellipse are R and $R\cos\alpha$. If the radius of the planet is R, the illuminated area is

$$\pi \frac{R^2}{2} + \frac{1}{2}\pi R \times R \cos\alpha = \frac{\pi}{2} R^2 (1 + \cos\alpha) \ ,$$

where α is the phase angle. The flux density is inversely proportional to the square of the distance Δ. Thus

$$F \propto \frac{1 + \cos\alpha}{\Delta^2} \ .$$

The cosine formula yields

$$M_\oplus^2 = r^2 + \Delta^2 - 2\Delta r \cos\alpha \ .$$

When $\cos\alpha$ is solved and inserted in the flux density we obtain

$$F \propto \frac{2\Delta r + r^2 + \Delta^2 - M_\oplus^2}{2r\Delta^3} \ .$$

The minimum of the equation yields the distance where Venus is brightest:

$$\frac{\partial F}{\partial \Delta} = -\frac{4r\Delta + 3r^2 - 3M_\oplus^2 + \Delta^2}{2r\Delta^4} = 0$$

$$\Rightarrow \Delta = -2r \pm \sqrt{r^2 + 3M_\oplus^2} \ .$$

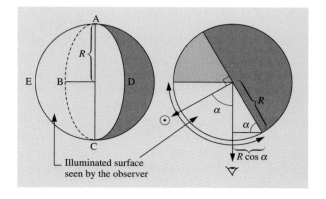

Illuminated surface
seen by the observer

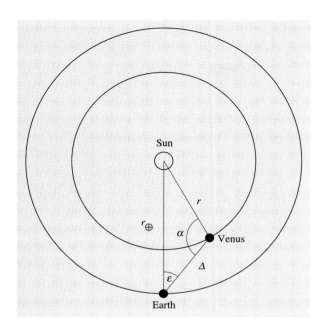

If $r = 0.723$ AU and $R_\oplus = 1$ AU, the distance is $\Delta = 0.43$ AU and the corresponding phase angle is $\alpha = 118°$.

Thus Venus is brightest shortly after the largest eastern elongation and before the largest western elongation. From the sine formula we obtain

$$\frac{\sin \varepsilon}{r} = \frac{\sin \alpha}{M_\oplus} .$$

The corresponding elongation is $\varepsilon = 40°$, and

$$\frac{1 + \cos \alpha}{2} \times 100\% = 27\%$$

of the surface is seen lit.

Example 7.5 *Size of an Asteroid*

The apparent visual magnitude V of a minor planet at the moment of opposition was observed to be $V_0 = 10.0$. Give an estimate of the asteroid's size when the geometric albedo is assumed to be $p = 0.15$. Give an error estimate if the geometric albedo is wrong by a factor of two. The visual magnitude of the Sun is $V_\odot = -26.8$, and the distance of the asteroid from the Earth was $\Delta = 3$ AU.

At opposition the distance of a body from the Sun is $r = \Delta + 1$. Assuming that $r_\oplus = 1$ AU, the radius R can be solved from (7.43)

$$R = \frac{r\Delta}{r_\oplus} \sqrt{\frac{10^{-0.4(V_0 - V_\odot)}}{p}}$$

$$= \frac{3 \cdot 4}{1} \sqrt{\frac{10^{-0.4(10+26.8)}}{0.15}} \times 1.5 \times 10^8 \text{ km}$$

$$= 200 \text{ km} .$$

If $p = 0.075$, the size of the asteroid is $R = 290$ km and if $p = 0.3$, the radius will be $R = 140$ km.

Example 7.6 *The Roche Limit*

A French mathematician *Edouard Roche* computed in 1848 a limit where a moon will be torn apart due to the tidal forces if it approaches its parent planet. Roche proposed that the Saturnian rings were formed in that way.

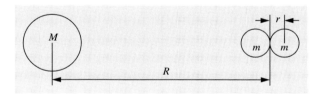

We can compute the Roche limit for a body at a distance R of a planet the mass of which is M approximating the body with two small spheres of radius r and mass m. The difference of gravitation affecting the small spheres by the planet is

$$\Delta F = GMm \left[\frac{1}{(R-r)^2} - \frac{1}{(R+r)^2} \right]$$

$$\overset{R \gg r}{\approx} GMm \frac{4r}{R^3} .$$

The gravitational force between the small spheres is

$$F' = \frac{Gm^2}{4r^2} .$$

If $\Delta F > F'$, the small spheres will be pulled apart. The forces are equal at the Roche limit:

$$GMm\frac{4r}{R^3} = \frac{Gm^2}{4r^2} .$$

Thus the distance of the Roche limit R is

$$R = \sqrt[3]{\frac{16r^3 M}{m}} .$$

Inserting the masses of the planet and the spheres in terms of the radii of the planet S and spheres r, and assuming that the densities ρ are equal, $m = \frac{4}{3}\pi r^3 \rho$, $M = \frac{4}{3}\pi S^3 \rho$, we obtain

$$R \approx 2.5 \times S .$$

Our result is valid only for a body without any internal strength. Smaller bodies with internal strength can survive inside the Roche limit. You, dear reader, act as an excellent example of this, because you read this example well inside the Roche limit of the Earth. A 100 km stony asteroid will survive even if it orbits the Earth just above the atmosphere but a sphere of water of the same size would break apart.

7.22 Exercises

Exercise 7.1 What is the greatest possible elongation of Mercury, Venus and Mars? How long before sunrise or after sunset is the planet visible? Assume that the declination of the planet and the Sun is $\delta = 0°$.

Exercise 7.2 a) What is the greatest possible geocentric latitude of Venus, i.e. how far from the Sun can the planet be at the inferior conjunction? Assume the orbits are circular. b) When is the situation possible? The longitude of the ascending node of Venus is $77°$.

Exercise 7.3 The interval between two oppositions of a planet was 398.9 d. The angular diameter of the planet at the opposition was $47.2''$. Find the sidereal period, semimajor axis, and the true diameter (in kilometres) of the planet. Which planet was it?

Exercise 7.4 a) Assume that three bodies move along circular orbits with angular velocities (mean motions) n_1, n_2 and n_3. Show that these bodies have a common synodic period if and only if there are nonzero integers k_1, k_2 and k_3 such that

$$k_1 n_1 + k_2 n_2 + k_3 n_3 = 0 , \qquad k_1 + k_2 + k_3 = 0 .$$

b) The resonance of the Galilean satellites can be expressed in terms of their mean motions as

$$n_{\text{Io}} - 3n_{\text{Europa}} + 2n_{\text{Ganymede}} = 0 .$$

Find the synodic period of these three moons.

Exercise 7.5 a) Find the daily retrograde apparent motion of an exterior planet at its opposition. Assume that the planet and the Earth have circular orbits. b) Pluto was found in 1930 from two plates, exposed 6 days apart during the opposition of the planet. On those plates one degree corresponded to 3 cm. How much (in cm) had Pluto moved between the exposures? How much does a typical main belt asteroid move in the same time?

Exercise 7.6 A planet is observed at the opposition or inferior conjunction. Due to the finite speed of light the apparent direction of the planet differs from the true place. Find this difference as a function of the radius of the orbit. You can assume the orbits are circular. Which planet has the largest deviation?

Exercise 7.7 The angular diameter of the Moon is $0.5°$. The full moon has an apparent magnitude of -12.5 and the Sun -26.7. Find the geometric and Bond albedos of the Moon, assuming that the reflected light is isotropic (into a solid angle 2π sterad).

Exercise 7.8 The eccentricity of the orbit of Mercury is 0.206. How much does the apparent magnitude of the Sun vary as seen from Mercury? How does the surface brightness of the Sun vary?

Exercise 7.9 An asteroid with a diameter of 100 m approaches the Earth at a velocity of $30\,\text{km}\,\text{s}^{-1}$. Find the apparent magnitude of the asteroid a) one week, b) one day before the collision. Assume that the phase angle is $\alpha = 0°$ and the geometric albedo of the asteroid is $p = 0.1$. What do you think about the chances of finding the asteroid well in advance the crash?

Exercise 7.10 Find the distance of a comet from the Sun when its temperature reaches $0\,°\text{C}$ and $100\,°\text{C}$. Assume the Bond albedo of the comet is 0.05.

8. Stellar Spectra

All our information about the physical properties of stars comes more or less directly from studies of their spectra. In particular, by studying the strength of various absorption lines, stellar masses, temperatures and compositions can be deduced. The line shapes contain detailed information about atmospheric processes.

As we have seen in Chap. 3, the light of a star can be dispersed into a spectrum by means of a prism or a diffraction grating. The distribution of the energy flux density over frequency can then be derived. The spectra of stars consist of a *continuous spectrum* or *continuum* with narrow *spectral lines* superimposed (Fig. 8.1). The lines in stellar spectra are mostly dark *absorption lines*, but in some objects bright *emission lines* also occur.

In a very simplified way the continuous spectrum can be thought of as coming from the hot surface of the star. Atoms in the atmosphere above the surface absorb certain characteristic wavelengths of this radiation, leaving dark "gaps" at the corresponding points in the spectrum. In reality there is no such sharp separation between surface and atmosphere. All layers emit and absorb radiation, but the net result of these processes is that less energy is radiated at the wavelengths of the absorption lines.

The spectra of stars are classified on the basis of the strengths of the spectral lines. *Isaac Newton* observed the solar spectrum in 1666, but, properly speaking, spectroscopy began in 1814 when *Joseph Fraunhofer* observed the dark lines in the spectrum of the Sun. He assigned capital letters, like D, G, H and K, to some of the stronger dark lines without knowing the elements responsible for the origin of the lines (Sect. 8.2). The absorption lines are also known as Fraunhofer lines. In 1860, *Gustav Robert Kirchhoff* and *Robert Bunsen* identified the lines as the characteristic lines produced by various elements in an incandescent gas.

8.1 Measuring Spectra

The most important methods of forming a spectrum are by means of an objective prism or a slit spectrograph. In the former case one obtains a photograph, where each stellar image has been spread into a spectrum. Up to several hundred spectra can be photographed on a single plate and used for spectral classification. The amount of detail that can be seen in a spectrum depends on its *dispersion*, the range of wavelengths per millimetre on the

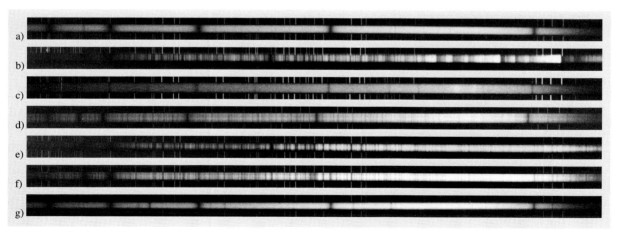

Fig. 8.1a–g. Typical stellar spectra. The spectrum of η Pegasi (**f**) is very similar to that of the Sun. The continuous spectrum is brightest at about 550 nm and gets fainter towards smaller and larger wavelengths. Dark absorption lines are superimposed on the continuum. See also Exercise 8.1. (Mt. Wilson Observatory)

plate (or per pixel on a CCD). The dispersion of an objective prism is a few tens of nanometres per millimetre. More detailed observations require a slit spectrograph, which can reach a dispersion 1–0.01 nm/mm. The detailed shape of individual spectral lines can then be studied.

The photograph of the spectrum is converted to an intensity tracing showing the flux density as a function of wavelength. This is done by means of a microdensitometer, measuring the amount of light transmitted by the recorded spectrum. Since the blackening of a photographic plate is not linearly related to the amount of energy it has received, the measured blackening has to be calibrated by comparison with known ex-

posures. In modern CCD spectrographs the intensity curve is determined directly without the intervening step of a photographic plate. For measurements of line strengths the spectrum is usually rectified by dividing by the continuum intensity.

Figure 8.2 shows a photograph of the spectrum of a star and the intensity curve obtained from a calibrated and rectified microdensitometer tracing. The second pair of pictures shows the intensity curve before and after the normalisation. The absorption lines appear as troughs of various sizes in the curve. In addition to the clear and deep lines, there are large numbers of weaker lines that can barely be discerned. The graininess of the photographic emulsion is a source of noise

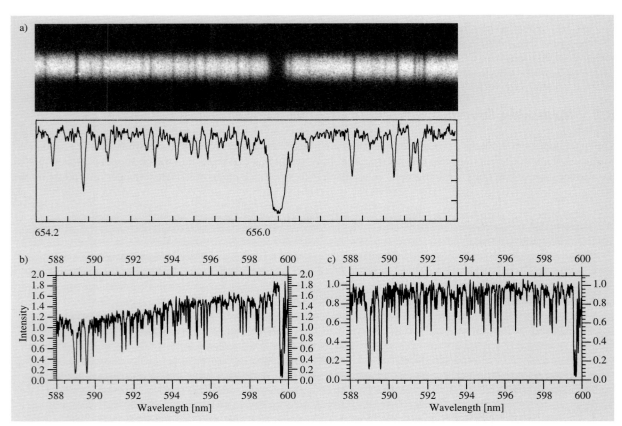

Fig. 8.2. (a) A section of a photograph of a stellar spectrum and the corresponding rectified microdensitometer intensity tracing. The original spectrum was taken at the Crimean Observatory. (b) A more extensive part of the spectrum. (c) The

picture the intensity curve of the first picture has been rectified by normalizing the value of the continuum intensity to one. (Pictures by J. Kyröläinen and H. Virtanen, Helsinki Observatory)

which appears as irregular fluctuations of the intensity curve. Some lines are so close together that they appear blended at this dispersion.

The detailed shape of a spectral line is called the *line profile* (Sect. 5.3). The true shape of the line reflects the properties of the stellar atmosphere, but the observed profile is also spread out by the measuring instrument. However, the total absorption in the line, usually expressed in terms of the *equivalent width*, is less sensitive to observational effects (see Fig. 5.6).

The equivalent width of a spectral line depends on how many atoms in the atmosphere are in a state in which they can absorb the wavelength in question. The more atoms there are, the stronger and broader the spectral line is. For example, a typical equivalent width of a metal line (Fe) in the solar spectrum is about 10 pm. Line widths are often expressed in ångströms $(1\,\text{Å} = 10^{-10}\,\text{m} = 0.1\,\text{nm})$.

Only in weak lines the equivalent width depends linearly on the number of absorbing atoms. The equivalent width as a function of the amount of absorbing atoms is known as the *curve of growth*. It is, however, beyond the scope of this book.

Line profiles are also broadened by the Doppler effect. In stellar atmospheres there are motions of small and large scale, like thermal motion of the atoms and convective flows.

The chemical composition of the atmosphere can be determined from the strengths of the spectral lines. With the introduction of large computers it has become feasible to construct quite detailed models of the structure of stellar atmospheres, and to compute the emergent spectrum for a given model. The computed synthetic spectrum can be compared with the observations and the theoretical model modified until a good fit is obtained. The theoretical models then give the number of absorbing atoms, and hence the element abundances, in the atmosphere. The construction of model atmospheres will be discussed in Sect. 8.6.

8.2 The Harvard Spectral Classification

The spectral classification scheme in present use was developed at Harvard Observatory in the United States in the early 20th century. The work was begun by *Henry Draper* who in 1872 took the first photograph of the spectrum of Vega. Later Draper's widow donated the observing equipment and a sum of money to Harvard Observatory to continue the work of classification.

The main part of the classification was done by *Annie Jump Cannon* using objective prism spectra. The *Henry Draper* Catalogue (HD) was published in 1918–1924. It contains 225,000 stars extending down to 9 magnitudes. Altogether more than 390,000 stars were classified at Harvard.

The Harvard classification is based on lines that are mainly sensitive to the stellar temperature, rather than to gravity or luminosity. Important lines are the hydrogen Balmer lines, the lines of neutral helium, the iron lines, the H and K doublet of ionized calcium at 396.8 and 393.3 nm, the G band due to the CH molecule and some metals around 431 nm, the neutral calcium line at 422.7 nm and the lines of titanium oxide (TiO).

The main types in the Harvard classification are denoted by capital letters. They were initially ordered in alphabetical sequence, but subsequently it was noticed that they could be ordered according to temperature. With the temperature decreasing towards the right the sequence is

$$
\begin{array}{c}
\text{C}\\
\text{O} - \text{B} - \text{A} - \text{F} - \text{G} - \text{K} - \text{M}\,.\\
\text{S}
\end{array}
$$

Additional notations are Q for novae, P for planetary nebulae and W for Wolf–Rayet stars. The class C consists of the earlier types R and N. The spectral classes C and S represent parallel branches to types G–M, differing in their surface chemical composition. There is a well-known mnemonic for the spectral classes, but due to its chauvinistic tone we refuse to tell it.

The spectral classes are divided into subclasses denoted by the numbers $0 \ldots 9$; sometimes decimals are used, e.g. B 0.5 (Figs. 8.3 and 8.4). The main characteristics of the different classes are:

O Blue stars, surface temperature 20,000–35,000 K. Spectrum with lines from multiply ionized atoms, e.g. He II, C III, N III, O III, Si V. He I visible, H I lines weak.

B Blue-white stars, surface temperature about 15,000 K. He II lines have disappeared, He I (403 nm) lines are strongest at B2, then get weaker and have disappeared at type B9. The K line of Ca II

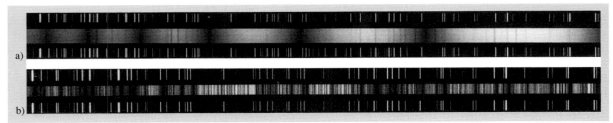

Fig. 8.3a,b. Spectra of early and late spectral type stars between 375 and 390 nm. (**a**) The upper star is Vega, of spectral type A0, and (**b**) the lower one is Aldebaran, of spectral type K5. The hydrogen Balmer lines are strong in the spectrum of Vega; in that of Aldebaran, there are many metal lines. (Lick Observatory)

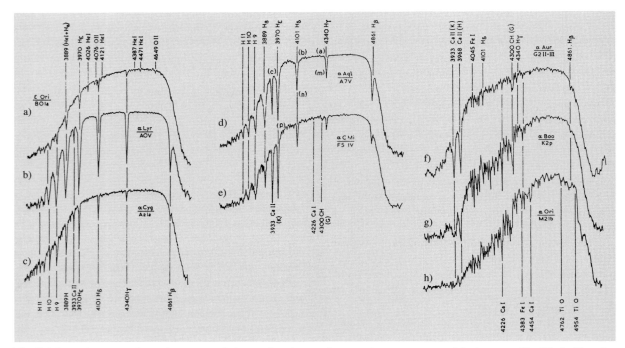

Fig. 8.4. Intensity curves for various spectral classes showing characteristic spectral features. The name of the star and its spectral and luminosity class are given next to each curve, and the most important spectral features are identified. (Drawing by J. Dufay)

becomes visible at type B3. H I lines getting stronger. O II, Si II and Mg II lines visible.

A White stars, surface temperature about 9000 K. The H I lines are very strong at A0 and dominate the whole spectrum, then get weaker. H and K lines of Ca II getting stronger. He I no longer visible. Neutral metal lines begin to appear.

F Yellow-white stars, surface temperature about 7000 K. H I lines getting weaker, H and K of Ca II getting stronger. Many other metal lines, e.g. Fe I, Fe II, Cr II, Ti II, clear and getting stronger.

G Yellow stars like the Sun, surface temperature about 5500 K. The H I lines still getting weaker, H and K lines very strong, strongest at G0. Metal lines getting stronger. G band clearly visible. CN lines seen in giant stars.

K Orange-yellow stars, surface temperature about 4000 K. Spectrum dominated by metal lines. H I lines insignificant. Ca I 422.7 nm clearly visible. Strong H and K lines and G band. TiO bands become visible at K5.

M Red stars, surface temperature about 3000 K. TiO bands getting stronger. Ca I 422.7 nm very strong. Many neutral metal lines.

C Carbon stars, previously R and N. Very red stars, surface temperature about 3000 K. Strong molecular bands, e. g. C_2, CN and CH. No TiO bands. Line spectrum like in the types K and M.

S Red low-temperature stars (about 3000 K). Very clear ZrO bands. Also other molecular bands, e. g. YO, LaO and TiO.

The main characteristics of the classification scheme can be seen in Fig. 8.5 showing the variations of some typical absorption lines in the different spectral classes. Different spectral features are mainly due to different effective temparatures. Different pressures and chemical compositions of stellar atmospheres are not very important factors in the spectral classification, execpt in some peculiar stars. The *early*, i. e. hot, spectral classes are characterised by the lines of ionized atoms, whereas the cool, or late, spectral types have lines of neutral atoms. In hot stars molecules dissociate into atoms; thus the absorption bands of molecules appear only in the spectra of cool stars of late spectral types.

To see how the strengths of the spectral lines are determined by the temperature, we consider, for example, the neutral helium lines at 402.6 nm and 447.2 nm. These are only seen in the spectra of hot stars. The rea-

son for this is that the lines are due to absorption by excited atoms, and that a high temperature is required to produce any appreciable excitation. As the stellar temperature increases, more atoms are in the required excited state, and the strength of the helium lines increases. When the temperature becomes even higher, helium begins to be ionized, and the strength of the neutral helium lines begins to decrease. In a similar way one can understand the variation with temperature of other important lines, such as the calcium H and K lines. These lines are due to singly ionized calcium, and the temperature must be just right to remove one electron but no more.

The hydrogen Balmer lines H_β, H_γ and H_δ are stongest in the spectral class A2. These lines correspond to transitions to the level the principal quantum number of which is $n = 2$. If the temprature is too high the hydrogen is ionized and such transitions are not possible.

8.3 The Yerkes Spectral Classification

The Harvard classification only takes into account the effect of the temperature on the spectrum. For a more precise classification, one also has to take into account the luminosity of the star, since two stars with the same effective temperature may have widely different luminosities. A two-dimensional system of spectral classification was introduced by *William W. Morgan*, *Philip C. Keenan* and *Edith Kellman* of Yerkes Obser-

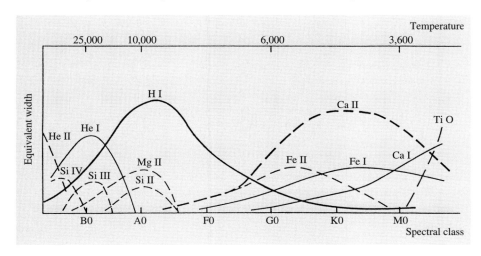

Fig. 8.5. Equivalent widths of some important spectral lines in the various spectral classes. [Struve, O. (1959): Elementary Astronomy (Oxford University Press, New York) p. 259]

vatory. This system is known as the MKK or *Yerkes classification*. (The MK classification is a modified, later version.) The MKK classification is based on the visual scrutiny of slit spectra with a dispersion of 11.5 nm/mm. It is carefully defined on the basis of standard stars and the specification of luminosity criteria. Six different *luminosity classes* are distinguished:

- Ia most luminous supergiants,
- Ib less luminous supergiants,
- II luminous giants,
- III normal giants,
- IV subgiants,
- V main sequence stars (dwarfs).

The luminosity class is determined from spectral lines that depend strongly on the stellar surface gravity, which is closely related to the luminosity. The masses of giants and dwarfs are roughly similar, but the radii of giants are much larger than those of dwarfs. Therefore

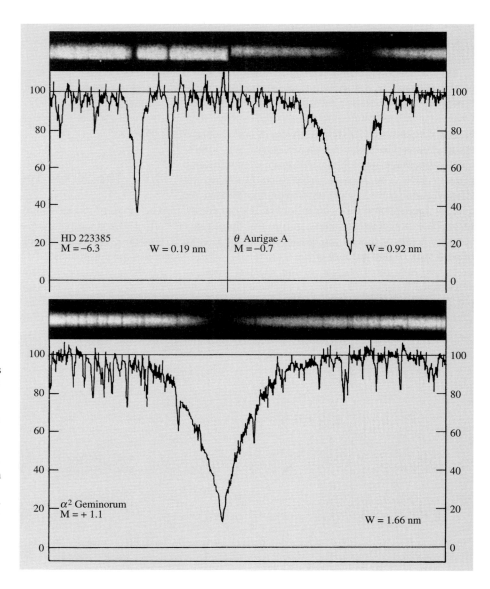

Fig. 8.6. Luminosity effects in the hydrogen H_γ line in A stars. The vertical axis gives the normalized intensity. HD 223385 (*upper left*) is an A2 supergiant, where the line is very weak, θ Aurigae A is a giant star and α^2 Geminorum is a main sequence star, where the line is very broad. [Aller, L.H. (1953): *Astrophysics. The Atmospheres of the Sun and Stars* (The Ronald Press Company, New York) p. 318]

the gravitational acceleration $g = GM/R^2$ at the surface of a giant is much smaller than for a dwarf. In consequence, the gas density and pressure in the atmosphere of a giant is much smaller. This gives rise to *luminosity effects* in the stellar spectrum, which can be used to distinguish between stars of different luminosities.

1. For spectral types B–F, the lines of neutral hydrogen are deeper and narrower for stars of higher luminosities. The reason for this is that the metal ions give rise to a fluctuating electric field near the hydrogen atoms. This field leads to shifts in the hydrogen energy levels (the Stark effect), appearing as a broadening of the lines. The effect becomes stronger as the density increases. Thus the hydrogen lines are narrow in absolutely bright stars, and become broader in main sequence stars and even more so in white dwarfs (Fig. 8.6).

2. The lines from ionized elements are relatively stronger in high-luminosity stars. This is because the higher density makes it easier for electrons and ions to recombine to neutral atoms. On the other hand, the rate of ionization is essentially determined by the radiation field, and is not appreciably affected by the gas density. Thus a given radiation field can maintain a higher degree of ionization in stars with more extended atmospheres. For example, in the spectral classes F–G, the relative strengths of the ionized strontium (Sr II) and neutral iron (Fe I) lines can be used as a luminosity indicator. Both lines depend on the temperature in roughly the same way, but the Sr II lines become relatively much stronger as the luminosity increases.

3. Giant stars are redder than dwarfs of the same spectral type. The spectral type is determined from the strengths of spectral lines, including ion lines. Since these are stronger in giants, a giant will be cooler, and thus also redder, than a dwarf of the same spectral type.

4. There is a strong cyanogen (CN) absorption band in the spectra of giant stars, which is almost totally absent in dwarfs. This is partly a temperature effect, since the cooler atmospheres of giants are more suitable for the formation of cyanogen.

8.4 Peculiar Spectra

The spectra of some stars differ from what one would expect on the basis of their temperature and luminosity (see, e. g., Fig. 8.7). Such stars are celled *peculiar*. The most common peculiar spectral types will now be considered.

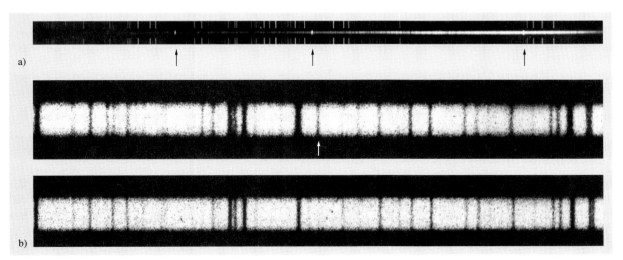

Fig. 8.7a,b. Peculiar spectra. (**a**) R Geminorum (above) is an emission line star, with bright emission lines, indicated by arrows, in its spectrum; (**b**) the spectrum of a normal star is compared with one in which the zirconium lines are unusually strong. (Mt. Wilson Observatory and Helsinki Observatory)

The *Wolf–Rayet stars* are very hot stars; the first examples were discovered by *Charles Wolf* and *Georges Rayet* in 1867. The spectra of Wolf–Rayet stars have broad emission lines of hydrogen and ionized helium, carbon, nitrogen and oxygen. There are hardly any absorption lines. The Wolf–Rayet stars are thought to be members of binary systems, where they have lost their outer layers to a companion star. This has exposed the stellar interior, which gives rise to a different spectrum than the normal outer layers.

In some O and B stars the hydrogen absorption lines have weak emission components either at the line centre or in its wings. These stars are called *Be* and *shell stars* (the letter e after the spectral type indicates that there are emission lines in the spectrum). The emission lines are formed in a rotationally flattened gas shell around the star. The shell and Be stars show irregular variations, apparently related to structural changes in the shell. About 15% of all O and B stars have emission lines in their spectra.

The strongest emission lines are those of the *P Cygni* stars, which have one or more sharp absorption lines on the short wavelength side of the emission line. It is thought that the lines are formed in a thick expanding envelope. The P Cygni stars are often variable. For example, P Cygni itself has varied between three and six magnitudes during the past centuries. At present its magnitude is about 5.

The peculiar A stars or *Ap stars* (p = peculiar) are usually strongly magnetic stars, where the lines are split into several components by the Zeeman effect. The lines of certain elements, such as magnesium, silicon, europium, chromium and strontium, are exceptionally strong in the Ap stars. Lines of rarer elements such as mercury, gallium or krypton may also be present. Otherwise, the Ap stars are like normal main sequence stars.

The *Am stars* (m = metallic) also have anomalous element abundances, but not to the same extent as the Ap stars. The lines of e. g. the rare earths and the heaviest elements are strong in their spectra; those of calcium and scandium are weak.

We have already mentioned the S and C stars, which are special classes of K and M giants with anomalous element abundances. In the S stars, the normal lines of titanium, scandium and vanadium oxide are replaced with oxides of heavier elements, zirconium, yttrium and barium. A large fraction of the S stars are irregular variables. The name of the C stars refers to carbon. The metal oxide lines are almost completely absent in their spectra; instead, various carbon compounds (CN, C_2, CH) are strong. The abundance of carbon relative to oxygen is 4–5 times greater in the C stars than in normal stars. The C stars are divided into two groups, hotter R stars and cooler N stars.

Another type of giant stars with abundance anomalies are the *barium stars*. The lines of barium, strontium, rare earths and some carbon compounds are strong in their spectra. Apparently nuclear reaction products have been mixed up to the surface in these stars.

8.5 The Hertzsprung–Russell Diagram

Around 1910, *Ejnar Hertzsprung* and *Henry Norris Russell* studied the relation between the absolute magnitudes and the spectral types of stars. The diagram showing these two variables is now known as the *Hertzsprung–Russell diagram* or simply the *HR diagram* (Fig. 8.8). It has turned out to be an important aid in studies of stellar evolution.

In view of the fact that stellar radii, luminosities and surface temperatures vary widely, one might have expected the stars to be uniformly distributed in the HR diagram. However, it is found that most stars are located along a roughly diagonal curve called the *main sequence*. The Sun is situated about the middle of the main sequence.

The HR diagram also shows that the yellow and red stars (spectral types G-K-M) are clustered into two clearly separate groups: the *main sequence* of *dwarf stars* and the *giants*. The giant stars fall into several distinct groups. The *horizontal branch* is an almost horizontal sequence, about absolute visual magnitude zero. The *red giant branch* rises almost vertically from the main sequence at spectral types K and M in the HR diagram. Finally, the *asymptotic branch* rises from the horizontal branch and approaches the bright end of the red giant branch. These various branches represent different phases of stellar evolution (c. f. Sects. 11.3 and 11.4): dense areas correspond to evolutionary stages in which stars stay a long time.

A typical horizontal branch giant is about a hundred times brighter than the Sun. Since giants and dwarfs of the same spectral class have nearly the same surface

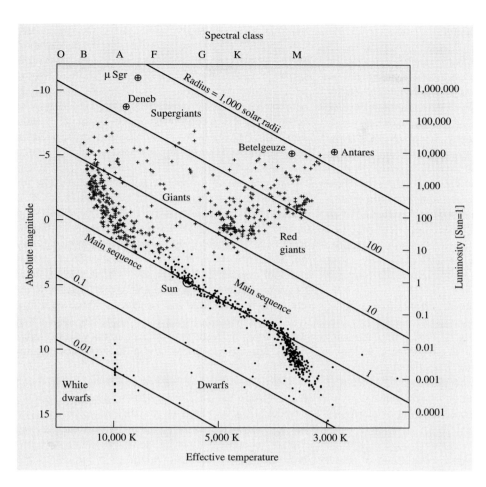

Fig. 8.8. The Hertzsprung–Russell diagram. The horizontal coordinate can be either the colour index $B-V$, obtained directly from observations, or the spectral class. In theoretical studies the effective temperature T_e is commonly used. These correspond to each other but the dependence varies somewhat with luminosity. The vertical axis gives the absolute magnitude. In a $(\lg(L/L_\odot), \lg T_e)$ plot the curves of constant radius are straight lines. The densest areas are the main sequence and the horizontal, red giant and asymptotic branches consisting of giant stars. The supergiants are scattered above the giants. To the lower left are some white dwarfs about 10 magnitudes below the main sequence. The apparently brightest stars ($m < 4$) are marked with crosses and the nearest stars ($r < 50$ ly) with dots. The data are from the Hipparcos catalogue

temperature, the difference in luminosity must be due to a difference in radius according to (5.21). For example Arcturus, which is one of the brightest stars in the sky, has a radius about thirty times that of the Sun.

The brightest red giants are the *supergiants* with magnitudes up to $M_V = -7$. One example is Betelgeuze in Orion, with a radius of 400 solar radii and 20,000 times more luminous than the Sun.

About 10 magnitudes below the main sequence are the *white dwarfs*. They are quite numerous in space, but faint and difficult to find. The best-known example is Sirius B, the companion of Sirius.

There are some stars in the HR diagram which are located below the giant branch, but still clearly above the main sequence. These are known as *subgiants*. Similarly, there are stars below the main sequence, but brighter than the white dwarfs, known as *subdwarfs*.

When interpreting the HR diagram, one has to take into account *selection effects*: absolutely bright stars are more likely to be included in the sample, since they can be discovered at greater distances. If only stars within a certain distance from the Sun are included, the distribution of stars in the HR diagram looks quite different. This can be seen in Fig. 8.8: there are no giant or bright main sequence stars among these.

The HR diagrams of star clusters are particularly important for the theory of stellar evolution. They will be discussed in Chap. 16.

8.6 Model Atmospheres

The stellar atmosphere consists of those layers of the star where the radiation that is transmitted directly to the observer originates. Thus in order to interpret stellar spectra, one needs to be able to compute the structure of the atmosphere and the emerging radiation.

In actual stars there are many factors, such as rotation and magnetic fields, that complicate the problem of computing the structure of the atmosphere. We shall only consider the classical problem of finding the structure, i.e. the distribution of pressure and temperature with depth, in a static, unmagnetized atmosphere. In that case a model atmosphere is completely specified by giving the chemical composition, the gravitational acceleration at the surface, g, and the energy flux from the stellar interior, or equivalently, the effective temperature T_e.

The basic principles involved in computing a model stellar atmosphere are the same as for stellar interiors and will be discussed in Chap. 10. Essentially, there are two differential equations to be solved: the equation of hydrostatic equilibrium, which fixes the distribution of pressure, and an equation of energy transport, which will have a different form depending on whether the atmosphere is radiative or convective, and which determines the temperature distribution.

The values of the various physical quantities in an atmosphere are usually given as functions of some suitably defined continuum optical depth τ. Thus pressure, temperature, density, ionization and the population numbers of various energy levels can all be obtained as functions of τ. When these are known, the intensity of radiation emerging from the atmosphere can be computed. In *The Intensity Emerging from a Stellar Atmosphere (p. 212), it is shown that approximately the emergent spectrum originates at unit optical depth, measured along each light ray. On this basis, one can predict whether a given spectral line will be present in the spectrum.

Consider a spectral line formed when an atom (or ion) in a given energy state absorbs a photon. From the model atmosphere, the occupation number of the absorbing level is known as a function of the (continuum) optical depth τ. If now there is a layer above the depth $\tau = 1$ where the absorbing level has a high occupancy, the optical depth in the line will become unity

before $\tau = 1$, i.e. the radiation in the line will originate higher in the atmosphere. Because the temperature increases inwards, the intensity in the line will correspond to a lower temperature, and the line will appear dark. On the other hand, if the absorbing level is unoccupied, the optical depth at the line frequency will be the same as the continuum optical depth. The radiation at the line frequency will then come from the same depth as the adjacent continuum, and no absorption line will be formed.

The expression for the intensity derived in *The Intensity Emerging from a Stellar Atmosphere (p. 212) also explains the phenomenon of *limb darkening* seen in the Sun (Sect. 12.2). The radiation that reaches us from near the edge of the solar disc emerges at a very oblique angle (θ near $90°$), i.e. $\cos \theta$ is small. Thus this radiation originates at small values of τ, and hence at low temperatures. In consequence, the intensity coming from near the edge will be lower, and the solar disc will appear darker towards the limb. The amount of limb darkening also gives an empirical way of determining the temperature distribution in the solar atmosphere.

Our presentation of stellar atmospheres has been highly simplified. In practice, the spectrum is computed numerically for a range of parameter values. The values of T_e and element abundances for various stars can then be found by comparing the observed line strengths and other spectral features with the theoretical ones. We shall not go into details on the procedures used.

8.7 What Do the Observations Tell Us?

To conclude this chapter, we shall give a summary of the properties of stars revealed by the observations. At the end of the book, there are tables of the brightest and of the nearest stars.

Of the brightest stars, four have a negative magnitude. Some of the apparently bright stars are absolutely bright supergiants, others are simply nearby.

In the list of the nearest stars, the dominance of faint dwarf stars, already apparent in the HR diagram, is worth noting. Most of these belong to the spectral types K and M. Some nearby stars also have very faint companions with masses about that of Jupiter, i.e. planets. They have not been included in the table.

Stellar spectroscopy offers an important way of determining fundamental stellar parameters, in particular mass and radius. However, the spectral information needs to be calibrated by means of direct measurements of these quantities. These will be considered next.

The *masses* of stars can be determined in the case of double stars orbiting each other. (The details of the method will be discussed in Chap. 9.) These observations have shown that the larger the mass of a main sequence star becomes, the higher on the main sequence it is located. One thus obtains an empirical *mass–luminosity relation*, which can be used to estimate stellar masses on the basis of the spectral type.

The observed relation between mass and luminosity is shown in Fig. 8.9. The luminosity is roughly proportional to the power 3.8 of the mass:

$$L \propto M^{3.8} . \tag{8.1}$$

The relations is only approximate. According to it, a ten solar mass star is about 6300 times brighter than the Sun, corresponding to 9.5 magnitudes.

The smallest observed stellar masses are about $1/20$ of the solar mass, corresponding to stars in the lower right-hand part of the HR diagram. The masses of white dwarfs are less than one solar mass. The masses of the most massive main sequence and supergiant stars are between 10 and 50 $M_\odot$.

Direct interferometric measurements of stellar angular diameters have been made for only a few dozen stars. When the distances are known, these immediately yield the value of the radius. In eclipsing binaries, the radius can also be directly measured (see Sect. 9.4). Altogether, close to a hundred stellar radii are known from direct measurements. In other cases, the radius must be estimated from the absolute luminosity and effective temperature.

In discussing stellar radii, it is convenient to use a version of the HR diagram with $\lg T_e$ on the horizontal and M_{bol} or $\lg(L/L_\odot)$ on the vertical axis. If the value of the radius R is fixed, then (5.21) yields a linear relation between the bolometric magnitude and $\lg T_e$. Thus lines of constant radius in the HR diagram are straight. Lines corresponding to various values of the radius are shown in Fig. 8.8. The smallest stars are the white dwarfs with radii of about one per cent of the solar radius, whereas the largest supergiants have radii sev-

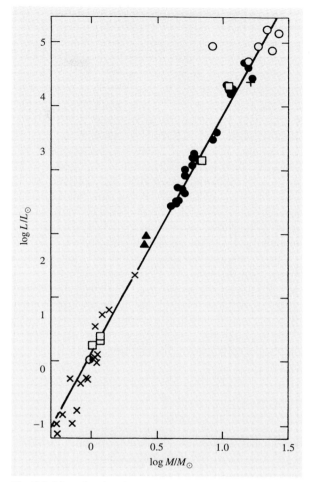

Fig. 8.9. Mass–luminosity relation. The picture is based on binaries with known masses. Different symbols refer to different kinds of binaries. (From Böhm-Vitense: *Introduction to Stellar Astrophysics*, Cambridge University Press (1989–1992))

eral thousand times larger than the Sun. Not included in the figure are the compact stars (neutron stars and black holes) with typical radii of a few tens of kilometres.

Since the stellar radii vary so widely, so do the densities of stars. The density of giant stars may be only 10^{-4} kg/m^3, whereas the density of white dwarfs is about 10^9 kg/m^3.

The range of values for stellar effective temperatures and luminosities can be immediately seen in the HR diagram. The range of effective temperature is 2,000–40,000 K, and that of luminosity 10^{-4}–10^6 $L_\odot$.

The *rotation* of stars appears as a broadening of the spectral lines. One edge of the stellar disc is approaching us, the other edge is receding, and the radiation from the edges is Doppler shifted accordingly. The rotational velocity observed in this way is only the component along the line of sight. The true velocity is obtained by dividing with $\sin i$, where i is the angle between the line of sight and the rotational axis. A star seen from the direction of the pole will show no rotation.

Assuming the axes of rotation to be randomly oriented, the distribution of rotational velocities can be statistically estimated. The hottest stars appear to rotate faster than the cooler ones. The rotational velocity at the equator varies from 200–250 km/s for O and B stars to about 20 km/s for spectral type G. In shell stars, the rotational velocity may reach 500 km/s.

The *chemical composition* of the outer layers is deduced from the strengths of the spectral lines. About three-fourths of the stellar mass is hydrogen. Helium comprises about one-fourth, and the abundance of other elements is very small. The abundance of heavy elements in young stars (about 2%) is much larger than in old ones, where it is less than 0.02%.

* The Intensity Emerging from a Stellar Atmosphere

The intensity of radiation emerging from an atmosphere is given by the expression (5.45), i. e.

$$I_\nu(0, \theta) = \int_0^\infty S_\nu(\tau_\nu) \, e^{-\tau_\nu \sec\theta} \sec\theta \, d\tau_\nu \,. \tag{8.2}$$

If a model atmosphere has been computed, the source function S_ν is known.

An approximate formula for the intensity can be derived as follows. Let us expand the source function as a Taylor series about some arbitrary point τ^*, thus

$$S_\nu = S_\nu(\tau^*) + (\tau_\nu - \tau^*) S_\nu'(\tau^*) + \ldots$$

where the dash denotes a derivative. With this expression, the integral in (8.2) can be evaluated, yielding

$$I_\nu(0, \theta) = S_\nu(\tau^*) + (\cos\theta - \tau^*) S_\nu'(\tau^*) + \ldots \,.$$

If we now choose $\tau^* = \cos\theta$, the second term will vanish. In local thermodynamic equilibrium the source function will be the Planck function $B_\nu(T)$. We thus obtain the *Eddington–Barbier approximation*

$$I_\nu(0, \theta) = B_\nu(T[\tau_\nu = \cos\theta]) \,.$$

According to this expression, the radiation emerging in a given direction originated at unit optical depth along that direction.

8.8 Exercise

Exercise 8.1 Arrange the spectra in Fig. 8.1 in the order of decreasing temperature.

9. Binary Stars and Stellar Masses

Quite often, two stars may appear to be close together in the sky, although they are really at very different distances. Such chance pairs are called *optical binary stars*. However, many close pairs of stars really are at the same distance and form a physical system in which two stars are orbiting around each other. Less than half of all stars are single stars like the Sun. More than 50% belong to systems containing two or more members. In general, the multiple systems have a hierarchical structure: a star and a binary orbiting around each other in triple systems, two binaries orbiting around each other in quadruple systems. Thus most multiple systems can be described as binaries with several levels.

Binaries are classified on the basis of the method of their discovery. *Visual binaries* can be seen as two separate components, i. e. the separation between the stars is larger than about 0.1 arc seconds. The relative position of the components changes over the years as they move in their orbits (Fig. 9.1). In *astrometric binary stars* only one component is seen, but its variable proper motion shows that a second invisible component must be present. The *spectroscopic binary stars* are discovered on the basis of their spectra. Either two sets of spectral lines are seen or else the Doppler shift of the observed lines varies periodically, indicating an invisible companion. The fourth class of binaries are the *photometric binary stars* or *eclipsing variables*. In these systems the components of the pair regularly pass in front of each other, causing a change in the total apparent magnitude.

Binary stars can also be classified on the basis of their mutual separation. In *distant binaries* the separation between the components is tens or hundreds of astronomical units and their orbital periods are from tens to thousands of years. In *close binaries* the separation is from about one AU down to the radius of the stars. The orbital period ranges from a few hours to a few years. The components of *contact binaries* are so close that they are touching each other.

The stars in a binary system move in an elliptical orbit around the centre of mass of the system. In Chap. 6 it was shown that the relative orbit, too, is an ellipse, and thus the observations are often described as if one component remained stationary and the other orbited around it.

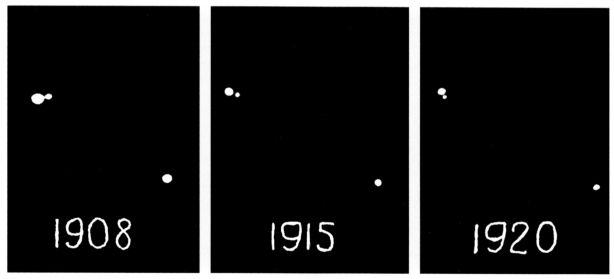

Fig. 9.1. When a visual binary is followed for a long time, the components can be seen to move with respect to each other. Picture of Krüger 60. (Yerkes Observatory)

9.1 Visual Binaries

We consider a visual binary, assuming initially that the brighter primary component is stationary and the fainter secondary component is orbiting around it. The angular separation of the stars and the angular direction to the secondary can be directly observed. Making use of observations extending over many years or decades, the relative orbit of the secondary can be determined. The first binary orbit to be determined was that of ξ UMa in 1830 (Fig. 9.2).

The observations of visual binaries only give the projection of the relative orbital ellipse on the plane of the sky. The shape and position of the true orbit are not known. However, they can be calculated if one makes use of the fact that the primary should be located at a focal point of the relative orbit. The deviation of the projected position of the primary from the focus of the projected relative orbit allows one to determine the orientation of the true orbit.

The absolute size of the orbit can only be found if the distance of the binary is known. Knowing this, the total mass of the system can be calculated from Kepler's third law.

The masses of the individual components can be determined by observing the motions of both components relative to the centre of mass (Fig. 9.3). Let the semimajor axes of the orbital ellipses of the primary and the secondary be a_1 and a_2. Then, according to the definition of the centre of mass,

$$\frac{a_1}{a_2} = \frac{m_2}{m_1} , \tag{9.1}$$

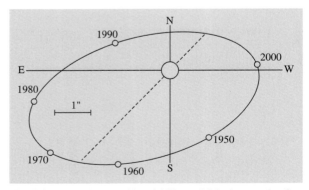

Fig. 9.2. In 1830 the orbit of ξ Ursae Majoris was the first binary orbit determined observationally

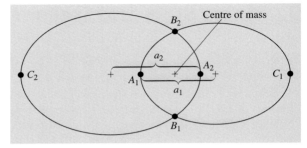

Fig. 9.3. The components of a binary system move around their common centre of mass. A_1, A_2 denote the positions of the stars at a given time A, and similarly for B and C

where m_1 and m_2 are the component masses. The semimajor axis of the relative orbit is

$$a = a_1 + a_2 . \tag{9.2}$$

For example, the masses of the components of ξ UMa have been found to be 1.3 and 1.0 solar masses.

9.2 Astrometric Binary Stars

In astrometric binaries, only the orbit of the brighter component about the centre of mass can be observed. If the mass of the visible component is estimated, e. g. from its luminosity, the mass of the invisible companion can also be estimated.

The first astrometric binary was Sirius, which in the 1830's was observed to have an undulating proper motion. It was concluded that it had a small companion, which was visually discovered a few decades later (Figs. 9.4 and 14.1). The companion, Sirius B, was a completely new type of object, a white dwarf.

The proper motions of nearby stars have been carefully studied in the search for planetary systems. Although e. g. Barnard's star may have unseen companions, the existence of planetary systems around other stars was not established by proper motion studies but with spectroscopic observations (see below).

9.3 Spectroscopic Binaries

The spectroscopic binaries (Fig. 9.5) appear as single stars in even the most powerful telescopes, but their

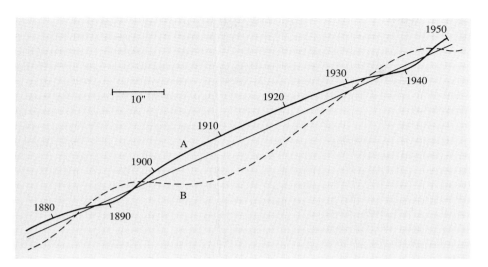

Fig. 9.4. The apparent paths of Sirius and its companion in the sky

spectra show a regular variation. The first spectroscopic binary was discovered in the 1880's, when it was found that the spectral lines of ζ UMa or Mizar split into two at regular intervals.

The Doppler shift of a spectral line is directly proportional to the radial velocity. Thus the separation of the spectral lines is largest when one component is directly approaching and the other is receding from the observer. The period of the variation is the orbital period of the stars. Unfortunately, there is no general way of determining the position of the orbit in space. The observed velocity v is related to the true velocity v_0 according to

$$v = v_0 \sin i \,, \tag{9.3}$$

where the inclination i is the angle between the line of sight and the normal of the orbital plane.

Consider a binary where the components move in circular orbits about the centre of mass. Let the radii of the orbits be a_1 and a_2. From the definition of the centre of mass $m_1 a_1 = m_2 a_2$, and writing $a = a_1 + a_2$, one obtains

$$a_1 = \frac{a m_2}{m_1 + m_2} \,. \tag{9.4}$$

The true orbital velocity is

$$v_{0,1} = \frac{2\pi a_1}{P} \,,$$

where P is the orbital period. The observed orbital velocity according to (9.3) is thus

$$v_1 = \frac{2\pi a_1 \sin i}{P} \,. \tag{9.5}$$

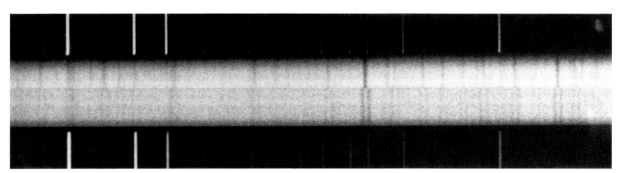

Fig. 9.5. Spectrum of the spectroscopic binary κ Arietis. In the upper spectrum the spectral lines are single, in the lower one doubled. (Lick Observatory)

Substituting (9.4), one obtains

$$v_1 = \frac{2\pi a}{P} \frac{m_2 \sin i}{m_1 + m_2} .$$

Solving for a and substituting it in Kepler's third law, one obtains the *mass function equation*:

$$\frac{m_2^3 \sin^3 i}{(m_1 + m_2)^2} = \frac{v_1^3 P}{2\pi G} . \qquad (9.6)$$

If one component in a spectroscopic binary is so faint that its spectral lines cannot be observed, only P and v_1 are observed. Equation (9.6) then gives the value of the mass function, which is the expression on the left-hand side. Neither the masses of the components nor the total mass can be determined. If the spectral lines of both components can be observed, v_2 is also known. Then (9.5) gives

$$\frac{v_1}{v_2} = \frac{a_1}{a_2}$$

and furthermore the definition of the centre of mass gives

$$m_1 = \frac{m_2 v_2}{v_1} .$$

When this is substituted in (9.6), the value of $m_2 \sin^3 i$, and correspondingly, $m_1 \sin^3 i$, can be determined. However, the actual masses cannot be found without knowing the inclination.

The size of the binary orbit (the semimajor axis a) is obtained from (9.5) apart from a factor $\sin i$. In general the orbits of binary stars are not circular and the preceding expressions cannot be applied as they stand. For an eccentric orbit, the shape of the velocity variation departs more and more from a simple sine curve as the eccentricity increases. From the shape of the velocity variation, both the eccentricity and the longitude of the periastron can be determined. Knowing these, the mass function or the individual masses can again be determined to within a factor $\sin^3 i$.

From accurate studies of the spectra of nearby stars, several planet-sized companions have been found. In the years 1995–2002, about 100 extrasolar planets were observed, with masses in the range of 0.1 up to 13 Jupiter masses.

9.4 Photometric Binary Stars

In the photometric binaries, a periodic variation in the total brightness is caused by the motions of the components in a double system. Usually the photometric binaries are *eclipsing variables*, where the brightness variations are due to the components passing in front of each other. A class of photometric binaries where there are no actual eclipses are the *ellipsoidal variables*. In these systems, at least one of the components has been distorted into an ellipsoidal shape by the tidal pull of the other one. At different phases of the orbit, the projected surface area of the distorted component varies. The surface temperature will also be lower at the ends of the tidal bulges. Together these factors cause a small variation in brightness.

The inclination of the orbit of an eclipsing binary must be very close to 90°. These are the only spectroscopic binaries for which the inclination is known and thus the masses can be uniquely determined.

The variation of the magnitude of eclipsing variables as a function of time is called the *lightcurve*. According to the shape of the lightcurve, they are grouped into three main types: *Algol*, *β Lyrae* and *W Ursae Majoris* type (Fig. 9.6).

Algol Stars. The Algol-type eclipsing variables have been named after *β Persei* or Algol. During most of the period, the lightcurve is fairly constant. This corresponds to phases during which the stars are seen separate from each other and the total magnitude remains constant. There are two different minima in the lightcurve, one of which, the primary minimum, is usually much deeper than the other one. This is due to the brightness difference of the stars. When the larger star, which is usually a cool giant, eclipses the smaller and hotter component, there is a deep minimum in the lightcurve. When the small, bright star passes across the disc of the giant, the total magnitude of the system does not change by much.

The shape of the minima depends on whether the eclipses are partial or total. In a partial eclipse the lightcurve is smooth, since the brightness changes smoothly as the depth of the eclipse varies. In a total eclipse there is an interval during which one component is completely invisible. The total brightness is then constant and the lightcurve has a flat bottomed minimum.

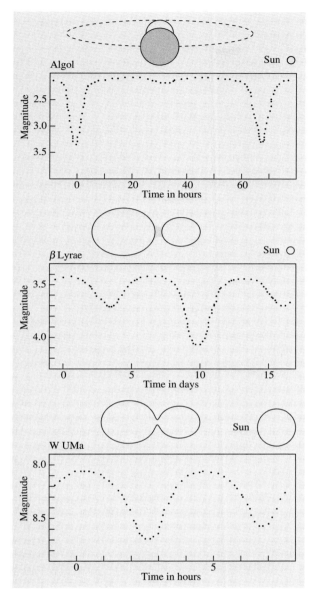

Fig. 9.6. Typical lightcurves and schematic views of Algol, β Lyrae and W Ursae Majoris type binary systems. The size of the Sun is shown for comparison

the orbit can be obtained. In that case the masses and the size of the orbit, and thus also the radii can be determined without having to know the distance of the system.

β Lyrae Stars. In the β Lyrae-type binaries, the total magnitude varies continuously. The stars are so close to each other that one of them has been pulled into ellipsoidal shape. Thus the brightness varies also outside the eclipses. The β Lyrae variables can be described as eclipsing ellipsoidal variables. In the β Lyrae system itself, one star has overfilled its Roche lobe (see Sect. 11.6) and is steadily losing mass to its companion. The mass transfer causes additional features in the lightcurve.

W UMa Stars. In W UMa stars, the lightcurve minima are almost identical, very round and broad. These are close binary systems where both components overfill their Roche lobes, forming a contact binary system.

The observed lightcurves of photometric binaries may contain many additional features that confuse the preceding classification.

– The shape of the star may be distorted by the tidal force of the companion. The star may be ellipsoidal or fill its Roche surface, in which case it becomes drop-like in shape.
– The limb darkening (Sects. 8.6 and 12.2) of the star may be considerable. If the radiation from the edges of the stellar disc is fainter than that from the centre, it will tend to round off the lightcurve.
– In elongated stars there is gravitational darkening: the parts most distant from the centre are cooler and radiate less energy.
– There are also reflection phenomena in stars. If the stars are close together, they will heat the sides facing each other. The heated part of the surface will then be brighter.
– In systems with mass transfer, the material falling onto one of the components will change the surface temperature.

All these additional effects cause difficulties in interpreting the lightcurve. Usually one computes a theoretical model and the corresponding lightcurve, which is then compared with the observations. The model is varied until a satisfactory fit is obtained.

The shape of the minima in Algol variables thus gives information on the inclination of the orbit.

The duration of the minima depends on the ratio of the stellar radii to the size of the orbit. If the star is also a spectroscopic binary, the true dimensions of

So far we have been concerned solely with the properties of binary systems in the optical domain. Recently many double systems that radiate strongly at other wavelengths have been discovered. Particularly interesting are the binary pulsars, where the velocity variation can be determined from radio observations. Many different types of binaries have also been discovered at X-ray wavelengths. These systems will be discussed in Chap. 14.

The binary stars are the only stars with accurately known masses. The masses for other stars are estimated from the mass-luminosity relation (Sect. 8.7), but this has to be calibrated by means of binary observations.

9.5 Examples

Example 9.1 *The Mass of a Binary Star*

The distance of a binary star is 10 pc and the largest angular separation of the components is $7''$ and the smallest is $1''$. The orbital period is 100 years. The mass of the binary is to be determined, assuming that the orbital plane is normal to the line of sight.

From the angular separation and the distance, the semimajor axis is

$$a = 4'' \times 10 \,\text{pc} = 40 \,\text{AU} .$$

According to Kepler's third law

$$m_1 + m_2 = \frac{a^3}{P^2} = \frac{40^3}{100^2} M_\odot = 6.4 \, M_\odot .$$

Let the semimajor axis of one component be $a_1 = 3''$ and for the other $a_2 = 1''$. Now the masses of the components can be determined separately:

$$m_1 a_1 = m_2 a_2 \quad \Rightarrow \quad m_1 = \frac{a_2}{a_1} m_2 = \frac{m_2}{3} ,$$

$$m_1 + m_2 = 6.4 \quad \Rightarrow \quad m_1 = 1.6 , \quad m_2 = 4.8 .$$

Example 9.2 *The Lightcurve of a Binary*

Let us suppose that the line of sight lies in the orbital plane of an Algol type binary, where both components have the same radius. The lightcurve is essentially as shown in the figure. The primary minimum occurs when the brighter component is eclipsed. The depth of the minima will be calculated.

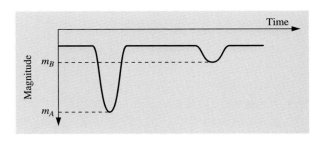

If the effective temperatures of the stars are T_A and T_B and their radius is R, their luminosities are given by

$$L_A = 4\pi R^2 \sigma T_A^4 , \quad L_B = 4\pi R^2 \sigma T_B^4 .$$

The flat part of the curve corresponds to the total luminosity

$$L_{\text{tot}} = L_A + L_B .$$

The luminosities may be expressed as absolute bolometric magnitudes by means of (4.13). Since the distance moduli of the components are the same, the apparent bolometric magnitude at the primary minimum will be

$$m_A - m_{\text{tot}} = M_A - M_{\text{tot}}$$

$$= -2.5 \lg \frac{L_A}{L_{\text{tot}}} = +2.5 \lg \frac{L_{\text{tot}}}{L_A}$$

$$= 2.5 \lg \frac{4\pi R^2 \sigma T_A^4 + 4\pi R^2 \sigma T_B^4}{4\pi R^2 \sigma T_A^4}$$

$$= 2.5 \lg \left(1 + \left(\frac{T_B}{T_A} \right)^4 \right) .$$

Similarly the depth of the secondary minimum is

$$m_B - m_{\text{tot}} = 2.5 \lg \left(1 + \left(\frac{T_A}{T_B} \right)^4 \right) .$$

Let the effective temperatures of the stars be $T_A = 5{,}000$ K and $T_B = 12{,}000$ K. The depth of the primary minimum is then

$$m_A - m_{\text{tot}} = 2.5 \lg \left(1 + \left(\frac{12{,}000}{5{,}000} \right)^4 \right)$$

$$\approx 3.8 \,\text{mag} .$$

The secondary minimum is

$$m_B - m_{\text{tot}} = 2.5 \lg \left(1 + \left(\frac{5000}{12000} \right)^4 \right) \approx 0.03 \text{ mag} .$$

9.6   Exercises

Exercise 9.1   The components of a binary move along circular orbits. The mutual distance is 1 AU, and the mass of each component is $1\,M_\odot$. An observer in the plane of the orbit will see periodic splitting of the spectral lines. What is the maximum separation of the components of the H_γ line.

Exercise 9.2   A planet (mass m) is orbiting a star (mass M) at a distance a. The distance of the star from the centre of gravity of the system is a'. Show that

$$MP = a\,(a - a') ,$$

where P is period in years, distances are in AU's and masses in solar masses.

Exercise 9.3   The distance of Barnard's star is 1.83 pc and mass $0.135\,M_\odot$. It has been suggested that it oscillates with an amplitude of $0.026''$ in 25 year periods. Assuming this oscillation is caused by a planet, find the mass and radius of the orbit of this planet.

10. Stellar Structure

The stars are huge gas spheres, hundreds of thousands or millions of times more massive than the Earth. A star such as the Sun can go on shining steadily for thousands of millions of years. This is shown by studies of the prehistory of the Earth, which indicate that the energy radiated by the Sun has not changed by much during the last four thousand million years. The equilibrium of a star must remain stable for such periods.

10.1 Internal Equilibrium Conditions

Mathematically the conditions for the internal equilibrium of a star can be expressed as four differential equations governing the distribution of mass, gas pressure and energy production and transport in the star. These equations will now be derived.

Hydrostatic Equilibrium. The force of gravity pulls the stellar material towards the centre. It is resisted by the pressure force due to the thermal motions of the gas molecules. The first equilibrium condition is that these forces be in equilibrium.

Consider a cylindrical volume element at the distance r from the centre of the star (Fig. 10.1). Its volume is $dV = dA\, dr$, where dA is its base area and dr its height; its mass is $dm = \rho\, dA\, dr$, where $\rho = \rho(r)$ is the gas density at the radius r. If the mass inside radius r is M_r, the gravitational force on the volume element will be

$$dF_g = -\frac{GM_r\, dm}{r^2} = -\frac{GM_r \rho}{r^2} dA\, dr \;,$$

where G is the gravitational constant. The minus sign in this expression means that the force is directed towards the centre of the star. If the pressure at the lower surface of the volume element is P and at its upper surface $P + dP$, the net force of pressure acting on the element is

$$dF_p = P\, dA - (P + dP)dA$$

$$= -dP\, dA \;.$$

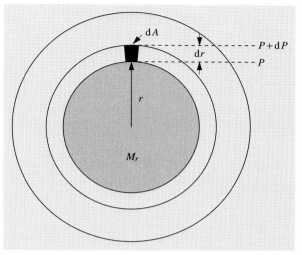

Fig. 10.1. In hydrostatic equilibrium the sum of the gravitational and pressure force acting on a volume element is zero

Since the pressure decreases outwards, dP will be negative and the force dF_p positive. The equilibrium condition is that the total force acting on the volume element should be zero, i. e.

$$0 = dF_g + dF_p$$

$$= -\frac{GM_r \rho}{r^2} dA\, dr - dP\, dA$$

or

$$\frac{dP}{dr} = -\frac{GM_r \rho}{r^2} \;. \tag{10.1}$$

This is the *equation of hydrostatic equilibrium.*

Mass Distribution. The second equation gives the mass contained within a given radius. Consider a spherical shell of thickness dr at the distance r from the centre (Fig. 10.2). Its mass is

$$dM_r = 4\pi r^2 \rho\, dr \;,$$

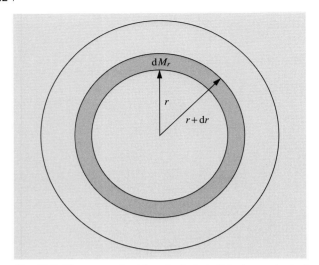

Fig. 10.2. The mass of a thin spherical shell is the product of its volume and its density

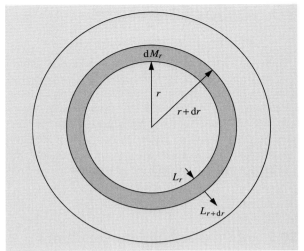

Fig. 10.3. The energy flowing out of a spherical shell is the sum of the energy flowing into it and the energy generated within the shell

giving the *mass continuity equation*

$$\frac{dM_r}{dr} = 4\pi r^2 \rho \ . \tag{10.2}$$

Energy Production. The third equilibrium condition expresses the conservation of energy, requiring that any energy produced in the star has to be carried to the surface and radiated away. We again consider a spherical shell of thickness dr and mass dM_r at the radius r (Fig. 10.3). Let L_r be the energy flux, i.e. the amount of energy passing through the surface r per unit time. If ε is the energy production coefficient, i.e. the amount of energy released in the star per unit time and mass, then

$$dL_r = L_{r+dr} - L_r = \varepsilon dM_r = 4\pi r^2 \rho \varepsilon \, dr \ .$$

Thus the *energy conservation equation* is

$$\frac{dL_r}{dr} = 4\pi r^2 \rho \varepsilon \ . \tag{10.3}$$

The rate at which energy is produced depends on the distance to the centre. Essentially all of the energy radiated by the star is produced in the hot and dense core. In the outer layers the energy production is negligible and L_r is almost constant.

The Temperature Gradient. The fourth equilibrium equation gives the temperature change as a function of the radius, i.e. the temperature gradient dT/dr. The form of the equation depends on how the energy is transported: by *conduction*, *convection* or *radiation*.

In the interiors of normal stars conduction is very inefficient, since the electrons carrying the energy can only travel a short distance without colliding with other particles. Conduction only becomes important in compact stars, white dwarfs and neutron stars, where the mean free path of photons is extremely short, but that of some electrons can be relatively large. In normal stars conductive energy transport can be neglected.

In radiative energy transport, photons emitted in hotter parts of the star are absorbed in cooler regions, which they heat. The star is said to be in radiative equilibrium, when the energy released in the stellar interior is carried outwards entirely by radiation.

The radiative temperature gradient is related to the energy flux L_r according to

$$\frac{dT}{dr} = \left(-\frac{3}{4ac}\right)\left(\frac{\kappa\rho}{T^3}\right)\left(\frac{L_r}{4\pi r^2}\right) \ , \tag{10.4}$$

where $a = 4\sigma/c = 7.564 \times 10^{-16} \, \mathrm{J\,m^{-3}\,K^{-4}}$ is the radiation constant, c the speed of light, and ρ the density. The *mass absorption coefficient* κ gives the amount of absorption per unit mass. Its value depends on the temperature, density and chemical composition.

In order to derive (10.4), we consider the equation of radiative transfer (5.44). In terms of the variables used in the present chapter, it may be written

$$\cos\theta \frac{dI_\nu}{dr} = -\kappa_\nu \rho I_\nu + j_\nu .$$

In this equation κ_ν is replaced with a suitable mean value κ. The equation is then multiplied with $\cos\theta$ and integrated over all directions and frequencies. On the left hand side, I_ν may be approximated with the Planck function B_ν. The frequency integral may then be evaluated by means of (5.16). On the right-hand side, the first term can be expressed in terms of the flux density according to (4.2) and the integral over directions of the second gives zero, since j_ν does not depend on θ. One thus obtains

$$\frac{4\pi}{3} \frac{d}{dr} \left(\frac{ac}{4\pi} T^4 \right) = -\kappa\rho F_r .$$

Finally, using the relation

$$F_r = \frac{L_r}{4\pi r^2} ,$$

between the flux density F_r and the energy flux L_r, one obtains (10.4).

The derivative dT/dr is negative, since the temperature increases inwards. Clearly there has to be a temperature gradient, if energy is to be transported by radiation: otherwise the radiation field would be the same in all directions and the net flux F_r would vanish.

If the radiative transfer of energy becomes inefficient, the absolute value of the radiative temperature gradient becomes very large. In that case motions are set up in the gas, which carry the energy outwards more efficiently than the radiation. In these convective motions, hot gas rises upwards into cooler layers, where it loses its energy and sinks again. The rising and sinking gas elements also mix the stellar material, and the composition of the convective parts of a star becomes homogeneous. Radiation and conduction, on the other hand, do not mix the material, since they move only energy, not gas.

In order to derive the temperature gradient for the convective case, consider a rising bubble. Assume that the gas moving with the bubble obeys the adiabatic equation of state

$$T \propto P^{1-\frac{1}{\gamma}} ,$$

(10.5)

where P is the pressure of the gas and γ, the *adiabatic exponent*

$$\gamma = C_P/C_V ,$$

(10.6)

is the ratio of the specific heats in constant pressure and constant volume. This ratio of the specific heats depends on the ionization of the gas, and can be computed when the temperature, density and chemical composition are known.

Taking the derivative of (10.5) we get the expression for the convective temperature gradient

$$\frac{dT}{dr} = \left(1 - \frac{1}{\gamma} \right) \frac{T}{P} \frac{dP}{dr} .$$

(10.7)

In the practical computation of stellar structure, one uses either (10.4) or (10.7), depending on which equation gives a less steep temperature gradient. In the outermost layers of stars heat exchange with the surroundings must be taken into account, and (10.7) is no longer a good approximation. An often used method for calculating the convective temperature gradient in that case is the mixing-length theory. The theory of convection is a difficult and still imperfectly understood problem, which is beyond the scope of this presentation.

The convective motions set in when the radiative temperature gradient becomes larger in absolute value than the adiabatic gradient, i. e. if either the radiative gradient becomes steep or if the convective gradient becomes small. From (10.4) it can be seen that a steep radiative gradient is expected, if either the energy flux density or the mass absorption coefficient becomes large. The convective gradient may become small, if the adiabatic exponent approaches 1.

Boundary Conditions. In order to obtain a well-posed problem, some boundary conditions have to be prescribed for the preceding differential equations:

– There are no sources of energy or mass at the centre inside the radius $r = 0$; thus $M_0 = 0$ and $L_0 = 0$.
– The total mass within the radius R of the star is fixed, $M_R = M$.

– The temperature and pressure at the stellar surface have some determinate values, T_R and P_R. These will be very small compared to those in the interior, and thus it is usually sufficient to take $T_R = 0$ and $P_R = 0$.

In addition to these boundary conditions one needs an expression for the pressure, which is given by the equation of state as well as expressions for the mass absorption coefficient and the energy generation rate, which will be considered later. The solution of the basic differential equations give the mass, temperature, density and energy flux as functions of the radius. The stellar radius and luminosity can then be calculated and compared with the observations.

The properties of a stellar equilibrium model are essentially determined once the mass and the chemical composition have been given. This result is known as the *Vogt–Russell theorem*.

10.2 Physical State of the Gas

Due to the high temperature the gas in the stars is almost completely ionized. The interactions between individual particles are small, so that, to a good approximation, the gas obeys the perfect gas equation of state,

$$P = \frac{k}{\mu m_H} \rho T \,, \tag{10.8}$$

where k is Boltzmann's constant, μ the mean molecular weight in units of m_H, and m_H the mass of the hydrogen atom.

The mean molecular weight can be approximately calculated assuming complete ionization. An atom with nuclear charge Z then produces $Z + 1$ free particles (the nucleus and Z electrons). Hydrogen gives rise to two particles per atomic mass unit; helium gives rise to three particles per four atomic mass units. For all elements heavier than hydrogen and helium it is usually sufficient to take $Z + 1$ to be half the atomic weight. (Exact values could easily be calculated, but the abundance of heavy elements is so small that this is usually not necessary.) In astrophysics the relative mass fraction of hydrogen is conventionally denoted by X, that of helium by Y and that of all heavier elements by Z, so that

$$X + Y + Z = 1 \,. \tag{10.9}$$

(The Z occuring in this equation should not be confused with the nuclear charge, which is unfortunately denoted by the same letter.) Thus the mean molecular weight will be

$$\mu = \frac{1}{2X + \frac{3}{4}Y + \frac{1}{2}Z} \,. \tag{10.10}$$

At high temperatures the radiation pressure has to be added to the gas pressure described by the perfect gas equation. The pressure exerted by radiation is (see p. 231)

$$P_{\text{rad}} = \frac{1}{3} a T^4 \,, \tag{10.11}$$

where a is the radiation constant. Thus the total pressure is

$$P = \frac{k}{\mu m_H} \rho T + \frac{1}{3} a T^4 \,. \tag{10.12}$$

The perfect gas law does not apply at very high densities.

The *Pauli exclusion principle* states that an atom with several electrons cannot have more than one electron with all four quantum numbers equal. This can also be generalized to a gas consisting of electrons (or other fermions). A *phase space* can be used to describe the electrons. The phase space is a 6-dimensional space, three coordinates of which give the position of the particle and the other three coordinates the momenta in x, y and z directions. A volume element of the phase space is

$$\Delta V = \Delta x \Delta y \Delta z \Delta p_x \Delta p_y \Delta p_z \,. \tag{10.13}$$

From the uncertainty principle it follows that the smallest meaningful volume element is of the order of h^3. According to the exclusion principle there can be only two electrons with opposite spins in such a volume element. When density becomes high enough, all volume elements of the phase space will be filled up to a certain limiting momentum. Such matter is called *degenerate*.

Electron gas begins to degenerate when the density is of the order 10^7 kg/m^3. In ordinary stars the gas is usually nondegenerate, but in white dwarfs and in neutron stars, degeneracy is of central importance. The pressure of a degenerate electron gas is (see p. 231)

$$P \approx \left(\frac{h^2}{m_e}\right)\left(\frac{N}{V}\right)^{5/3} \,, \tag{10.14}$$

where m_e is the electron mass and N/V the number of electrons per unit volume. This equation may be written in terms of the density

$$\rho = N\mu_e m_H/V \,,$$

where μ_e is the mean molecular weight per free electron in units of m_H. An expression for μ_e may be derived in analogy with (10.10):

$$\mu_e = \frac{1}{X + \frac{2}{4}Y + \frac{1}{2}Z} = \frac{2}{X+1} \,. \tag{10.15}$$

For the solar hydrogen abundance this yields

$$\mu_e = 2/(0.71+1) = 1.17 \,.$$

The final expression for the pressure is

$$P \approx \left(\frac{h^2}{m_e}\right)\left(\frac{\rho}{\mu_e m_H}\right)^{5/3} \,. \tag{10.16}$$

This is the equation of state of a degenerate electron gas. In contrast to the perfect gas law the pressure no longer depends on the temperature, only on the density and on the particle masses.

In normal stars the degenerate gas pressure is negligible, but in the central parts of giant stars and in white dwarfs, where the density is of the order of 10^8 kg/m^3, the degenerate gas pressure is dominant, in spite of the high temperature.

At even higher densities the electron momenta become so large that their velocities approach the speed of light. In this case the formulas of the special theory of relativity have to be used. The pressure of a relativistic degenerate gas is

$$P \approx hc\left(\frac{N}{V}\right)^{4/3} = hc\left(\frac{\rho}{\mu_e m_H}\right)^{4/3} \,. \tag{10.17}$$

In the relativistic case the pressure is proportional to the density to the power 4/3, rather than 5/3 as for the nonrelativistic case. The transition to the relativistic situation takes place roughly at the density 10^9 kg/m^3.

In general the pressure inside a star depends on the temperature (except for a completely degenerate gas), density and chemical composition. In actual stars the gas will never be totally ionized or completely degenerate. The pressure will then be given by more complicated expressions. Still it can be calculated for each case of interest. One may then write

$$P = P(T, \rho, X, Y, Z) \,, \tag{10.18}$$

giving the pressure as a known function of the temperature, density and chemical composition.

The *opacity* of the gas describes how difficult it is for radiation to propagate through it. The change dI of the intensity in a distance dr can be expressed as

$$dI = -I\alpha\, dr \,,$$

where α is the opacity (Sect. 4.5). The opacity depends on the chemical composition, temperature and density of the gas. It is usually written as $a = \kappa\rho$, where ρ is the density of the gas and κ the *mass absorption coefficient* ($[\kappa] = $ m^2/kg).

The inverse of the opacity represents the mean free path of radiation in the medium, i.e. the distance it can propagate without being scattered or absorbed. The different types of absorption processes (bound–bound, bound–free, free–free) have been described in Sect. 5.1. The opacity of the stellar material due to each process can be calculated for relevant values of temperature and density.

10.3 Stellar Energy Sources

When the equations of stellar structure were derived, the character of the source of stellar energy was left unspecified. Knowing a typical stellar luminosity, one can calculate how long different energy sources would last. For instance, normal chemical burning could produce energy for only a few thousand years. The energy released by the contraction of a star would last slightly longer, but after a few million years this energy source would also run out.

Terrestrial biological and geological evidence shows that the solar luminosity has remained fairly constant for at least a few thousand million years. Since the age of the Earth is about 5000 million years, the Sun has presumably existed at least for that time. Since the solar luminosity is 4×10^{26} W, it has radiated about 6×10^{43} J in 5×10^9 years. The Sun's mass is 2×10^{30} kg; thus it must be able to produce at least 3×10^{13} J/kg.

The general conditions in the solar interior are known, regardless of the exact energy source. Thus, in Example 10.5, it will be estimated that the temperature at half the radius is about 5 million degrees. The central temperature must be about ten million kelvins, which is high enough for *thermonuclear fusion reactions* to take place.

In fusion reactions light elements are transformed into heavier ones. The final reaction products have a smaller total mass than the initial nuclei. This mass difference is released as energy according to Einstein's relation $E = mc^2$. Thermonuclear reactions are commonly referred to as burning, although they have no relation to the chemical burning of ordinary fuels.

The atomic nucleus consists of protons and neutrons, together referred to as nucleons. We define

m_p = proton mass ,

m_n = neutron mass ,

Z = nuclear charge = atomic number ,

N = neutron number ,

$A = Z + N$ = atomic weight ,

$m(Z, N)$ = mass of the nucleus .

The mass of the nucleus is smaller than the sum of the masses of all its nucleons. The difference is called the *binding energy*. The binding energy per nucleon is

$$Q = \frac{1}{A}(Zm_p + Nm_n - m(Z, N))c^2 . \qquad (10.19)$$

It turns out that Q increases towards heavier elements up to iron ($Z = 26$). Beyond iron the binding energy again begins to decrease (Fig. 10.4).

It is known that the stars consist mostly of hydrogen. Let us consider how much energy would be released by the fusion of four hydrogen nuclei into a helium nucleus. The mass of a proton is 1.672×10^{-27} kg and that of a helium nucleus is 6.644×10^{-27} kg. The mass difference, 4.6×10^{-29} kg, corresponds to an energy difference $E = 4.1 \times 10^{-12}$ J. Thus 0.7% of the mass is turned into energy in the reaction, corresponding to an energy release of 6.4×10^{14} J per one kilogram of hydrogen. This should be compared with our previous estimate that 3×10^{13} J/kg is needed.

Already in the 1930's it was generally accepted that stellar energy had to be produced by nuclear fusion. In 1938 *Hans Bethe* and independently *Carl Friedrich von Weizsäcker* put forward the first detailed mechanism for energy production in the stars, the *carbon–nitrogen–oxygen (CNO) cycle*. The other important energy generation processes (the *proton–proton chain* and the *triple-alpha reaction*) were not proposed until the 1950's.

The Proton–Proton Chain (Fig. 10.5). In stars with masses of about that of the Sun or smaller, the energy

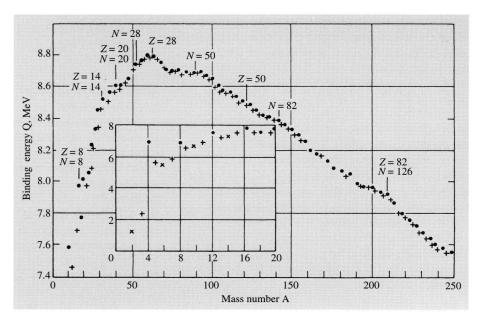

Fig. 10.4. The nuclear binding energy per nucleon as a function of the atomic weight. Among isotopes with the same atomic weight the one with the largest binding energy is shown. The points correspond to nuclei with even proton and neutron numbers, the crosses to nuclei with odd mass numbers. Preston, M.A. (1962): *Physics of the Nucleus* (Addison-Wesley Publishing Company, Inc., Reading, Mass.)

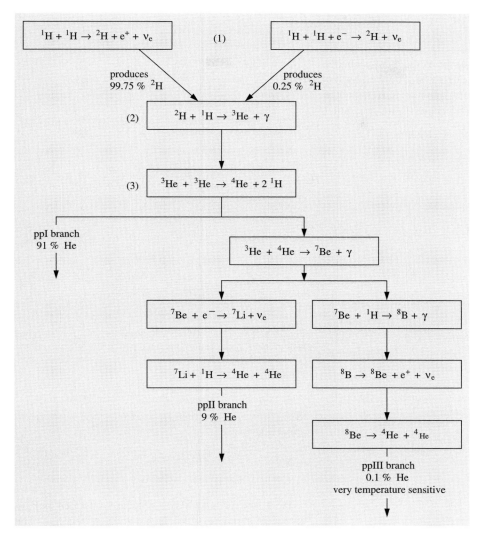

Fig. 10.5. The proton–proton chain. In the ppI branch, four protons are transformed into one helium nucleus, two positrons, two neutrinos and radiation. The relative weights of the reactions are given for conditions in the Sun. The pp chain is the most important energy source in stars with mass below 1.5 $M_\odot$

is produced by the proton–proton (pp) chain. It consists of the following steps:

ppI: (1) $\quad {}^1H + {}^1H \quad\quad \rightarrow {}^2H + e^+ + \nu_e$,

$\quad\quad\quad {}^1H + {}^1H + e^- \rightarrow {}^2H + \nu_e$,

(2) $\quad {}^2H + {}^1H \quad\quad \rightarrow {}^3He + \gamma$,

(3) $\quad {}^3He + {}^3He \quad \rightarrow {}^4He + 2\,{}^1H$.

For each reaction (3) the reactions (1) and (2) have to take place twice. The first reaction step has a very small probability, which has not been measured in the

laboratory. At the central density and temperature of the Sun, the expected time for a proton to collide with another one to form a deuteron is 10^{10} years on the average. It is only thanks to the slowness of this reaction that the Sun is still shining. If it were faster, the Sun would have burnt out long ago. The neutrino produced in the reaction (1) can escape freely from the star and carries away some of the energy released. The positron e^+ is immediately annihilated together with an electron, giving rise to two gamma quanta.

The second reaction, where a deuteron and a proton unite to form the helium isotope ${}^3He$, is very fast compared to the preceding one. Thus the abundance of deuterons inside stars is very small.

The last step in the pp chain can take three different forms. The ppI chain shown above is the most probable one. In the Sun 91% of the energy is produced by the ppI chain. It is also possible for ^{3}He nuclei to unite into ^{4}He nuclei in two additional branches of the pp chain.

ppII: (3) ^{3}He $+ \, ^4$He $\to \, ^7$Be $+ \gamma$,

          (4) ^{7}Be $+ \mathrm{e}^- \to \, ^7$Li $+ \nu_\mathrm{e}$,

          (5) ^{7}Li $+ \, ^1$H $\to \, ^4$He $+ \, ^4$He ,

ppIII: (3) ^{3}He $+ \, ^4$He $\to \, ^7$Be $+ \gamma$,

          (4) ^{7}Be $+ \, ^1$H $\to \, ^8$B $+ \gamma$,

          (5) ^{8}B $\to \, ^8$Be $+ \mathrm{e}^+ + \nu_\mathrm{e}$,

          (6) ^{8}Be $\to \, ^4$He $+ \, ^4$He .

The Carbon Cycle (Fig. 10.6). At temperatures below 20 million degrees the pp chain is the main energy production mechanism. At higher temperatures corresponding to stars with masses above 1.5 $M_\odot$, the carbon (CNO) cycle becomes dominant, because its reaction rate increases more rapidly with temperature. In the CNO cycle carbon, oxygen and nitrogen act as catalysts. The reaction cycle is the following:

(1) ^{12}C $+ \, ^1$H $\to \, ^{13}$N $+ \gamma$,

(2) ^{13}N $\to \, ^{13}$C $+ \mathrm{e}^+ + \nu_\mathrm{e}$,

(3) ^{13}C $+ \, ^1$H $\to \, ^{14}$N $+ \gamma$,

(4) ^{14}N $+ \, ^1$H $\to \, ^{15}$O $+ \gamma$,

(5) ^{15}O $\to \, ^{15}$N $+ \gamma + \nu_\mathrm{e}$,

(6) ^{15}N $+ \, ^1$H $\to \, ^{12}$C $+ \, ^4$He .

Reaction (4) is the slowest, and thus determines the rate of the CNO cycle. At a temperature of 20 million degrees the reaction time for the reaction (4) is a million years.

The fraction of energy released as radiation in the CNO cycle is slightly smaller than in the pp chain, because more energy is carried away by neutrinos.

The Triple Alpha Reaction. As a result of the preceding reactions, the abundance of helium in the stellar interior increases. At a temperature above 10^8 degrees

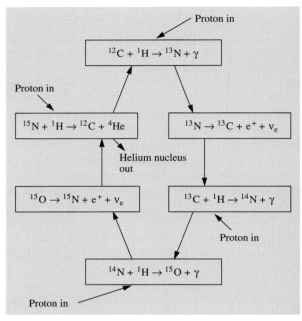

Fig. 10.6. The CNO cycle is catalysed by ^{12}C. It transforms four protons into a helium nucleus, two positrons, two neutrinos and radiation. It is the dominant energy source for stars more massive than 1.5 $M_\odot$

the helium can be transformed into carbon in the triple alpha reaction:

(1) ^{4}He $+ \, ^4$He $\leftrightarrow \, ^8$Be ,

(2) ^{8}Be $+ \, ^4$He $\to \, ^{12}$C $+ \gamma$.

Here ^{8}Be is unstable and decays into two helium nuclei or alpha particles in 2.6×10^{-16} seconds. The production of carbon thus requires the almost simultaneous collision of three particles. The reaction is often written

$$3 \, ^4\mathrm{He} \to \, ^{12}\mathrm{C} + \gamma .$$

Once helium burning has been completed, at higher temperatures other reactions become possible, in which heavier elements up to iron and nickel are built up. Examples of such reactions are various alpha reactions and oxygen, carbon and silicon burning.

Alpha Reactions. During helium burning some of the carbon nuclei produced react with helium nuclei to form oxygen, which in turn reacts to form neon, etc. These reactions are fairly rare and thus are not important as stellar energy sources. Examples are

$$^{12}\text{C} + {}^4\text{He} \rightarrow {}^{16}\text{O} + \gamma \,,$$
$$^{16}\text{O} + {}^4\text{He} \rightarrow {}^{20}\text{Ne} + \gamma \,,$$
$$^{20}\text{Ne} + {}^4\text{He} \rightarrow {}^{24}\text{Mg} + \gamma \,.$$

Carbon Burning. After the helium is exhausted, carbon burning sets in at the temperature $(5\text{–}8) \times 10^{10}$ K:

$$^{12}\text{C} + {}^{12}\text{C} \rightarrow {}^{34}\text{Mg} + \gamma$$
$$\rightarrow {}^{23}\text{Na} + {}^1\text{H}$$
$$\rightarrow {}^{20}\text{Ne} + {}^4\text{He}$$
$$\rightarrow {}^{23}\text{Mg} + \text{n}$$
$$\rightarrow {}^{16}\text{O} + 2\,{}^4\text{He} \,.$$

Oxygen Burning. Oxygen is consumed at slightly higher temperatures in the reactions

$$^{16}\text{O} + {}^{16}\text{O} \rightarrow {}^{32}\text{S} + \gamma$$
$$\rightarrow {}^{31}\text{P} + {}^1\text{H}$$
$$\rightarrow {}^{28}\text{Si} + {}^4\text{He}$$
$$\rightarrow {}^{31}\text{S} + \text{n}$$
$$\rightarrow {}^{24}\text{Mg} + 2\,{}^4\text{He} \,.$$

Silicon Burning. After several intermediate steps the burning of silicon produces nickel and iron. The total process may be expressed as

$$^{28}\text{Si} + {}^{28}\text{Si} \rightarrow {}^{56}\text{Ni} + \gamma \,,$$
$$^{56}\text{Ni} \qquad \rightarrow {}^{56}\text{Fe} + 2\,\text{e}^+ + 2\,\nu_\text{e} \,.$$

When the temperature becomes higher than about 10^9 K, the energy of the photons becomes large enough to destroy certain nuclei. Such reactions are called *photonuclear reactions* or *photodissociations*.

The production of elements heavier than iron requires an input of energy, and therefore such elements cannot be produced by thermonuclear reactions. Elements heavier than iron are almost exclusively produced by *neutron capture* during the final violent stages of stellar evolution (Sect. 11.5).

The rates of the reactions presented above can be determined by laboratory experiments or by theoretical calculations. Knowing them, one can calculate the rate at which energy is released per unit mass and time as a function of the density, temperature and chemical composition:

$$\varepsilon = \varepsilon(T, \rho, X, Y, Z) \,. \tag{10.20}$$

In reality the relative abundance of each of the heavier nuclei needs to be known, not just their total abundance Z.

10.4 Stellar Models

A theoretical stellar model is obtained if one solves the differential equations for stellar structure. As we have already noted, the model is uniquely defined once the chemical composition and the mass of the star have been given.

Stars just formed out of the interstellar medium are chemically homogeneous. When stellar models for homogeneous stars are plotted in the HR diagram, they fall along the lower edge of the main sequence. The theoretical sequence obtained in this way is called the *zero age main sequence*, ZAMS. The exact position of the ZAMS depends on the initial chemical composition. For stars with an initial abundance of heavy elements like that in the Sun, the computed ZAMS is in good agreement with the observations. If the initial value of Z is smaller, the theoretical ZAMS falls below the main sequence in the subdwarf region of the HR diagram. This is related to the classification of stars into populations I and II, which is discussed in Sect. 17.2.

The theoretical models also provide an explanation for the mass–luminosity relation. The computed properties of zero age main sequence stars of different masses are given in Table 10.1. The chemical composition assumed is $X = 0.71$ (hydrogen mass fraction), $Y = 0.27$ (helium) and $Z = 0.02$ (heavier elements), except for the 30 $M_\odot$ star, which has $X = 0.70$ and $Y = 0.28$. The luminosity of a one solar mass star is $0.74\,L_\odot$ and the radius $0.87\,R_\odot$. Thus the Sun has brightened and expanded to some extent during its evolution. However, these changes are small and do not conflict with the evidence for a steady solar energy output. In addition the

Table 10.1. Properties of zero age main sequence stars. (T_c = central temperature; ρ_c = central density; M_{ci} = relative mass of convective interior; M_{ce} = relative mass of convective envelope)

M [$M_\odot$]	R [$R_\odot$]	L [$L_\odot$]	T_e [K]	T_c [10^6 K]	ρ_c [kg/m^3]	M_{ci} [M]	M_{ce} [M]
30	6.6	140,000	44,000	36	3,000	0.60	0
15	4.7	21,000	32,000	34	6,200	0.39	0
9	3.5	4,500	26,000	31	7,900	0.26	0
5	2.2	630	20,000	27	26,000	0.22	0
3	1.7	93	14,000	24	42,000	0.18	0
1.5	1.2	5.4	8,100	19	95,000	0.06	0
1.0	0.87	0.74	5,800	14	89,000	0	0.01
0.5	0.44	0.038	3,900	9.1	78,000	0	0.41

biological evidence only goes back about 3000 million years.

The model calculations show that the central temperature in the smallest stars ($M \approx 0.08\,M_\odot$) is about 4×10^6 K, which is the minimum temperature required for the onset of thermonuclear reactions. In the biggest stars ($M \approx 50\,M_\odot$), the central temperature reaches 4×10^7 K.

The changes in chemical composition caused by the nuclear reactions can be computed, since the rates of the various reactions at different depths in the star are known. For example, the change ΔX of the hydrogen abundance in the time interval Δt is proportional to the rate of energy generation ε and to Δt:

$$\Delta X \propto -\varepsilon \Delta t \ . \tag{10.21}$$

The constant of proportionality is clearly the amount of hydrogen consumed per unit energy [kg/J]. The value of this constant of proportionality is different for the pp chain and the CNO cycle. Therefore the contribution from each reaction chain must be calculated separately in (10.21). For elements that are produced by the nuclear reactions, the right-hand side contribution in (10.21) is positive. If the star is convective, the change in composition is obtained by taking the average of (10.21) over the convection zone.

* Gas Pressure and Radiation Pressure

Let us consider noninteracting particles in a rectangular box. The particles may also be photons. Let the sides of the box be Δx, Δy and Δz, and the number of particles, N. The pressure is caused by the collisions of the particles with the sides of the box. When a particle

hits a wall perpendicular to the x axis, its momentum in the x direction, p_x, changes by $\Delta p = 2p_x$. The particle will return to the same wall after the time $\Delta t = 2\Delta x/v_x$. Thus the pressure exerted by the particles on the wall (surface area $A = \Delta y \Delta z$) is

$$P = \frac{F}{A} = \frac{\sum \Delta p / \Delta t}{A} = \frac{\sum p_x v_x}{\Delta x \Delta y \Delta z} = \frac{N \langle p_x v_x \rangle}{V} \ ,$$

where $V = \Delta x \Delta y \Delta z$ is the volume of the box and the angular brackets represent the average value. The momentum is $p_x = m v_x$ (where $m = h\nu/c^2$ for photons), and hence

$$P = \frac{Nm \langle v_x^2 \rangle}{V} \ .$$

Suppose the velocities of the particles are isotropically distributed. Then $\langle v_x^2 \rangle = \langle v_y^2 \rangle = \langle v_z^2 \rangle$, and thus

$$\langle v^2 \rangle = \langle v_x^2 \rangle + \langle v_y^2 \rangle + \langle v_z^2 \rangle = 3 \langle v_x^2 \rangle$$

and

$$P = \frac{Nm \langle v^2 \rangle}{3V} \ .$$

If the particles are gas molecules, the energy of a molecule is $\varepsilon = \frac{1}{2} m v^2$. The total energy of the gas is $E = N \langle \varepsilon \rangle = \frac{1}{2} Nm \langle v^2 \rangle$, and the pressure may be written

$$P = \frac{2}{3} \frac{E}{V} \quad \text{(gas)} \ .$$

If the particles are photons, they move with the speed of light and their energy is $\varepsilon = mc^2$. The total energy of a photon gas is thus $E = N \langle \varepsilon \rangle = Nmc^2$ and the pressure is

$$P = \frac{1}{3} \frac{E}{V} \quad \text{(radiation)} \ .$$

According to (4.7), (4.4) and (5.16) the energy density of blackbody radiation is

$$\frac{E}{V} = u = \frac{4\pi}{c} I = \frac{4}{c} F = \frac{4}{c} \sigma T^4 \equiv a T^4 ,$$

where $a = 4\sigma/c$ is the radiation constant. Thus the radiation pressure is

$$P_{\mathrm{rad}} = a \, T^4/3 .$$

* The Pressure of a Degenerate Gas

A gas where all available energy levels up to a limiting momentum p_0, known as the Fermi momentum, are filled is called degenerate. We shall determine the pressure of a completely degenerate electron gas.

Let the volume of the gas be V. We consider electrons with momenta in the range $[p, p+\mathrm{d}p]$. Their available phase space volume is $4\pi p^2 \mathrm{d}p V$. According to the Heisenberg uncertainty relation the elementary volume in phase space is h^3 and, according to the Pauli exclusion principle, this volume can contain two electrons with opposite spins. Thus the number of electrons in the momentum interval $[p, p+\mathrm{d}p]$ is

$$\mathrm{d}N = 2 \frac{4\pi p^2 \mathrm{d}p \, V}{h^3} .$$

The total number of electrons with momenta smaller than p_0 is

$$N = \int \mathrm{d}N = \frac{8\pi V}{h^3} \int_0^{p_0} p^2 \, \mathrm{d}p = \frac{8\pi V}{3h^3} p_0^3 .$$

Hence the Fermi momentum p_0 is

$$p_0 = \left(\frac{3}{\pi}\right)^{1/3} \frac{h}{2} \left(\frac{N}{V}\right)^{1/3} .$$

Nonrelativistic Gas. The kinetic energy of an electron is $\varepsilon = p^2/2m_{\mathrm{e}}$. The total energy of the gas is

$$E = \int \varepsilon \, \mathrm{d}N = \frac{4\pi V}{m_{\mathrm{e}} h^3} \int_0^{p_0} p^4 \, \mathrm{d}p$$

$$= \frac{4\pi V}{5 m_{\mathrm{e}} h^3} p_0^5 .$$

Introducing the expression for the Fermi momentum p_0, one obtains

$$E = \frac{\pi}{40} \left(\frac{3}{\pi}\right)^{5/3} \frac{h^2}{m_{\mathrm{e}}} V \left(\frac{N}{V}\right)^{5/3} .$$

The pressure of the gas was derived in *Gas Pressure and Radiation Pressure:

$$P = \frac{2}{3} \frac{E}{V}$$

$$= \frac{1}{20} \left(\frac{3}{\pi}\right)^{2/3} \frac{h^2}{m_{\mathrm{e}}} \left(\frac{N}{V}\right)^{5/3} \quad \text{(nonrelativistic)} .$$

Here N/V is the number density of electrons.

Relativistic Gas. If the density becomes so large that the electron kinetic energy ϵ corresponding to the Fermi momentum exceeds the rest energy $m_{\mathrm{e}} c^2$, the relativistic expression for the electron energy has to be used. In the extreme relativistic case $\varepsilon = cp$ and the total energy

$$E = \int \varepsilon \, \mathrm{d}N = \frac{8\pi c V}{h^3} \int_0^{p_0} p^3 \, \mathrm{d}p$$

$$= \frac{2\pi c V}{h^3} p_0^4 .$$

The expression for the Fermi momentum remains unchanged, and hence

$$E = \frac{\pi}{8} \left(\frac{3}{\pi}\right)^{4/3} h c V \left(\frac{N}{V}\right)^{4/3} .$$

The pressure of the relativistic electron gas is obtained from the formula derived for a photon gas in *Gas Pressure and Radiation Pressure:

$$P = \frac{1}{3} \frac{E}{V}$$

$$= \frac{1}{8} \left(\frac{3}{\pi}\right)^{1/3} h c \left(\frac{N}{V}\right)^{4/3} \quad \text{(relativistic)} .$$

We have obtained the nonrelativistic and extreme relativistic approximations to the pressure. In intermediate cases the exact expression for the electron energy,

$$\varepsilon = (m_{\mathrm{e}}^2 c^4 + p^2 c^2)^{1/2} ,$$

has to be used.

The preceding derivations are rigorously valid only at zero temperature. However, the densities in compact stars are so high that the effects of a nonzero temperature are negligible, and the gas may be considered completely degenerate.

10.5 Examples

Example 10.1   *The Gravitational Acceleration at the Solar Surface*

The expression for the gravitational acceleration is

$$g = \frac{GM_\odot}{R^2} \,.$$

Using the solar mass $M = 1.989 \times 10^{30}$ kg and radius $R = 6.96 \times 10^8$ m, one obtains

$$g = 274 \, \mathrm{m \, s^{-2}} \approx 28 \, g_0 \,,$$

where $g_0 = 9.81 \, \mathrm{m \, s^{-2}}$ is the gravitational acceleration at the surface of the Earth.

Example 10.2   *The Average Density of the Sun*

The volume of a sphere with radius R is

$$V = \frac{4}{3}\pi R^3 \,;$$

thus the average density of the Sun is

$$\overline{\rho} = \frac{M}{V} = \frac{3M}{4\pi R^3} \approx 1410 \, \mathrm{kg \, m^{-3}} \,.$$

Example 10.3   *Pressure at Half the Solar Radius*

The pressure can be estimated from the condition for the hydrostatic equilibrium (10.1). Suppose the density is constant and equal to the average density $\overline{\rho}$. Then the mass within the radius r is

$$M_r = \frac{4}{3}\pi \overline{\rho} r^3 \,,$$

and the hydrostatic equation can be written

$$\frac{dP}{dr} = -\frac{GM_r\overline{\rho}}{r^2} = -\frac{4\pi G \overline{\rho}^2 r}{3} \,.$$

This can be integrated from half the solar radius, $r = R_\odot/2$, to the surface, where the pressure vanishes:

$$\int_P^0 dP = -\frac{4}{3}\pi G \overline{\rho}^2 \int_{R_\odot/2}^{R_\odot} r \, dr \,,$$

which gives

$$P = \frac{1}{2}\pi G \overline{\rho}^2 R_\odot^2$$

$$\approx \frac{1}{2}\pi \, 6.67 \times 10^{-11} \times 1410^2 \times (6.96 \times 10^8)^2 \mathrm{N/m^2}$$

$$\approx 10^{14} \, \mathrm{Pa} \,.$$

This estimate is extremely rough, since the density increases strongly inwards.

Example 10.4   *The Mean Molecular Weight of the Sun*

In the outer layers of the Sun the initial chemical composition has not been changed by nuclear reactions. In this case one can use the values $X = 0.71$, $Y = 0.27$ and $Z = 0.02$. The mean molecular weight (10.10) is then

$$\mu = \frac{1}{2 \times 0.71 + 0.75 \times 0.27 + 0.5 \times 0.02} \approx 0.61 \,.$$

When the hydrogen is completely exhausted, $X = 0$ and $Y = 0.98$, and hence

$$\mu = \frac{1}{0.75 \times 0.98 + 0.5 \times 0.02} \approx 1.34 \,.$$

Example 10.5   *The Temperature of the Sun at $r = R_\odot/2$*

Using the density from Example 10.2 and the pressure from Example 10.3, the temperature can be estimated from the perfect gas law (10.8). Assuming the surface value for the mean molecular weight (Example 10.4), one obtains the temperature

$$T = \frac{\mu m_\mathrm{H} P}{k\rho} = \frac{0.61 \times 1.67 \times 10^{-27} \times 1.0 \times 10^{14}}{1.38 \times 10^{-23} \times 1410}$$

$$\approx 5 \times 10^6 \, \mathrm{K} \,.$$

Example 10.6 *The Radiation Pressure in the Sun at $r = R_\odot/2$*

In the previous example we found that the temperature is $T \approx 5 \times 10^6$ K. Thus the radiation pressure given by (10.11) is

$$P_{\text{rad}} = \frac{1}{3} aT^4 = \frac{1}{3} \times 7.564 \times 10^{-16} \times (5 \times 10^6)^4$$

$$\approx 2 \times 10^{11}\ \text{Pa}\ .$$

This is about a thousand times smaller than the gas pressure estimated in Example 10.3. Thus it confirms that the use of the perfect gas law in Example 10.5 was correct.

Example 10.7 *The Path of a Photon from the Centre of a Star to Its Surface*

Radiative energy transport can be described as a random walk, where a photon is repeatedly absorbed and re-emitted in a random direction. Let the step length of the walk (the mean free path) be d. Consider, for simplicity, the random walk in a plane. After one step the photon is absorbed at

$$x_1 = d \cos \theta_1\ , \quad y_1 = d \sin \theta_1\ ,$$

where θ_1 is an angle giving the direction of the step. After N steps the coordinates are

$$x = \sum_{i=1}^{N} d \cos \theta_i\ , \quad y = \sum_{i=1}^{N} d \sin \theta_i\ ,$$

and the distance from the starting point is

$$r^2 = x^2 + y^2$$

$$= d^2 \left[\left(\sum_{1}^{N} \cos \theta_i \right)^2 + \left(\sum_{1}^{N} \sin \theta_i \right)^2 \right]\ .$$

The first term in square brackets can be written

$$\left(\sum_{1}^{N} \cos \theta_i \right)^2 = (\cos \theta_1 + \cos \theta_2 + \ldots + \cos \theta_N)^2$$

$$= \sum_{1}^{N} \cos^2 \theta_i + \sum_{i \neq j} \cos \theta_i \cos \theta_j\ .$$

Since the directions θ_i are randomly distributed and independent,

$$\sum_{i \neq j} \cos \theta_i \cos \theta_j = 0\ .$$

The same result applies for the second term in square brackets. Thus

$$r^2 = d^2 \sum_{1}^{N} (\cos^2 \theta_i + \sin^2 \theta_i) = Nd^2\ .$$

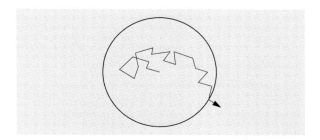

After N steps the photon is at the distance $r = d\sqrt{N}$ from the starting point. Similarly, a drunkard taking a hundred one-metre steps at random will have wandered ten metres from his/her starting point. The same result applies in three dimensions.

The time taken by a photon to reach the surface from the centre depends on the mean free path $d = 1/\alpha = 1/\kappa\rho$. The value of κ at half the solar radius can be estimated from the values of density and temperature obtained in Example 10.2 and 10.5. The mass absorption coefficient in these conditions is found to be $\kappa = 10\ \text{m}^2/\text{kg}$. (We shall not enter on how it is calculated.) The photon mean free path is then

$$d = \frac{1}{\kappa\rho} \approx 10^{-4}\ \text{m}\ .$$

This should be a reasonable estimate in most of the solar interior. Since the solar radius $r = 10^9$ m, the number of steps needed to reach the surface will be $N = (r/d)^2 = 10^{26}$. The total path travelled by the photon is $s = Nd = 10^{22}$ m, and the time taken is $t = s/c = 10^6$ years; a more careful calculation gives $t = 10^7$ years. Thus it takes 10 million years for the energy generated at the centre to radiate into space. Of course the radiation that leaves the surface does not consist of the same gamma photons that were produced near the centre. The intervening scattering, emission and absorption processes have transformed the radiation into visible light (as can easily be seen).

10.6 Exercises

Exercise 10.1 How many hydrogen atoms are there in the Sun per each helium atom?

Exercise 10.2 a) How many pp reactions take place in the Sun every second? The luminosity of the Sun is 3.9×10^{26} W, the mass of a proton is 1.00728 amu, and that of the α particle 4.001514 amu (1 amu is 1.6604×10^{-27} kg).

b) How many neutrinos produced in these pp reactions will hit the Earth in one second?

Exercise 10.3 The mass absorption coefficient of a neutrino is $\kappa = 10^{-21}$ m^2 kg^{-1}. Find the mean free path at the centre of the Sun.

11. Stellar Evolution

In the preceding chapter we have seen how one can compute the evolution of a star by starting from a homogeneous model representing a newly formed system. When the chemical composition of the star changes with time, a new model is computed each time. In this chap-

ter we shall consider the theoretical evolutionary paths of systems with various masses and see how the computed evolution explains the observational data. The following discussion is rather qualitative, since the details of the theoretical calculations are too involved for the present book.

11.1 Evolutionary Time Scales

Changes in a star may take place on quite different time scales at different evolutionary phases. There are three important basic time scales: the nuclear time scale t_n, the thermal time scale t_t and the dynamical or freefall time scale t_d.

The Nuclear Time Scale. The time in which a star radiates away all the energy that can be released by nuclear reactions is called the nuclear time scale. An estimate of this time can be obtained if one calculates the time in which all available hydrogen is turned into helium. On the basis of theoretical considerations and evolutionary computations it is known that only just over 10% of the total mass of hydrogen in the star can be consumed before other, more rapid evolutionary mechanisms set in. Since 0.7% of the rest mass is turned into energy in hydrogen burning, the nuclear time scale will be

$$t_n \approx \frac{0.007 \times 0.1 \, Mc^2}{L} \, . \tag{11.1}$$

For the Sun one obtains the nuclear time scale 10^{10} years, and thus

$$t_n \approx \frac{M/M_\odot}{L/L_\odot} \times 10^{10} \, \text{a} \, . \tag{11.2}$$

This gives the nuclear time scale as a function of the mass M and luminosity L of a given star. For example, if the mass is $30 \, M_\odot$, one obtains t_n about 2 million years. The reason for the shorter time scale is that the stellar luminosity strongly increases for higher masses (Table 10.1).

The Thermal Time Scale. The time in which a star would radiate away all its thermal energy if the nuclear energy production were suddenly turned off is called the thermal time scale. This is also the time it takes for radiation from the centre to reach the surface. The thermal time scale may be estimated as

$$t_t \approx \frac{0.5 \, G M^2/R}{L}$$
$$\approx \frac{(M/M_\odot)^2}{(R/R_\odot)(L/L_\odot)} \times 2 \times 10^7 \, \text{a} \, , \tag{11.3}$$

Table 11.1. Stellar lifetimes (unit 10^6 years)

Mass [$M_\odot$]	Spectral type on the main sequence	Contraction to main sequence	Main sequence	Main sequence to red giant	Red giant
30	O5	0.02	4.9	0.55	0.3
15	B0	0.06	10	1.7	2
9	B2	0.2	22	0.2	5
5	B5	0.6	68	2	20
3	A0	3	240	9	80
1.5	F2	20	2,000	280	
1.0	G2	50	10,000	680	
0.5	M0	200	30,000		
0.1	M7	500	10^7		

where G is the constant of gravity and R the stellar radius. For the Sun the thermal time scale is about 20 million years or $1/500$ of the nuclear time scale.

The Dynamical Time Scale. The third and shortest time scale is the time it would take a star to collapse if the pressure supporting it against gravity were suddenly removed. It can be estimated from the time it would take for a particle to fall freely from the stellar surface to the centre. This is half of the period given by Kepler's third law, where the semimajor axis of the orbit corresponds to half the stellar radius R:

$$t_d = \frac{2\pi}{2}\sqrt{\frac{(R/2)^3}{GM}} \approx \sqrt{\frac{R^3}{GM}} . \qquad (11.4)$$

The dynamical time scale of the Sun is about half an hour.

The ordering of the time scales is normally like that in the Sun, i.e. $t_d \ll t_t \ll t_n$.

11.2 The Contraction of Stars Towards the Main Sequence

The formation and subsequent gravitational collapse of condensations in the interstellar medium will be considered in a later chapter. Here we shall follow the behaviour of such a *protostar*, when it is already in the process of contraction.

When a cloud contracts, gravitational potential energy is released and transformed into thermal energy of the gas and into radiation. Initially the radiation can propagate freely through the material, because the density is low and the opacity small. Therefore most of the liberated energy is radiated away and the temperature does not increase. The contraction takes place on the dynamical time scale; the gas is falling freely inwards.

The density and the pressure increase most rapidly near the centre of the cloud. As the density increases, so does the opacity. A larger fraction of the released energy is then turned into heat, and the temperature begins to rise. This leads to a further increase in the pressure that is resisting the free fall. The contraction of the central part of the cloud slows down. The outer parts, however, are still falling freely.

At this stage, the cloud may already be considered a protostar. It consists mainly of hydrogen in molecular form. When the temperature reaches 1800 K, the

hydrogen molecules are dissociated into atoms. The dissociation consumes energy, and the rise in temperature is slowed down. The pressure then also grows more slowly and this in turn means that the rate of contraction increases. The same sequence of events is repeated, first when hydrogen is ionized at 10^4 K, and then when helium is ionized. When the temperature has reached about 10^5 K, the gas is essentially completely ionized.

The contraction of a protostar only stops when a large fraction of the gas is fully ionized in the form of *plasma*. The star then settles into hydrostatic equilibrium. Its further evolution takes place on the thermal time scale, i.e. much more slowly. The radius of the protostar has shrunk from its original value of about 100 AU to about 1/4 AU. It will usually be located inside a larger gas cloud and will be accreting material from its surroundings. Its mass therefore grows, and the central temperature and density increase.

The temperature of a star that has just reached equilibrium is still low and its opacity correspondingly large. Thus it will be convective in its centre. The convective energy transfer is quite efficient and the surface of the protostar will therefore be relatively bright.

We now describe the evolution in the HR diagram. Initially the protostar will be faint and cool, and it will reside at the lower far right in the HR diagram (outside Fig. 11.1). During the collapse its surface rapidly heats up and brightens and it moves to the upper right of Fig. 11.1. At the end of the collapse the star will settle at a point corresponding to its mass on the *Hayashi track*. The Hayashi track (Fig. 11.1) gives the location in the HR diagram of completely convective stars. Stars to its right cannot be in equilibrium and will collapse on the dynamic time scale.

The star will now evolve almost along the Hayashi track on the thermal time scale. In the HR diagram it moves almost vertically downwards, its radius decreases and its luminosity drops (Fig. 11.1). As the temperature goes on increasing in its centre, the opacity diminishes and energy begins to be transported by radiation. The mass of the radiative region will gradually grow until finally most of the star is radiative. By then the central temperature will have become so large that nuclear reactions begin. Previously all the stellar energy had been released potential energy, but now the nuclear reactions make a growing contribution and the luminosity increases. The stellar surface temperature will also in-

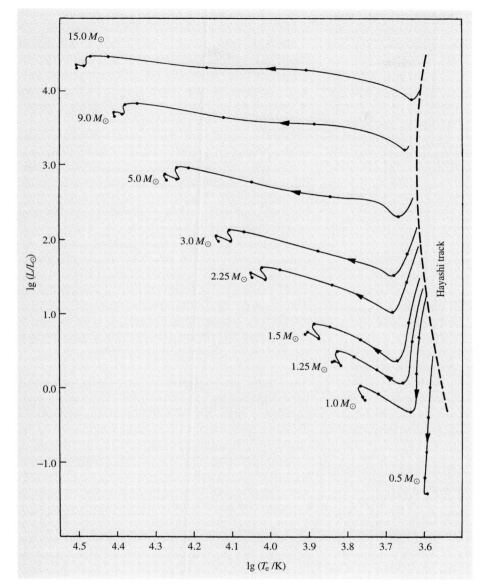

Fig. 11.1. The paths in the HR diagram of stars contracting to the main sequence on the thermal time scale. After a rapid dynamical collapse the stars settle on the Hayashi track and evolve towards the main sequence on the thermal time scale. (Models by Iben, I. (1965): Astrophys. J. **141**, 993)

crease and the star will move slightly upwards to the left in the HR diagram. In massive stars, this turn to the left occurs much earlier, because their central temperatures are higher and the nuclear reactions are initiated earlier.

For solar mass stars, the rapid collapse of the protostellar cloud only lasts for a few hundred years. The final stage of condensation is much slower, lasting several tens of millions of years. This length of time strongly depends on the stellar mass because of the luminosity dependence of the thermal time scale. A 15 $M_\odot$ star condenses to the main sequence in 60,000 years, whereas for a 0.1 $M_\odot$ star, the time is hundreds of millions of years.

Some of the hydrogen burning reactions start already at a few million degrees. For example, lithium, beryllium and boron burn to helium in the ppII and ppIII branches of the pp chain long before the complete chain has turned on. Because the star is convective and thus well mixed during the early stages, even its surface material will have been processed in the centre. Although the abundances of the above-mentioned elements are small, they give important information on the central temperature.

The beginning of the main sequence phase is marked by the start of hydrogen burning in the pp chain at a temperature of about 4 million degrees. The new form of energy production completely supersedes the energy release due to contraction. As the contraction is halted, the star makes a few oscillations in the HR diagram, but soon settles in an equilibrium and the long, quiet main sequence phase begins.

It is difficult to observe stars during contraction, because the new-born stars are usually hidden among dense clouds of dust and gas. However, some condensations in interstellar clouds have been discovered and near them, very young stars. One example are the *T Tauri stars*. Their lithium abundance is relatively high, which indicates that they are newly formed stars in which the central temperature has not yet become large enough to destroy lithium. Near the T Tauri stars, small, bright, star-like nebulae, *Herbig–Haro objects*, have been discovered. These are thought to be produced in the interaction between a stellar wind and the surrounding interstellar medium.

11.3 The Main Sequence Phase

The *main sequence phase* is that evolutionary stage in which the energy released by the burning of hydrogen in the core is the only source of stellar energy. During this stage, the star is in stable equilibrium, and its structure changes only because its chemical composition is gradually altered by the nuclear reactions. Thus the evolution takes place on a nuclear time scale, which means that the main sequence phase is the longest part of the life of a star. For example, for a solar mass star, the main sequence phase lasts for about $10,000$ million years. More massive stars evolve more rapidly, because they radiate much more energy. Thus the main sequence phase of a 15 solar mass star is only about 10 million years. On the other hand, less massive stars have

a longer main sequence lifetime: a $0.25\,M_\odot$ star spends about $70,000$ million years on the main sequence.

Since stars are most likely to be found in the stage of steady hydrogen burning, the main sequence in the HR diagram is richly populated, in particular at its low-mass end. The more massive upper main sequence stars are less abundant because of their shorter main sequence lifetimes.

If the mass of a star becomes too large, the force of gravity can no longer resist the radiation pressure. Stars more massive than this upper limit cannot form, because they cannot accrete additional mass during the contraction phase. Theoretical computations give a limiting mass of about $100\,M_\odot$; the most massive stars observed are about $70\,M_\odot$.

There is also a lower-mass limit of the main sequence. Stars below $0.08\,M_\odot$ never become hot enough for hydrogen burning to begin. The smallest protostars therefore contract to planet-like dwarfs. During the contraction phase they radiate because potential energy is released, but eventually they begin to cool. In the HR diagram such stars first move almost vertically downwards and then further downwards to the right.

The Upper Main Sequence. The stars on the *upper main sequence* are so massive and their central temperature so high that the CNO cycle can operate. On the *lower main sequence* the energy is produced by the pp chain. The pp chain and the CNO cycle are equally efficient at a temperature of 18 million degrees, corresponding to the central temperature of a $1.5\,M_\odot$ star. The boundary between the upper and the lower main sequence corresponds roughly to this mass.

The energy production in the CNO cycle is very strongly concentrated at the core. The outward energy flux will then become very large, and can no longer be maintained by radiative transport. Thus the upper main sequence stars have a *convective core*, i.e. the energy is transported by material motions. These keep the material well mixed, and thus the hydrogen abundance decreases uniformly with time within the entire convective region.

Outside the core, there is *radiative equilibrium*, i.e. the energy is carried by radiation and there are no nuclear reactions. Between the core and the envelope, there is a transition region where the hydrogen abundance decreases inwards.

The mass of the convective core will gradually diminish as the hydrogen is consumed. In the HR diagram the star will slowly shift to the upper right as its luminosity grows and its surface temperature decreases (Fig. 11.2). When the central hydrogen supply becomes exhausted, the core of the star will begin to shrink rapidly. The surface temperature will increase and the star will quickly move to the upper left. Because of the contraction of the core, the temperature in the hydrogen shell just outside the core will increase. It rapidly becomes high enough for hydrogen burning to set in again.

The Lower Main Sequence. On the lower main sequence, the central temperature is lower than for mas-

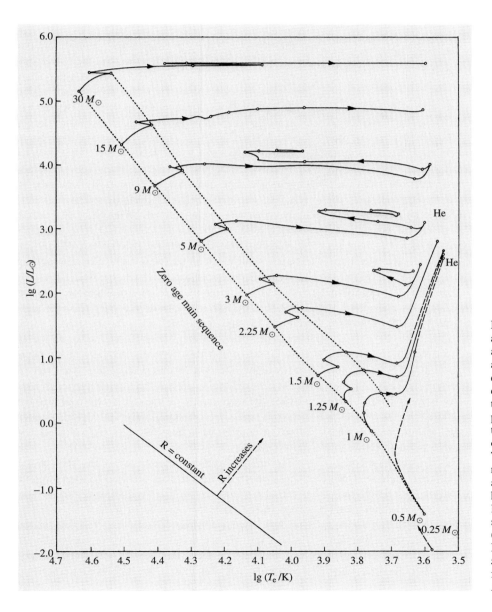

Fig. 11.2. Stellar evolutionary paths in the HR diagram at the main sequence phase and later. On the main sequence, bounded by dashed curves, the evolution is on the nuclear time scale. The post-main sequence evolution to the red giant phase is on the thermal time scale. The point marked He corresponds to helium ignition and in low-mass stars the helium flash. The straight line shows the location of stars with the same radius. (Iben, I. (1967): Annual Rev. Astron. Astrophys. **5**, 571; data for 30 $M_\odot$ from Stothers, R. (1966): Astrophys. J. **143**, 91)

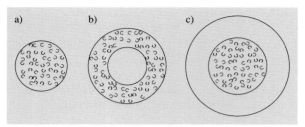

Fig. 11.3a–c. Energy transport in the main sequence phase. (**a**) The least massive stars ($M < 0.26\,M_\odot$) are convective throughout. (**b**) For $0.26\,M_\odot < M < 1.5\,M_\odot$ the core is radiative and the envelope convective. (**c**) Massive stars ($M > 1.5\,M_\odot$) have a convective core and a radiative envelope

sive stars, and the energy is generated by the pp chain. Since the rate of the pp chain is not as sensitive to temperature as that of the CNO cycle, the energy production is spread over a larger region than in the more massive stars (Fig. 11.3). In consequence, the core never becomes convectively unstable, but remains radiative.

In the outer layers of lower main sequence stars, the opacity is high because of the low temperature. Radiation can then no longer carry all the energy, and convection will set in. The structure of lower main sequence stars is thus opposite to that of the upper main sequence: the centre is radiative and the envelope is convective. Since there is no mixing of material in the core, the hydrogen is most rapidly consumed at the very centre, and the hydrogen abundance increases outwards.

As the amount of hydrogen in the core decreases, the star will slowly move upwards in the HR diagram, almost along the main sequence (Fig. 11.2). It becomes slightly brighter and hotter, but its radius will not change by much. The evolutionary track of the star will then bend to the right, as hydrogen in the core nears its end. Eventually the core is almost pure helium. Hydrogen will continue to burn in a thick shell around the core.

Stars with masses between $0.08\,M_\odot$ and $0.26\,M_\odot$ have a very simple evolution. During their whole main sequence phase they are fully convective, which means that their entire hydrogen content is available as fuel. These stars evolve very slowly toward the upper left in the HR diagram. Finally, when all their hydrogen has burned to helium, they contract to become white dwarfs.

11.4 The Giant Phase

The main-sequence phase of stellar evolution ends when hydrogen is exhausted at the centre. The star then settles in a state in which hydrogen is burning in a shell surrounding a helium core. As we have seen, the transition takes place gradually in low-mass stars, while the high-mass stars make a rapid jump in the HR diagram at this point.

The mass of the helium core is increased by the hydrogen burning in the shell. This leads to an expansion of the envelope of the star, which moves almost horizontally to the right in the HR diagram. As the convective envelope becomes more extensive, the star approaches the Hayashi track. Since it cannot pass further to the right, and since its radius continues to grow, the star has to move upwards along the Hayashi track towards larger luminosities (Fig. 11.2). The star has become a red giant.

In low-mass stars, as the mass of the core grows, its density will eventually become so high that it becomes degenerate. The central temperature will continue to rise. The whole helium core will have a uniform temperature because of the high conductivity of the degenerate gas. If the mass of the star is larger than $0.26\,M_\odot$ the central temperature will eventually reach about 100 million degrees, which is enough for helium to burn to carbon in the triple alpha process.

Helium burning will set in simultaneously in the whole central region and will suddenly raise its temperature. Unlike a normal gas, the degenerate core cannot expand, although the temperature increases (c. f. (10.16)), and therefore the increase in temperature will only lead to a further acceleration of the rate of the nuclear reactions. When the temperature increases further, the degeneracy of the gas is removed and the core will begin to expand violently. Only a few seconds after the ignition of helium, there is an explosion, the *helium flash*.

The energy from the helium flash is absorbed by the outer layers, and thus it does not lead to the complete disruption of the star. In fact the luminosity of the star drops in the flash, because when the centre expands, the outer layers contract. The energy released in the flash is turned into potential energy of the expanded core. Thus after the helium flash, the star will settle into a new state, where helium is steadily burning to carbon in a nondegenerate core.

After the helium flash the star finds itself on the horizontal giant branch in the HR diagram. However, the route leading there is not straight. The star may make several oscillations back and forth in the HR diagram before finally settling down.

When the helium is exhausted in the core, there are two nuclear burning shells in the star. In the inner one helium is burning; in the outer one hydrogen. Such a configuration is unstable and the stellar material may become mixed or matter may be ejected into space in a shell, like that of a planetary nebula.

In high-mass stars ($M \geq 1.5 \, M_\odot$), the central temperature is higher and the central density lower, and the core will therefore not be degenerate. Thus helium burning can set in noncatastrophically as the central regions contract. When the central helium supply is exhausted, helium will continue to burn in a shell.

The core will go on contracting and becoming hotter. First carbon burning and subsequently oxygen and silicon burning (see Sect. 10.3) will be ignited. As each nuclear fuel is exhausted in the centre, the burning will continue in a shell. The star will thus contain several nuclear burning shells. At the end the star will consist of a sequence of layers differing in composition, in massive stars (more massive than $15 \, M_\odot$) all the way up to iron.

The End of the Giant Phase. The evolution that follows helium burning depends strongly on the stellar mass. The mass determines how high the central temperature can become and the degree of degeneracy, when heavier nuclear fuels are ignited.

Stars less massive than $3 \, M_\odot$ never become hot enough to ignite carbon burning in the core. At the end of the giant phase the radiation pressure expels the outer layers, which form a *planetary nebula*. The hot core remains as a white dwarf.

In stars with masses in the range $3–15 \, M_\odot$ either carbon or oxygen is ignited explosively just like helium in low-mass stars: there is a *carbon* or *oxygen flash*. This is much more powerful than the helium flash, and will make the star explode as a supernova (Sects. 11.5 and 13.3). The star will probably be completely disrupted by the explosion.

The central parts of the most massive stars with masses larger than $15 \, M_\odot$ burn all the way to iron ^{56}Fe. All nuclear sources of energy will then be completely exhausted. The structure of a 30 solar mass star at this stage is schematically shown in Fig. 11.4. The star is made up of a nested sequence of zones bounded by shells burning silicon ^{28}Si, oxygen ^{16}O and carbon ^{12}C, helium ^{4}He and hydrogen ^{1}H. However, this is not a stable state, since the end of nuclear reactions in the core means that the central pressure will fall, and the core will collapse. The energy released in the collapse goes into dissociating the iron nuclei first to helium and then to protons and neutrons. This will further speed up the collapse, just like the dissociation of molecules speeds up the collapse of a protostar. The collapse takes place on a dynamical time scale, which, in the dense stellar core, is only a fraction of a second. The outer parts will also collapse, but more slowly. In consequence, the temperature will increase in layers containing unburnt nuclear fuel. This will burn explosively, releasing immense amounts of energy in a few seconds.

The final result is that the outer layers will explode as a supernova. In the dense central core, the protons and electrons combine to form neutrons. The core will finally consist almost entirely of neutrons, which become degenerate because of the high density. The degeneracy pressure of the neutrons will stop the collapse of a small mass core. However, if the mass of the core is large enough, a black hole will probably be formed.

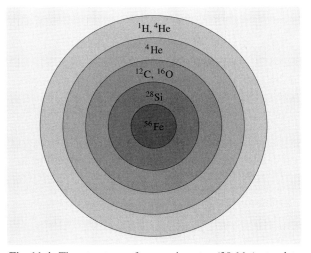

Fig. 11.4. The structure of a massive star ($30 \, M_\odot$) at a late evolutionary stage. The star consists of layers with different composition separated by nuclear burning shells

11.5 The Final Stages of Evolution

The endpoints of stellar evolution can be seen from Fig. 11.6. This shows the relation between mass and central density for a body at zero temperature, i. e. the final equilibrium when a massive body has cooled. There are two maxima on the curve. The mass corresponding to the left-hand maximum is called the *Chandrasekhar mass*, $M_{Ch} \approx 1.2$–$1.4\,M_\odot$, and that corresponding to the right-hand one, the *Oppenheimer–Volkoff mass*, $M_{OV} \approx 1.5$–$2\,M_\odot$.

Let us first consider a star with mass less than M_{Ch}. Suppose the mass does not change. When the nuclear fuel is exhausted, the star will become a white dwarf, which will gradually cool down and contract. In Fig. 11.6 it moves horizontally to the right. Finally it will reach zero temperature and end up on the left-hand rising part of the equilibrium curve. Its final equilibrium is a completely degenerate black dwarf.

If the mass of the star is larger than M_{Ch} but smaller than M_{OV}, it can continue cooling until it reaches the right-hand rising section of the curve. Again

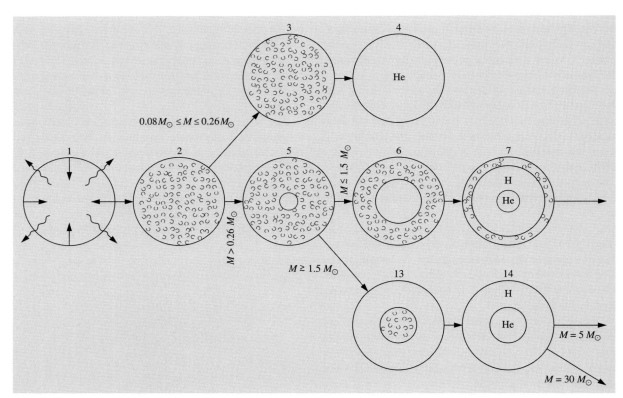

Fig. 11.5. Evolution schemes for stars with different masses. The radius is scaled to be the same in all drawings. In reality, there are vast differences in the sizes of different stars and different phases of evolution. In the beginning (*1*) a gas cloud is contracting rapidly in free fall. Because the gas is quite rarefied, radiation escapes easily from the cloud. As the density increases, radiation transport becomes more difficult, and the released energy tends to warm up the gas. The contraction

lasts until the gas is completely ionized, and the star, which has become a protostar, is in hydrostatic equilibrium (*2*). The star is convective throughout its interior.

Now evolution continues on a thermal time scale. The contraction is much slower than in the free-fall phase. The phases of further evolution are determined by the mass M of the star. For $M < 0.08\,M_\odot$ the temperature in the centre does not rise high enough for hydrogen burning, and these stars contract to planetlike brown dwarfs. Stars with $M \geq 0.08\,M_\odot$ start hydrogen burning when the temperature has reached about 4×10^6 K. This is the beginning of the main sequence phase. In the main sequence, the lowest-mass stars with $0.08\,M_\odot \leq M \leq 0.26\,M_\odot$ are entirely convective, and thus they remain homogeneous (*3*). Their evolution is very slow, and after all the hydrogen has been burnt to helium, they contract to white dwarfs (*4*).

The increasing temperature makes the stars with $M > 0.26\,M_\odot$ radiative in the centre as the opacity decreases (*5*). The low-mass stars with $0.26\,M_\odot \leq M \leq 1.5\,M_\odot$ remain radiative in the centre during the main sequence phase (*6*) as they burn their hydrogen through the pp chain. The outer part is convective. At the end of the main sequence phase, hydrogen burning continues in a shell surrounding the helium core (*7*).

The outer part expands, and the giant phase begins. The contracting helium core is degenerate and warms up. At about 10^8 K, the triple alpha process begins and leads immediately to the helium flash (*8*). The explosion is damped by the outer parts, and helium burning goes on in the core (*9*). Hydrogen is still burning in an outer shell. As the central helium is exhausted, helium burning moves over to a shell (*10*). At the same time, the outer part expands and the star loses some of its mass. The expanding envelope forms a planetary nebula (*11*). The star in the centre of the nebula becomes a white dwarf (*12*).

In the upper main sequence with $M \geq 1.5\,M_\odot$ energy is released through the CNO cycle, and the core becomes convective, while the outer part is radiative (*13*). The main sequence phase ends as the hydrogen in the core is exhausted, and shell burning begins (*14*). The helium core remains convective and nondegenerate, and helium burning begins without perturbations (*15* and *19*). Afterwards, helium burning moves over to a shell (*16* and *20*). For stars with $3\,M_\odot \leq M \leq 15\,M_\odot$ the carbon in the core is degenerate, and a carbon flash occurs (*17*). This leads to a supernova explosion (*18*) and possibly to the complete destruction of the star.

For the most massive stars with $M \geq 15\,M_\odot$ the carbon core remains convective, and carbon burns to oxygen and magnesium. Finally, the star consists of an iron core surrounded by shells with silicon, oxygen, carbon, helium and hydrogen (*21*). The nuclear fuel is now exhausted, and the star collapses on a dynamical time scale. The result is a supernova (*22*). The outer parts explode, but the remaining core continues to contract to a neutron star or a black hole

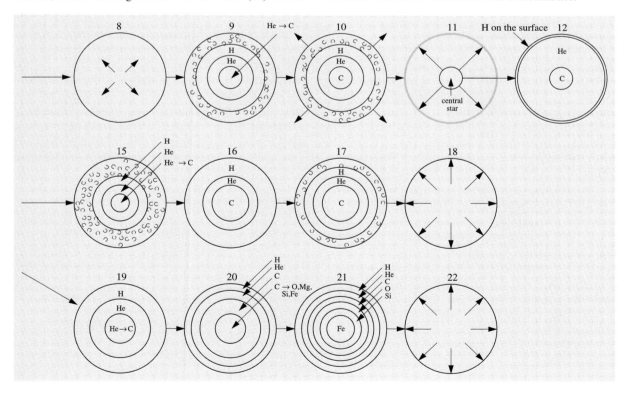

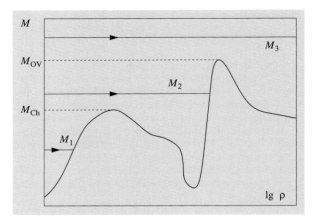

Fig. 11.6. The evolutionary end points of stars with different masses shown as a function of central density. The curve shows the behaviour of the central density of completely degenerate ($T = 0\,\mathrm{K}$) bodies. The Chandrasekhar mass M_{Ch} and the Oppenheimer–Volkoff mass M_{OV} correspond to maxima on this curve

there is a stable final state, this time corresponding to a completely degenerate neutron star.

An even more massive star with mass larger than M_{OV} will go on contracting past the density corresponding to a neutron star. There is then no longer any known possible stable equilibrium, and the star must go on contracting to form a black hole.

The only endpoints of stellar evolution predicted by theory are the two stable states of Fig. 11.6 and the two extreme possibilities, collapse to a black hole or explosive disruption.

The preceding considerations are purely theoretical. The final evolutionary stages of real stars involve many imperfectly known factors, which may affect the final equilibrium. Perhaps most important is the question of mass loss, which is very difficult to settle either observationally or theoretically. For example, in a supernova explosion the whole star may be disrupted and it is very uncertain whether what remains will be a neutron star, a black hole or nothing at all. (The structure of compact stars will be discussed in Chap. 14.)

A summary of the variuos evolutionary paths is given in Fig. 11.5.

11.6 The Evolution of Close Binary Stars

If the components of a binary star are well separated, they do not significantly perturb one another. When studying their evolution, one can regard them as two single stars evolving independently, as described above. However, in close binary pairs, this will no longer be the case.

Close binary stars are divided into three classes, as shown in Fig. 11.7: *detached*, *semidetached* and *contact binaries*. The figure-eight curve drawn in the figure is an equipotential surface called the *Roche surface*. If the star becomes larger than this surface, it begins to lose mass to its companion through the waist of the Roche surface.

During the main sequence phase the stellar radius does not change much, and each component will remain within its own Roche lobe. When the hydrogen is exhausted, the stellar core will rapidly shrink and the outer layers expand, as we have seen. At this stage a star may exceed its Roche lobe and mass transfer may set in.

Close binary stars are usually seen as eclipsing binaries. One example is Algol in the constellation Perseus. The components in this binary system are a normal main sequence star and a subgiant, which is much less massive than the main sequence star. The subgiant has a high luminosity and thus has apparently already left the main sequence. This is unexpected, since the components were presumably formed at the same time, and the more massive star should evolve more

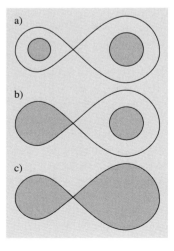

Fig. 11.7a–c. The types of close binary systems: (**a**) detached, (**b**) semidetached and (**c**) contact binary

rapidly. The situation is known as the *Algol paradox*: for some reason, the less massive star has evolved more rapidly.

In the 1950's a solution to the paradox proposed that the subgiant was originally more massive, but that it had lost mass to its companion during its evolution. Since the 1960's mass transfer in close binary systems has been much studied, and has turned out be a very significant factor in the evolution of close binaries.

As an example, let us consider a close binary, where the initial masses of the components are 1 and 2 solar masses and the initial orbital period 1.4 days (Fig. 11.8). After evolving away from the main sequence the more massive component will exceed the Roche limit and begin to lose mass to its companion. Initially the mass

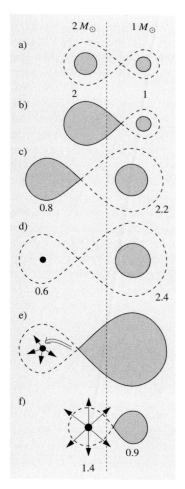

Fig. 11.8a–f. Evolution of a low-mass binary: (**a**) both components on the main sequence; (**b**) mass transfer from the more massive component; (**c**) light subgiant and massive main sequence star; (**d**) white dwarf and main sequence star; (**e**) mass transferred to the white dwarf from the more massive component leads to nova outbursts; (**f**) the white dwarf mass exceeds the Chandrasekhar mass and explodes as a type I supernova

will be transferred on the thermal time scale, and after a few million years the roles of the components will be changed: the initially more massive component has become less massive than its companion.

The binary is now semidetached and can be observed as an Algol-type eclipsing binary. The two components are a more massive main sequence star and a less massive subgiant filling its Roche surface. The mass transfer will continue, but on the much slower nuclear time scale. Finally, mass transfer will cease and the less massive component will contract to a 0.6 $M_\odot$ white dwarf.

The more massive 2.4 $M_\odot$ star now evolves and begins to lose mass, which will accumulate on the surface of the white dwarf. The accumulated mass may give rise to *nova outbursts*, where material is ejected into space by large explosions. Despite this, the mass of the white dwarf will gradually grow and may eventually exceed the Chandrasekhar mass. The white dwarf will then collapse and explode as a type I supernova.

As a second example, we can take a massive binary with the initial masses 20 and 8 $M_\odot$ and the initial period 4.7 days (Fig. 11.9). The more massive component evolves rapidly, and at the end of the main sequence phase, it will transfer more than 15 $M_\odot$ of its material to the secondary. The mass transfer will occur on the thermal time scale, which, in this case, is only a few ten thousand years. The end result is a *helium star*, having as a companion an unevolved main sequence star. The properties of the helium star are like those of a *Wolf–Rayet star*.

Helium continues to burn to carbon in the core of the helium star, and the mass of the carbon core will grow. Eventually the carbon will be explosively ignited, and the star will explode as a supernova. The consequences of this explosion are not known, but let us suppose that a 2 $M_\odot$ compact remnant is left. As the more massive star expands, its stellar wind will become stronger, giving rise to strong X-ray emission as it hits the compact star. This X-ray emission will only cease when the more massive star exceeds its Roche surface.

The system will now rapidly lose mass and angular momentum. A steady state is finally reached when the system contains a 6 $M_\odot$ helium star in addition to the 2 $M_\odot$ compact star. The helium star is seen as a Wolf–Rayet star, which, after about a million years, explodes as a supernova. This will probably lead to the breakup of the binary system. However, for certain values of

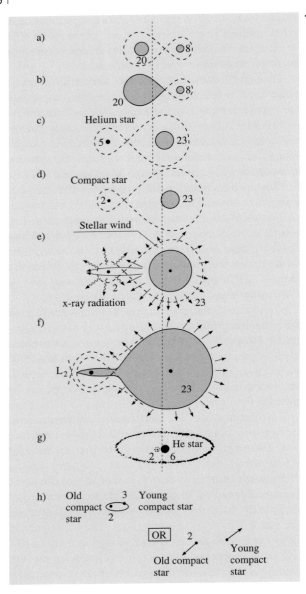

a)

b)

20

8

Helium star

c)

5

23

Compact star

d)

2

23

Stellar wind

e)

2

x-ray radiation   23

f)

L₂

23

g)

He star

2   6

h)   Old   3   Young
     compact   compact star
     star   2

     OR   2

     Old compact   Young
     star   compact
                        star

◄ Fig. 11.9a–h.
Evolution of a massive binary. It has been assumed that the supernova explosion of a $5\,M_\odot$ helium star leaves a $2\,M_\odot$ compact remnant (neutron star or black hole). (a) Main sequence phase; (b) beginning of the first mass transfer phase; (c) end of the first mass transfer phase; the first Wolf–Rayet phase begins; (d) the helium star (Wolf–Rayet star) has exploded as a supernova; (e) the $23\,M_\odot$ component becomes a supergiant; the compact component is a strong X-ray source; (f) beginning of the second mass transfer phase; the X-ray source is throttled and large-scale mass loss begins; (g) second Wolf–Rayet phase; (h) the $6\,M_\odot$ helium star has exploded as a supernova; the binary may or may not be disrupted, depending on the remaining mass

observed HR diagrams. If the theoretical models are correct, the observed number of stars should reflect the duration of the various evolutionary phases. These are given for stars of different masses in Table 11.1. The stars are most numerous along the main sequence. Giants are also common and, in addition to these, there are white dwarfs, subgiants, etc. The sparsely populated region to the right of the main sequence, the *Hertzsprung gap*, is explained by the rapid transition from the main sequence to the giant phase.

The cepheids provide an important test for the evolutionary models. The pulsations and the relation between period and luminosity for the cepheids can be understood on the basis of theoretical stellar models.

The evolutionary models can also explain the HR diagrams of star clusters. Let us assume that all the stars in a cluster were formed at the same time. In the youngest systems, the associations, the stars will mainly be found on the upper main sequence, since the most massive stars evolve most rapidly. To the right of the main sequence, there will be less massive T Tauri stars, which are still contracting. In intermediate age open clusters, the main sequence will be well developed and its upper end should bend to the right, since the most massive stars will already have begun to evolve off the main sequence. In the old globular clusters, the giant branch should increase in importance in the older clusters. These predictions are confirmed by the observations, which will be further discussed in Chap. 16 on star clusters.

Of course, the most detailed observations can be made of the Sun, which is therefore a crucial point of comparison for the theoretical models. If a star of one solar mass with an initial composition of 71% hydrogen, 27% helium and 2% heavier elements is allowed

the mass, the binary may remain bound. Thus a binary neutron star may be formed.

11.7 Comparison with Observations

The most important direct support for the theoretical evolutionary models is obtained from the properties of

to evolve for 5000 million years, it will be very similar to our present Sun. In particular, it will have the same radius, surface temperature and luminosity. According to calculations, about half of the Sun's supply of hydrogen fuel has been consumed. The Sun will go on shining like a normal main sequence star for another 5000 million years, before there will be any dramatic change.

Some problems remain in regard to the observations. One is the solar neutrino problem. The neutrinos produced by solar nuclear reactions have been observed since the beginning of the 1970's by the techniques described in Sect. 3.7. Only the neutrinos formed in the relatively rare ppIII reaction are energetic enough to be observed in this way. Their observed number is too small: whereas the models predict about 5 units, the observations have consistently only registered 1–2.

The discrepancy may be due to a fault in the observational technique or to some unknown properties of the neutrinos. However, if the solar models are really in error, the central temperature of the Sun would have to be about 20% lower than thought, which would be in serious conflict with the observed solar luminosity. One possibility is that some of the electron neutrinos change to other, unobservable particles during their passage to Earth. (See also Sect. 12.1.)

A second problem is the observed abundance of lithium and beryllium. The solar surface contains a normal abundance of beryllium, but very little lithium. This should mean that during its contraction, the Sun was still fully convective when the central temperature was high enough to destroy lithium (3×10^6 K), but not beryllium (4×10^6 K). However, according to the standard solar evolution models, convection ceased in the centre already at a temperature of 2×10^6 K. One suggested explanation is that the convection has later carried down lithium to layers where the temperature is high enough to destroy it.

11.8 The Origin of the Elements

There are just under a hundred naturally occurring elements, and about 300 isotopes in the solar system (Fig. 11.10). In Sect. 11.4, we have seen how the elements up to iron are produced when hydrogen burns

to helium and helium further to carbon, oxygen and heavier elements.

Almost all nuclei heavier than helium were produced in nuclear reactions in stellar interiors. In the oldest stars, the mass fraction of heavy elements is only about 0.02%, whereas in the youngest stars it is a few per cent. Nevertheless, most of the stellar material is hydrogen and helium. According to the standard cosmological model, those were formed in the early stages of the Universe, when the temperature and density were suitable for nuclear reactions. (This will be discussed in Chap. 19.) Although helium is produced during the main sequence stellar evolution, very little of it is actually returned into space to be incorporated into later stellar generations. Most of it is either transformed into heavier elements by further reactions, or else remains locked up inside white dwarf remnants. Therefore the helium abundance does not increase by much due to stellar processes.

The most important nuclear reactions leading to the build-up of the heavy nuclei up to iron were presented in Sect. 10.3. The probabilities of the various reactions are determined either by experiments or by theoretical calculations. When they are known, the relative abundances of the various nuclei produced can be calculated.

The formation of elements heavier than iron requires an input of energy, and thus they cannot be explained in the same manner. Still heavy nuclei are continually produced. In 1952 technetium was discovered in the atmosphere of a red giant. The half-life of the most longlived isotope ^{98}Tc is about 1.5×10^6 years, so that the observed technetium must have been produced in the star.

Most of the nuclei more massive than iron are formed by *neutron capture*. Since the neutron does not have an electric charge, it can easily penetrate into the nucleus. The probability for neutron capture depends both on the kinetic energy of the incoming neutron and on the mass number of the nucleus. For example, in the solar system the abundances of isotopes show maxima at the mass numbers $A = 70$–90, 130, 138, 195 and 208. These mass numbers correspond to nuclei with closed neutron shells at the neutron numbers $N = 50$, 82, and 126. The neutron capture probability for these nuclei is very small. The closed shell nuclei thus react more slowly and are accumulated in greater abundances.

Fig. 11.10. Element abundances in the solar system as a function of the nuclear mass number. The abundance of Si has been normalized as 10^6

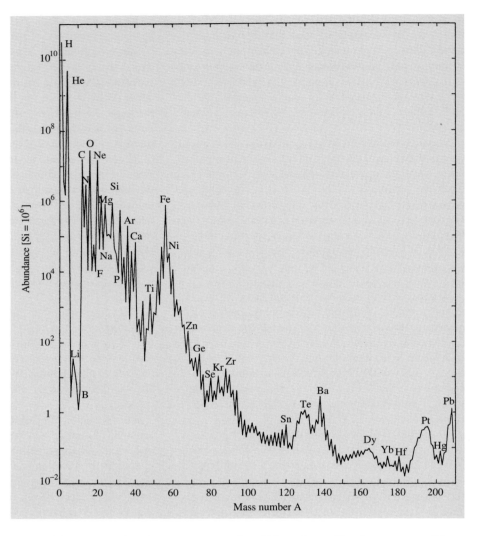

In a neutron capture, a nucleus with mass number A is transformed into a more massive nucleus:

$$(Z, A) + n \rightarrow (Z, A+1) + \gamma .$$

The newly formed nucleus may be unstable to β *decay*, where one neutron is transformed into a proton:

$$(Z, A+1) \rightarrow (Z+1, A+1) + e^- + \bar{\nu}_e .$$

Two kinds of neutron capture processes are encountered, depending on the value of the neutron flux. In the slow *s-process*, the neutron flux is so small that any β decays have had time to occur before the next neutron capture reaction takes place. The most stable nuclei up to mass number 210 are formed by the s-process. These nuclei are said to correspond to the β stability valley. The s-process explains the abundance peaks at the mass numbers 88, 138 and 208.

When the neutron flux is large, β decays do not have time to happen before the next neutron capture. One then speaks of the rapid *r-process*, which gives rise to more neutron-rich isotopes. The abundance maxima produced by the r-process lie at mass numbers about ten units smaller than those of the s-process.

A neutron flux sufficient for the s-process is obtained in the course of normal stellar evolution. For example, some of the carbon and oxygen burning reactions produce free neutrons. If there is convection between the hydrogen and helium burning shells, free protons may be carried into the carbon-rich layers. Then the following neutron-producing reaction chain becomes important:

$$^{12}C + p \rightarrow {}^{13}N + \gamma ,$$
$$^{13}N \rightarrow {}^{13}C + e^+ + \nu_e ,$$
$$^{13}C + {}^4He \rightarrow {}^{16}O + n .$$

The convection can also carry the reaction products nearer to the surface.

The neutron flux required for the r-process is about 10^{22} cm^{-3}, which is too large to be produced during normal stellar evolution. The only presently known site where a large enough neutron flux is expected is near a neutron star forming in a supernova explosion. In this case, the rapid neutron capture leads to nuclei that cannot capture more neutrons without becoming strongly unstable. After one or more rapid β decays, the process continues.

The r-process stops when the neutron flux decreases. The nuclei produced then gradually decay by the β-process towards more stable isotopes. Since the path of the r-process goes about ten mass units below the stability valley, the abundance peaks produced will fall about ten units below those of the s-process. This is shown in Fig. 11.11. The most massive naturally occurring elements, such as uranium, thorium and plutonium, are formed by the r-process.

There are about 40 isotopes on the proton-rich side of the β stability valley that cannot be produced by neutron capture processes. Their abundances are very small, relative to the neighbouring isotopes. They are formed in supernova explosions at temperatures higher than 10^9 K by reactions known as the *p-process*. At this temperature, pair formation can take place:

$$\gamma \rightarrow e^+ + e^- .$$

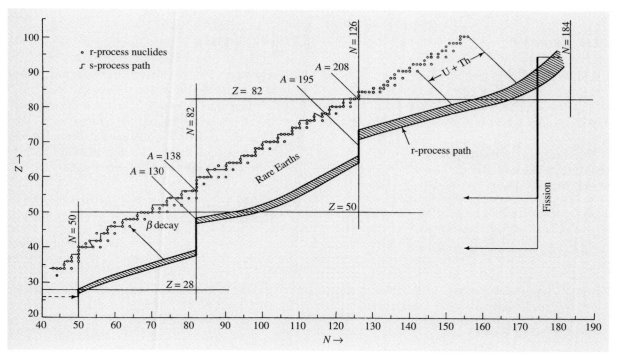

Fig. 11.11. Neutron capture paths for the s-process and r-process (*from left to right*). The s-process follows a path along the line of beta stability. The stable r-process nuclei (*small circles*) result from beta decay of their neutron rich progenitors on the shaded path shown lower. Beta decay occurs along straight lines $A = $ const. The closed neutron shells in nuclei at $N = 50$, 82 and 126 correspond to abundance peaks in s-process nuclei at $A = 88$, 138 and 208, and in r-process nuclei at $A = 80$, 130 and 195. (Seeger, P.A., Fowler, W.A., Clayton, D.D. (1965): Astrophys. J. Suppl. **11**, 121)

The positron may either be annihilated immediately or be consumed in the reaction

$$e^+ + (Z, A) \rightarrow (Z+1, A) + \bar{\nu}_e .$$

Another reaction in the p-process is

$$(Z, A) + p \rightarrow (Z+1, A+1) + \gamma .$$

Finally, the *fission* of some heavier isotopes may give rise to p-process nuclei. Examples of this are the isotopes ^{184}W, ^{190}Pt and ^{196}Hg formed by the fission of lead.

All the preceding reaction products are ejected into the interstellar medium in the supernova explosion. Collisions between cosmic rays and heavy nuclei then finally give rise to the light elements lithium, beryllium and boron. Thus the abundances of essentially all naturally occurring isotopes can be explained.

During succeeding generations of stars the relative abundance of heavy elements increases in the interstellar medium. They can then be incorporated into new stars, planets – and living beings.

11.9 Example

Example 11.1   An interstellar cloud has a mass of one solar mass and density of 10^{10} hydrogen atoms per cm^3. Its rotation period is 1000 years. What is the rotation period after the cloud has condensed into a star of solar size?

The angular momentum is $L = I\omega$, where ω is the angular velocity and I is the moment of inertia. For a homogeneous sphere

$$I = \frac{2}{5}MR^2 ,$$

where M is the mass and R the radius. From the conservation of the angular momentum we get

$$L = I_1\omega_1 = I_2\omega_2$$
$$\Rightarrow \frac{I_1 2\pi}{P_1} = \frac{I_2 2\pi}{P_2}$$
$$\Rightarrow P_2 = P_1 \frac{I_2}{I_1} = P_1 \frac{\frac{2}{5}MR_2^2}{\frac{2}{5}MR_1^2} = P_1 \left(\frac{R_2}{R_1}\right)^2 ,$$

where P_1 and P_2 are the rotation periods before and after the collapse. The mass of the cloud is

$$M = \frac{4}{3}\pi R^3 \rho$$
$$= \frac{4}{3}\pi R^3 \times 10^{16} \times 1.6734 \times 10^{-27} \text{ kg}$$
$$= 1 M_\odot = 1.989 \times 10^{30} \text{ kg} .$$

Solving for the radius we get $R = 3 \times 10^{13}$ m. The rotation period after the collapse is

$$P_2 = 1000 \text{ a} \times \left(\frac{6.96 \times 10^8 \text{ m}}{3 \times 10^{13} \text{ m}}\right)^2$$
$$= 5.4 \times 10^{-7} \text{ a} = 17 \text{ s} .$$

This is several orders of magnitude shorter than the actual period. Somehow the star has to get rid of most of its angular momentum during the process.

11.10 Exercises

Exercise 11.1   Find the free fall time scale for a hydrogen cloud, if the density of H_2 molecules is 3000 cm^{-3}. Assume that stars condense from such clouds, there are 100 clouds in the Galaxy, the mass of each cloud is $5 \times 10^4 M_\odot$, and 10% of the mass is converted into stars. Also assume that the average mass of a star is $1 M_\odot$. How many stars are born in one year?

Exercise 11.2   The mass of Vega (spectral class A0 V) is $2 M_\odot$, radius $3 R_\odot$, and luminosity $60 L_\odot$. Find its thermal and nuclear time scales.

Exercise 11.3   Assume that a star remains 10^9 years in the main sequence and burns 10% of its hydrogen. Then the star will expand into a red giant, and its luminosity will increase by a factor of 100. How long is the red giant stage, if we assume that the energy is produced only by burning the remaining hydrogen?

12. The Sun

The Sun is our nearest star. It is important for astronomy because many phenomena which can only be studied indirectly in other stars can be directly observed in the Sun (e.g. stellar rotation, starspots, the structure of the stellar surface). Our present picture of the Sun is based both on observations and on theoretical calculations. Some observations of the Sun disagree with the theoretical solar models. The details of the models will have to be changed, but the general picture should remain valid.

12.1 Internal Structure

The Sun is a typical main sequence star. Its principal properties are:

mass	$m = M_\odot$	=	1.989×10^{30} kg
radius	$R = R_\odot$	=	6.960×10^8 m
mean density	$\bar{\rho}$	=	1409 kg/m^3
central density	ρ_c	=	1.6×10^5 kg/m^3
luminosity	$L = L_\odot$	=	3.9×10^{26} W
effective temperature	T_e	=	5785 K
central temperature	T_c	=	1.5×10^7 K
absolute bolometric magnitude	M_bol	=	4.72
absolute visual magnitude	M_V	=	4.79
spectral class			G2 V
colour indices	$B - V$	=	0.62
	$U - B$	=	0.10
surface chemical composition	X	=	0.71
	Y	=	0.27
	Z	=	0.02
rotational period			
at the equator			25 d
at latitude 60°			29 d

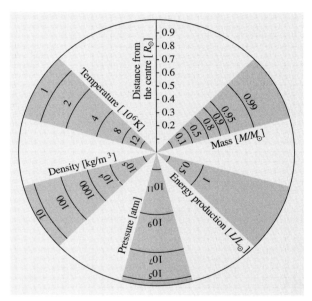

Fig. 12.1. The distribution of temperature, pressure, energy production and mass as functions of radius in the Sun

On the basis of these data, the solar model shown in Fig. 12.1 has been calculated. The energy is produced by the pp chain in a small central region. 99% of the solar energy is produced within a quarter of the solar radius.

The Sun produces energy at the rate of 4×10^{26} W, which is equivalent to changing about four million tonnes of mass into energy every second. The mass of the Sun is so large, about 330,000 times that of the Earth, that during the whole main sequence lifetime of the Sun less than 0.1% of its mass is turned into energy.

When the Sun formed about 5000 million years ago, its composition was the same everywhere as its present surface composition. Since energy production is concentrated at the very centre, hydrogen is consumed most rapidly there. At about a quarter of the radius the hydrogen abundance is still the same as in the surface layers, but going inwards from that point it rapidly decreases. In the central core only 40% of the material is hydrogen. About 5% of the hydrogen in the Sun has been turned into helium.

The radiative central part of the Sun extends to about 70% of the radius. At that radius the temperature has dropped so much that the gas is no longer completely ionized. The opacity of the solar material then strongly increases, preventing the propagation of radiation. In consequence, convection becomes a more efficient means of energy transport. Thus the Sun has a convective envelope (Fig. 12.2).

Fig. 12.2. The interior and surface of the Sun. The various kinds of solar phenomena are schematically indicated. (Based on Van Zandt, R.P. (1977): *Astronomy for the Amateur*, Planetary Astronomy, Vol. 1, 3rd ed. (published by the author, Peoria, Ill.))

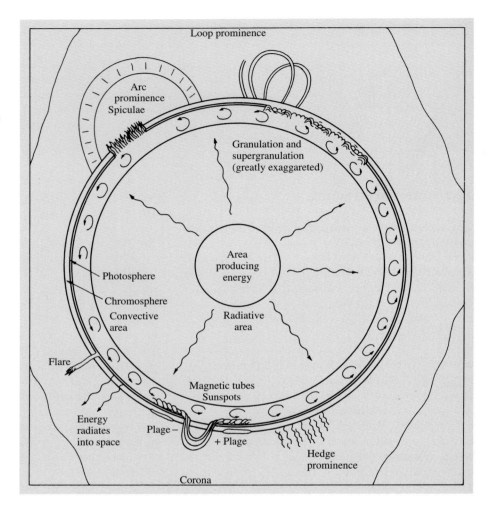

The Solar Neutrino Problem. The central nuclear reactions produce neutrinos at several of the steps in the pp chain (see Fig. 10.5). These neutrinos can propagate freely through the outer layers, and thus give direct information about conditions near the centre of the Sun. When neutrinos from the Sun were first observed in the 1970's, their number was found to be only about a third of what was predicted. This disagreement is called the *solar neutrino problem*.

In the first experiments only neutrinos from the ppII and ppIII branches were observed (Sect. 10.3). Since only a small fraction of the solar luminosity is produced in these reactions, it was not clear what were the consequences of these results for solar models. In the 1990's

neutrinos produced in the ppI branch, the main branch of the pp chain, have been observed. Although the disagreement with the standard models is slightly smaller in these observations (about 60% of the predicted flux has been observed), the neutrino problem still remains.

Perhaps the most popular explanation for the solar neutrino problem is based on neutrino oscillations. According to this explanation, if neutrinos have a small mass (about 10^{-2} eV), an electron neutrino could change into a μ or a τ neutrino as it passes through the outer parts of the Sun. Since only electron neutrinos have been observed in the experiments, the small observed flux of neutrinos could thus be understood. Since there are no direct measurements of neutrino masses, it is also possible that there is some flaw in the standard solar models.

The Solar Rotation. As soon as telescopes were introduced, it was observed from the motions of sunspots that the Sun is rotating with a rotational period of about 27 days. As early as 1630 *Christoph Scheiner* showed that there was *differential rotation*: the rotational period near the poles was more than 30 days, while it was only 25 days at the equator. The rotational axis of the Sun is inclined at 7° with respect to the plane of the ecliptic, so that the North Pole of the Sun is best visible from the Earth in September.

The motions of sunspots still give the best information on the rotation near the surface of the Sun. Other surface features also have been used for this purpose. The rotational velocity has also been measured directly from the Doppler effect. The angular velocity is usually written

$$\Omega = A - B \sin^2 \psi , \qquad (12.1)$$

where ψ is the latitude with respect to the equator. The measured values of the coefficients are $A = 14.5$ and $B = 2.9$ degrees/day.

The rotational velocity deeper down in the Sun cannot be directly observed. In the 1980's a method to estimate the rotation in the interior became available, when it became possible to measure the frequencies of solar oscillations from the variations in spectral lines. These oscillations are essentially sound waves produced by turbulent gas motions in the convection zone. These sound waves have calculable oscillation periods (about 3–12 minutes), which depend on the conditions in the solar interior. By comparing the observed and theoretical values one can get information about the conditions deep inside the Sun. The idea of the method is the same as that used when studying the interior of the Earth by means of waves from earthquakes, and it is therefore called *helioseismology*.

Using helioseismology, models for the solar rotation throughout the convection zone have been deduced. It appears that the angular velocity in the whole convection zone is almost the same as at the surface, although it decreases slightly with radius near the equator, and increases near the poles. The angular velocity of the radiative core is still uncertain, but there are indica-

tions that the core is rotating as a solid body with approximately the average surface angular velocity.

The solar differential rotation is maintained by gas motions in the convection zone. Explaining the observed behaviour is a difficult problem that is not yet completely understood.

12.2 The Atmosphere

The solar atmosphere is divided into the *photosphere* and the *chromosphere*. Outside the actual atmosphere, the *corona* extends much further outwards.

The Photosphere. The innermost layer of the atmosphere is the photosphere, which is only about 300–500 km thick. The photosphere is the visible surface of the Sun, where the density rapidly increases inwards, hiding the interior from sight. The temperature at the inner boundary of the photosphere is 8000 K and at the outer boundary 4500 K. Near the edge of the solar disc, the line of sight enters the photosphere at a very small angle and never penetrates to large depths. Near the edges one therefore only sees light from the cooler, higher layers. For this reason, the edges appear darker; this phenomenon is known as *limb darkening*. Both the continuous spectrum and the absorption lines are formed in the photosphere, but the light in the absorption lines comes from higher layers and therefore the lines appear dark.

The solar convection is visible on the surface as the *granulation* (Fig. 12.3), an uneven, constantly changing granular pattern. At the bright centre of each granule, gas is rising upward, and at the darker granule boundaries, it is sinking down again. The size of a granule seen from the Earth is typically $1''$, corresponding to about 1000 km on the solar surface. There is also a larger scale convection called *supergranulation* in the photosphere. The cells of the supergranulation may be about $1'$ in diameter. The observed velocities in the supergranulation are mainly directed along the solar surface.

The Chromosphere. Outside the photosphere there is a layer, perhaps about 500 km thick, where the temperature increases from 4500 K to about 6000 K, the chromosphere. Outside this layer, there is a transition region of a few thousand kilometres, where the chromo-

Fig. 12.3. The granulation of the solar surface. The granules are produced by streaming gas. Their typical diameter is 1000 km. (Photograph Mt. Wilson Observatory)

Fig. 12.4. Flash spectrum of the solar chromosphere, showing bright emission lines

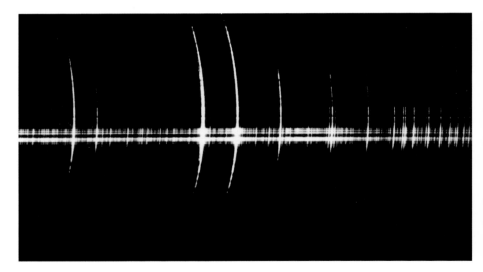

sphere gradually goes over into the corona. In the outer parts of the transition region, the kinetic temperature is already about 10^6 K.

Normally the chromosphere is not visible, because its radiation is so much weaker than that of the photosphere. However, during total solar eclipses, the chromosphere shines into view for a few seconds at both ends of the total phase, when the Moon hides the photosphere completely. The chromosphere then appears as a thin reddish sickle or ring.

During eclipses the chromospheric spectrum, called the *flash spectrum*, can be observed (Fig. 12.4). It is an emission line spectrum with more than 3000 identified lines. Brightest among these are the lines of hydrogen, helium and certain metals.

One of the strongest chromospheric emission lines is the hydrogen Balmer α line (Fig. 12.5) at a wavelength of 656.3 nm. Since the H_α line in the normal solar spectrum is a very dark absorption line, a photograph taken at this wavelength will show the solar chromosphere. For this purpose, one uses narrow-band filters letting through only the light in the H_α line. The resulting pictures show the solar surface as a mottled, wavy disc. The bright regions are usually the size of

Fig. 12.5. The solar surface in the hydrogen H_α line. Active regions near the equator appear bright; the dark filaments are prominences

a supergranule, and are bounded by *spicules* (Fig. 12.6). These are flamelike structures rising up to 10,000 km above the chromosphere, and lasting for a few minutes. Against the bright surface of the Sun, they look like dark streaks; at the edges, they look like bright flames.

The Corona. The chromosphere gradually goes over into the corona. The corona is also best seen during total solar eclipses (Fig. 12.7). It then appears as a halo of light extending out to a few solar radii. The surface brightness of the corona is about that of the full moon, and it is therefore difficult to see next to the bright photosphere.

The inner part of the corona, the K corona, has a continuous spectrum formed by the scattering of the photospheric light by electrons. Further out, a few solar radii from the surface, is the F corona, which has a spec-

trum showing Fraunhofer absorption lines. The light of the F corona is sunlight scattered by dust.

In the latter part of the 19th century strong emission lines, which did not correspond to those of any known element, were discovered in the corona (Fig. 12.8). It was thought that a new element, called coronium, had been found – a little earlier, helium had been discovered in the Sun before it was known on Earth. About 1940, it was established that the coronal lines were due to highly ionized atoms, e. g. thirteen times ionized iron. Much energy is needed to remove so many electrons from the atoms. The entire corona has to have a temperature of about a million degrees.

A continuous supply of energy is needed in order to maintain the high temperature of the corona. According to earlier theories, the energy came in the form of acoustic or magnetohydrodynamic shock waves generated at the solar surface by the convection. Most recently, heat-

Fig. 12.6. Spicules, flamelike uprisings near the edge of the solar disc. (Photograph Big Bear Solar Observatory)

Fig. 12.7. Previously, the corona could be studied only during total solar eclipses. The picture is from the eclipse on March 7, 1970. Nowadays the corona can be studied continuously using a device called the coronagraph

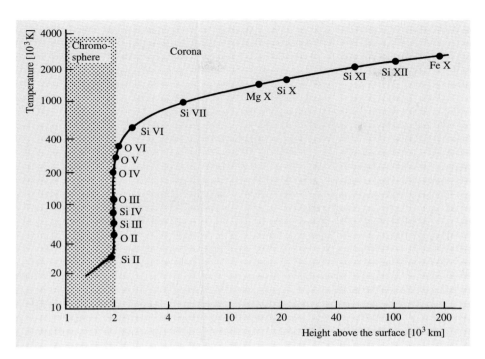

Fig. 12.8. The presence of lines from highly ionized atoms in the coronal spectrum shows that the temperature of the corona has to be very high

ing by electric currents induced by changing magnetic fields has been suggested. Heat would then be generated in the corona almost like in an ordinary light bulb.

In spite of its high temperature the coronal gas is so diffuse that the total energy stored in it is small. It is constantly streaming outwards, gradually becoming a *solar wind*, which carries a flux of particles away from the Sun. The gas lost in this way is replaced with new material from the chromosphere. Near the Earth the density of the solar wind is typically 5–10 particles/cm^3 and its velocity about 500 km/s. The mass loss of the Sun due to the solar wind is about $10^{-13}\,M_\odot$ per year.

12.3 Solar Activity

Sunspots. The clearest visible sign of solar activity are the *sunspots*. The existence of sunspots has been known for long (Fig. 12.9), since the largest ones can be seen with the naked eye by looking at the Sun through a suitably dense layer of fog. More precise observations became available beginning in the 17th century, when Galilei started to use the telescope for astronomical observations.

A sunspot looks like a ragged hole in the solar surface. In the interior of the spot there is a dark *umbra* and around it, a less dark *penumbra*. By looking at spots near the edge of the solar disc, it can be seen that the spots are slightly depressed with respect to the rest of the surface. The surface temperature in a sunspot is about 1500 K below that of its surroundings, which explains the dark colour of the spots.

The diameter of a typical sunspot is about $10,000$ km and its lifetime is from a few days to several months, depending on its size. The larger spots are more likely to be long-lived. Sunspots often occur in pairs or in larger groups. By following the motions of the spots, the period of rotation of the Sun can be determined.

The variations in the number of sunspots have been followed for almost 250 years. The frequency of spots is described by the Zürich sunspot number Z:

$$Z = C(S + 10\,G)\,, \tag{12.2}$$

where S is the number of spots and G the number of spot groups visible at a particular time. C is a con-

Fig. 12.9. The sunspots are the form of solar activity that has been known for the longest time. (Photograph Mt. Wilson Observatory)

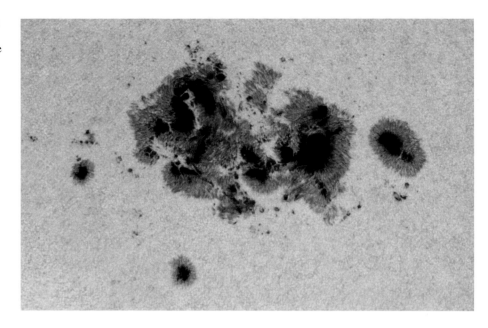

stant depending on the observer and the conditions of observation.

In Fig. 12.10, the variations in the Zürich sunspot number between the 18th century and the present are shown. Evidently the number of spots varies with an average period of about 11 years. The actual period may be between 7 and 17 years. In the past decades, it has been about 10.5 years. Usually the activity rises to its maximum in about 3–4 years, and then falls off slightly more slowly. The period was first noted by *Samuel Heinrich Schwabe* in 1843.

The variations in the number of sunspots have been fairly regular since the beginning of the 18th century.

However, in the 17th century there were long intervals when there were essentially no spots at all. This quiescent period is called the *Maunder minimum*. The similar *Spörer minimum* occurred in the 15th century, and other quiet intervals have been inferred at earlier epochs. The mechanism behind these irregular variations in solar activity is not yet understood.

The magnetic fields in sunspots are measured on the basis of the Zeeman effect, and may be as large as 0.45 tesla. (The magnetic field of the Earth is 0.03 mT.) The strong magnetic field inhibits convective energy transport, which explains the lower temperature of the spots.

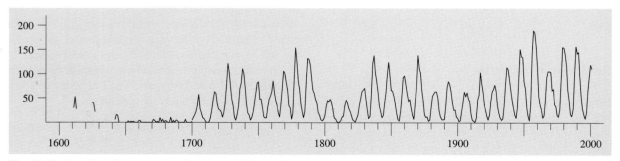

Fig. 12.10. The Zürich sunspot number from 1700 to 2001. Prior to 1700 there are only occasional observations. The number of sunspots and spot groups varies with a period of about 11 years

Fig. 12.11. In pairs of sunspots the magnetic field lines form a loop outside the solar surface. Material streaming along the field lines may form loop prominences. (Photograph Mt. Wilson Observatory)

Sunspots often occur in pairs where the components have opposite polarity. The structure of such *bipolar groups* can be understood if the field rises into a loop above the solar surface, connecting the components of the pair. If gas is streaming along such a loop, it becomes visible as a *loop prominence* (Fig. 12.11).

The periodic variation in the number of sunspots reflects a variation in the general solar magnetic field. At the beginning of a new activity cycle spots first begin to appear at latitudes of about $\pm 40°$. As the cycle advances, the spots move closer to the equator. The characteristic pattern in which spots appear, shown in Fig. 12.12, is known as the *butterfly diagram*. Spots of

the next cycle begin to appear while those of the old one are still present near the equator. Spots belonging to the new cycle have a polarity opposite to that of the old ones. (Spots in opposite hemispheres also have opposite polarity.) Since the field is thus reversed between consecutive 11 year cycles the complete period of solar magnetic activity is 22 years.

The following general qualitative description of the mechanism of the solar cycle was proposed by *Horace W. Babcock*. Starting at a solar minimum, the field will be of a generally dipolar character. Because a conducting medium, such as the outer layers of the Sun, cannot move across the field lines, these will be

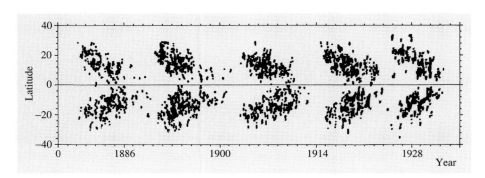

Fig. 12.12. At the beginning of an activity cycle, sunspots appear at high latitudes. As the cycle advances the spots move towards the equator. (Diagram by H. Virtanen, based on Greenwich Observatory observations)

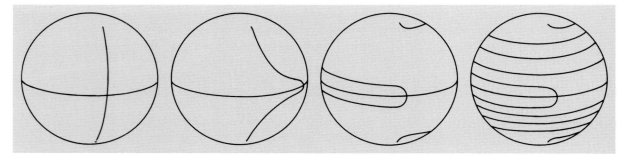

Fig. 12.13. Because the Sun rotates faster at the equator than at the poles, the field lines of the solar magnetic field are drawn out into a tight spiral

frozen into the plasma and carried along by it. Thus the differential rotation will draw the field into a tight spiral (Fig. 12.13). In the process the field becomes stronger, and this amplification will be a function of latitude.

When the subsurface field becomes strong enough, it gives rise to a "magnetic buoyancy" that lifts ropes of magnetic flux above the surface. This happens first at a latitude about 40°, and later at lower latitudes. These protruding flux ropes expand into loops forming bipolar groups of spots. As the loops continue expanding they make contact with the general dipolar field, which still remains in the polar regions. This leads to a rapid

reconnection of the field lines neutralising the general field. The final result when activity subsides is a dipolar field with a polarity opposite the initial one.

Thus the Babcock model accounts for the butterfly diagram, the formation of bipolar magnetic regions and the general field reversal between activity maxima. Nevertheless, it remains an essentially phenomenological model, and alternative scenarios have been proposed. In *dynamo theory* quantitative models for the origin of magnetic fields in the Sun and other celestial bodies are studied. In these models the field is produced by convection and differential rotation of the gas. A completely satisfactory dynamo model for the solar magnetic cycle

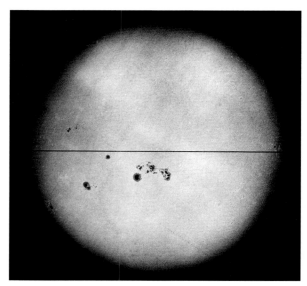

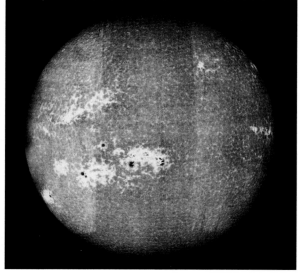

Fig. 12.14. The Sun seen in visible light (*left*) and in the hydrogen H_α line (*right*). The bright regions around the sunspots and sunspot groups are called plages. (Photographs Yerkes Observatory)

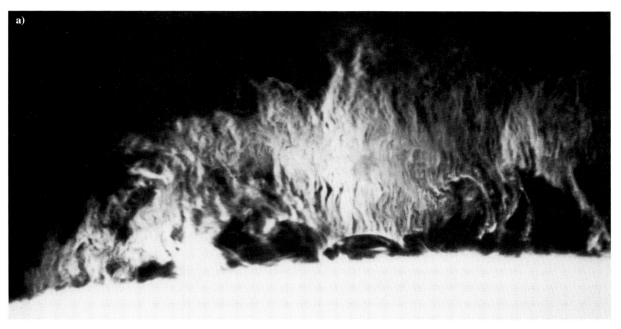

Fig. 12.15. (a) Quiescent "hedgerow" prominence (Photograph Sacramento Peak Observatory). **(b)** Larger eruptive prominence (Photograph Big Bear Solar Observatory)

has not yet been found. For example, it is not yet known whether the field is produced everywhere in the convection zone, or just in the boundary layer between the convective and radiative regions, as some indications suggest.

Other Activity. The Sun shows several other types of surface activity: *faculae* and *plages*; *prominences*; *flares*.

The faculae and plages are local bright regions in the photosphere and chromosphere, respectively. Observations of the plages are made in the hydrogen H_α or the calcium K lines (Fig. 12.14). The plages usually occur where new sunspots are forming, and disappear when the spots disappear. Apparently they are caused by the enhanced heating of the chromosphere in strong magnetic fields.

The prominences are among the most spectacular solar phenomena. They are glowing gas masses in the corona, easily observed near the edge of the Sun. There are several types of prominences (Fig. 12.15): the quiescent prominences, where the gas is slowly sinking along the magnetic field lines; loop prominences, connected with magnetic field loops in sunspots; and the rarer eruptive prominences, where gas is violently thrown outwards.

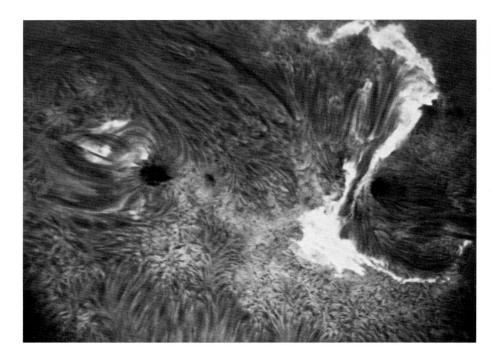

Fig. 12.16. A violent flare near some small sunspots. (Photograph Sacramento Peak Observatory)

The temperature of prominences is about 10,000–20,000 K. In H$_\alpha$ photographs of the chromosphere, the prominences appear as dark filaments against the solar surface.

The flare outbursts are among the most violent forms of solar activity (Fig. 12.16). They appear as bright flashes, lasting from one second to just under an hour. In the flares a large amount of energy stored in the magnetic field is suddenly released. The detailed mechanism is not yet known.

Flares can be observed at all wavelengths. The hard X-ray emission of the Sun may increase hundredfold during a flare. Several different types of flares are observed at radio wavelengths (Fig. 12.17). The emission of solar cosmic ray particles also rises.

The flares give rise to disturbances on the Earth. The X-rays cause changes in the ionosphere, which affect short-wave radio communications. The flare particles give rise to strong auroras when they enter the Earth's magnetic field a few days after the outburst.

Solar Radio Emission. The Sun is the strongest radio source in the sky and has been observed since the 1940's. In contrast to optical emission the radio picture of the Sun shows a strong *limb brightening*. This is because the radio radiation comes from the upper layers of the atmosphere. Since the propagation of radio waves is obstructed by free electrons, the high electron density near the surface prevents radio radiation from getting out. Shorter wavelengths can propagate more easily, and thus millimetre-wavelength observations give a picture of deeper layers in the atmosphere, whereas the long wavelengths show the upper layers. (The 10 cm emission originates in the upper layers of the chromosphere and the 1 m emission, in the corona.)

The Sun looks different at different wavelengths. At long wavelengths the radiation is coming from the largest area, and its electron temperature is about 10^6 K, since it originates in the corona.

The radio emission of the Sun is constantly changing according to solar activity. During large storms the total emission may be 100,000 times higher than normal.

X-ray and UV Radiation. The X-ray emission of the Sun is also related to active regions. Signs of activity are bright *X-ray regions* and smaller *X-ray bright points*, which last for around ten hours. The inner solar corona

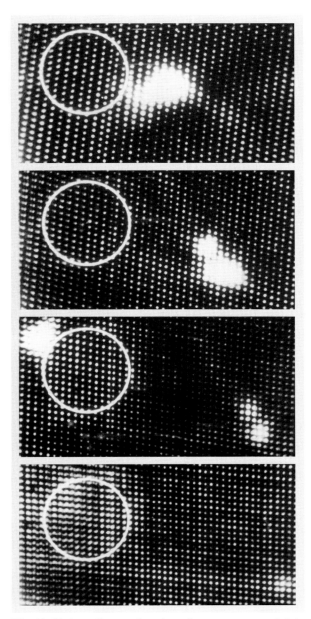

Fig. 12.17. At radio wavelengths a flare appears as a bright region, rapidly moving away from the Sun. (Picture CSIRO)

also emits X-rays. Near the solar poles there are *coronal holes*, where the X-ray emission is weak.

Ultraviolet pictures of the solar surface show it as much more irregular than it appears in visible light. Most of the surface does not emit much UV radiation,

but there are large active regions that are very bright in the ultraviolet.

Several satellites have made observations of the Sun at UV and X-ray wavelengths, for example Soho (Solar and Heliospheric Observatory, 1995–). These observations have made possible detailed studies of the outer layers of the Sun. Observations of other stars have revealed coronae, chromospheres and magnetic variations similar to those in the Sun. Thus the new observational techniques have brought the physics of the Sun and the stars nearer to each other.

12.4 Example

Example 12.1 Assume that the Sun converts 0.8% of its mass into energy. Find an upper limit for the age of the Sun, assuming that its luminosity has remained constant.

The total amount of energy released is

$$
\begin{aligned}
E = mc^2 &= 0.008\, M_\odot c^2 \\
&= 0.008 \times 2 \times 10^{30}\ \text{kg} \times (3 \times 10^8\ \text{ms}^{-1})^2 \\
&= 1.4 \times 10^{45}\ \text{J}\,.
\end{aligned}
$$

The time needed to radiate this energy is

$$
\begin{aligned}
t = \frac{E}{L_\odot} &= \frac{1.4 \times 10^{45}\ \text{J}}{3.9 \times 10^{26}\ \text{W}} \\
&= 3.6 \times 10^{18}\ \text{s} \approx 10^{11}\ \text{years}\,.
\end{aligned}
$$

12.5 Exercises

Exercise 12.1 The solar constant, i. e. the flux density of the solar radiation at the distance of the Earth is $1370\ \text{W m}^{-2}$.

a) Find the flux density on the surface of the Sun, when the apparent diameter of the Sun is $32''$.

b) How many square metres of solar surface is needed to produce 1000 megawatts?

Exercise 12.2 Some theories have assumed that the effective temperature of the Sun 4.5 billion years ago was 5000 K and radius 1.02 times the current radius. What was the solar constant then? Assume that the orbit of the Earth has not changed.

13. Variable Stars

Stars with changing magnitudes are called *variables* (Fig. 13.1). Variations in the brightness of stars were first noted in Europe at the end of the 16th century, when *Tycho Brahe's supernova* lit up (1572) and the regular light variation of the star o Ceti (Mira) was observed (1596). The number of known variables has grown steadily as observational precision has improved (Fig. 13.2). The most recent catalogues contain about 40,000 stars known or suspected to be variable.

Strictly speaking, all stars are variable. As was seen in Chap. 11, the structure and brightness of a star change as it evolves. Although these changes are usually slow, some evolutionary phases may be extremely rapid. In certain evolutionary stages, there will also be periodic variations, for example pulsations of the outer layers of a star.

Small variations in stellar brightness are also caused by hot and cool spots on a star's surface, appearing and disap-pearing as it rotates about its axis. The luminosity of the Sun changes slightly because of the sunspots. Probably there are similar spots on almost all stars.

Initially stellar brightnesses were determined visually by comparing stars near each other. Later on, comparisons were made on photographic plates. At present the most accurate observations are made photoelectrically or using a CCD camera. The magnitude variation as a function of time is called the *lightcurve* of a star (Fig. 13.3). From it one obtains the *amplitude* of the magnitude variation and its *period*, if the variation is periodic.

The basic reference catalogue of variable stars is the *General Catalogue of Variable Stars* by the Soviet astronomer *Boris Vasilyevich Kukarkin*. New, supplemented editions appear at times; the fourth edition published in 1985−1987, edited by P.N. Kholopov, contains about 32,000 variables of the Milky Way galaxy.

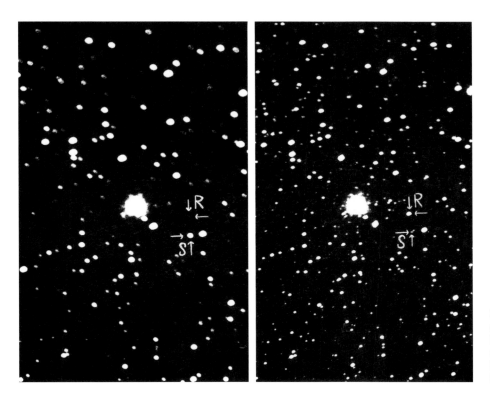

Fig. 13.1. The variables are stars changing in brightness. Two variables in Scorpius, R and S Sco (Photograph Yerkes Observatory)

Fig. 13.2. The location of variables in the HR diagram

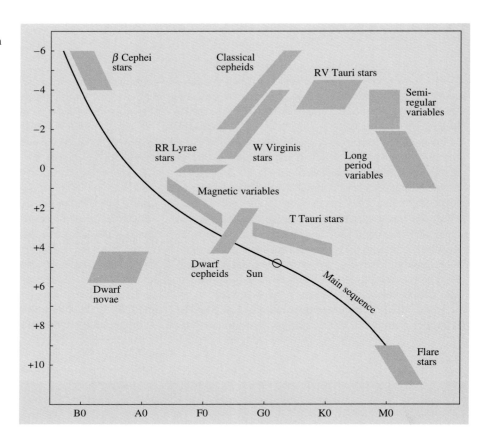

13.1 Classification

When a new variable is discovered, it is given a name according to the constellation in which it is located. The name of the first variable in a given constellation is R, followed by the name of the constellation (in the genitive case). The symbol for the second variable is S, and so on, to Z. After these, the two-letter symbols RR, RS, ... to ZZ are used, and then AA to QZ (omitting I). This is only enough for 334 variables, a number that has long been exceeded in most constellations. The numbering therefore continues: V335, V336, etc. (V stands for variable). For some stars the established Greek letter symbol has been kept, although they have later been found to be variable (e. g. δ Cephei).

The classification of variables is based on the shape of the lightcurve, and on the spectral class and observed radial motions. The spectrum may also contain dark absorption lines from material around the star. Observations can be made outside the optical region as well. Thus the radio emission of some variables (e. g. flare stars) increases strongly, simultaneously with their optical brightness. Examples of radio and X-ray variables are the radio and X-ray pulsars, and the X-ray bursters.

Variables are usually divided into three main types: *pulsating*, *eruptive* and *eclipsing variables*. The eclipsing variables are binary systems in which the components periodically pass in front of each other. In these variables the light variations do not correspond to any physical change in the stars. They have been treated in connection with the binary stars. In the other variables the brightness variations are intrinsic to the stars. In the pulsating variables the variations are due to the expansion and contraction of the outer layers. These variables are giants and supergiants that have reached an unstable stage in their evolution. The eruptive variables are usually faint stars ejecting mass.

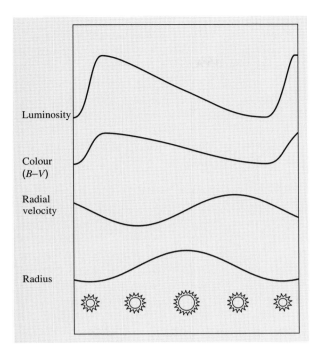

Fig. 13.3. The variation of brightness, colour and size of a cepheid during its pulsation

Table 13.1. The main properties of pulsating variables (N, number of stars of the given type in Kukarkin's catalogue, P, pulsation period in days, Δm, pulsation amplitude in magnitudes)

Variable	N	P	Spectrum	Δm
Classical cepheids (δ Cep, W Vir)	800	1–135	F–K I	$\lesssim 2$
RR Lyrae	6100	< 1	A–F8	$\lesssim 2$
Dwarf cepheids (δ Scuti)	200	0.05–7	A–F	$\lesssim 1$
β Cephei	90	0.1–0.6	B1–B3 III	$\gtrsim 0.3$
Mira variables	5800	80–1000	M–C	$\gtrsim 2.5$
RV Tauri	120	30–150	G–M	$\lesssim 4$
Semiregular	3400	30–1000	K–C	$\lesssim 4.5$
Irregular	2300	–	K–M	$\lesssim 2$

They are mostly members of close binary systems in which mass is transferred from one component to the other.

In addition a few *rotating variables* are known, where the brightness variations are due to an uneven temperature distribution on the surface, starspots coming into sight when the star rotates. Such stars may be quite common – after all, our Sun is a weak rotating variable. The most prominent group of rotating variables are the magnetic A stars (e. g. the α^2 Canum Venaticorum stars). These stars have strong magnetic fields that may be giving rise to starspots. The periods of rotating variables range from about 1 day to 25 d, and the amplitudes are less than 0.1 mag.

13.2 Pulsating Variables

The wavelengths of the spectral lines of the pulsating variables change along with the brightness variations (Table 13.1). These changes are due to the Doppler effect, showing that the outer layers of the star are indeed pulsating. The observed gas velocities are in the range of 40–200 km/s.

The period of pulsation corresponds to a *proper frequency* of the star. Just like a tuning fork vibrates with a characteristic frequency when hit, a star has a fundamental frequency of vibration. In addition to the fundamental frequency other frequencies, "overtones", are possible. The observed brightness variation can be understood as a superposition of all these modes of vibration. Around 1920, the English astrophysicist *Sir Arthur Eddington* showed that the period of pulsation P is inversely proportional to the square root of the mean density,

$$P \propto \frac{1}{\sqrt{\rho}} . \tag{13.1}$$

The diameter of the star may double during the pulsation, but usually the changes in size are minor. The main cause of the light variation is the periodic variation of the surface temperature. We have seen in Sect. 5.6 that the luminosity of a star depends sensitively on its effective temperature, $L \propto T_e^4$. Thus a small change in effective temperature leads to a large brightness variation.

Normally a star is in stable hydrostatic equilibrium. If its outer layers expand, the density and temperature decrease. The pressure then becomes smaller and the force of gravity compresses the gas again. However, unless energy can be transferred to the gas motions, these oscillations will be damped.

The flux of radiative energy from the stellar interior could provide a source of energy for the stellar oscillations, if it were preferentially absorbed in regions of higher gas density. Usually this is not the case but in the *ionization zones*, where hydrogen and helium are partially ionized, the opacity in fact becomes larger when the gas is compressed. If the ionization zones are at a suitable depth in the atmosphere, the energy absorbed during compression and released during expansion of an ionization zone can drive an oscillation. Stars with surface temperatures of 6000–9000 K are liable to this instability. The corresponding section of the HR diagram is called the cepheid instability strip.

Cepheids. Among the most important pulsating variables are the cepheids, named after δ Cephei (Fig. 13.3). They are population I supergiants (stellar populations are discussed in Sect. 17.2) of spectral class F–K. Their periods are 1–50 days and their amplitudes, 0.1–2.5 magnitudes. The shape of the light curve is regular, showing a fairly rapid brightening, followed by a slower fall off. There is a relationship between the period of a cepheid and its absolute magnitude (i.e. luminosity), discovered in 1912 by *Henrietta Leavitt* from cepheids in the Small Magellanic Cloud. This *period–luminosity relation* (Fig. 13.4) can be used to measure distances of stars and nearby galaxies.

We have already noted that the pulsation period is related to the mean density. On the other hand the size of a star, and hence its mean density, is related to its total luminosity. Thus one can understand why there should be a relation between the period and the luminosity of a pulsating star.

The magnitudes M and periods P of classical cepheids are shown in Fig. 13.4. The relation between M and $\log P$ is linear. However, to some extent, the cepheid luminosities also depend on colour: bluer stars are brighter. For accurate distance determinations, this effect needs to be taken into consideration.

W Virginis Stars. In 1952 *Walter Baade* noted that there are in fact two types of cepheids: the classical cepheids and the W Virginis stars. Both types obey a period–luminosity relation, but the W Vir stars of a given period are 1.5 magnitudes fainter than the corresponding classical cepheids. This difference is due to the fact that the classical cepheids are young population I objects, whereas the W Vir stars are old stars of population II. Otherwise, the two classes of variables are similar.

Earlier, the W Vir period–luminosity relation had been used for both types of cepheids. Consequently the calculated distances to classical cepheids were too small. For example, the distance to the Andromeda Galaxy had been based on classical cepheids, since only these were bright enough to be visible at that distance. When the correct period–luminosity relation was used, all extragalactic distances had to be doubled. Distances within the Milky Way did not have to be changed, since their measurements were based on other methods.

RR Lyrae Stars. The third important class of pulsating variables are the *RR Lyrae stars*. Their brightness variations are smaller than those of the cepheids, usually less than a magnitude. Their periods are also shorter, less than a day. Like the W Vir stars, the RR Lyrae stars are old population II stars. They are very common in the globular star clusters and were therefore previously called cluster variables.

The absolute magnitudes of the RR Lyrae stars are about $M_V = 0.6 \pm 0.3$. They are all of roughly the same age and mass, and thus represent the same evolutionary phase, where helium is just beginning to burn in the core. Since the absolute magnitudes of the RR Lyrae variables are known, they can be used to determine distances to the globular clusters.

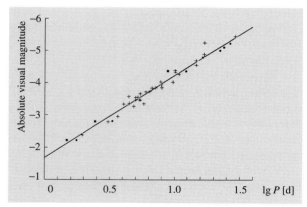

Fig. 13.4. The period–luminosity relation for cepheids. The black points and squares are theoretically calculated values, the crosses and the straight line represent the observed relation. (Drawing from Novotny, E. (1973): *Introduction to Stellar Atmospheres and Interiors* (Oxford University Press, New York) p. 359)

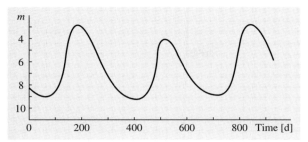

Fig. 13.5. The lightcurve of a long period Mira variable

Mira Variables (Fig. 13.5). The Mira variables (named after Mira Ceti) are supergiants of spectral classes M, S or C, usually with emission lines in their spectrum. They are losing gas in a steady stellar wind. Their periods are normally 100–500 days, and for this reason, they are also sometimes called long period variables. The amplitude of the light variations is typically about 6 magnitudes in the visual region. The period of Mira itself is about 330 days and its diameter is about 2 AU. At its brightest, Mira has the magnitude 2–4, but at light minimum, it may be down to 12. The effective temperature of the Mira variables is only about 2000 K. Thus 95% of their radiation is in the infrared, which means that a very small change in temperature can cause a very large change in visual brightness.

Other Pulsating Variables. One additional large group of pulsating stars are the *semiregular* and *irregular variables*. They are supergiants, often very massive young stars with unsteady pulsations in their extended outer layers. If there is some periodicity in the pulsations, these variables are called semiregular; otherwise they are irregular. An example of a semiregular variable is Betelgeuze (α Orionis). The pulsation mechanism of these stars is not well understood, since their outer layers are convective, and the theory of stellar convection is still poorly developed.

In addition to the main types of pulsating variables, there are some smaller separate classes, shown in Fig. 13.2.

The *dwarf cepheid* and the δ Scuti stars, which are sometimes counted as a separate type, are located below the RR Lyrae stars in the cepheid instability strip in the HR diagram. The dwarf cepheids are fainter and more rapidly varying than the classical cepheids. Their light curves often show a beating due to interference between the fundamental frequency and the first overtone.

The *β Cephei* stars are located in a different part of the HR diagram than the other variables. They are hot massive stars, radiating mainly in the ultraviolet. The variations are rapid and of small amplitude. The pulsation mechanism of the *β* Cephei stars is unknown.

The *RV Tauri* stars lie between the cepheids and the Mira variables in the HR diagram. Their period depends slightly on the luminosity. There are some unexplained features in the light curves of the RV Tauri stars, e. g. the minima are alternately deep and shallow.

13.3 Eruptive Variables

In the eruptive variables there are no regular pulsations. Instead sudden outbursts occur in which material is ejected into space. Nowadays such stars are divided into two main categories, *eruptive* and *cataclysmic variables*. Brightness changes of eruptive variables are caused by sudden eruptions in the chromosphere or corona, the contributions of which are, however, rather small in the stellar scale. These stars are usually surrounded by a gas shell or interstellar matter participating in the eruption. This group includes e. g. *flare stars*, various kinds of *nebular variables*, and *R Coronae Borealis* stars. Eruptions of the cataclysmic variables are due to nuclear reactions on the stellar surface or interior. Explosions are so violent that they can even destroy the whole star. This group includes *novae* and nova-like stars, *dwarf novae* and *supernovae* (Table 13.2).

Flare Stars. The *flare* or *UV Ceti stars* are dwarf stars of spectral class M. They are young stars, mostly found in young star clusters and associations. At irregular intervals there are flare outbursts on the surface of the stars similar to those on the Sun. The flares are related to disturbances in the surface magnetic fields. The energy of the outbursts of the flare stars is apparently about the same as in solar flares, but because the stars are much fainter than the Sun, a flare can cause a brightening by up to 4–5 magnitudes. A flare lights up in a few seconds and then fades away in a few minutes (Fig. 13.6). The same star may flare several times in one day. The optical flare is accompanied by a radio outburst, like in

Table 13.2. Main properties of eruptive variables (N, number of stars of the given type in Kukarkin's catalogue, Δm, change in brightness in magnitudes. The velocity is the expansion velocity in km/s, based on the Doppler shifts of the spectral lines)

Variable	N	Δm	Velocity
Supernovae	7	$\gtrsim 20$	4000–10000
Ordinary novae		7–18	200–3500
	210		
Recurrent novae		$\lesssim 10$	600
Nova-like stars	80	$\lesssim 2$	30–100
(P Cygni, symbiotic)			
Dwarf novae	330	2–6	(700)
(SS Cyg = U Gem, ZZ Cam)			
R Coronae Borealis	40	1–9	–
Irregular	1450	$\lesssim 4$	(300)
(nebular variables,			
T Tau, RW Aur)			
Flare stars	750	$\lesssim 6$	2000
(UV Ceti)			

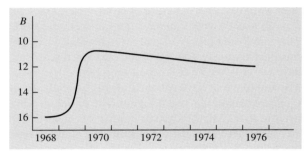

Fig. 13.7. Light curve of a T Tauri variable

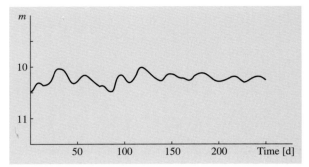

Fig. 13.8. In 1969–1970, the star V1057 Cygni brightened by almost 6 magnitudes

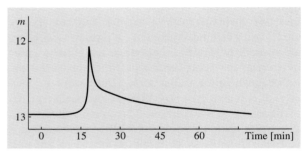

Fig. 13.6. The outbursts of typical flare stars are of short duration

the Sun. In fact, the flare stars were the first stars to be detected as radio sources.

Nebular Variables. In connection with bright and dark interstellar clouds e.g. in the constellations of Orion, Taurus and Auriga, there are variable stars. The *T Tauri stars* are the most interesting of them. These stars are newly formed or just contracting towards the main sequence. The brightness variations of the T Tauri stars are irregular (Fig. 13.7). Their spectra contain bright emission lines, formed in the stellar chromosphere, and forbidden lines, which can only be formed at extremely low densities. The spectral lines also show that matter is streaming out from the stars.

Since the T Tauri stars are situated inside dense gas clouds, they are difficult to observe. However, this situation has improved with the development of radio and infrared techniques.

Stars in the process of formation may change in brightness very rapidly. For example, in 1937, FU Orionis brightened by 6 magnitudes. This star is a strong source of infrared radiation, which shows that it is still enveloped by large quantities of interstellar dust and gas. A similar brightening by six magnitudes was observed in 1969 in V1057 Cygni (Fig. 13.8). Before its brightening, it was an irregular T Tauri variable; since then, it has remained a fairly constant tenth-magnitude AB star.

Stars of the *R Coronae Borealis* type have "inverse nova" light curves. Their brightness may drop by almost ten magnitudes and stay low for years, before the star brightens to its normal luminosity. For example, R CrB itself is of magnitude 5.8, but may fade to 14.8 magnitudes. Figure 13.9 shows its recent decline, based on observations by Finnish and French amateurs. The R CrB stars are rich in carbon and the decline is pro-

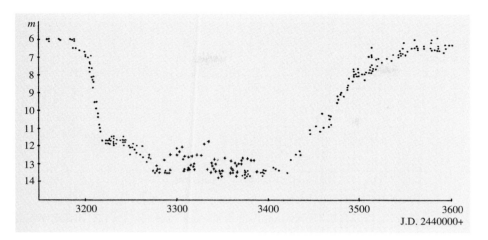

Fig. 13.9. The decline of R Coronae Borealis in 1977–1978; observations by Finnish and French amateur astronomers. (Kellomäki, Tähdet ja Avaruus 5/1978)

duced when the carbon condenses into a circumstellar dust shell.

One very interesting variable is *η Carinae* (Fig. 13.10). At present it is a six magnitude star surrounded by a thick, extensive envelope of dust and gas. In the early 19th century *η* Carinae was the second brightest star in the sky after Sirius. Around the middle of the century it rapidly dimmed to magnitude 8, but during the 20th century it has brightened somewhat. The circumstellar dust cloud is the brightest infrared source in the sky outside the solar system. The energy radiated by *η* Carinae is absorbed by the nebula and re-radiated at infrared wavelengths. It is not known whether *η* Carinae is related to the novae or whether it is a very young star that cannot evolve in the normal way because of the very thick cloud surrounding it.

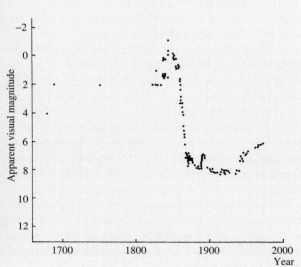

Fig. 13.10. In the 19th century, *η* Carinae was one of the brightest stars in the sky; since then it has dimmed considerably. In an outburst in 1843 the star ejected an expanding nebula, which has been called "Homunculus". (Photograph NASA/HST)

Fig. 13.11. The light curve of the dwarf nova SS Cygni in the beginning of 1966. (Drawing by Martti Perälä is based on observations by Nordic amateurs)

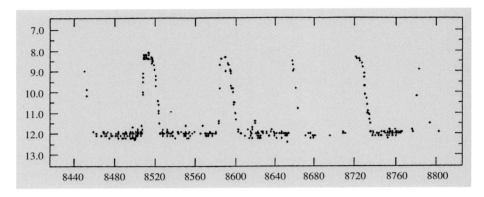

Fig. 13.11. The light curve of the dwarf nova SS Cygni in the beginning of 1966. (Drawing by Martti Perälä is based on observations by Nordic amateurs)

Novae. One of the best known types of eruptive variables are the *novae*. They are classified into several subtypes: *ordinary novae*, *recurrent novae* and *nova-like variables*. The *dwarf novae* (Fig. 13.11) are nova-like rather frequently eruptive stars with some special characteristics. They are nowadays divided into three subgroups.

The outbursts of all novae are rapid. Within a day or two the brightness rises to a maximum, which may be 7–18 magnitudes brighter than the normal luminosity. This is followed by a gradual decline, which may go on for months or years. The light curve of a typical nova is shown in Fig. 13.12. This light curve of Nova Cygni 1975 has been composed from hundreds of observations, mostly by amateurs.

In recurrent novae, the brightening is somewhat less than 10 magnitudes and in dwarf novae, 2–6 magnitudes. In both types there are repeated outbursts. For recurrent novae the time between outbursts is a few decades and for the dwarf novae 20–600 days. The interval depends on the strength of the outburst: the stronger the outburst, the longer the time until the next one. The brightening in magnitudes is roughly proportional to the logarithm of the recharging interval. It is possible that ordinary novae obey the same relationship. However, their amplitude is so much larger that the time between outbursts should be thousands or millions of years.

Observations have shown all novae and dwarf novae to be members of close binary systems. One component of the system is a normal star and the other is a white dwarf surrounded by a gas ring. (The evolution of close binary systems was considered in Sect. 11.6, where it was seen how this kind of system might have been formed.) The normal star fills its Roche surface, and

material from it streams over to the white dwarf. When enough mass has collected on the surface of the white dwarf, the hydrogen is explosively ignited and the outer shell is ejected. The brightness of the star grows rapidly. As the ejected shell expands, the temperature of the star drops and the luminosity gradually decreases. However, the outburst does not stop the mass transfer from the companion star, and gradually the white dwarf accretes new material for the next explosion (Fig. 13.13).

The emission and absorption lines from the expanding gas shell can be observed in the spectrum of a nova. The Doppler shifts correspond to an expansion velocity of about 1000 km/s. As the gas shell disperses, its spectrum becomes that of a typical diffuse emission nebula. The expanding shell around a nova can also sometimes be directly seen in photographs.

A considerable fraction of the novae in our Galaxy are hidden by interstellar clouds and their number is therefore difficult to estimate. In the Andromeda Galaxy, observations indicate 25–30 nova explosions per year. The number of dwarf novae is much larger. In addition there are nova-like variables, which share many of the properties of novae, such as emission lines from circumstellar gas and rapid brightness variations. These variables, some of which are called *symbiotic stars*, are close binaries with mass transfer. Gas streaming from the primary hits a gas disc around the secondary in a hot spot, but there are no nova outbursts.

Supernovae. The supernovae are stars with the largest brightness variations. In a few days their brightness may rise more than 20 magnitudes, i. e. the luminosity may increase by a factor of a hundred million. After the maximum, there is a slow decline lasting several years.

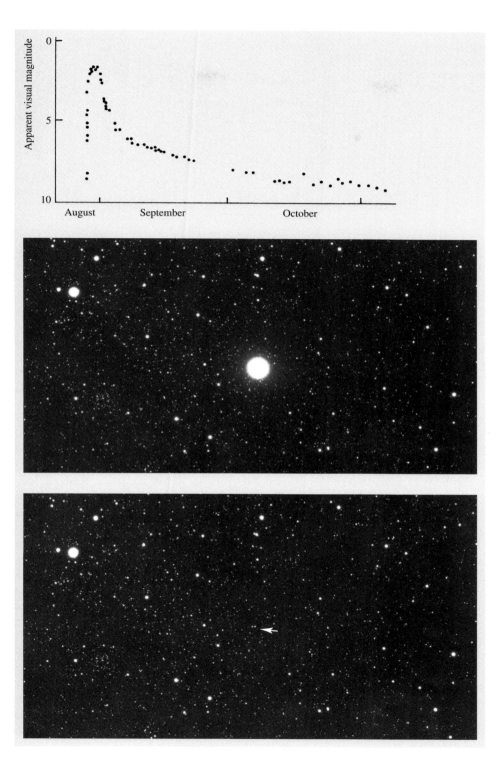

Fig. 13.12. In 1975 a new variable, Nova Cygni or V1500 Cygni, was discovered in Cygnus. In the upper photograph the nova is at its brightest (about 2 magnitudes), and in the lower photograph it has faded to magnitude 15. (Photographs Lick Observatory)

Fig. 13.13. The novae are thought to be white dwarfs accreting matter from a nearby companion star. At times, nuclear reactions burning the accreted hydrogen are ignited, and this is seen as the flare-up of a nova

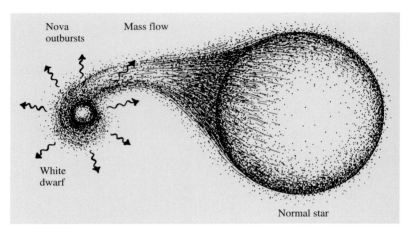

The supernovae are exploding stars. In the explosion a gas shell expanding with a velocity around 10,000 km/s is ejected. The expanding gas shell remains visible for thousands of years. A few tens of such supernova remnants have been discovered in the Milky Way. The remnant of the actual star may be a neutron star or a black hole.

The supernovae are classified as type I and II on the basis of their light curves (Fig. 13.14). *Type I* supernovae fade away in a regular manner, almost exponentially. The decline of a *type II* supernova is less regular and its maximum luminosity is smaller. The reason for this difference is that the exploding stars

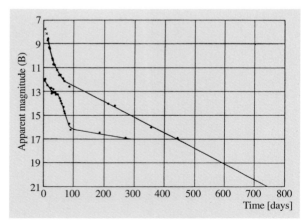

Fig. 13.14. The lightcurves of the two types of supernovae. Above SN 1972e type I and below SN 1970g type II. (Kirchner, R.P. (1976): Sci. Am. **235**, No. 6, 88)

are of quite different character. Type I supernovae are produced by old, low-mass stars; those of type II, by young, massive stars. As observations have improved, further subdivisions have been introduced. The classical supernovae are now usually designated as type Ia.

In Chap. 11 it was mentioned that there are several possible ways in which a star may come to explode at the end of its evolution. The type II supernova is the natural endpoint of the evolution of a single star. The type I supernovae, on the other hand, have about the mass of the Sun and should end up as white dwarfs. However, if a star is accreting mass from a binary companion, it will undergo repeated nova outbursts. Some of the accreted material will then be turned into helium or carbon and oxygen, and will collect on the star and increase its mass. Finally the mass may exceed the Chandrasekhar limit. The star will then collapse and explode as a supernova.

At least six supernova explosions have been observed in the Milky Way. Best known are the "guest star" seen in China in 1054 (whose remnant is the Crab nebula), Tycho Brahe's supernova in 1572 and Kepler's supernova in 1604. On the basis of observations of other Sb–Sc-type spiral galaxies, the interval between supernova explosions in the Milky Way is predicted to be about 50 years. Some will be hidden by obscuring material, but the 400 years' interval since the last observed supernova explosion is unusually long.

On February 23, 1987 the first burst of light from a supernova in the Large Magellanic Cloud, the small companion galaxy of the Milky Way, reached the Earth (Fig. 13.15). This supernova, SN 1987A, was of type II,

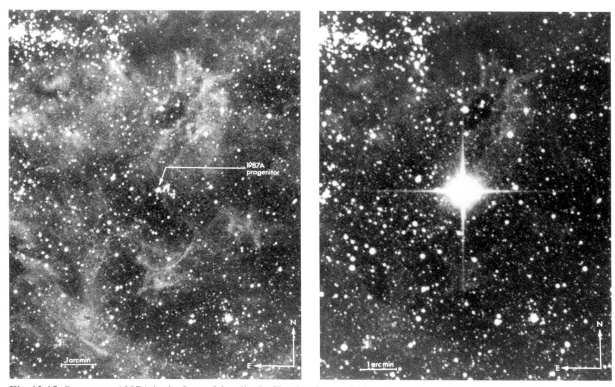

Fig. 13.15. Supernova 1987A in the Large Magellanic Cloud before and after the explosion. (Photographs ESO)

and was the brightest supernova for 383 years. After its first detection, SN 1987A was studied in great detail by all available means. Although the general ideas of Sects. 11.4 and 11.5 on the final stages of stellar evolution have been confirmed, there are complications. Thus e. g. the progenitor star was a blue rather than a red giant as expected, perhaps because of the lower abundance of heavy elements in the Large Magellanic Cloud compared to that in the Milky Way. The collapse of its core released a vast amount of energy as a pulse of neutrinos, which was detected in Japan and the USA. The amount of energy released indicates that the remnant is a neutron star. As of June 2002 the remnant has however not been seen yet.

13.4 Examples

Example 13.1 The observed period of a cepheid is 20 days and its mean apparent magnitude $m = 20$. From Fig. 13.4, its absolute magnitude is $M \approx -5$. According to (4.11), the distance of the cepheid is

$$r = 10 \times 10^{(m - M)/5} = 10 \times 10^{(20 + 5)/5}$$
$$= 10^6 \text{ pc} = 1 \text{ Mpc} .$$

Example 13.2 The brightness of a cepheid varies 2 mag. If the effective temperature is 6000 K at the maximum and 5000 K at the minimum, how much does the radius change?

The luminosity varies between

$$L_{\text{max}} = 4\pi R_{\text{max}}^2 \sigma T_{\text{max}}^4 ,$$
$$L_{\text{min}} = 4\pi R_{\text{min}}^2 \sigma T_{\text{min}}^4 .$$

In magnitudes the difference is

$$\Delta m = -2.5 \lg \frac{L_{\text{min}}}{L_{\text{max}}} = -2.5 \lg \frac{4\pi R_{\text{min}}^2 \sigma T_{\text{min}}^4}{4\pi R_{\text{max}}^2 \sigma T_{\text{max}}^4}$$
$$= -5 \lg \frac{R_{\text{min}}}{R_{\text{max}}} - 10 \lg \frac{T_{\text{min}}}{T_{\text{max}}} .$$

This gives

$$\lg \frac{R_{min}}{R_{max}} = -0.2\Delta m - 2\lg \frac{T_{min}}{T_{max}}$$

$$= -0.4 - 2\lg \frac{5000}{6000} = -0.24 \, ,$$

whence

$$\frac{R_{min}}{R_{max}} = 0.57 \, .$$

13.5 Exercises

Exercise 13.1 The absolute visual magnitude of RR Lyrae variables is 0.6 ± 0.3. What is the relative error of distances due to the deviation in the magnitude?

Exercise 13.2 The bolometric magnitude of a long period variable varies by one magnitude. The effective temperature at the maximum is 4500 K.
a) What is the temperature at the minimum, if the variation is due to temperature change only?
b) If the temperature remains constant, what is the relative variation in the radius?

Exercise 13.3 In 1983 the radius of the Crab nebula was about $3'$. It is expanding $0.21''$ a year. Radial velocities of $1300 \, km \, s^{-1}$ with respect to the central star have been observed in the nebula.

a) What is the distance of the nebula, assuming its expansion is symmetric?
b) A supernova explosion has been observed in the direction of the nebula. Estimate, how long time ago?
c) What was the apparent magnitude of the supernova, if the absolute magnitude was a typical -18?

14. Compact Stars

In astrophysics those stars in which the density of matter is much larger than in ordinary stars are known as compact objects. These include white dwarfs, neutron stars, and black holes. In addition to a very high density, the compact objects are characterised by the fact that nuclear reactions have completely ceased in their interiors. Conse- quently they cannot support themselves against gravity by thermal gas pressure. In the white dwarfs and neutron stars, gravity is resisted by the pressure of a degenerate gas. In the black holes the force of gravity is completely dominant and compresses the stellar material to infinite density.

14.1 White Dwarfs

As was mentioned in Sect. 10.2, in ordinary stars the pressure of the gas obeys the equation of state of an ideal gas. In stellar interiors the gas is fully ionized, i. e. it is plasma consisting of ions and free electrons. The partial pressures of the ions and electrons together with the radiation pressure important in hot stars comprise the total pressure balancing gravitation. When the star runs out of its nuclear fuel, the density in the interior increases but the temperature does not change much. The electrons become degenerate, and the pressure is mainly due to the pressure of the degenerate electron gas, the pressure due to the ions and radiation being negligible. The star becomes a *white dwarf*.

As will be explained in *The Radius of White Dwarfs and Neutron Stars (p. 282) the radius of a degenerate star is inversely proportional to the cubic root of the mass. Unlike in a normal star the radius decreases as the mass increases.

The first white dwarf to be discovered was Sirius B, the companion of Sirius (Fig. 14.1). Its exceptional na- ture was realized in 1915, when it was discovered that its effective temperature was very high. Since it is faint, this meant that its radius had to be very small, slightly smaller than that of the Earth. The mass of Sirius B was known to be about equal to that of the Sun, so its density had to be extremely large.

The high density of Sirius B was confirmed in 1925, when the gravitational redshift of its spectral lines was measured. This measurement also provided early obser- vational support to Einstein's general theory of relativity.

White dwarfs occur both as single stars and in binary systems. Their spectral lines are broadened by the strong gravitational field at the surface. In some white dwarfs

Fig. 14.1. One of the best-known white dwarfs is Sirius B, the small companion of Sirius. In this picture taken with a 3 m telescope, it can be seen as a faint spot to the left of the primary star. (Photograph Lick Observatory)

the spectral lines are further broadened by rapid rotation. Strong magnetic fields have also been observed.

White dwarfs have no internal sources of energy, but further gravitational contraction is prevented by the pressure of the degenerate electron gas. Radiating away the remaining heat, white dwarfs will slowly cool, changing in colour from white to red and finally to black. There ought to be a large number of invisible black dwarfs in the Milky Way.

White dwarfs that are members of binary systems can often be seen as novae, as was explained previosly in Sect. 13.3.

14.2 Neutron Stars

If the mass of a star is large enough, the density of matter may grow even larger than in normal white dwarfs. The equation of state of a classical degenerate electron gas then has to be replaced with the corresponding relativistic formula. In this case decreasing the radius of the star no longer helps in resisting the gravitational attraction. Equilibrium is possible only for one particular value of the mass, the Chandrasekhar mass M_{Ch}, already introduced in Sect. 11.5. The value of M_{Ch} is about $1.4 M_{\odot}$, which is thus the upper limit to the mass of a white dwarf. If the mass of the star is larger than M_{Ch}, gravity overwhelms the pressure and the star will rapidly contract towards higher densities. The final stable state reached after this collapse will be a *neutron star* (Fig. 14.2). On the other hand, if the mass is smaller than M_{Ch}, the pressure dominates. The star will then expand until the density is small enough to allow an equilibrium state with a less relativistic equation of state.

When a massive star reaches the end of its evolution and explodes as a supernova, the simultaneous collapse of its core will not necessarily stop at the density of a white dwarf. If the mass of the collapsing core is larger than the Chandrasekhar mass ($\gtrsim 1.4 M_{\odot}$), the collapse continues to a neutron star.

An important particle reaction during the final stages of stellar evolution is the *URCA process*, which was put forward by *Schönberg* and *Gamow* in the 1940's and which produces a large neutrino emission without otherwise affecting the composition of matter. (The URCA process was invented in Rio de Janeiro and named after a local casino. Apparently money disappeared at URCA just as energy disappeared from stellar interiors in the form of neutrinos. It is claimed that the casino was closed by the authorities when this similarity became known.) The URCA process consists of the reactions

$$(Z, A) + e^{-} \rightarrow (Z - 1, A) + \nu_e \,,$$
$$(Z - 1, A) \rightarrow (Z, A) + e^{-} + \bar{\nu}_e \,,$$

where Z is the number of protons in a nucleus; A the mass number; e^{-} an electron; and ν_e and $\bar{\nu}_e$ the electron neutrino and antineutrino. When the electron gas is degenerate, the latter reaction is suppressed by the Pauli exclusion principle. In consequence the protons in the nuclei are transformed into neutrons. As the number of neutrons in the nuclei grows, their binding energies decrease. At densities of about 4×10^{14} kg/m^3 the neutrons begin to leak out of the nucleus, and at 10^{17} kg/m^3 the nuclei disappear altogether. Matter then consists of a neutron "porridge", mixed with about 0.5% electrons and protons.

Neutron stars are supported against gravity by the pressure of the degenerate neutron gas, just as white dwarfs are supported by electron pressure. The equation of state is the same, except that the electron mass is replaced by the neutron mass, and that the mean molecular weight is defined with respect to the number of free neutrons. Since the gas consists almost entirely of neutrons, the mean molecular weight is approximately one.

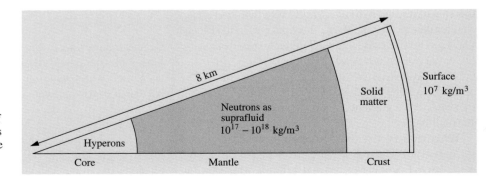

Fig. 14.2. The structure of a neutron star. The crust is rigid solid material and the mantle a freely streaming superfluid

The typical diameters of neutron stars are about 10 km. Unlike ordinary stars they have a well-defined solid surface. The atmosphere above it is a few centimetres thick. The upper crust is a metallic solid with the density growing rapidly inwards. Most of the star is a neutron superfluid, and in the centre, where the density exceeds 10^{18} kg/m^3, there may be a solid nucleus of heavier particles (hyperons).

A neutron star formed in the explosion and collapse of a supernova will initially rotate rapidly, because its angular momentum is unchanged while its radius is much smaller than before. In a few hours the star will settle in a flattened equilibrium, rotating several hundred times per second. The initial magnetic field of the neutron star will also be compressed in the collapse, so that there will be a strong field coupling the star to the surrounding material. The angular momentum of the neutron star is steadily decreased by the emission of electromagnetic radiation, neutrinos, cosmic ray particles and possibly gravitational radiation. Thus the angular velocity decreases. The rotation can also break the star into several separate objects. They will eventually recombine when the energy of the system is reduced. In some cases the stars can remain separated, resulting e. g. in a binary neutron star.

The theory of neutron stars was developed in the 1930's, but the first observations were not made until the 1960's. At that time *the pulsars*, a new type of rapidly pulsating radio sources, were discovered and identified as neutron stars. In the 1970's neutron stars were also seen as *X-ray pulsars* and *X-ray bursters* (Figs. 14.7 and 14.8).

Pulsars. The pulsars were discovered in 1967, when *Anthony Hewish* and *Jocelyn Bell* in Cambridge, England, detected sharp, regular radio pulses coming from the sky. Since then about four hundred pulsars have been discovered (Fig. 14.4). Their periods range from 0.0016 s (for the pulsar 1937+214) to several seconds.

In addition to the steady slowing down of the rotation, sometimes small sudden jumps in the period are observed. These might be a sign of rapid mass movements in the neutron star crust ("starquakes") or in its surroundings.

The origin of the radio pulses can be understood if the magnetic field is tilted at an angle of $45° - 90°$ with respect to the rotation axis. There will then be

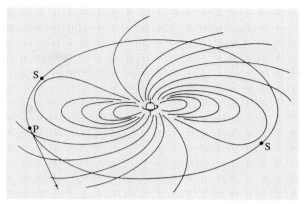

Fig. 14.3. The magnetic field around a rotating neutron star carries plasma with it. At a certain distance the speed of the plasma approaches the speed of light. At this distance the radiating regions S emit radiation in a narrow forward beam. Radiation from the point P hits the observer located in the direction of the arrow. (Drawing from Smith, F.G. (1977): Pulsars (Cambridge University Press, Cambridge) p. 189)

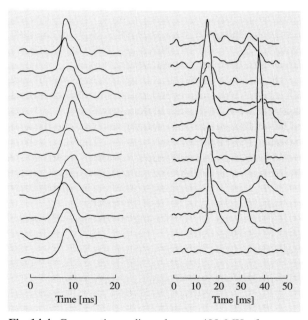

Fig. 14.4. Consecutive radio pulses at 408 MHz from two pulsars. To the left PSR 1642-03 and to the right PSR 1133+16. Observations made at Jodrell Bank. (Picture from Smith, F.G. (1977): Pulsars (Cambridge University Press, Cambridge) pp. 93, 95)

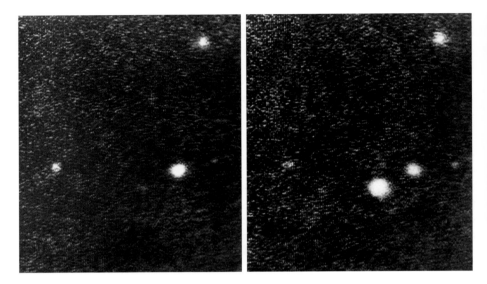

a magnetosphere around the star, where the particles are tied to the magnetic field and rotate with it (Fig. 14.3). At a certain distance from the star, the speed of rotation approaches the speed of light. The rapidly moving particles will there emit radiation in a narrow cone in their direction of motion. As the star rotates, this cone sweeps around like a lighthouse beam, and is seen as rapid pulses. At the same time relativistic particles stream out from the neutron star.

The best-known pulsar is located in the Crab nebula (Fig. 14.5). This small nebula in the constellation Taurus was noted by the French astronomer *Charles Messier* in the middle of the 18th century and became the first object in the Messier catalogue, M1. The Crab nebula was found to be a strong radio source in 1948 and an X-ray source in 1964. The pulsar was discovered in 1968. In the following year it was observed optically and was also found to be an X-ray emitter.

Neutron stars are difficult to study optically, since their luminosity in the visible region is very small (typically about $10^{-6} L_\odot$). For instance the Vela pulsar has been observed at a visual magnitude of about 25. In the radio region, it is a very strong pulsating source.

A few pulsars have been discovered in binary systems; the first one, PSR 1913+16, in 1974. In 1993 *Joseph Taylor* and *Russell Hulse* were awarded the Nobel prize for the detection and studies of this pulsar. The pulsar orbits about a companion, presumably another neutron star, with the orbital eccentricity 0.6 and the pe-

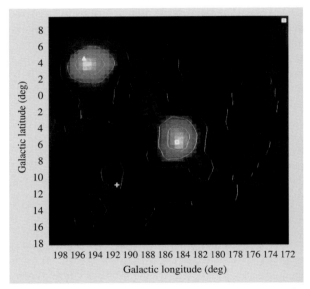

Fig. 14.6. Some pulsars shine brightly in gamma-rays. In the center the Crab pulsar and on the upper left the gamma source Geminga, which was identified in 1992 to be the nearest pulsar with a distance of about 100 pc from the Sun. (Photo by Compton Gamma Ray Observatory)

riod 8 hours. The observed period of the pulses is altered by the Doppler effect, and this allows one to determine the velocity curve of the pulsar. These observations can be made very accurately, and it has therefore been possible to follow the changes in the orbital elements of the system over a period of several years. For example, the periastron direction of the binary pulsar has been found to rotate about 4° per year. This phenomenon can be explained by means of the general theory of relativity; in the solar system, the corresponding rotation (the minor fraction of the rotation not explained by the Newtonian mechanics) of the perihelion of Mercury is 43 arc seconds per century.

The binary pulsar PSR 1913+16 has also provided the first strong evidence for the existence of gravitational waves. During the time of observation the orbital period of the system has steadily decreased. This shows that the system is losing orbital energy at a rate that agrees exactly with that predicted by the general theory of relativity. The lost energy is radiated as gravitational waves.

X-ray Pulsars. Many kinds of variable X-ray sources were discovered in the 1970's. Among these, the X-ray pulsars and the X-ray bursters are related to neutron stars. In contrast to radio pulsars, the period of the pulsed emission of X-ray pulsars decreases with time. Since the origin of the X-ray pulses is thought to be similar to that of radio pulses, this means that the rotation rate of the neutron stars must increase. The pulse periods

of X-ray pulsars are significantly longer than those of radio pulsars, from a few seconds to tens of minutes.

X-ray pulsars always belong to binary systems. Typically, the companion is an OB supergiant. The characteristic properties of X-ray pulsars can be understood from their binary nature. A neutron star formed in a binary system is first seen as a normal radio pulsar. Initially, the strong radiation of the pulsar prevents gas from falling onto it. However, as it slows down, its energy decreases, and eventually the stellar wind from the companion can reach its surface. The incoming gas is channelled to the magnetic polar caps of the neutron star, where it emits strong X-ray radiation as it hits the surface. This produces the observed pulsed emission. At the same time, the angular momentum of the incoming gas speeds up the rotation of the pulsar.

The emission curve of a typical fast X-ray pulsar, Harcules X1, is shown in Fig. 14.7. The period of the pulses is 1.24 s. This neutron star is part of an eclipsing binary system, known from optical observations as HZ Herculis. The orbital properties of the system can therefore be determined. Thus e. g. the mass of the pulsar is about one solar mass, reasonable for a neutron star.

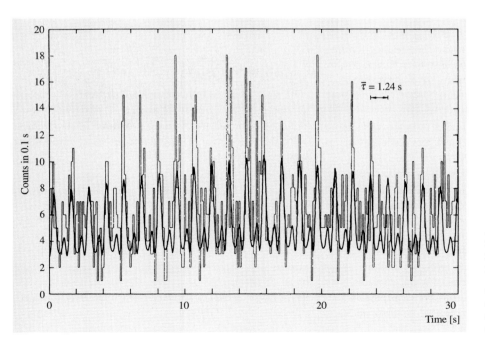

Fig. 14.7. The pulses of the X-ray pulsar Harcules X1 have the period 1.24 s. The best-fitting curve has been superimposed on the observations. (Tananbaum, H. et al. (1972): Astrophys. J. (Lett.) **174**, L143)

Fig. 14.8. The variations of the rapid X-ray burster MXB 1730–335. An 100 second interval is marked in the diagram. (Lewin, W.H.G. (1977): Ann. N.Y. Acad. Sci. **302**, 310)

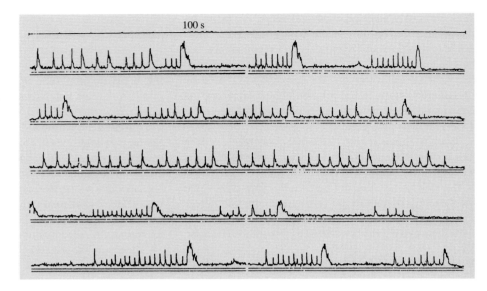

100 s

X-ray Bursters. X-ray bursters are irregular variables, showing random outbursts like dwarf novae (Fig. 14.8). The typical interval between outbursts is a few hours or days, but more rapid bursters are also known. The strength of the outburst seems to be related to the recharging time.

The source of radiation in X-ray bursters cannot be the ignition of hydrogen, since the maximum emission is in the X-ray region. One explanation is that the bursters are binary systems like the dwarf novae, but containing a neutron star rather than a white dwarf. Gas from the companion settles on the surface of the neutron star, where hydrogen burns to helium. When the growing shell of helium reaches a critical temperature, it burns to carbon in a rapid helium flash. Since, in this case, there are no thick damping outer layers, the flash appears as a burst of X-ray radiation.

It used to be believed that also the *gamma ray bursts*, very short and sharp gamma ray pulses, come from neutron stars. Satellite observations with the Compton Gamma Ray Observatory have now shown that the gamma ray bursts are almost uniformly distributed in the sky, unlike the known neutron stars. Some bursts have now been detected optically, and their redshifts show them to be at cosmological distances. At least some of these bursts were probably produced when a supermassive star exploded to form a black hole.

It was seen in Sect. 11.5 that there is an upper limit to the mass of a neutron star, the Oppenheimer–Volkoff mass, above which not even the degenerate neutron gas pressure can resist the gravitational force. The precise value of this upper mass limit is difficult to calculate theoretically, since it depends on imperfectly known particle interaction probabilities, but it is estimated to be in the range 1.5–2 $M_\odot$.

* The Radius of White Dwarfs and Neutron Stars

The mass of a white dwarf or a neutron star determines its radius. This follows from the equation of hydrostatic equilibrium and from the pressure-density relation for a degenerate gas. Using the hydrostatic equilibrium equation (10.1)

$$\frac{dP}{dr} = -\frac{GM_r\rho}{r^2}$$

one can estimate the average pressure P:

$$\left|\frac{dP}{dr}\right| \approx \frac{P}{R} \propto \frac{M \times M/R^3}{R^2} = \frac{M^2}{R^5} \, .$$

Here we have used $\rho \propto M/R^3$. Thus the pressure obeys

$$P \propto M^2/R^4 \, . \tag{14.1}$$

In the nonrelativistic case, the pressure of a degenerate electron gas is given by (10.16):

$$P \approx (h^2/m_\mathrm{e})(\mu_\mathrm{e} m_\mathrm{H})^{-5/3} \rho^{5/3}$$

and hence

$$P \propto \frac{\rho^{5/3}}{m_\mathrm{e} \mu_\mathrm{e}^{5/3}} \,. \tag{14.2}$$

By combining (14.1) and (14.2) we obtain

$$\frac{M^2}{R^4} \propto \frac{M^{5/3}}{R^5 m_\mathrm{e} \mu_\mathrm{e}^{5/3}}$$

or

$$R \propto \frac{1}{M^{1/3} m_\mathrm{e} \mu_\mathrm{e}^{5/3}} \propto M^{-1/3} \,.$$

Thus the smaller the radius of a white dwarf is, the larger its mass will be. If the density becomes so large that the relativistic equation of state (10.17) has to be used, the expression for the pressure is

$$P \propto \rho^{4/3} \propto \frac{M^{4/3}}{R^4} \,.$$

As the star contracts, the pressure grows at the same rate as demanded by the condition for hydrostatic support (14.1). Once contraction has begun, it can only stop when the state of matter changes: the electrons and protons combine into neutrons. Only a star that is massive enough can give rise to a relativistic degenerate pressure.

The neutrons are fermions, just like the electrons. They obey the Pauli exclusion principle, and the degenerate neutron gas pressure is obtained from an expression analogous to (14.2):

$$P_\mathrm{n} \propto \frac{\rho^{5/3}}{m_\mathrm{n} \mu_\mathrm{n}^{5/3}} \,,$$

where m_n is the neutron mass and μ_n, the molecular weight per free neutron. Correspondingly, the radius of a neutron star is given by

$$R_\mathrm{ns} \propto \frac{1}{M^{1/3} m_\mathrm{n} \mu_\mathrm{n}^{5/3}} \,.$$

If a white dwarf consists purely of helium, $\mu_\mathrm{e} = 2$; for a neutron star, $\mu_\mathrm{n} \approx 1$. If a white dwarf and a neutron star have the same mass, the ratio of their radii is

$$\frac{R_\mathrm{wd}}{R_\mathrm{ns}} = \left(\frac{M_\mathrm{ns}}{M_\mathrm{wd}} \right)^{1/3} \left(\frac{\mu_\mathrm{n}}{\mu_\mathrm{e}} \right)^{5/3} \frac{m_\mathrm{n}}{m_\mathrm{e}}$$

$$\approx 1 \times \left(\frac{1}{2} \right)^{5/3} \times 1840 \approx 600 \,.$$

Thus the radius of a neutron star is about 1/600 of that of a white dwarf. Typically R_ns is about 10 km.

14.3 Black Holes

If the mass of a star exceeds M_OV, and if it does not lose mass during its evolution it can no longer reach any stable final state. The force of gravity will dominate over all other forces, and the star will collapse to a black hole. A black hole is black because not even light can escape from it. Already at the end of the 18th century *Laplace* showed that a sufficiently massive body would prevent the escape of light from its surface. According to classical mechanics, the escape velocity from a body of radius R and mass M is

$$v_\mathrm{e} = \sqrt{\frac{2GM}{R}} \,.$$

This is greater than the speed of light, if the radius is smaller than the critical radius

$$R_\mathrm{S} = 2GM/c^2 \,. \tag{14.3}$$

The same value for the critical radius, the *Schwarzschild radius*, is obtained from the general theory of relativity. For example, for the Sun, R_S is about 3 km; however, the Sun's mass is so small that it cannot become a black hole by normal stellar evolution. Because the mass of a black hole formed by stellar collapse has to be larger than M_OV the radius of the smallest black holes formed in this way is about 5–10 km.

The properties of black holes have to be studied on the basis of the general theory of relativity, which is beyond the scope of this book. Thus only some basic properties are discussed qualitatively.

An *event horizon* is a surface through which no information can be sent out, even in principle. A black hole is surrounded by an event horizon at the Schwarzschild radius (Fig. 14.9). In the theory of relativity each observer carries with him his own local measure of time. If two observers are at rest with respect to each other at the same point their clocks go at the same rate. Otherwise their clock rates are different, and the apparent course of events differs, too.

Near the event horizon the different time definitions become significant. An observer falling into a black hole reaches the centre in a finite time, according to his

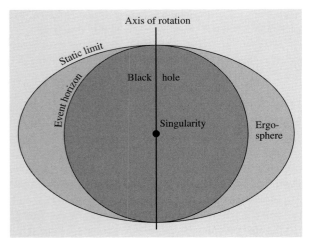

Fig. 14.9. A black hole is surrounded by a spherical event horizon. In addition to this a rotating black hole is surrounded by a flattened surface inside which no matter can remain stationary. This region is called the ergosphere

own clock, and does not notice anything special as he passes through the event horizon. However, to a distant observer he never seems to reach the event horizon; his velocity of fall seems to decrease towards zero as he approaches the horizon.

The slowing down of time also appears as a decrease in the frequency of light signals. The formula for the gravitational redshift can be written in terms of the Schwarzschild radius as (Appendix B)

$$\nu_\infty = \nu\sqrt{1 - \frac{2GM}{rc^2}} = \nu\sqrt{1 - \frac{R_S}{r}} \, . \qquad (14.4)$$

Here, ν is the frequency of radiation emitted at a distance r from the black hole and ν_∞ the frequency observed by an infinitely distant observer. It can be seen that the frequency at infinity approaches zero for radiation emitted near the event horizon.

Since the gravitational force is directed towards the centre of the hole and depends on the distance, different parts of a falling body feel a gravitational pull that is different in magnitude and direction. The tidal forces become extremely large near a black hole so that any material falling into the hole will be torn apart. All atoms and elementary particles are destroyed near the central point, and the final state of matter is unknown to present-day physics. The observable properties of a black hole do not depend on how it was made.

Not only all information on the material composition disappears as a star collapses into a black hole; any magnetic field, for example, also disappears behind the event horizon. A black hole can only have three observable properties: mass, angular momentum and electric charge.

It is improbable that a black hole could have a significant net charge. Rotation, on the other hand, is typical to stars, and thus black holes, too, must rotate. Since the angular momentum is conserved, stars collapsed to black holes must rotate very fast.

In 1963 *Roy Kerr* managed to find a solution of the field equations for a rotating black hole. In addition to the event horizon a rotating hole has another limiting surface, an ellipsoidal *static limit* (Fig. 14.9). Objects inside the static limit cannot be kept stationary by any force, but they must orbit the hole. However, it is possible to escape from the region between the static limit and the event horizon, called the *ergosphere*. In fact it is possible to utilize the rotational energy of a black hole by dropping an object to the ergosphere in such a way that part of the object falls into the hole and another part is slung out. The outcoming part may then have considerably more kinetic energy than the original object.

At present the only known way in which a black hole could be directly observed is by means of the radiation from gas falling into it. For example, if a black hole is part of a binary system, gas streaming from the companion will settle into a disc around the hole. Matter at the inner edge of the disc will fall into the hole. The accreting gas will lose a considerable part of its energy (up to 40% of the rest mass) as radiation, which should be observable in the X-ray region.

Some rapidly and irregularly varying X-ray sources of the right kind have been discovered in the sky. The most promising black hole candidate is probably Cygnus X-1 (Fig. 14.11). Its luminosity varies on the time scale of 0.001 s, which means that the emitting region must be only 0.001 light-seconds or a few hundred kilometres in size. Only neutron stars and black holes are small and dense enough to give rise to such high-energy processes. Cygnus X-1 is the smaller component of the double system HDE 226868. The larger component is an optically visible supergiant with a mass 20–25 $M_\odot$. The mass of the unseen component has been calculated to be 10–15 $M_\odot$. If this is correct, the mass

Fig. 14.10. The arrow shows the variable star V1357 Cyg. Its companion is the suspected black hole Cygnus X-1. The bright star to the lower right of V1357 is η Cygni, one of the brightest stars in the constellation Cygnus

of the secondary component is much larger than the upper limit for a neutron star, and thus it has to be a black hole.

Many frightening stories about black holes have been invented. It should therefore be stressed that they obey the same dynamical laws as other stars – they are not lurking in the darkness of space to attack innocent passers-by. If the Sun became a black hole, the planets would continue in their orbits as if nothing had happened.

Fig. 14.11. No periodicity is evident in the X-ray radiation of the black hole candidate Cygnus X-1, and thus it is probably not a neutron star. (Schreier, E. et al. (1971): Astrophys. J. (Lett.) **170**, L24)

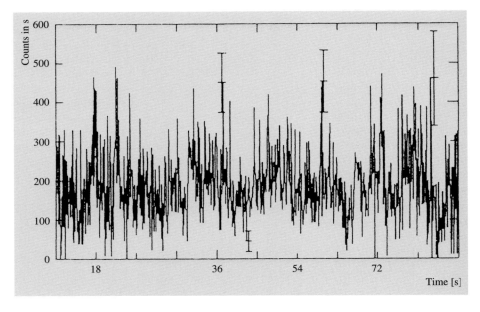

So far we have discussed only black holes with masses in the range of stellar masses. There is however no upper limit to the mass of a black hole. Many active phenomena in the nuclei of galaxies can be explained with supermassive black holes with masses of millions or thousands of millions solar masses (see Sect. 18.4 and 19.9).

14.4 Examples

Example 14.1 Assume that the Sun collapses into a neutron star with a radius of 20 km. a) What will be the mean density of the neutron star? b) What would be it rotation period?

a) The mean density is

$$\rho = \frac{M_\odot}{\frac{4}{3}\pi R^3} = \frac{2 \times 10^{30} \text{ kg}}{\frac{4}{3}\pi (20 \times 10^3)^3 \text{ m}^3}$$

$$\approx 6 \times 10^{16} \text{ kg/m}^3 \ .$$

One cubic millimetre of this substance would weigh 60 million kilos.

b) To obtain an exact value, we should take into account the mass distributions of the Sun and the resulting neutron star. Very rough estimates can be found assuming that both are homogeneous. Then the moment of inertia is $I = \frac{2}{5}MR^2$, and the angular momentum is $L = I\omega$. The rotation period is then obtained as in Example 11.1:

$$P = P_\odot \left(\frac{R}{R_\odot}\right)^2$$

$$= 25 \text{ d} \left(\frac{20 \times 10^3 \text{ m}}{6.96 \times 10^8 \text{ m}}\right)^2 = 2.064 \times 10^{-8} \text{ d}$$

$$\approx 0.0018 \text{ s} \ .$$

The Sun would make over 550 revolutions per second.

Example 14.2 What should be the radius of the Sun if the escape velocity from the surface were to exceed the speed of light?

The escape velocity exceeds the speed of light if

$$\sqrt{\frac{2GM}{R}} > c$$

or

$$R < \frac{2GM}{c^2} = R_\mathrm{S} \ .$$

For the Sun we have

$$R_\mathrm{S} = \frac{2 \times 6.67 \times 10^{-11} \text{ m}^3 \text{ s}^{-2} \text{ kg}^{-1} \times 1.989 \times 10^{30} \text{ kg}}{(2.998 \times 10^8 \text{ m s}^{-1})^2}$$

$$= 2950 \text{ m} \ .$$

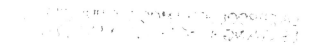

14.5 Exercises

Exercise 14.1 The mass of a pulsar is 1.5 $M_\odot$, radius 10 km, and rotation period 0.033 s. What is the angular momentum of the pulsar? Variations of 0.0003 s are observed in the period. If they are due to radial oscillations ("starquakes"), how large are these oscillations?

Exercise 14.2 In *Dragon's Egg* by Robert L. Forward a spaceship orbits a neutron star at a distance of 406 km from the centre of the star. The orbital period is the same as the rotation period of the star, 0.1993 s.

a) Find the mass of the star and the gravitational acceleration felt by the spaceship.

b) What is the effect of the gravitation on a 175 cm tall astronaut, if (s)he stands with her/his feet pointing towards the star? And if (s)he is lying tangential to the orbit?

Exercise 14.3 A photon leaves the surface of star at a frequency ν_e. An infinitely distant observer finds that its frequency is ν. If the difference is due to gravitation only, the change in the energy of the photon, $h\Delta\nu$, equals the change in its potential energy. Find the relation between ν and ν_e, assuming the mass and radius of the star are M and R. How much will the solar radiation redshift on its way to the Earth?

15. The Interstellar Medium

Although most of the mass of the Milky Way Galaxy is condensed into stars, interstellar space is not completely empty. It contains *gas* and *dust* in the form both of individual clouds and of a diffuse medium. Interstellar space typically contains about one gas atom per cubic centimetre and 100 dust particles per cubic kilometre.

Altogether, about 10 % of the mass of the Milky Way consists of interstellar gas. Since the gas is strongly concentrated in the galactic plane and the spiral arms, in these regions there are many places where the quantities of stars and interstellar matter are about equal. The dust (a better name would be "smoke", since the particle sizes are much smaller than in terrestrial dust) constitutes about one percent of the gas. High-energy cosmic ray particles are mixed with the gas and dust. There is also a weak, but still very important, galactic magnetic field.

At present the most important observations of the interstellar medium are made at radio and infrared wavelengths, since the peak of the emission often lies at these wavelengths. But many forms of interstellar matter (such as solid bodies with diameters larger than 1 mm) would be almost impossible to detect on the basis of their emission or absorption. In principle, the mass of these forms of matter might be larger than the observed mass of all other forms put together. However, an upper limit on the total mass of interstellar matter, regardless of its form, can be derived on the basis of its gravitational effects. This is the *Oort limit*. The galactic gravitational field is determined by the distribution of matter. By observing the motions of stars perpendicular to the galactic plane, the vertical gravitational force and hence the amount of mass in the galactic plane can be determined. The result is that the local density within 1 kpc of the Sun is $(7.3-10.0) \times 10^{-21}$ kg m^{-3}. The density of known stars is $(5.9-6.7) \times 10^{-21}$ kg m^{-3} and that of known interstellar matter about 1.7×10^{-21} kg m^{-3}. Thus there is very little room for unknown forms of mass in the solar neighbourhood. However, the limit concerns only the dark matter concentrated in the galactic plane. There are indications that the Milky Way is surrounded by a spherical halo of dark matter (Chap. 17).

15.1 Interstellar Dust

The first clear evidence for the existence of interstellar dust was obtained around 1930. Before that, it had been generally thought that space is completely transparent and that light can propagate indefinitely without extinction.

In 1930 *Robert Trumpler* published his study of the space distribution of the open clusters. The absolute magnitudes M of the brightest stars could be estimated on the basis of the spectral type. Thus the distance r to the clusters could be calculated from the observed apparent magnitudes m of the bright stars:

$$m - M = 5 \lg \frac{r}{10 \, \text{pc}} . \qquad (15.1)$$

Trumpler also studied the diameters of the clusters. The linear diameter D is obtained from the apparent angular diameter d by means of the formula

$$D = dr , \qquad (15.2)$$

where r is the distance of the cluster.

It caught Trumpler's attention that the more distant clusters appeared to be systematically larger than the nearer ones (Fig. 15.1). Since this could hardly be true, the distances of the more distant clusters must have been overestimated. Trumpler concluded that space is not completely transparent, but that the light of a star is dimmed by some intervening material. To take this into account, (15.1) has to be replaced with (4.17)

$$m - M = 5 \lg \frac{r}{10 \, \text{pc}} + A , \qquad (15.3)$$

where $A \geq 0$ is the extinction in magnitudes due to the intervening medium. If the opacity of the medium is assumed to be the same at all distances and in all directions, A can be written

$$A = ar , \qquad (15.4)$$

where a is a constant. Trumpler obtained for the average value of a in the galactic plane, $a_{\text{pg}} = 0.79$ mag/kpc, in photographic magnitudes. At present, a value of 2 mag/kpc is used for the average extinction. Thus the extinction over a 5 kpc path is already 10 magnitudes.

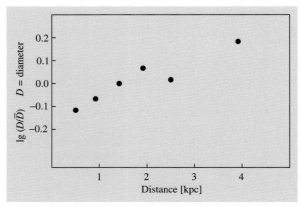

Fig. 15.1. The diameters of open star clusters calculated with the distance given by the formula (15.1) according to Trumpler (1930). The increase of the diameter with distance is not a real phenomenon, but an effect of interstellar extinction, which was discovered in this way

Extinction due to dust varies strongly with direction. For example, visible light from the galactic centre (distance 8–9 kpc) is dimmed by 30 magnitudes. Therefore the galactic centre cannot be observed at optical wavelengths.

Extinction is due to dust grains that have diameters near the wavelength of the light. Such particles scatter light extremely efficiently. Gas can also cause extinction by scattering, but its scattering efficiency per unit mass is much smaller. The total amount of gas allowed by the Oort limit is so small that scattering by gas is negligible in interstellar space. (This is in contrast with the Earth's atmosphere, where air molecules make a significant contribution to the total extinction).

Interstellar particles can cause extinction in two ways:

1. In *absorption* the radiant energy is transformed into heat, which is then re-radiated at infrared wavelengths corresponding to the temperature of the dust particles.

2. In *scattering* the direction of light propagation is changed, leading to a reduced intensity in the original direction of propagation.

An expression for interstellar extinction will now be derived. The size, index of refraction and number density of the particles are assumed to be known. For simplicity we shall assume that all particles are spheres with the same radius a and the geometrical cross section πa^2. The true extinction cross section of the particles C_{ext} will be

$$C_{ext} = Q_{ext}\pi a^2 , \qquad (15.5)$$

where Q_{ext} is the extinction efficiency factor.

Let us consider a volume element with length dl and cross section dA, normal to the direction of propagation (Fig. 15.2). It is assumed that the particles inside the element do not shadow each other. If the particle density is n, there are $n\,dl\,dA$ particles in the volume element

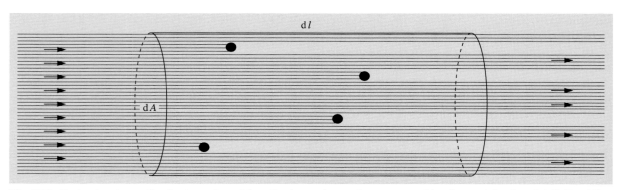

Fig. 15.2. Extinction by a distribution of particles. In the volume element with length dl and cross section dA, there are $n\,dA\,dl$ particles, where n is the particle density in the medium. If the extinction cross section of one particle is C_{ext},

the total area covered by the particles is $n\,dA\,dl\,C_{ext}$. Thus the fractional decrease in intensity over the distance dl is $dI/I = -n\,dA\,dl\,C_{ext}/dA = -n\,C_{ext}\,dl$

and they will cover the fraction $d\tau$ of the area dA, where

$$d\tau = \frac{n\, dA\, dl\, C_{ext}}{dA} = n\, C_{ext}\, dl\ .$$

In the length dl the intensity is thus changed by

$$dI = -I\, d\tau\ . \tag{15.6}$$

On the basis of (15.6) $d\tau$ can be identified as the optical depth.

The total optical depth between the star and the Earth is

$$\tau(r) = \int_0^r d\tau = \int_0^r n\, C_{ext}\, dl = C_{ext}\bar{n}r\ ,$$

where $\bar{n}$ is the average particle density along the given path. According to (4.18) the extinction in magnitudes is

$$A = (2.5\lg e)\tau\ ,$$

and hence

$$A(r) = (2.5\lg e)C_{ext}\bar{n}r\ . \tag{15.7}$$

This formula can also be inverted to calculate $\bar{n}$, if the other quantities are known.

The extinction efficiency factor Q_{ext} can be calculated exactly for spherical particles with given radius a and refractive index m. In general,

$$Q_{ext} = Q_{abs} + Q_{sca}\ ,$$

where

$$Q_{abs} = \text{absorption efficiency factor}\ ,$$
$$Q_{sca} = \text{scattering efficiency factor}\ .$$

If we define

$$x = 2\pi a/\lambda\ , \tag{15.8}$$

where λ is the wavelength of the radiation, then

$$Q_{ext} = Q_{ext}(x, m)\ . \tag{15.9}$$

The exact expression for Q_{ext} is a series expansion in x that converges more slowly for larger values of x. When $x \ll 1$, the process is called *Rayleigh scattering*; otherwise it is known as *Mie scattering*. Figure 15.3 shows Q_{ext} as a function of x for $m = 1.5$ and $m = 1.33$. For very large particles, $(x \gg 1)$ $Q_{ext} = 2$, as appears

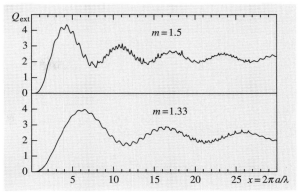

Fig. 15.3. Mie scattering: the extinction efficiency factor for spherical particles for the refractive indices $m = 1.5$ and $m = 1.33$ (refractive index of water). The horizontal axis is related to the size of the particle according to $x = 2\pi a/\lambda$, where a is the particle radius and λ, the wavelength of the radiation

from Fig. 15.3. Purely geometrically one would have expected $Q_{ext} = 1$; the two times larger scattering efficiency is due to the diffraction of light at the edges of the particle.

Other observable phenomena, apart from extinction, are also caused by interstellar dust. One of these is the *reddening* of the light of stars. (This should not be confused with the redshift of spectral lines.) Reddening is due to the fact that the amount of extinction becomes larger for shorter wavelengths. Going from red to ultraviolet, the extinction is roughly inversely proportional to wavelength. For this reason the light of distant stars is redder than would be expected on the basis of their spectral class. The spectral class is defined on the basis of the relative strengths of the spectral lines which are not affected by extinction.

According to (4.20), the observed colour index $B - V$ of a star is

$$\begin{aligned} B - V &= M_B - M_V + A_B - A_B \\ &= (B - V)_0 + E_{B-V}\ , \end{aligned} \tag{15.10}$$

where $(B - V)_0$ is the *intrinsic colour* of the star and E_{B-V} the *colour excess*. As noted in Sect. 4.5 the ratio between the visual extinction A_V and the colour excess is approximately constant:

$$R = \frac{A_V}{E_{B-V}} = \frac{A_V}{A_B - A_V} \approx 3.0\ . \tag{15.11}$$

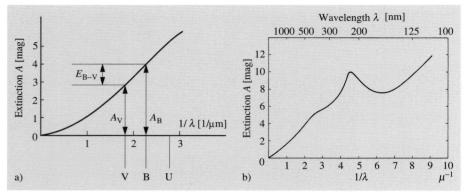

Fig. 15.4. (a) Schematic representation of the interstellar extinction. As the wavelength increases, the extinction approaches zero. (Drawing based on Greenberg, J. M. (1968): "Interstellar Grains", in *Nebulae and Interstellar Matter*, ed. by Middlehurst, B.M., Aller, L.H., Stars and Stellar Systems, Vol. VII (The University of Chicago Press, Chicago) p. 224). (b) Measured extinction curve, normalized to make $E_{B-V} = 1$. (Hoyle, F., Narlikar, J. (1980): *The Physics-Astronomy Frontier* (W. H. Freeman and Company, San Francisco) p. 156. Used by permission)

R does not depend on the properties of the star or the amount of extinction. This is particularly important in photometric distance determinations because of the fact that the colour excess E_{B-V} can be directly determined from the difference between the observed colour index $B - V$ and the intrinsic colour $(B - V)_0$ known from the spectral class. One can then calculate the extinction

$$A_V \approx 3.0 \, E_{B-V} \qquad (15.12)$$

and finally the distance. Since the interstellar medium is far from homogeneous, the colour excess method gives a much more reliable value than using some average value for the extinction in (4.18).

The wavelength dependence of the extinction, $A(\lambda)$, can be studied by comparing the magnitudes of stars of the same spectral class in different colours. These measurements have shown that $A(\lambda)$ approaches zero as λ becomes very large. In practice $A(\lambda)$ can be measured up to a wavelength of about two micrometres. The extrapolation to zero inverse wavelength is then fairly reliable. Figure 15.4a shows $A(\lambda)$ as a function of inverse wavelength. It also illustrates how the quantities A_V and E_{B-V}, which are needed in order to calculate the value of R, are obtained from this *extinction* or *reddening curve*. Figure 15.4b shows the observed extinction curve. The points in the ultraviolet ($\lambda \leq 0.3$ m) are based on rocket measurements.

It is clear from Fig. 15.4b that interstellar extinction is largest at short wavelengths in the ultraviolet and decreases for longer wavelengths. In the infrared it is only about ten percent of the optical extinction and in the radio region it is vanishingly small. Objects that are invisible in the optical region can therefore be studied at infrared and radio wavelengths.

Another observed phenomenon caused by dust is the *polarization* of the light of the stars. Since spherical particles cannot produce any polarization, the interstellar dust particles have to be nonspherical in shape.

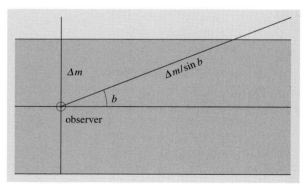

Fig. 15.5. In a homogeneous medium the extinction in magnitudes is proportional to the pathlength traversed. If the extinction in the direction of the galactic pole is Δm, then the extinction at the galactic latitude b will be $\Delta m / \sin b$

If the particles in a cloud are aligned by the interstellar magnetic field, they will polarize the radiation passing through the cloud. The degree of polarization and its wavelength dependence give information on the properties of the dust particles. By studying the direction of polarization in various directions, one can map the structure of the galactic magnetic field.

In the Milky Way interstellar dust is essentially confined to a very thin, about 100 pc, layer in the galactic plane. The dust in other spiral galaxies has a similar distribution and is directly visible as a dark band in the disc of the galaxy (Fig. 18.12b). The Sun is located near the central plane of the galactic dust layer, and thus the extinction in the direction of the galactic plane is very large, whereas the total extinction towards the galactic

Fig. 15.6. The Coalsack is a dark nebula next to the Southern Cross. (Photograph K. Mattila, Helsinki University)

poles may be less than 0.1 magnitudes. This is apparent in the distribution of galaxies in the sky: at high galactic latitudes, there are many galaxies, while near the galactic plane, there is a 20° zone where hardly any galaxies are seen. This empty region is called the *zone of avoidance*.

If a homogeneous dust layer gives rise to a total extinction of Δm magnitudes in the vertical direction, then according to Fig. 15.5, the total extinction at galactic latitude b will be

$$\Delta m(b) = \Delta m / \sin b . \tag{15.13}$$

If the galaxies are uniformly distributed in space, then in the absence of extinction, the number of galaxies per square degree brighter than the magnitude m would be

$$\lg N_0(m) = 0.6 m + C , \tag{15.14}$$

where C is a constant (see Example 17.1). However, due to extinction, a galaxy that would otherwise have the apparent magnitude m_0 will have the magnitude

$$m(b) = m_0 + \Delta m(b) = m_0 + \Delta m / \sin b , \tag{15.15}$$

where b is the galactic latitude. Thus the observable number of galaxies at latitude b will be

$$\lg N(m, b) = \lg N_0(m - \Delta m(b))$$
$$= 0.6(m - \Delta m(b)) + C$$
$$= \lg N_0(m) - 0.6 \Delta m(b)$$

or

$$\lg N(m, b) = C' - 0.6 \frac{\Delta m}{\sin b} , \tag{15.16}$$

where $C' = \lg N_0(m)$ does not depend on the galactic latitude. By making galaxy counts at various latitudes b, the extinction Δm can be determined. The value obtained from galaxy counts made at Lick Observatory is $\Delta m_{pg} = 0.51$ mag.

The total vertical extinction of the Milky Way has also been determined from the colour excesses of stars. These investigations have yielded much smaller extinction values, about 0.1 mag. In the direction of the north pole, extinction is only 0.03 mag. The disagreement between the two extinction values is probably largely due to the fact that the dust layer is not really homogeneous. If the Sun is located in a local region of low dust content, the view towards the galactic poles might be almost unobstructed by dust.

Dark Nebulae. Observations of other galaxies show that the dust is concentrated in the spiral arms, in particular at their inner edge. In addition dust is concentrated in individual clouds, which appear as star-poor regions or *dark nebulae* against the background of the Milky Way. Examples of dark nebulae are the Coalsack in the southern sky (Fig. 15.6) and the Horsehead nebula in Orion. Sometimes the dark nebulae form extended winding bands, and sometimes small, almost spherical, objects. Objects of the latter type are most easy to see against a bright background, e. g. a gas nebula (see Fig. 15.19). These objects have been named *globules* by *Bart J. Bok*, who put forward the hypothesis that they are clouds that are just beginning to contract into stars.

The extinction by a dark nebula can be illustrated and studied by means of a *Wolf diagram*, shown schematically in Fig. 15.7. The diagram is constructed on the basis of star counts. The number of stars per square degree in some magnitude interval (e. g. between magnitudes 14 and 15) in the cloud is counted and compared with the number outside the nebula. In the comparison area, the number of stars increases monotonically towards fainter magnitudes. In the dark nebula the num-

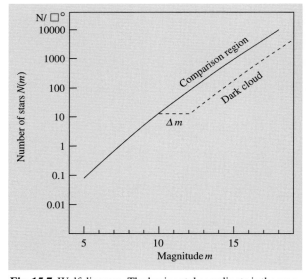

Fig. 15.7. Wolf diagram. The horizontal coordinate is the magnitude and the vertical coordinate is the number of stars per square degree in the sky brighter than that magnitude. A dark nebula diminishes the brightness of stars lying behind it by the amount Δm

bers first increase in the same way, but beyond some limiting magnitude (10 in the figure) the number of stars falls below that outside the cloud. The reason for this is that the fainter stars are predominantly behind the nebula, and their brightness is reduced by some constant amount Δm (2 magnitudes in the figure). The brighter stars are mostly in front of the nebula and suffer no extinction.

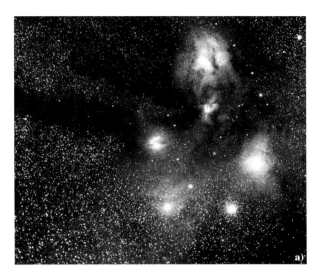

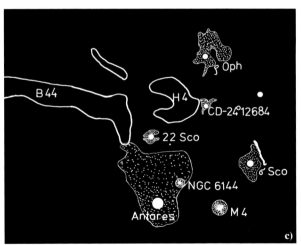

Fig. 15.8a–c. Bright and dark nebulae in Scorpius and Ophiuchus. Photograph (**a**) was taken in the blue colour region, $\lambda = 350$–500 nm, and (**b**) in the red colour region, $\lambda = 600$–680 nm. (The sharp rings in (**b**) are reflections of Antares in the correction lens of the Schmidt camera.) The nebulae located in the area are identified in drawing (**c**). B44 and H4 are dark nebulae. There is a large reflection nebula around Antares, which is faintly visible in the blue (**a**), but bright in the red (**b**) regions. Antares is very red (spectral class M1) and therefore the reflec-

tion nebula is also red. In contrast, the reflection nebulae around the blue stars ρ Ophiuchi (B2), CD-24° 12684 (B3), 22 Scorpii (B2) and σ Scorpii (B1) are blue and are visible only in (**a**). In (**b**) there is an elongated nebula to the right of σ Scorpii, which is invisible in (**a**). This is an emission nebula, which is very bright in the red hydrogen H_α line (656 nm). In this way reflection and emission nebulae can be distinguished by means of pictures taken in different wavelength regions. (Photograph (**a**) E. Barnard, and (**b**) K. Mattila)

Fig. 15.9. The reflection nebula NGC 2068 (M78) in Orion. In the middle of the nebula there are two stars of about magnitude 11. The northern one (at the top) is the illuminating star, while the other one probably lies in the foreground. (Photography Lunar and Planetary Laboratory, Catalina Observatory)

Reflection Nebulae. If a dust cloud is near a bright star, it will scatter, i. e. reflect the light of the star. Thus individual dust clouds can sometimes be observed as bright *reflection nebulae*. Some 500 reflection nebulae are known.

The regions in the sky richest in reflection nebulae are the areas around the Pleiades and around the giant star Antares. Antares itself is surrounded by a large red reflection nebula. This region is shown in Fig. 15.8. Figure 15.9 shows the reflection nebula NGC 2068, which is located near a large, thick dust cloud a few degrees northwest of Orion's belt. It is one of the brightest reflection nebulae and the only one included in the Messier catalogue (M78). In the middle of the nebula there are two stars of about 11 magnitudes. The northern star illuminates the nebula, while the other one is probably in front of the nebula. Figure 15.10 shows the reflection nebula NGC 1435 around Merope in the Pleiades. Another bright and much-studied reflection nebula is NGC 7023 in Cepheus. It, too, is connected with a dark nebula. The illuminating star has emission lines in its spectrum (spectral type Be). Infrared stars have also been discovered in the area of the nebula, probably a region of star formation.

In 1922 *Edwin Hubble* published a fundamental investigation of bright nebulae in the Milky Way. On the basis of extensive photographic and spectroscopic observations, he was able to establish two interesting relationships. First he found that emission nebulae only occur near stars with spectral class earlier than B0, whereas reflection nebulae may be found near stars of spectral class B1 and later. Secondly Hubble discovered a relationship between the angular size R of the nebula and the apparent magnitude m of the illuminating star:

$$5 \lg R = -m + \text{const} . \tag{15.17}$$

Thus the angular diameter of a reflection nebula is larger for a brighter illuminating star. Since the measured size of a nebula generally increases for longer exposures, i. e. fainter limiting surface brightness, the value of R should be defined to correspond to a fixed limiting surface brightness. The value of the constant in the Hubble relation depends on this limiting surface

Fig. 15.10. The reflection nebula NGC 1435 around Merope (23 Tau, spectral class B6) in the Pleiades. This figure should be compared with Fig. 16.1, where Merope is visible as the lowest of the bright stars in the Pleiades. (National Optical Astronomy Observatories, Kitt Peak National Observatory)

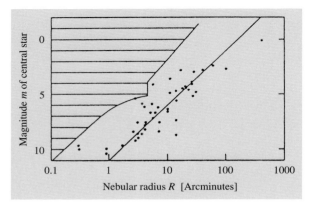

brightness. The Hubble relation for reflection nebulae is shown in Fig. 15.11, based on measurements by *Sidney van den Bergh* from Palomar Sky Atlas plates. Each point corresponds to a reflection nebula and the straight line represents the relation (15.17), where the value of the constant is 12.0 (R is given in arc minutes).

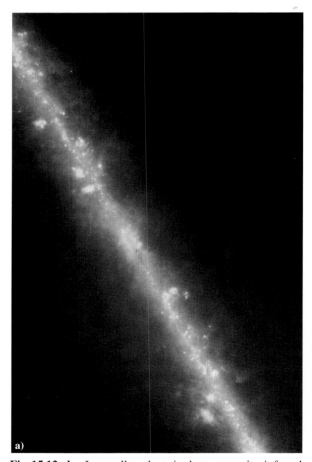

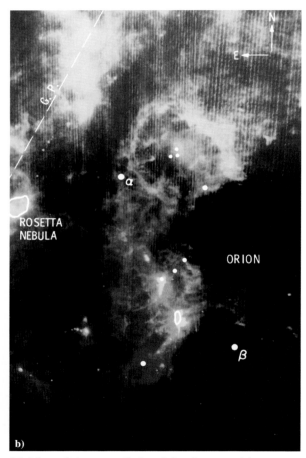

Fig. 15.12a,b. Interstellar dust is best seen in infrared wavelengths. Two examples of the images by the IRAS satellite. (**a**) In a view towards the Galactic centre, the dust is seen to be concentrated in a narrow layer in the galactic plane. Several separate clouds are also seen.

(**b**) Most of the constellation Orion is covered by a complex area of interstellar matter. The densest concentrations of dust below the centre of the image, are in the region of the Horsehead nebula and the Orion nebula. (Photos NASA)

The Hubble relation can be derived theoretically, if it is assumed that the illumination of a dust cloud is inversely proportional to the square of the distance to the illuminating star, and that the dust clouds are uniformly distributed in space. The theoretical Hubble relation also gives an expression for the constant on the right-hand side, which involves the albedo and the phase function of the grains.

The observations of reflection nebulae show that the albedo of interstellar grains must be quite high. It has not yet been possible to obtain its precise numerical value in this way, since the distances between the nebulae and their illuminating stars are not known well enough.

One may also consider the surface brightness of dark nebulae that are not close enough to a star to be visible as reflection nebulae. These nebulae will still reflect the diffuse galactic light from all the stars in the Milky Way. Calculations show that if the dust grains have a large albedo, then the reflected diffuse light should be bright enough to be observable, and it has indeed been observed. Thus the dark nebulae are not totally dark. The diffuse galactic light constitutes about 20–30% of the total brightness of the Milky Way.

Dust Temperature. In addition to scattering the interstellar grains also absorb radiation. The absorbed energy is re-radiated by the grains at infrared wavelengths corresponding to their temperatures. The temperature of dust in interstellar space (including dark nebulae) is about 10–20 K. The corresponding wavelength according to Wien's displacement law (5.21) is 300–150 μm. Near a hot star the temperature of the dust may be 100–600 K and the maximum emission is then at 30–5 μm. In H II regions the dust temperature is about 70–100 K.

The rapid development of infrared astronomy in the 1970's has brought all the above-mentioned dust sources within the reach of observations (Fig. 15.12). In addition infrared radiation from the nuclei of normal and active galaxies is largely thermal radiation from dust. Thermal dust emission is one of the most important sources of infrared radiation in astronomy.

One of the strongest infrared sources in the sky is the nebula around the star η Carinae. The nebula consists of ionized gas, but infrared radiation from dust is also clearly visible in its spectrum (Fig. 15.13). In even more extreme cases, the central star may be completely ob-

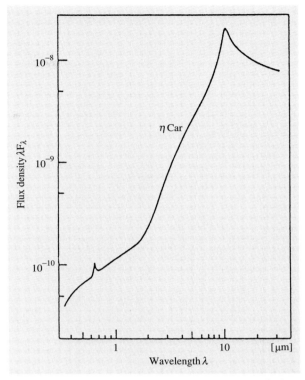

Fig. 15.13. More than 99% of the radiation from the η Carinae nebula (Fig. 13.10) is in the infrared. The peak in the visual region is from the hydrogen H_α line (0.66 μm). In the infrared, the silicate emission from dust is evident at 10 μm. (Allen, D.A. (1975): *Infrared, the New Astronomy* (Keith Reid Ltd., Shaldon) p. 103)

scured, but revealed by the infrared emission from hot dust.

Composition and Origin of the Dust (Table 15.1). From the peaks in the extinction curve, it may be concluded that interstellar dust contains water ice and silicates, and probably graphite as well. The sizes of the grains can be deduced from their scattering properties; usually they are smaller than one micrometre. The strongest scattering is due to grains of about 0.3 μm but smaller particles must also be present.

Dust grains are formed in the atmospheres of stars of late spectral types (K, M). Gas condenses into grains just as water in the Earth's atmosphere may condense into snow and ice. The grains are then expelled into interstellar space by the radiation pressure. Grains may

Table 15.1. Main properties of interstellar gas and dust

Property	Gas	Dust
Mass fraction	10%	0.1%
Composition	H I, H II, H_2 (70%) He (28%) C, N, O, Ne, Na, Mg, Al, Si, S, ... (2%)	Solid particles $d \approx 0.1$–$1\,\mu m$ H_2O (ice), silicates, graphite + impurities
Particle density	$1\,/cm^3$	$10^{-13}\,/cm^3 = 100\,/km^3$
Mass density	$10^{-21}\,kg/m^3$	$10^{-23}\,kg/m^3$
Temperature	100 K (H I), 10^4 K (H II) 50 K (H_2)	10–20 K
Method of study	Absorption lines in stellar spectra. Optical: Ca I, Ca II, Na I, K I, Ti II, Fe I, CN, CH, CH^+ Ultraviolet: H_2, CO, HD Radio lines: hydrogen 21 cm emission and absorption; H II, He II, C II recombination lines; molecular emission and absorption lines OH, H_2CO, NH_3, H_2O, CO, H_2C_2HCN, C_2H_5OH	Absorption and scattering of starlight. Interstellar reddening Interstellar polarization Thermal infrared emission

also form in connection with star formation and possibly directly from atoms and molecules in interstellar clouds as well.

15.2 Interstellar Gas

The mass of gas in interstellar space is a hundred times larger than that of dust. Although there is more gas, it is less easily observed, since the gas does not cause a general extinction of light. In the optical region it can only be observed on the basis of a small number of spectral lines.

The existence of interstellar gas began to be suspected in the first decade of the 20th century, when in 1904 *Johannes Hartmann* observed that some absorption lines in the spectra of certain binary stars were not Doppler shifted by the motions of the stars like the other lines. It was concluded that these absorption lines were formed in gas clouds in the space between the Earth and the stars. In some stars there were several lines, apparently formed in clouds moving with different velocities. The strongest lines in the visible region are those of neu-

tral sodium and singly ionized calcium (Fig. 15.14). In the ultraviolet region, the lines are more numerous. The strongest one is the hydrogen Lyman α line (121.6 nm).

On the basis of the optical and ultraviolet lines, it has been found that many atoms are ionized in interstellar space. This ionization is mainly due to ultraviolet radiation from stars and, to some extent, to ionization by cosmic rays. Since the density of interstellar matter is very low, the free electrons only rarely encounter ions, and the gas remains ionized.

About thirty elements have been discovered by absorption line observations in the visible and ultraviolet region. With a few exceptions, all elements from hydrogen to zinc (atomic number 30) and a few additional heavier elements have been detected (Table 15.3). Like in the stars, most of the mass is hydrogen (about 70%) and helium (almost 30%). On the other hand, heavy elements are significantly less abundant than in the Sun and other population I stars. It is thought that they have been incorporated into dust grains, where they do not produce any absorption lines. The element abundances in the interstellar medium (gas + dust) would then be normal, although the interstellar gas is depleted in heavy

Observable phenomenon	Cause
Interstellar extinction and polarization	Non-spherical dust grains aligned by magnetic field
Dark nebulae, uneven distribution of stars and galaxies	Dust clouds
Interstellar absorption lines in stellar spectra	Atoms and molecules in the interstellar gas
Reflection nebulae	Interstellar dust clouds illuminated by nearby stars
Emission nebulae or H II regions (optical, infrared and radio emission)	Interstellar gas and dust cloud, where a nearby hot star ionizes the gas and heats the dust to 50–100 K
Optical galactic background (diffuse galactic light)	Interstellar dust illuminated by the integrated light of all stars
Galactic background radiation: a) short wavelength ($\lesssim 1$ m) b) long wavelength ($\gtrsim 1$ m)	Free–free emission from hot interstellar gas Synchrotron radiation from cosmic ray electrons in the magnetic field
Galactic 21 cm emission	Cold (100 K) interstellar neutral hydrogen clouds (H I regions)
Molecular line emission (extended)	Giant molecular clouds (masses even 10^5–$10^6 \, M_\odot$), dark nebulae
Point-like OH, H_2O and SiO sources	Maser sources near protostars and long-period variables

Table 15.2. Phenomena caused by the interstellar medium

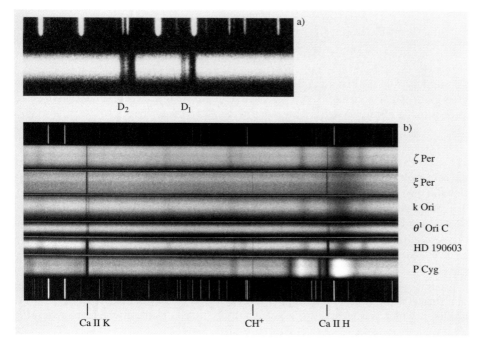

Fig. 15.14. (a) The D lines D_1 and D_2 of interstellar sodium (rest wavelengths 589.89 and 589.00 nm) in the spectrum of the star HD 14134. Both lines consist of two components formed in the gas clouds of two spiral arms. The radial velocity difference of the arms is about 30 km/s. (Mt. Wilson Observatory). (b) The interstellar absorption lines of ionized calcium Ca II and ionized methylidyne CH^+ in the spectra of several stars. The emission spectrum of iron is shown for comparison in (a) and (b). (Lick Observatory)

Table 15.3. Element abundances in the interstellar medium towards ζ Ophiuchi and in the Sun. The abundances are given relative to that of hydrogen, which has been defined to be 1,000,000. An asterisk (*) means that the abundance has been determined from meteorites. The last column gives the ratio of the abundances in the interstellar medium and in the Sun

Atomic number	Name	Chemical symbol	Interstellar abundance	Solar abundance	Abundance ratio
1	Hydrogen	H	1,000,000	1,000,000	1.00
2	Helium	He	85,000	85,000	≈ 1
3	Lithium	Li	0.000051	0.00158*	0.034
4	Beryllium	Be	< 0.000070	0.000012	< 5.8
5	Boron	B	0.000074	0.0046*	0.016
6	Carbon	C	74	370	0.20
7	Nitrogen	N	21	110	0.19
8	Oxygen	O	172	660	0.26
9	Fluorine	F	–	0.040	–
10	Neon	Ne	–	83	–
11	Sodium	Na	0.22	1.7	0.13
12	Magnesium	Mg	1.05	35	0.030
13	Aluminium	Al	0.0013	2.5	0.00052
14	Silicon	Si	0.81	35	0.023
15	Phosphorus	P	0.021	0.27	0.079
16	Sulfur	S	8.2	16	0.51
17	Chlorine	Cl	0.099	0.45	0.22
18	Argon	Ar	0.86	4.5	0.19
19	Potassium	K	0.010	0.11	0.094
20	Calcium	Ca	0.00046	2.1	0.00022
21	Scandium	Sc	–	0.0017	–
22	Titanium	Ti	0.00018	0.055	0.0032
23	Vanadium	V	< 0.0032	0.013	< 0.25
24	Chromium	Cr	< 0.002	0.50	< 0.004
25	Manganese	Mn	0.014	0.26	0.055
26	Iron	Fe	0.28	25	0.011
27	Cobalt	Co	< 0.19	0.032	< 5.8
28	Nickel	Ni	0.0065	1.3	0.0050
29	Copper	Cu	0.00064	0.028	0.023
30	Zinc	Zn	0.014	0.026	0.53

elements. This interpretation is supported by the observation that in regions where the amount of dust is smaller than usual, the element abundances in the gas are closer to normal.

Atomic Hydrogen. Ultraviolet observations have provided an excellent way of studying interstellar *neutral hydrogen*. The strongest interstellar absorption line, as has already been mentioned, is the hydrogen Lyman α line (Fig. 15.15). This line corresponds to the transition of the electron in the hydrogen atom from a state with principal quantum number $n = 1$ to one with $n = 2$. The conditions in interstellar space are such that almost all hydrogen atoms are in the ground state with $n = 1$. Therefore the Lyman α line is a strong absorption line, whereas the Balmer absorption lines, which arise from the excited initial state $n = 2$, are unobservable. (The Balmer lines are strong in stellar atmospheres with temperatures of about 10,000 K, where a large number of atoms are in the first excited state.)

The first observations of the interstellar Lyman α line were made from a rocket already in 1967. More extensive observations comprising 95 stars were obtained by the OAO 2 satellite. The distances of the observed stars are between 100 and 1000 parsecs.

Comparison of the Lyman α observations with observations of the 21 cm neutral hydrogen line have been especially useful. The distribution of neutral hydrogen over the whole sky has been mapped by means of the 21 cm line. However, the distances to nearby hydrogen clouds are difficult to determine from these observations. In the Lyman α observations one usually knows the distance to the star in front of which the absorbing clouds must lie.

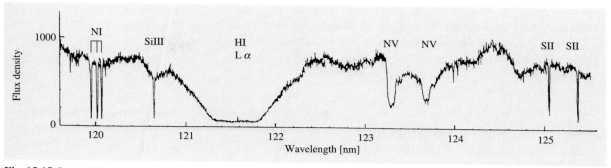

Fig. 15.15. Interstellar absorption lines in the ultraviolet spectrum of ζ Ophiuchi. The strongest line is the hydrogen Lyman α line (equivalent width, more than 1 nm). The observations were made with the Copernicus satellite. (Morton, D.C. (1975): Astrophys. J. **197**, 85)

The average gas density within about 1 kpc of the Sun derived from the Lyman α observations is 0.7 atoms/cm³. Because the interstellar Lyman α line is so strong, it can be observed even in the spectra of very nearby stars. For example, it has been detected by the Copernicus satellite in the spectrum of Arcturus, whose distance is only 11 parsecs. The deduced density of neutral hydrogen between the Sun and Arcturus is 0.02–0.1 atoms/cm³. Thus the Sun is situated in a clearing in the interstellar medium, where the density is less than one tenth of the average density.

If a hydrogen atom in its ground state absorbs radiation with a wavelength smaller than 91.2 nm, it will be ionized. Knowing the density of neutral hydrogen, one can calculate the expected distance a 91.2 nm photon can propagate before being absorbed in the ionization of a hydrogen atom. Even in the close neighbourhood of the Sun, where the density is exceptionally low, the mean free path of a 91.2 nm photon is only about a parsec and that of a 10 nm photon a few hundred parsecs. Thus only the closest neighbourhood of the Sun can be studied in the extreme ultraviolet (XUV) spectral region.

The Hydrogen 21 cm Line. The spins of the electron and proton in the neutral hydrogen atom in the ground state may be either parallel or opposite. The energy difference between these two states corresponds to the frequency of 1420.4 MHz. Thus transitions between these two hyperfine structure energy levels will give rise to a spectral line at the wavelength of 21.049 cm (Fig. 5.8). The existence of the line was theoretically predicted by *Hendrick van de Hulst* in 1944, and was first observed by *Harold Ewen* and *Edward Purcell* in 1951. Studies of this line have revealed more about the properties of the interstellar medium than any other

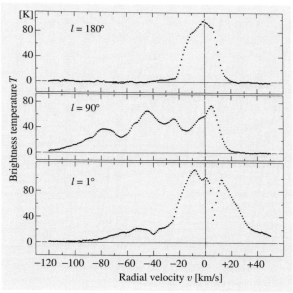

Fig. 15.16. Hydrogen 21 cm emission line profiles in the galactic plane at longitude 180°, 90° and 1° (in the direction $l = 0°$ there is strong absorption). The horizontal axis gives the radial velocity according to the Doppler formula, the vertical axis gives the brightness temperature. (Burton, W. B. (1974): "The Large Scale Distribution of Neutral Hydrogen in the Galaxy", in *Galactic and Extra-Galactic Radio Astronomy*, ed. by Verschuur, G.L., Kellermann, K.I. (Springer, Berlin, Heidelberg, New York) p. 91)

method – one might even speak of a special branch of 21 cm astronomy. The spiral structure and rotation of the Milky Way and other galaxies can also be studied by means of the 21 cm line.

Usually the hydrogen 21 cm line occurs in emission. Because of the large abundance of hydrogen, it can be observed in all directions in the sky. Some observed 21 cm line profiles are shown in Fig. 15.16. Rather than frequency or wavelength, the radial velocity calculated from the Doppler formula is plotted on the horizontal axis. This is because the broadening of the 21 cm spectral line is always due to gas motions either within the cloud (turbulence) or of the cloud as a whole. The vertical axis is mostly plotted in terms of the antenna temperature T_A (see Chap. 5), the usual radio astronomical measure of intensity. The brightness temperature of an extended source is then $T_b = T_A/\eta_B$, where η_B is the beam efficiency of the antenna.

For the 21 cm line $h\nu/k = 0.07$ K, and thus $h\nu/kT \ll 1$ for all relevant temperatures. One may therefore use the Rayleigh–Jeans approximation (5.24)

$$I_\nu = \frac{2\nu^2 kT}{c^2} \,. \tag{15.18}$$

In the solution of the equation of radiative transfer (5.42) the intensity can thus be directly related to a corresponding temperature. By definition, I_ν is related to the brightness temperature T_b, and the source function S_ν is related to the excitation temperature T_{exc}, i.e.

$$T_b = T_{\mathrm{exc}}(1 - e^{-\tau_\nu}) \,. \tag{15.19}$$

In certain directions in the Milky Way there is so much hydrogen along the line of sight that the 21 cm line is optically thick, $\tau_\nu \gg 1$. In that case

$$T_b = T_{\mathrm{exc}} \,, \tag{15.20}$$

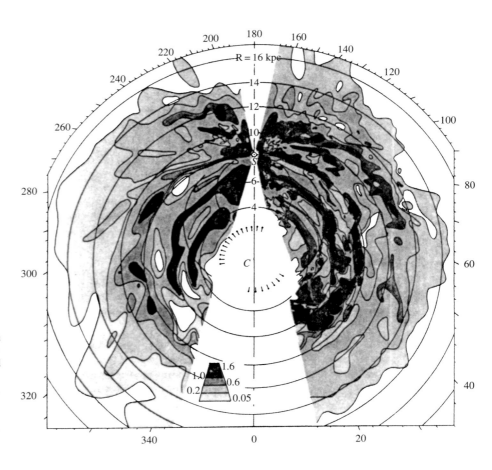

Fig. 15.17. The distribution of neutral hydrogen in the galaxy from the Leiden and Parkes surveys. The density is given in atoms/cm^3. (Oort, J.H., Kerr, P.T., Westerhout, G.L. (1958): Mon. Not. R. Astron. Soc. **118**, 379)

i.e. the brightness temperature immediately yields the excitation temperature of the cloud. This is often referred to as the *spin temperature* T_S.

The excitation temperature need not always agree with the kinetic temperature of the gas. However, in the present case the population numbers of the hyperfine levels are determined by mutual collisions of hydrogen atoms: the time between collisions is 400 years on the average, whereas the time for spontaneous radiative transitions is 11 million years; thus the excitation temperature will be the same as the kinetic temperature. The observed temperature is $T \approx 125$ K.

The distance to a source cannot be obtained directly from the observed emission. Thus one can only study the number of hydrogen atoms in a cylinder with a 1 cm^2 base area extending from the observer to outside the Milky Way along the line of sight. This is called the *projected* or *column density* and is denoted by N. One may also consider the column density $N(v)\,\mathrm{d}v$ of atoms with velocities in the interval $[v, v + \mathrm{d}v]$.

It can be shown that if the gas is optically thin, the brightness temperature in a spectral line is directly proportional to the column density N of atoms with the corresponding radial velocity. Hence, if the diameter L of a cloud along the line of sight is known, the gas density can be determined from the observed line profile:

$$n = N/L .$$

The diameter L can be obtained from the apparent diameter, if the distance and shape of the cloud are assumed known.

The distances of clouds can be determined from their radial velocities by making use of the rotation of the Milky Way (Sect. 17.3). Thus if the observed peaks in the 21 cm line profiles (Fig. 15.16) are due to individual clouds, their distances and densities can be obtained. Since radio observations are not affected by extinction, it has been possible in this way to map the density distribution of neutral hydrogen in the whole galactic plane. The resulting distribution, based on observations at Leiden and Parkes, is shown in Fig. 15.17. It appears that the Milky Way is a spiral galaxy and that the interstellar hydrogen is concentrated in the spiral arms. The average density of interstellar hydrogen is 1 atom/cm^3, but the distribution is very inhomogeneous. Typically the hydrogen forms denser regions, a few parsecs in size, where the densities may be 10–100 atoms/cm^3. Regions where the hydrogen is predominantly neutral are known as *H I regions* (in contrast to H II regions of ionized hydrogen).

The hydrogen 21 cm line may also occur in absorption, when the light from a bright radio source, e.g. a quasar, passes through an intervening cloud. The same cloud may give rise to both an absorption and an emission spectrum. In that case the temperature, optical thickness and hydrogen content of the cloud can all be derived.

Like interstellar dust hydrogen is concentrated in a thin disc in the galactic plane. The thickness of the hydrogen layer is about twice that of the dust or about 200 pc.

H II Regions. In many parts of space hydrogen does not occur as neutral atoms, but is ionized. This is true in particular around hot O stars, which radiate strongly in the ultraviolet. If there is enough hydrogen around such a star, it will be visible as an emission nebula of ionized hydrogen. Such nebulae are known as *H II region* (Figs. 15.18 and 15.19).

A typical emission nebula is the great nebula in Orion, M42. It is visible even to the unaided eye, and is a beautiful sight when seen through a telescope. In the middle of the nebula there is a group of four hot stars known as the Trapezium, which can be distinguished inside the bright nebula, even with a small telescope. The Trapezium stars emit strong ultraviolet radiation, which keeps the gas nebula ionized.

Unlike a star a cloud of ionized gas has a spectrum dominated by a few narrow emission lines. The continuous spectrum of H II regions is weak. In the visible region the hydrogen Balmer emission lines are particularly strong. These are formed when a hydrogen atom recombines into an excited state and subsequently returns to the ground state via a sequence of radiative transitions. Typically a hydrogen atom in a H II region remains ionized for several hundred years. Upon recombination it stays neutral for some months, before being ionized again by a photon from a nearby star.

The number of recombinations per unit time and volume is proportional to the product of the densities of electrons and ions,

$$n_{\mathrm{rec}} \propto n_{\mathrm{e}} n_{\mathrm{i}} . \tag{15.21}$$

Fig. 15.18. The great nebula in Orion (M42, NGC 1976). The nebula gets its energy from newly formed hot stars. The dark regions are opaque dust clouds in front of the nebula. Radio and infrared observations have revealed a rich molecular cloud behind the nebula (Fig. 15.20). In the upper part of this picture is the gas nebula NGC 1977, in the lower part the bright star ι Orionis. (Lick Observatory)

In completely ionized hydrogen, $n_e = n_i$, and hence

$$n_{\text{rec}} \propto n_e^2 \,. \tag{15.22}$$

Most recombinations will include the transition $n = 3 \rightarrow 2$, i.e. will lead to the emission of a H_α photon.

Thus the surface brightness of a nebula in the H_α line will be proportional to the *emission measure*,

$$EM = \int n_e^2 \, dl \,, \tag{15.23}$$

where the integral is along the line of sight through the nebula.

The ionization of a helium atom requires more energy than that of a hydrogen atom, and thus regions of ionized helium are formed only around the hottest stars. In these cases, a large H II region will surround a smaller central

Fig. 15.19. The Lagoon nebula (M8, NGC 6523) in Sagittarius. This H II region contains many stars of early spectral types and stars that are still contracting towards the main sequence. Small, round dark nebulae, globules, are also visible against the bright background. These are presumably gas clouds in the process of condensation into stars. (National Optical Astronomy Observatories, Kitt Peak National Observatory)

He$^+$ or He^{++} region. The helium lines will then be strong in the spectrum of the nebula.

Although hydrogen and helium are the main constituents of clouds, their emission lines are not always strongest in the spectrum. At the beginning of this century it was even suggested that some strong unidentified lines in the spectra of nebulae were due to the new element nebulium. However, in 1927 *Ira S. Bowen* showed that they were *forbidden lines* of ionized oxygen and nitrogen, O$^+$, O^{++} and N$^+$. Forbidden lines are extremely difficult to observe in the laboratory, because their transition probabilities are so small that at laboratory densities the ions are de-excited by collisions before they have had time to radiate. In the extremely diffuse interstellar gas, collisions are much less frequent, and thus there is a chance that an excited ion will make the transition to a lower state by emitting a photon.

Because of interstellar extinction, only the nearest H II regions can be studied in visible light. At infrared and radio wavelengths much more distant regions can be studied. The most important lines at radio wavelengths are recombination lines of hydrogen and helium; thus the hydrogen transition between energy levels 110 and 109 at 5.01 GHz has been much studied. These lines are also important because with their help radial velocities, and hence (using the galactic rotation law), distances of H II regions can be determined, just as for neutral hydrogen.

The physical properties of H II regions can also be studied by means of their continuum radio emission. The radiation is due to bremsstrahlung or free–free emission from the electrons. The intensity of the radiation is proportional to the emission measure *EM* defined in (15.23). H II regions also have a strong infrared continuum emission. This is thermal radiation from dust inside the nebula.

H II regions are formed when a hot O or B star begins to ionize its surrounding gas. The ionization steadily propagates away from the star. Because neutral hydrogen absorbs ultraviolet radiation so efficiently, the boundary between the H II region and the neutral gas is very sharp. In a homogeneous medium the H II region around a single star will be spherical, forming a *Strömgren sphere*. For a B0 V star the radius of the Strömgren sphere is 50 pc and for an A0 V star only 1 pc.

The temperature of a H II region is higher than that of the surrounding gas, and it therefore tends to expand. After millions of years, it will have become extremely diffuse and will eventually merge with the general interstellar medium.

15.3 Interstellar Molecules

The first *interstellar molecules* were discovered in 1937–1938, when molecular absorption lines were found in the spectra of some stars. Three simple diatomic molecules were detected: *methylidyne* CH, its positive ion CH$^+$ and *cyanogen* CN. A few other molecules were later discovered by the same method in the ultraviolet. Thus *molecular hydrogen* H$_2$ was discovered in the early 1970's, and *carbon monoxide*, which had been discovered by radio observations, was also detected in the ultraviolet. Molecular hydrogen is the most abundant interstellar molecule, followed by carbon monoxide.

Molecular Hydrogen. The detection and study of molecular hydrogen has been one of the most important achievements of UV astronomy. Molecular hydrogen has a strong absorption band at 105 nm, which was first observed in a rocket experiment in 1970 by *George R. Carruthers*, but more extensive observations could only be made with the Copernicus satellite. The observations showed that a significant fraction of interstellar hydrogen is molecular, and that this fraction increases strongly for denser clouds with higher extinction. In clouds with visual extinction larger than one magnitude essentially all the hydrogen is molecular.

Hydrogen molecules are formed on the surface of interstellar grains, which thus act as a chemical catalyst. Dust is also needed to shield the molecules from the stellar UV radiation, which would otherwise destroy them. Molecular hydrogen is thus found where dust is abundant. It is of interest to know whether gas and dust are well mixed or whether they form separate clouds and condensations.

UV observations have provided a reliable way of comparing the distribution of interstellar gas and dust. The amount of dust between the observer and a star is obtained from the extinction of the stellar light. Furthermore, the absorption lines of atomic and molecular hydrogen in the ultraviolet spectrum of the same star can be observed. Thus the total amount of hydrogen (atomic + molecular) between the observer and the star can also be determined.

Observations indicate that the gas and dust are well mixed. The amount of dust giving rise to one magnitude visual extinction corresponds to 1.9×10^{21} hydrogen atoms (one molecule is counted as two atoms). The mass ratio of gas and dust obtained in this way is 100.

Radio Spectroscopy. Absorption lines can only be observed if there is a bright star behind the molecular cloud. Because of the large dust extinction, no observations of molecules in the densest clouds can be made in the optical and ultraviolet spectral regions. Thus only radio observations are possible for these objects, where molecules are especially abundant.

Radio spectroscopy signifies an immense step forward in the study of interstellar molecules. In the early 1960's, it was still not believed that there might be more complicated molecules than diatomic ones in interstellar space. It was thought that the gas was too diffuse for molecules to form and that any that formed would be destroyed by ultraviolet radiation. The first molecular

radio line, the hydroxyl radical OH, was discovered in 1963. Many other molecules have been discovered since then. By 2002, about 130 molecules had been detected, the heaviest one being the 13-atom molecule $HC_{11}N$.

Molecular lines in the radio region may be observed either in absorption or in emission. Radiation from diatomic molecules like CO (see Fig. 15.20) may correspond to three kinds of transitions. (1) *Electron transitions* correspond to changes in the electron cloud of the molecule. These are like the transitions in single atoms, and their wavelengths lie in the optical or ultraviolet region. (2) *Vibrational transitions* correspond to changes in the vibrational energy of the molecule. Their energies are usually in the infrared region. (3) Most important for radio spectroscopy are the *rotational transitions*, which are changes in the rotational energy of the molecule. Molecules in their ground state do not rotate, i.e. their angular momentum is zero, but they may be excited and start rotating in collisions with other molecules. For example, carbon sulfide CS returns to its ground state in a few hours by emitting a millimetre region photon.

Fig. 15.20. Radio map of the distribution of carbon monoxide $^{13}C^{16}O$ in the molecular cloud near the Orion nebula. The curves are lines of constant intensity. (Kutner, M. L., Evans 11, N. J., Tucker K. D. (1976): Astrophys. J. **209**, 452)

Table 15.4. Some molecules observed in the interstellar medium

Molecule	Name	Year of discovery
Discovered in the optical and ultraviolet region:		
CH	methylidyne	1937
CH^+	methylidyne ion	1937
CN	cyanogen	1938
H_2	hydrogen molecule	1970
CO	carbon monoxide	1971
Discovered in the radio region:		
OH	hydroxyl radical	1963
CO	carbon monoxide	1970
CS	carbon monosulfide	1971
SiO	silicon monoxide	1971
SO	sulfur monoxide	1973
H_2O	water	1969
HCN	hydrogen cyanide	1970
NH_3	ammonia	1968
H_2CO	formaldehyde	1969
HCOOH	formic acid	1975
HCCNC	isocyanoacetylene	1991
C_2H_4O	vinyl alcohol	2001
H_2CCCC	cumulene carbene	1991
$(CH_3)_2O$	dimethyl ether	1974
C_2H_5OH	ethanol	1975
$HC_{11}N$	cyanopentacetylene	1981

A number of interstellar molecules are listed in Table 15.4. Many of them have only been detected in the densest clouds (mainly the Sagittarius B2 cloud at the galactic centre), but others are very common. The most abundant molecule H_2 cannot be observed at radio wavelengths, because it has no suitable spectral lines. The next most abundant molecules are carbon monoxide CO, the hydroxyl radical OH and ammonia NH_3, although their abundance is only a small fraction of that of hydrogen. However, the masses of interstellar clouds are so large that the number of molecules is still considerable. (The Sagittarius B2 cloud contains enough ethanol, C_2H_5OH, for 10^{28} bottles of vodka.)

Both the formation and survival of interstellar molecules requires a higher density than is common in interstellar clouds; thus they are most common in dense clouds. Molecules are formed in collisions of atoms or simpler molecules or catalysed on dust grains. Molecular clouds must also contain a lot of dust to absorb the ultraviolet radiation entering from outside that otherwise would disrupt the molecules. The most suit-

able conditions are thus found inside dust and molecular clouds near dense dark nebulae and H II regions.

Most of the molecules in Table 15.4 have only been detected in dense molecular clouds occurring in connection with H II regions. Almost every molecule yet discovered has been detected in Sagittarius B2 near the galactic centre. Another very rich molecular cloud has been observed near the H II region Orion A. In visible light this region has long been known as the Orion nebula M42 (Fig. 15.18). Inside the actual H II regions there are no molecules, since they would be rapidly dissociated by the high temperature and strong ultraviolet radiation. Three types of molecular sources have been found near H II regions (Fig. 15.21):

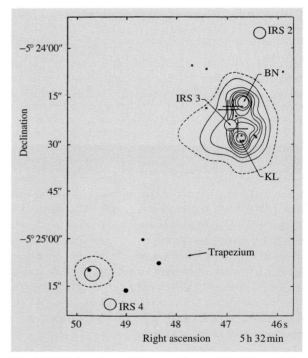

Fig. 15.21. Infrared map of the central part of the Orion nebula. In the lower part are the four Trapezium stars. Above is an infrared source of about 0.5″ diameter, the Kleinmann–Low nebula (KL). BN is an infrared point source, the Becklin–Neugebauer object. Other infrared sources are denoted IRS. The large crosses indicate OH masers and the small crosses H_2O masers. On the scale of Fig. 15.18 this region would only be a few millimetres in size. (Goudis, C. (1982): The Orion Complex: A Case Study of Interstellar Matter (Reidel, Dordrecht) p. 176)

Table 15.5. The five phases of interstellar gas

		T [K]	n [cm^{-3}]
1.	Very cold molecular gas clouds (mostly hydrogen H$_2$)	20	$\gtrsim 10^3$
2.	Cold gas clouds (mostly atomic neutral hydrogen)	100	20
3.	Warm neutral gas enveloping the cooler clouds	6000	0.05–0.3
4.	Hot ionized gas (mainly H II regions around hot stars)	8000	> 0.5
5.	Very hot and diffuse ionized coronal gas, ionized and heated by supernova explosions	10^6	10^{-3}

1. Large gas and dust envelopes around the H II region.
2. Small dense clouds inside these envelopes.
3. Very compact OH and H$_2$O maser sources.

The large envelopes have been discovered primarily by CO observations. OH and H$_2$CO have also been detected. Like in the dark nebulae the gas in these clouds is probably mainly molecular hydrogen. Because of the large size and density ($n \approx 10^3$–10^4 molecules/cm^3) of these clouds, their masses are very large, 10^5 or even 10^6 solar masses (Sgr B2). They are among the most massive objects in the Milky Way. The dust in molecular clouds can be observed on the basis of its thermal radiation. Its peak falls at wavelengths of 10–100 μm, corresponding to a dust temperature of 30–300 K.

Some interstellar clouds contain very small *maser sources*. In these the emission lines of OH, H$_2$O and SiO may be many million times stronger than elsewhere. The diameter of the radiating regions is only about 5–10 AU. The conditions in these clouds are such that radiation in some spectral lines is amplified by stimulated emission as it propagates through the cloud. Hydroxyl and water masers occur in connection with dense H II regions and infrared sources, and appear to be related to the formation of protostars. In addition maser emission (OH, H$_2$O and SiO) occurs in connection with Mira variables and some red supergiant stars. This maser emission comes from a molecule and dust envelope around the star, which also gives rise to an observable infrared excess.

15.4 The Formation of Protostars

The mass of the Milky Way is about 10^{11} solar masses. Since its age is about 10^{10} years, stars have been forming at the average rate of 10 $M_\odot$ per year. This estimate is only an upper limit for the present rate, because earlier the rate of star formation must have been much higher. Since the lifetime of O stars is only about a million years, a better estimate of the star formation rate can be made, based on the observed number of O stars. Accordingly, it has been concluded that at present, new stars are forming in the Milky Way at a rate of about three solar masses per year.

Stars are now believed to form inside large dense interstellar clouds mostly located in the spiral arms of the Galaxy. Under its own gravity, a cloud begins to contract and fragment into parts that will become protostars. The observations seem to indicate that stars are not formed individually, but in larger groups. Young stars are found in open clusters and in loose associations, typically containing a few hundred stars which must have formed simultaneously.

Theoretical calculations confirm that the formation of single stars is almost impossible. An interstellar cloud can contract only if its mass is large enough for gravity to overwhelm the pressure. As early as in the 1920's, James Jeans calculated that a cloud with a certain temperature and density can condense only if its mass is high enough. If the mass is too small the pressure of the gas is sufficient to prevent the gravitational contraction. The limiting mass is the Jeans mass (Sect. 6.11), given by

$$M_{\rm J} \approx 3 \times 10^4 \sqrt{\frac{T^3}{n}} \, M_\odot \, ,$$

where n is the density in atoms/m^3 and T the temperature.

In a typical interstellar neutral hydrogen cloud $n = 10^6$ and $T = 100$ K, giving the Jeans mass, 30,000 $M_\odot$. In the densest dark clouds $n = 10^{12}$ and $T = 10$ K and hence, $M_{\rm J} = 1 \, M_\odot$.

It is thought that star formation begins in clouds of a few thousand solar masses and diameters of about 10 pc. The cloud begins to contract, but does not heat up because the liberated energy is carried away by radiation. As the density increases, the Jeans mass thus decreases. Because of this, separate condensation nuclei are formed in the cloud, which go on contracting independently: the cloud *fragments*. Fragmentation is further advanced by the increasing rotation velocity. The original cloud has a certain angular momentum which is conserved during the contraction; thus the angular velocity must increase.

This contraction and fragmentation continues until the density becomes so high that the individual fragments become optically thick. The energy liberated by the contraction can then no longer escape, and the tem-

perature will begin to rise. In consequence the Jeans mass begins to increase, further fragmentation ceases and the rising pressure in existing fragments stops their contraction. Some of the protostars formed in this way may still be rotating too rapidly. These may split into two, thus forming double systems. The further evolution of protostars has been described in Sect. 11.2.

Although the view that stars are formed by the collapse of interstellar clouds is generally accepted, many details of the fragmentation process are still highly conjectural. Thus the effects of rotation, magnetic fields and energy input are very imperfectly known. Why a cloud begins to contract is also not certain; one theory

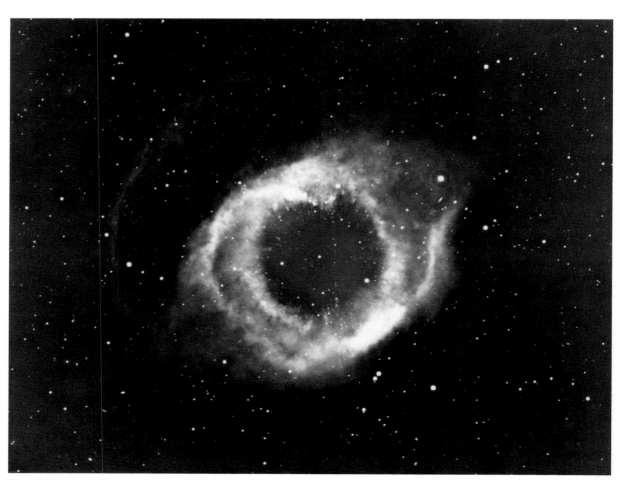

Fig. 15.22. The Helix nebula (NGC 7293). The planetary nebulae are formed during the final stages of evolution of solar-type stars. The centrally visible star has ejected its outer layers into space. (National Optical Astronomy Observatories, Kitt Peak National Observatory)

is that passage through a spiral arm compresses clouds and triggers contraction (see Sect. 17.4). This would explain why young stars are predominantly found in the spiral arms of the Milky Way and other galaxies. The contraction of an interstellar cloud might also be initiated by a nearby expanding H II region or supernova explosion.

Star formation can be observed particularly well in the infrared, since the temperatures of of the condensing clouds and protostars are of the order 100–1000 K and the infrared radiation can escape even the densest dust clouds. For example, in connection with the Orion nebula there is a large cloud of hydrogen, found in radio observations, containing small infrared sources. E.g. the Becklin–Neugebauer object has a temperature of a couple of hundred kelvins but a luminosity that is thousandfold compared with the Sun. It is a strong H_2O maser source, located next to a large H II region.

15.5 Planetary Nebulae

Bright regions of ionized gas do not occur only in connection with newly formed stars, but also around stars in late stages of their evolution. The *planetary nebulae* are gas shells around small hot blue stars. As we

Fig. 15.23. The Crab nebula (M1, NGC 1952) in Taurus is the remnant of a supernova explosion observed in 1054. The photograph was taken at red wavelengths. The nebula is also a strong radio source. Its energy source is the central rapidly rotating neutron star, pulsar, which is the collapsed core of the original star. (Palomar Observatory)

have seen in connection with stellar evolution, instabilities may develop at the stage of helium burning. Some stars begin to pulsate, while in others the whole outer atmosphere may be violently ejected into space. In the latter case, a gas shell expanding at 20–30 km/s will be formed around a small and hot (50,000–100,000 K) star, the core of the original star.

The expanding gas in a planetary nebula is ionized by ultraviolet radiation from the central star, and its spectrum contains many of the same bright emission lines as that of an H II region. Planetary nebulae are, however, generally much more symmetrical in shape than most H II regions, and they expand more rapidly. For example, the well-known Ring nebula in Lyra (M57) has expanded visibly in photographs taken at 50-year intervals. In a few ten thousand years, the planetary nebulae disappear in the general interstellar medium and their central stars cool to become white dwarfs.

The planetary nebulae were given their name in the 19th century, because certain small nebulae visually look quite like planets such as Uranus. The apparent diameter of the smallest known planetary nebulae is only a few arc seconds, whereas the largest ones (like the Helix nebula) may be one degree in diameter (Fig. 15.22).

The brightest emission lines are often due to forbidden transitions, like in H II regions. For example, the green colour of the central parts of the Ring nebula in Lyra is due to the forbidden lines of doubly ionized oxygen at 495.9 and 500.7 nm. The red colour of the outer parts is due to the hydrogen Balmer α line (656.3 nm) and the forbidden lines of ionized nitrogen (654.8 nm, 658.3 nm).

The total number of planetary nebulae in the Milky Way has been estimated to be 50,000. About 2000 have actually been observed.

15.6 Supernova Remnants

In Chap. 11 we have seen that massive stars end their evolution in a supernova explosion. The collapse of the stellar core leads to the violent ejection of the outer layers, which then remain as an expanding gas cloud.

About 120 *supernova remnants* (SNR's) have been discovered in the Milky Way. Some of them are optically visible as a ring or an irregular nebula (e.g. the Crab nebula; see Fig. 15.23), but most are detectable only

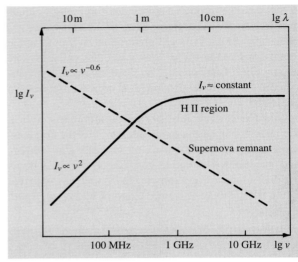

Fig. 15.24. The radio spectra of typical H II regions and supernova remnants. The radiation of H II regions is thermal and obeys the Rayleigh–Jeans law, $I \propto \nu^2$, at wavelengths larger than 1 m. In supernova remnants the intensity decreases with increasing frequency. (After Scheffler, H., Elsässer, H. (1987): *Physics of the Galaxy and the Interstellar Matter* (Springer, Berlin, Heidelberg, New York))

in the radio region (because radio emission suffers no extinction).

In the radio region the SNR's are extended sources similar to H II regions. However, unlike H II regions the radiation from SNR's is often polarized. Another characteristic difference between these two kinds of sources is that whereas the radio brightness of H II regions grows or remains constant as the frequency increases, that of SNR's falls off almost linearly (in a $\log I_\nu - \log \nu$ diagram) with increasing frequency (Fig. 15.24).

These differences are due to the different emission processes in H II regions and in SNR's. In an H II region, the radio emission is free-free radiation from the hot plasma. In a SNR it is *synchrotron radiation* from relativistic electrons moving in spiral orbits around the magnetic field lines. The synchrotron process gives rise to a continuous spectrum extending over all wave-

Fig. 15.25. The Veil nebula (NGC 6960 at the *right*, NGC 6992 at the *left*) in Cygnus is the remnant of a supernova explosion which occurred several ten thousand years ago. (Mt. Wilson Observatory)

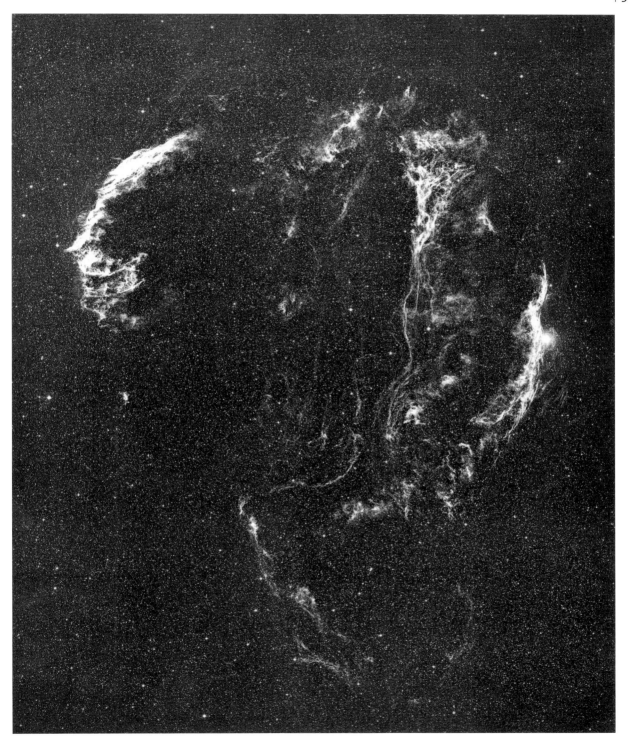

length regions. For example, the Crab nebula looks blue or green in colour photographs because of optical synchrotron radiation.

In the Crab nebula red filaments are also visible against the bright background. Their emission is principally in the hydrogen H_α line. The hydrogen in a SNR is not ionized by a central star as in the H II regions, but by the ultraviolet synchrotron radiation.

The supernova remnants in the Milky Way fall into two classes. One type has a clearly ring-like structure (e. g. Cassiopeia A or the Veil nebula in Cygnus; see Fig. 15.25); another is irregular and bright at the middle (like the Crab nebula). In the remnants of the Crab nebula type there is always a rapidly rotating pulsar at the centre. This pulsar provides most of the energy of the remnant by continuously injecting relativistic electrons into the cloud. The evolution of this type of SNR reflects that of the pulsar and for this reason has a time scale of a few ten thousand years.

Ring-like SNR's do not contain an energetic pulsar; their energy comes from the actual supernova explosion. After the explosion, the cloud expands at a speed of 10,000–20,000 km/s. About 50–100 years after the explosion the remnant begins to form a spherical shell as the ejected gas starts to sweep up interstellar gas and to slow down in its outer parts. The swept-up shell expands with a decreasing velocity and cools until, after about 100,000 years, it merges into the interstellar medium. The two types of supernova remnants may be related to the two types (I and II) of supernovae.

* Synchrotron Radiation

Synchrotron radiation was first observed in 1948 by *Frank Elder*, *Robert Langmuir* and *Herbert Pollack*, who were experimenting with an electron synchrotron, in which electrons were accelerated to relativistic energies in a magnetic field. It was observed that the electrons radiated visible light in a narrow cone along their momentary direction of motion. In astrophysics synchrotron radiation was first invoked as an explanation of the radio emission of the Milky Way, discovered by *Karl Jansky* in 1931. This radiation had a spectrum and a large metre-wave brightness temperature (more than 10^5 K) which were inconsistent with ordinary thermal free-free emission from ionized gas. In 1950 *Hannes Alfvén* and *Nicolai Herlofson* as well as

Karl-Otto Kiepenheuer proposed that the galactic radio background was due to synchrotron radiation. According to Kiepenheuer the high-energy cosmic ray electrons would emit radio radiation in the weak galactic magnetic field. This explanation has turned out to be correct. Synchrotron radiation is also an important emission process in supernova remnants, radio galaxies and quasars. It is *a non-thermal* radiation process, i. e. the energy of the radiating electrons is not due to thermal motions.

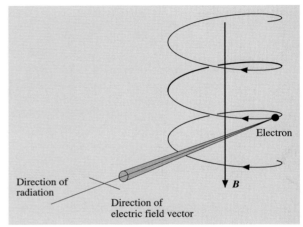

The emission of synchrotron radiation. A charged particle (electron) propagating in a magnetic field moves in a spiral. Because of the centripetal acceleration, the particle emits electromagnetic radiation

The origin of synchrotron radiation is schematically shown in the figure. The magnetic field forces the electron to move in a spiral orbit. The electron is thus constantly accelerated and will emit electromagnetic radiation. According to the special theory of relativity, the emission from a relativistic electron will be concentrated in a narrow cone. Like the beam from a lighthouse, this cone sweeps across the observer's field of vision once for each revolution. Thus the observer sees a sequence of radiation flashes of very short duration compared with their interval. (In the total emission of a large number of electrons, separate flashes cannot be distinguished.) When this series of pulses is represented as a sum of different frequency components (Fourier transform), a broad spectrum is obtained with a maximum at

$$\nu_{\max} = aB_\perp E^2 \,,$$

where $B_\perp$ is the magnetic field component perpendicular to the velocity of the electron, and E its energy, a is a constant of proportionality.

The table gives the frequency and wavelength of the maximum as functions of the electron energy for the typical galactic field strength 0.5 nT:

λ_{max}	ν_{max} [Hz]	E [eV]
300 nm	10^{15}	6.6×10^{12}
30 μm	10^{13}	6.6×10^{11}
3 mm	10^{11}	6.6×10^{10}
30 cm	10^{9}	6.6×10^{9}
30 m	10^{7}	6.6×10^{8}

To produce even radio synchrotron radiation, very energetic electrons are required, but these are known to be present in the cosmic radiation. In the optical galactic background radiation, the contribution from synchrotron radiation is negligible, but, for example, in the Crab nebula, a significant part of the optical emission is due to this mechanism.

15.7 The Hot Corona of the Milky Way

As early as 1956 *Lyman Spitzer* showed that the Milky Way has to be surrounded by a large envelope of very hot gas. Almost two decades later the Copernicus satellite, whose scientific program was directed by Spitzer, found evidence for this kind of gas, which began to be called *galactic coronal gas*, in analogy with the solar corona. The satellite observed emission lines of e. g. five times ionized oxygen (O VI), four times ionized nitrogen (N V) and triply ionized carbon (C IV). The formation of these lines requires a high temperature (100,000–1,000,000 K), and a high temperature is also indicated by the broadening of the lines.

Galactic coronal gas is distributed through the whole Milky Way and extends several thousand parsecs from the galactic plane. Its density is only of the order of 10^{-3} atoms/cm^3 (recall that the mean density in the galactic plane is 1 atom/cm^3). Thus coronal gas forms a kind of background sea, from which the denser and cooler forms of interstellar matter, such as neutral hydrogen and molecular clouds, rise as islands. In the early 1980's the IUE satellite also detected similar coronae in the Large Magellanic Cloud and in the spiral galaxy M100. Coronal gas is probably quite a common and important form of matter in galaxies.

Supernova explosions are probably the source of both coronal gas and its energy. When a supernova explodes, it forms a hot bubble in the surrounding medium. The bubbles from neighbouring supernovae will expand and merge, forming a foamlike structure. In addition to supernovae, stellar winds from hot stars may provide some of the energy of the coronal gas.

15.8 Cosmic Rays and the Interstellar Magnetic Field

Cosmic Rays. Elementary particles and atomic nuclei reaching the Earth from space are called *cosmic rays*. They occur throughout interstellar space with an energy density of the same order of magnitude as that of the radiation from stars. Cosmic rays are therefore important for the ionization and heating of interstellar gas.

Since cosmic rays are charged, their direction of propagation in space is constantly changed by the magnetic field. Their direction of arrival therefore gives no information about their place of origin. The most important properties of cosmic rays that can be observed from the Earth are their particle composition and energy distribution. As noted in Sect. 3.6, these observations have to be made in the upper atmosphere or from satellites, since cosmic ray particles are destroyed in the atmosphere.

The main constituent of the cosmic rays (about 90%) is hydrogen nuclei or protons. The second most important constituent (about 9%) is helium nuclei or α particles. The rest of the particles are electrons and nuclei more massive than helium.

Most cosmic rays have an energy smaller than 10^9 eV. The number of more energetic particles drops rapidly with increasing energy. The most energetic protons have an energy of 10^{20} eV, but such particles are very rare – the energy of one such proton could lift this book about one centimetre. (The largest particle accelerators reach "only" energies of 10^{12} eV.)

The distribution of low-energy (less than 10^8 eV) cosmic rays cannot be reliably determined from the Earth, since solar "cosmic rays", high-energy protons and electrons formed in solar flares fill the solar system and strongly affect the motion of low-energy cosmic rays.

The distribution of cosmic rays in the Milky Way can be directly inferred from gamma-ray and radio observations. The collisions of cosmic ray protons with interstellar hydrogen atoms gives rise to pions which then decay to form a gamma-ray background. The radio background is formed by cosmic ray electrons which emit synchrotron radiation in the interstellar magnetic field.

Both radio and gamma-ray emission are strongly concentrated in the galactic plane. From this it has been concluded that the sources of cosmic rays must also be located in the galactic plane. In addition there are individual peaks in the backgrounds around known supernova remnants. In the gamma-ray region such peaks are observed at e.g. the Crab nebula and the Vela pulsar; in the radio region the North Polar Spur is a large, nearby ring-like region of enhanced emission.

Apparently a large fraction of cosmic rays have their origin in supernovae. An actual supernova explosion will give rise to energetic particles. If a pulsar is formed, observations show that it will accelerate particles in its surroundings. Finally the shock waves formed in the expanding supernova remnant will also give rise to relativistic particles.

On the basis of the relative abundances of various cosmic ray nuclei, it can be calculated how far they have travelled before reaching the Earth. It has been found that typical cosmic ray protons have travelled for a period of a few million years (and hence also a distance of a few million light-years) from their point of origin. Since the diameter of the Milky Way is about 100,000 light-years, the protons have crossed the Milky Way tens of times in the galactic field.

The Interstellar Magnetic Field. The strength and direction of the *interstellar magnetic field* are difficult to determine reliably. Direct measurements are impossible, since the magnetic fields of the Earth and the Sun are much stronger. However, using various sources it has been possible to deduce the existence and strength of the field.

We have already seen that interstellar grains give rise to interstellar polarization. In order to polarize light, the dust grains have to be similarly oriented; this can only be achieved by a general magnetic field. Figure 15.26 shows the distribution of interstellar polarization over the sky. Stars near each other generally have the same polarization. At low galactic latitudes the polarization is almost parallel to the galactic plane, except where one is looking along a spiral arm.

More precise estimates of the strength of the magnetic field can be obtained from the rotation of the plane of polarization of the radio radiation from distant sources. This Faraday rotation is proportional to

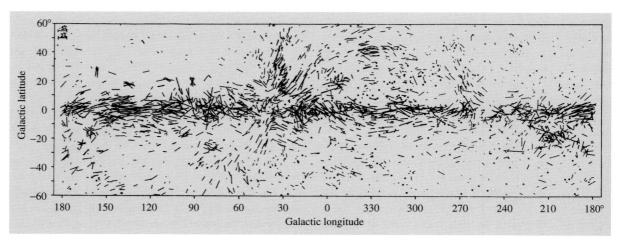

Fig. 15.26. The polarization of starlight. The dashes give the direction and degree of the polarization. The thinner dashes correspond to stars with polarization smaller than 0.6%; the thicker dashes to stars with larger polarization. The scale is shown in the upper left-hand corner. Stars with polarization smaller than 0.08% are indicated by a small circle. (Mathewson, D.S., Ford, V.L. (1970): Mem. R.A.S. **74**, 139)

the strength of the magnetic field and to the electron density. Another method is to measure the Zeeman splitting of the 21 cm radio line. These measurements have fairly consistently given a value of 10^{-10}–10^{-9} T for the strength of the interstellar magnetic field. This is about one millionth of the interplanetary field in the solar system.

15.9 Examples

Example 15.1 Estimate the dust grain size and number density in the galactic plane.

Let us compare the interstellar extinction curve in Fig. 15.4b with the Mie scattering curves in Fig. 15.3. We see that the leftmost parts of the curves may correspond to each other: the interval $0 < x < 5$ with $m = 1.5$ in Fig. 15.3 matches the interval $0 < 1/\lambda < 5\,\mu m^{-1}$ in Fig. 15.4b. Remembering that $x = 2\pi a/\lambda$, this suggests a constant grain radius a, given by $2\pi a \approx 1\,\mu m$, or $a \approx 0.16\,\mu m$.

In the blue wavelength region ($\lambda = 0.44\,\mu m$), $x = 2.3$ and, according to the upper Fig. 15.3, $Q_{ext} \approx 2$. Using $A = 2$ mag for the interstellar extinction at $r = 1$ kpc, we get, substituting (15.5) into (15.7), $\bar{n} \approx 4 \times 10^{-7}\,m^{-3}$. This should give the order of magnitude of the interstellar dust density.

As a summary we could say that a considerable fraction of the interstellar extinction might be due to grains of diameter $0.3\,\mu m$ and particle density of the order of $10^{-7}\,m^{-3} = 100\,km^{-3}$.

Example 15.2 Estimate the time interval between successive collisions of a hydrogen atom in interstellar gas.

Two atoms will collide if the separation between their centres becomes less than $2r$, where r is the radius of the atom. Thus, the microscopic cross section for the collision is $\sigma = \pi(2r)^2 = 4\pi r^2$. The macroscopic cross section, or the number of collisions of an H atom per unit length, is then $\Sigma = n\sigma$, where n is the number density of the H atoms. The mean free path l of an atom is the inverse of the macroscopic cross section, $l = 1/\Sigma$, and the time between two collisions is $t = l/v$, where v is the velocity of the atom.

Considering the numerical values, the Bohr radius of an H atom is $r = 5.3 \times 10^{-11}$ m. Taking $n = 1\,cm^{-3}$ we

get $l = 2.8 \times 10^{13}$ m ≈ 0.0009 pc. The average velocity is not far from the root mean square velocity at $T = 125$ K, given by (5.33):

$$v = \sqrt{\frac{3kT}{m}} = 1760\,m\,s^{-1}\,.$$

These values of l and v give $t = l/v = 510$ years for the collision interval. Taking into account the velocity distribution in the gas, the mean free path appears to be shorter by a factor of $1/\sqrt{2}$, which reduces the time to about 400 years.

Example 15.3 Consider the lowest rotational transition of the CO molecule. For ^{12}CO the frequency of this line is $\nu(^{12}CO) = 115.27$ GHz, and for ^{13}CO, $\nu(^{13}CO) = 110.20$ GHz. Estimate the optical thickness of each line in a molecular cloud, where the observed brightness temperatures of the lines are $T_b(^{12}CO) = 40$ K and $T_b(^{13}CO) = 9$ K.

For the ^{12}CO line, $h\nu/k = 5.5$ K. Thus, the Rayleigh–Jeans approximation is valid if the temperature is considerably higher than 5 K. This is not always the case, but the measured value of $T_b(^{12}CO)$ suggests that the approximation can be used.

Ignoring the background, (15.19) gives

$$T_b = T_{exc}\left(1 - e^{-\tau_\nu}\right)\,.$$

The optical thickness τ_ν is proportional to the opacity or the absorption coefficient α_ν [see (4.15)], and α_ν is evidently proportional to the number of CO molecules present. Other differences between the lines are small, so we can write

$$\frac{\tau_\nu(^{12}CO)}{\tau_\nu(^{13}CO)} \approx \frac{n(^{12}CO)}{n(^{13}CO)}\,.$$

Adopting the terrestrial value $n(^{12}CO)/n(^{13}CO) = 89$, we set

$$\tau_\nu(^{12}CO) = 89\tau_\nu(^{13}CO)\,.$$

Assuming the excitation temperatures equal and denoting $\tau_\nu(^{12}CO)$ by τ, we get

$$T_{exc}\left(1 - e^{-\tau}\right) = 40\,,$$

$$T_{exc}\left(1 - e^{-\tau/89}\right) = 9\,.$$

The solution of this pair of equations is

$$\tau_\nu(^{12}\text{CO}) = 23, \quad \tau_\nu(^{13}\text{CO}) = 0.25, \quad T_{\text{exc}} = 40\,\text{K} .$$

Thus, the ^{12}CO line seems to be optically thick, and $T_{\text{exc}} = T_{\text{b}}(^{12}\text{CO})$. If also the ^{13}CO line were optically thick, the brightness temperatures would be practically equal, and the optical thicknesses could not be determined.

15.10 Exercises

Exercise 15.1 Two open clusters, which are seen near each other in the galactic plane, have angular diameters α and 3α, and distance moduli 16.0 and 11.0, respectively. Assuming their actual diameters are equal, find their distances and the interstellar extinction coefficient a in (15.4).

Exercise 15.2 Estimate the free fall velocity on the surface of a spherical gas cloud contracting under the influence of its own gravity. Assume $n(\text{H}_2) = 10^3\,\text{cm}^{-3}$ and $R = 5\,\text{pc}$.

Exercise 15.3 The force F exerted by a magnetic field B on a charge q moving with velocity v is $F = qv \times B$. If v is perpendicular to B, the path of the charge is circular. Find the radius of the path of an interstellar proton with a kinetic energy of 1 MeV. Use $B = 0.1\,\text{nT}$ for the galactic magnetic field.

16. Star Clusters and Associations

Several collections of stars can be picked out in the sky, even with the naked eye. Closer study reveals that they really do form separate clusters in space. E.g. the Pleiades in Taurus and the Hyades around Aldebaran, the brightest star in Taurus, are such *open star clusters*. Almost the whole of the constellation Coma Berenices is also an open star cluster. Many objects appearing as nebulous patches to the unaided eye, when looked at with a telescope, turn out to be star clusters, like Praesepe in the constellation Cancer, or Misan, the double cluster in Perseus (Fig. 16.1). In addition to open clusters some apparently nebulous objects are very dense *globular clusters*, such as those in Hercules and in Canes Venatici (Fig. 16.2).

The first catalogue of star clusters was prepared by the French astronomer *Charles Messier* in 1784. Among his 103 objects there were about 30 globular clusters and the same number of open clusters, in addition to gas nebulae and galaxies. A larger catalogue, published in 1888, was the *New General Catalogue of Nebulae and Clusters of Stars* prepared by the Danish astronomer *John Louis Emil Dreyer*. The catalogue numbers of objects in this list are preceded by the initials NGC. For example, the large globular cluster in Hercules is object M13 in the Messier catalogue, and it is also known as NGC 6205. The NGC catalogue was supplemented with the *Index Catalogue* in 1895 and 1910. The objects of this catalogue are given the initials IC.

16.1 Associations

In 1947 the Soviet astronomer *Viktor Amazaspovich Ambartsumyan* discovered that there are groups of young stars scattered over so large regions of the sky that they would be very difficult to identify merely on the basis of their appearance. These *associations* may have a few tens of members. One association is found around the star ζ Persei, and in the region of Orion, there are several associations.

Associations are groups of very young stars. They are usually identified on the basis either of absolutely bright main sequence stars or of T Tauri stars. According to the type, one speaks of OB associations and T Tauri associations. The most massive stars of spectral class O stay on the main sequence for only a few million years, and therefore associations containing them are necessarily young. The T Tauri stars are even younger stars that are in the process of contracting towards the main sequence.

Studies of the internal motions in associations show that they are rapidly dispersing. There are so few stars in an association that their gravity cannot hold them together for any length of time. The observed motions have often confirmed that the stars in an association were very close together a few million years ago (Fig. 16.3).

Large amounts of interstellar matter, gas and dust nebulae often occur in connection with associations, supplying information about the connection between star formation and the interstellar medium. Infrared observations have shown that stars are now forming or have recently formed in many dense interstellar clouds.

Associations are strongly concentrated in the spiral arms in the plane of the Milky Way. Both in the Orion region and in the direction of Cepheus, three generations of associations have been identified, the oldest ones being most extended and the youngest ones, most dense.

16.2 Open Star Clusters

Open clusters usually contain from a few tens to a few hundreds of stars. The kinetic energy of the cluster members, the differential rotation of the Milky Way (Sect. 17.3) and external gravitational disturbances tend to gradually disperse the open clusters. Still, many of them are fairly permanent; for example, the Pleiades is many hundreds of millions of years old, but nevertheless, quite a dense cluster.

The distances of star clusters – and also of associations – can be obtained from the photometric or spectroscopic distances of their brightest members. For the nearest clusters, in particular for the Hyades, one can use the method of *kinematic parallaxes*, which is based on the fact that the stars in a cluster all have the same average space velocity with respect to the Sun. The proper motions in the Hyades are shown in Fig. 16.4a. They all appear to be directed to the same point. Figure 16.4b explains how this convergence can be understood as an

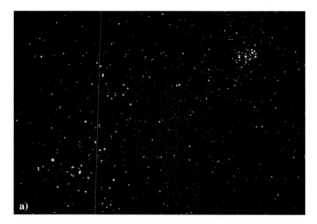

Fig. 16.1a–c. Open clusters. (**a**) The Hyades slightly to the lower left in the photograph. Above them to the right the Pleiades (Photograph M. Korpi). (**b**) The Pleiades photographed with the Metsähovi Schmidt camera. The diameter of the cluster is about 1°. Reflection nebulae are visible around some of the stars (Photograph M. Poutanen and H. Virtanen, Helsinki University). (**c**) Misan or h and χ Persei, the double cluster in Perseus. The separation between the clusters is about 25′. Picture taken with the Metsähovi 60-cm Ritchey Chrétien telescope (Photograph T. Markkanen, Helsinki University)

effect of perspective, if all cluster members have the same velocity vector with respect to the observer. Let θ be the angular distance of a given star from the convergence point. The angle between the velocity of the star and the line of sight will then also be θ. The velocity components along the line of sight and at right angles to it, v_r and v_t, are therefore given by

$$v_r = v \cos \theta \,,$$
$$v_t = v \sin \theta \,. \tag{16.1}$$

The radial velocity v_r can be measured from the Doppler shift of the stellar spectrum. The tangential velocity v_t is related to the proper motion μ and the distance r:

$$v_t = \mu r \,. \tag{16.2}$$

Thus the distance can be calculated:

$$r = \frac{v_t}{\mu} = \frac{v \sin \theta}{\mu} = \frac{v_r}{\mu} \tan \theta \,. \tag{16.3}$$

By means of this method, the distances of the individual stars can be determined from the motion of the cluster as a whole. Since the method of (ground-based) trigonometric parallaxes is reliable only out to a distance of 30 pc, the moving cluster method is an indispensable way of determining stellar distances. The distance of the Hyades obtained in this way is about 46 pc. The Hyades is the nearest open cluster.

The observed HR diagram or the corresponding colour–magnitude diagram of the Hyades and other nearby star clusters show a very well-defined and narrow main sequence (Fig. 16.5). Most of the cluster mem-

Fig. 16.2. The globular cluster ω Centauri. The picture was taken with the Danish 1.5-m telescope at La Silla, Chile. Thanks to the excellent seeing, one can see through the entire cluster in some places. (Photograph T. Korhonen, Turku University)

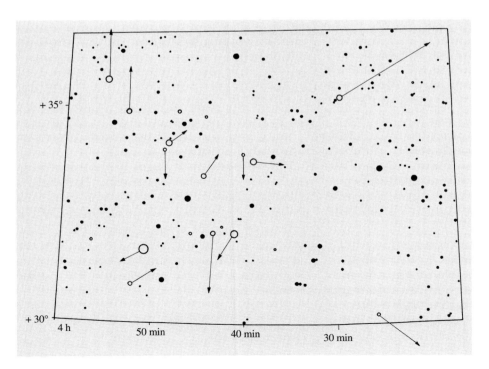

Fig. 16.3. ζ Persei association. O and B stars are shown as open circles. The proper motion vectors show the movements of the stars in the next 500,000 years

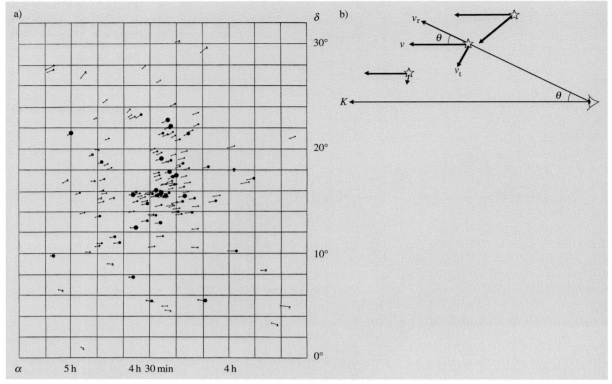

Fig. 16.4. (a) Proper motions of the Hyades. The vectors show the movement of the stars in about 10,000 years. (van Bueren, H. G. (1952): Bull. Astr. Inst. Neth. **11**). **(b)** If all stars move in the same direction, their tangential velocity components appear to be directed towards the convergence point K

bers are main sequence stars; there are only a few giants. There are quite a few stars slightly less than one magnitude above the main sequence. These are apparently binary stars whose components have not been resolved. To see this, let us consider a binary, where both components have the same magnitude m and the same colour index. If this system is unresolved, the colour index will still be the same, but the observed magnitude will be $m - 0.75$, i.e. slightly less than one magnitude brighter.

The main sequences of open clusters are generally located in the same section of the HR or colour-magnitude diagram (Fig. 16.6). This is because the material from which the clusters formed has not varied much, i.e. their initial chemical composition has been fairly constant. In younger clusters the main sequence extends to brighter and hotter stars and earlier spectral types. Usually one can clearly see the point in the diagram where the main

sequence ends and bends over towards the giant branch. This point will depend very strongly on the age of the cluster. It can therefore be used in determining the ages of open clusters. Star clusters are of central importance in the study of stellar evolution.

The colour–magnitude diagrams of star clusters can also be used to determine their distances. The method is called *main sequence fitting*. By means of multicolour photometry the reddening due to interstellar dust can be removed from the observed colours $B - V$ of the stars, yielding the intrinsic colours $(B - V)_0$. Most star clusters are so far away from us that all cluster members can be taken to be at the same distance. The distance modulus

$$m_{V_0} - M_V = 5 \lg \frac{r}{10 \, \text{pc}} \tag{16.4}$$

will then be the same for all members. In (16.4), m_{V_0} is the apparent, M_V the absolute visual magnitude of a star,

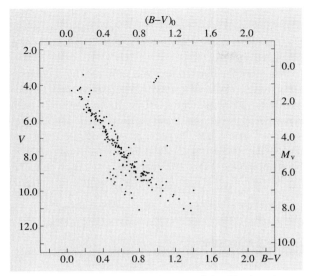

Fig. 16.5. Colour–magnitude diagram of the Hyades. Apparent visual magnitude on the left-hand vertical axis; absolute visual magnitude on the right-hand one

and r the distance. It has been assumed that the extinction due to interstellar dust A_V has been determined from multicolour photometry and its effect removed from the observed visual magnitude m_V:

$$m_{V_0} = m_V - A_V .$$

When the observed colour–magnitude diagram of the cluster is plotted using the apparent magnitude m_{V_0} rather than the absolute magnitude M_V on the vertical axis, the only change will be that the position of the main sequence is shifted vertically by an amount corresponding to the distance modulus. The observed (m_{V_0}, $(B-V)_0$) diagram may now be compared with the Hyades (M_V, $(B-V)_0$) diagram used as a standard. By demanding that the main sequences of the two diagrams agree, the distance modulus and hence the distance can be determined. The method is very accurate and efficient. It can be used to determine cluster distances out to many kiloparsecs.

16.3 Globular Star Clusters

Globular star clusters usually contain about 10^5 stars. The distribution of the stars is spherically symmetric, and the central densities are about ten times larger than in open clusters. Stars in globular clusters are among the oldest in the Milky Way, and therefore they are of great importance for studies of stellar evolution. There are about 150–200 globular clusters in the Milky Way.

The colour-magnitude diagram of a typical globular cluster is shown in Fig. 16.7. The main sequence only contains faint red stars; there is a prominent giant branch, and the horizontal and asymptotic branches are clearly seen. The main sequence is lower than that of the open clusters, because the metal abundance is much lower in the globular clusters.

The horizontal branch stars have a known absolute magnitude, which has been calibrated using principally RR Lyrae type variables. Because the horizontal branch stars are bright, they can be observed even in distant clusters, and thus using them the distances of globular clusters can be well determined.

Using the known distances, the linear sizes of globular clusters can be calculated. It is found that most of the mass is concentrated to a central core with a radius of about 0.3–10 pc. Outside this there is an extended en-

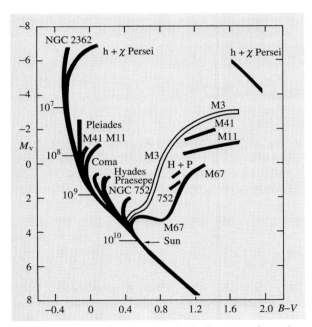

Fig. 16.6. Schematic colour–magnitude diagrams of star clusters. M3 is a globular cluster; the others are open clusters. Cluster ages are shown along the main sequence. The age of a cluster can be told from the point where its stars begin to turn off the main sequence. (Sandage, A. (1956): Publ. Astron. Soc. Pac. **68**, 498)

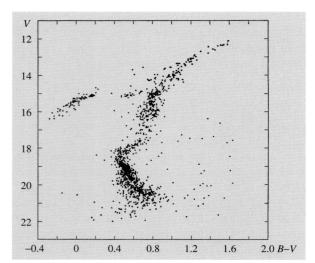

Fig. 16.7. Colour–magnitude diagram of the globular cluster M5. In addition to the main sequence one can see the giant branch bending to the right and to its left the horizontal brunch. (Arp, H. (1962): Astrophys. J. **135**, 311)

velope with a radius that may be 10–100 times larger. At even larger radii stars will escape from the cluster because of the tidal force of the Galaxy.

The masses of globular clusters can be roughly estimated from the virial theorem, if the stellar velocities in the cluster have been measured. More precise values are calculated by fitting theoretical models to the observed density and velocity distributions. In this way masses in the range 10^4–$10^6\,M_\odot$ have been obtained.

The globular clusters in the Milky Way fall into two classes. In the classification given in Table 17.1 these correspond to intermediate and halo population II. The disc globular clusters are concentrated towards the centre and the plane of the Milky Way and they form a system that is rotating with the general rotation of the Milky Way. In contrast, the halo clusters are almost spherically distributed in an extensive distribution reaching out to at least 35 kpc. The system of halo clusters does not rotate, but instead the velocities of individual clusters are uniformly distributed in all directions. The abundance of heavy elements is also different in the two classes of clusters. For disc clusters it is typically about 30% of the solar value, for halo clusters it is only about 1%. The smallest known heavy element abundances, about 10^{-3} times the solar value, have been

detected in some halo globular clusters. They therefore give important information about the production of elements in the early Universe and during the formation of the Milky Way.

All globular clusters are old, and the halo clusters are among the oldest known astronomical objects. Determining a precise age is difficult, and requires both accurate observations of the turn-off point of the main sequence in the HR diagram, as well detailed theoretical stellar evolution models. The ages obtained have been about $(13–16) \times 10^9$ years. This age is close to the age of the Universe calculated from its rate of expansion (see Chap. 19).

16.4 Example

Example 16.1 Assume that a globular cluster has a diameter of 40 pc and contains 100,000 stars of one solar mass each.

a) Use the virial theorem to find the average velocity of the stars. You can assume that the average distance between stars equals the radius of the cluster.

b) Find the escape velocity.

c) Comparing these velocities, can you tell something about the stability of the cluster?

a) First, we have to estimate the potential energy. There are $n(n-1)/2 \approx n^2/2$ pairs of stars in the cluster, and the average distance of each pair is R. Thus the potential energy is about

$$U = -G\frac{m^2}{R}\frac{n^2}{2} ,$$

where $m = 1\,M_\odot$. The kinetic energy is

$$T = \frac{1}{2}mv^2 n ,$$

where v is the root mean square velocity. According to the virial theorem we have $T = -1/2U$, whence

$$\frac{1}{2}mv^2 n = \frac{1}{2}G\frac{m^2}{R}\frac{n^2}{2} .$$

Solving for the velocity we get

$$v^2 = \frac{Gmn}{2R}$$

$$= \frac{6.7 \times 10^{-11}\,\mathrm{m^3\,kg^{-1}\,s^{-2}} \times 2.0 \times 10^{30}\,\mathrm{kg} \times 10^5}{40 \times 3.1 \times 10^{16}\,\mathrm{m}}$$

$$= 1.1 \times 10^7\,\mathrm{m^2\,s^{-2}}\,,$$

which gives $v \approx 3\,\mathrm{km\,s^{-1}}$.

b) The escape velocity from the edge of the cluster is

$$v_e = \sqrt{\frac{2Gmn}{R}}$$

$$= \sqrt{4v^2} = 2v = 6\,\mathrm{km\,s^{-1}}\,.$$

c) No. The average velocity seems to be smaller than the escape velocity, but it was derived from the virial theorem assuming that the cluster is stable.

16.5 Exercises

Exercise 16.1 A globular cluster consists of 100,000 stars of the solar absolute magnitude. Calculate the total apparent magnitude of the cluster, if its distance is 10 kpc.

Exercise 16.2 The Pleiades open cluster contains 230 stars within 4 pc. Estimate the velocities of the stars in the cluster using the virial theorem. For simplicity, let the mass of each star be replaced by $1\,M_\odot$.

17. The Milky Way

On clear, moonless nights a nebulous band of light can be seen stretching across the sky. This is the Milky Way (Fig. 17.1). The name is used both for the phenomenon in the sky and for the large stellar system causing it. The Milky Way system is also called the Galaxy — with a capital letter. The general term galaxy is used to refer to the countless stellar systems more or less like our Milky Way.

The band of the Milky Way extends round the whole celestial sphere. It is a huge system consisting mostly of stars, among them the Sun. The stars of the Milky Way form a flattened disc-like system. In the direction of the plane of the disc huge numbers of stars are visible, whereas relatively few are seen in the perpendicular direction. The faint light of distant stars merges into a uniform glow, and therefore the Milky Way appears as a nebulous band to the naked eye. A long-exposure photograph reveals hundreds of thousands of stars (Fig. 17.2).

In the early 17th century *Galileo Galilei*, using his first telescope, discovered that the Milky Way consists of innumerable stars. In the late 18th century *William Herschel* attempted to determine the size and shape of the Milky Way by means of star counts. Only early in the 20th century did the Dutch astronomer *Jacobus Kapteyn* obtain the first estimate of the size of the Milky Way. The true size of the Milky Way and the Sun's position in it became clear in the 1920's from *Harlow Shapley's* studies of the space distribution of globular clusters.

In studying the structure of the Milky Way, it is convenient to choose a spherical coordinate system so that the fundamental plane is the symmetry plane of the Milky Way. This is defined to be the symmetry plane of the distribution of neutral hydrogen, and it agrees quite closely with the symmetry plane defined by the distribution of stars in the solar neighbourhood (within a few kpc).

Fig. 17.1. The nebulous band of the Milky Way stretches across the entire sky. (Photograph M. and T. Kesküla, Lund Observatory)

The basic direction in the fundamental plane has been chosen to be the direction of the centre of the Milky Way. This is located in the constellation Sagittarius ($\alpha = 17\,\mathrm{h}\,45.7\,\mathrm{min}$, $\delta = -29°\,00'$, epoch 2000.0) at a distance of about 8.5 kpc. The galactic latitude is counted from the plane of the Galaxy to its pole, going from 0° to $+90°$, and to the galactic south pole, from 0° to $-90°$. The galactic coordinate system is shown in Fig. 17.3 (see also Sect. 2.8).

Fig. 17.2. A section of about 40° of the Milky Way between the constellations of Cygnus and Aquila. The brightest star at the upper right is Vega (α Lyrae). (Photograph Palomar Observatory)

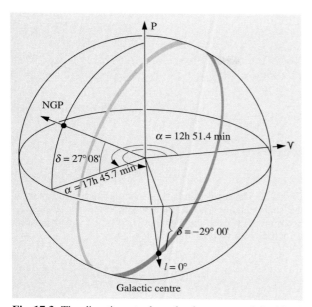

Fig. 17.3. The directions to the galactic centre and the North galactic pole (NGP) in equatorial coordinates. The galactic longitude l is measured from the galactic centre along the galactic plane

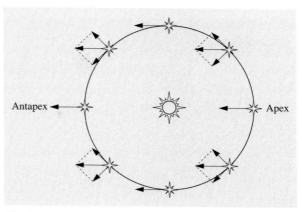

Fig. 17.4. Because of the motion of the Sun towards the apex, the average radial velocity of the nearby stars appears largest in the apex and antapex directions

17.1 Methods of Distance Measurement

In order to study the structure of the Milky Way, one needs to know how various kinds of objects, such as stars, star clusters and interstellar matter, are distributed in space. The most important ways of measuring the distances will first be considered.

Trigonometric Parallaxes. The method of *trigonometric parallaxes* is based on the apparent yearly back-and-forth movement of stars in the sky, caused by the orbital motion of the Earth. From Earth-based observations the trigonometric parallaxes can be reliably measured out to a distance of about 30 pc; beyond 100 pc this method is no longer useful. The situation is, however, changing. The limit has already been pushed to a few hundred parsecs by the Hipparcos satellite, and Gaia will mean another major leap in the accuracy.

The Motion of the Sun with Respect to the Neighbouring Stars. The Local Standard of Rest. The motion of the Sun with respect to the neighbouring

stars is reflected in their proper motions and radial velocities (Fig. 17.4). The point towards which the Sun's motion among the stars seems to be directed is called the *apex*. The opposite point is the *antapex*. The stars near the apex appear to be approaching; their (negative) radial velocities are smallest, on the average. In the direction of the antapex the largest (positive) radial velocities are observed. On the great circle perpendicular to the apex-antapex direction, the radial velocities are zero on the average, but the proper motions are large. The average proper motions decrease towards the apex and the antapex, but always point from the apex towards the antapex.

In order to study the true motions of the stars, one has to define a coordinate system with respect to which the motions are to be defined. The most practical frame of reference is defined so that the stars in the solar neighbourhood are at rest, on the average. More precisely, this *local standard of rest* (LSR) is defined as follows:

Let us suppose the velocities of the stars being considered are distributed at random. Their velocities with respect to the Sun, i.e. their radial velocities, proper motions and distances, are assumed to be known. The local standard of rest is then defined so that the mean value of the velocity vectors is opposite to the velocity of the Sun with respect to the LSR. Clearly the mean velocity of the relevant stars with respect to the LSR will then be zero. The motion of the Sun with respect to the LSR is found to be:

Apex coordinates	$\alpha = 18\,\mathrm{h}\,00\,\mathrm{min} = 270°$	$l = 56°$
	$\delta = +30°$	$b = +23°$
Solar velocity	$v_0 = 19.7\,\mathrm{kms}^{-1}$	

The apex is located in the constellation of Hercules. When the sample of stars used to determine the LSR is restricted to a subset of all the stars in the solar neighbourhood, e.g. to stars of a given spectral class, the sample will usually have slightly different kinematic properties, and the coordinates of the solar apex will change correspondingly.

The velocity of an individual star with respect to the local standard of rest is called the *peculiar motion* of the star. The peculiar velocity of a star is obtained by adding the velocity of the Sun with respect to the LSR to the measured velocity. Naturally the velocities should be treated as vectors.

The local standard of rest is at rest only with respect to a close neighbourhood of the Sun. The Sun and the nearby stars, and thus also the LSR, are moving round the centre of the Milky Way at a speed that is ten times greater than the typical peculiar velocities of stars in the solar neighbourhood (Fig. 17.5).

Statistical Parallaxes. The velocity of the Sun with respect to neighbouring stars is about $20\,\mathrm{km\,s}^{-1}$. This means that in one year, the Sun moves about 4 AU with respect to the stars.

Let us consider a star S (Fig. 17.6), whose angular distance from the apex is ϑ and which is at a distance r from the Sun. In a time interval t the star will move away from the apex at the angular velocity $u/t = \mu_\mathrm{A}$ because of the solar motion. In the same time interval, the Sun will move the distance s. The sine theorem for triangles yields

$$r \approx r' = \frac{s \sin \vartheta}{\sin u} \approx \frac{s \sin \vartheta}{u} , \tag{17.1}$$

because the distance remains nearly unchanged and the angle u is very small. In addition to the component μ_A due to solar motion, the observed proper motion has a component due to the peculiar velocity of the star. This can be removed by taking an average of (17.1) for a sample of stars, since the peculiar velocities of the stars in the solar neighbourhood can be assumed to be randomly distributed. By observing the average proper motion of objects known to be at the same distance one

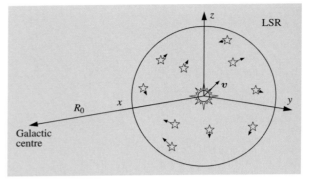

Fig. 17.5. The local standard of rest, defined by the stars in the solar neighbourhood, moves with respect to the galactic centre. However, the average value of the stellar peculiar velocities with respect to the LSR is zero

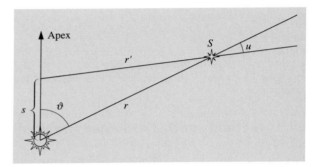

Fig. 17.6. When the Sun has moved the distance s towards the apex, the direction to the star S appears to have changed by the angle u

thus obtains their actual distance. A similar statistical method can be applied to radial velocities.

Objects that are at the same distance can be found as follows. We know that the distance modulus $m - M$ and the distance r are related according to:

$$m - M = 5 \lg(r/10\,\mathrm{pc}) + A(r) , \tag{17.2}$$

where A is the interstellar extinction. Thus objects that have the same apparent and the same absolute magnitude will be at the same distance. It should be noted that we need not know the absolute magnitude as long as it is the same for all stars in the sample. Suitable classes of stars are main sequence A4 stars, RR Lyrae variables and classical cepheids with some given period. The stars in a cluster are also all at the same distance. This method has been used, for example, to determine the distance to the Hyades as explained in Sect. 16.2.

Parallaxes based on the peculiar or apex motion of the Sun are called statistical or secular parallaxes.

Main Sequence Fitting. If the distance of a cluster is known, it is possible to plot its HR diagram with the absolute magnitude as the vertical coordinate. Another cluster, whose distance is to be determined, can then be plotted in the same diagram using the apparent magnitudes as the vertical coordinate. Now the vertical distance of the main sequences tells how much the apparent magnitudes differ from the absolute ones. Thus the distance modulus $m - M$ can be measured. This method, known as the *main sequence fitting*, works for clusters whose stars are roughly at the same distance; if the distances vary too much, a clear main sequence cannot be distinguished.

Photometric Parallaxes. The determination of the distance directly from (17.2) is called the photometric method of distance determination and the corresponding parallax, the *photometric parallax*. The most difficult task when using this method usually involves finding the absolute magnitude; there are many ways of doing this. For example, the two-dimensional MKK spectral classification allows one to determine the absolute magnitude from the spectrum. The absolute magnitudes of cepheids can be obtained from their periods. A specially useful method for star clusters is the procedure of main sequence fitting. A condition for the photometric method is that the absolute magnitude scale first be calibrated by some other method.

Trigonometric parallaxes do not reach very far. For example, even with the Hipparcos satellite, only a few cepheid distances have been accurately measured by this method. The method of statistical parallaxes is indispensable for calibrating the absolute magnitudes of bright objects. When this has been done, the photometric method can be used to obtain distances of objects even further away.

Other examples of indicators of brightness, luminosity criteria, are characteristic spectral lines or the periods of cepheids. Again, their use requires that they first be calibrated by means of some other method. It is a characteristic feature of astronomical distance determinations that the measurement of large distances is based on knowledge of the distances to nearer objects.

17.2 Stellar Statistics

The Stellar Luminosity Function. By systematically observing all stars in the solar neighbourhood, one can find the distribution of their absolute magnitudes. This is given by the luminosity function $\Phi(M)$, which gives the relative number of main sequence stars with absolute magnitudes in the range $[M - 1/2, M + 1/2]$. No stars appear to be forming at present in the region of space where the luminosity function has been determined. The age of the Milky Way is 10–15 Ga, which means that all stars less massive than $0.9\,M_\odot$, will still be on the main sequence. On the other hand, more massive stars, formed early in the history of the Milky Way, will have completed their evolution and disappeared. Low-mass stars have accumulated in the luminosity function for many generations of star formation, whereas bright, high-mass stars are the result of recent star formation.

By taking into account the different main sequence lifetimes of stars of different masses and hence of different magnitudes, one can determine the initial luminosity function $\Psi(M)$, which gives the brightness distribution at the time of star formation, the zero age main sequence luminosity function. The relation between the function Ψ and the observed luminosity function is

$$\Psi(M) = \Phi(M)T_0/t_E(M) , \qquad (17.3)$$

where T_0 is the age of the Milky Way and $t_E(M)$ is the main sequence lifetime of stars of magnitude M. Here we assume that the birth rate of stars of magnitude M has remained constant during the lifetime of the Milky Way. The initial luminosity function is shown in Fig. 17.7.

The Fundamental Equation of Stellar Statistics. The Stellar Density. A crucial problem for studies of the structure of the Milky Way is to find out how the density of stars varies in space. The number of stars per unit volume at a distance r in the direction (l, b) from the Sun is given by the stellar density $D = D(r, l, b)$.

The stellar density cannot be directly observed except in the immediate neighbourhood of the Sun. However, it

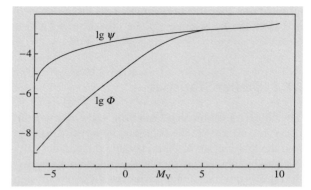

Fig. 17.7. The observed luminosity function $\Phi(M_V)$ and the initial luminosity function $\Psi(M_V)$ for main sequence stars in the solar neighbourhood. The functions give the number of stars per cubic parsec in the magnitude interval $[M_V - 1/2, M_V + 1/2]$; they are actually the products $D\Phi$ and $D\Psi$, where D is the stellar density function (in the solar neighbourhood)

can be calculated if one knows the luminosity function and the interstellar extinction as a function of distance in a given direction. In addition the number of stars per unit solid angle (e.g. per square arc second) can be determined as a function of limiting apparent magnitude by means of star counts (Fig. 17.8).

Let us consider the stars within the solid angle ω in the direction (l, b) and in the distance range $[r, r + dr]$. We let their luminosity function $\Phi(M)$ be the same as in the solar neighbourhood and their unknown stellar density D. The absolute magnitude M of the stars of apparent magnitude m is, as usual,

$$M = m - 5\lg(r/10\,\text{pc}) - A(r) .$$

The number of stars in the apparent magnitude interval $[m - 0.5, m + 0.5]$ in the volume element $dV = \omega r^2 dr$ at distance r is (Fig. 17.9)

$$dN(m) = D(r, l, b)$$
$$\times \Phi\left[m - 5\lg\frac{r}{10\,\text{pc}} - A(r)\right]dV . \tag{17.4}$$

The stars of apparent magnitude m in the given area of the sky will in reality be at many different distances. In order to obtain their total number $N(m)$, one has to

Fig. 17.8. The stellar density is determined by means of star counts. In practice, the counting is done on photographic plates. (Cartoon S. Harris)

integrate $dN(m)$ over all distances r:

$$N(m) = \int_0^\infty D(r, l, b)$$
$$\times \Phi\left[m - 5\lg\frac{r}{10\,\text{pc}} - A(r)\right]\omega r^2\,dr . \tag{17.5}$$

Equation (17.5) is called the *fundamental equation of stellar statistics*. Its left-hand side, the number of stars in the apparent magnitude interval $[m - 0.5, m + 0.5]$ in the solid angle ω, is obtained from the observations: one counts the stars of different magnitudes in a chosen area of a photographic plate. The luminosity function is known from the solar neighbourhood. The extinction $A(r)$ can be determined for the chosen areas, for instance, by means of multicolour photometry. In order to solve the integral equation (17.5) for $D(r, l, b)$, several methods have been developed, but we shall not go into them here.

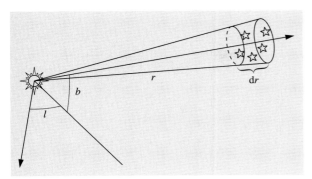

Fig. 17.9. The size of the volume element at distance r in the direction (l, b) is $\omega r^2 \mathrm{d}r$

Figure 17.10a shows the stellar density in the solar neighbourhood in the plane of the Milky Way, and Fig. 17.10b in the direction perpendicular to the plane. There are several individual concentrations, but e. g. spiral structure cannot be observed in such a limited region of space.

The Distribution of Bright Objects. Using stellar statistical methods, one can only study the close neighbourhood of the Sun, out to about 1 kpc at the most. Absolutely faint objects cannot be observed very far. Since the solar neighbourhood appears to be fairly representative of the general properties of the Milky Way, its study is naturally important, giving information e. g.

on the distributions and luminosity functions of stars of various spectral types. However, in order to get an idea of the larger-scale structure of the Milky Way, one has to make use of objects that are as absolutely bright as possible, and which can be observed even at large distances.

Examples of suitable objects are stars of early spectral types, H II regions, OB associations, open star clusters, cepheids, RR Lyrae stars, supergiants and giants of late spectral types, and globular clusters. Some of these objects differ greatly in age, such as the young OB associations, on the one hand, and the old globular clusters, on the other. Any differences in their space distribution tell us about changes in the general structure of the Milky Way.

The young optical objects, the H II regions, OB associations and open clusters, are strongly concentrated in the plane of the Milky Way (Table 17.1). Figure 17.11 shows that they also appear to be concentrated in three drawn-out bands, at least within the observed region. Since these types of objects in other galaxies are known to be part of a spiral structure, the observed bands in the Milky Way have been interpreted as portions of three spiral arms passing through the solar neighbourhood. Stars of later spectral types seem to be much more evenly distributed. Apart from a few special directions, interstellar dust limits observations in the galactic plane to within 3–4 kpc.

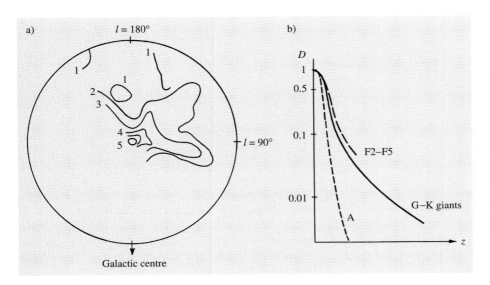

Fig. 17.10a,b. The stellar density near the Sun. (**a**) The stellar density of spectral classes A2–A5 in the galactic plane, according to S.W. McCuskey. The numbers next to the isodensity curves give the number of stars in $10,000\,\mathrm{pc}^3$. (**b**) The distribution of different spectral classes perpendicularly to the galactic plane according to T. Elvius. The density in the galactic plane has been normalized to one

Table 17.1. Populations of the Milky Way; z is the vertical distance from the galactic plane, and v_r the velocity component perpendicular to the galactic plane

Population	Typical objects	Average age $[10^9 \text{ a}]$	z [pc]	v_r [k/s]	Metal abundance
Halo population II	Subdwarfs, globular clusters RR Lyr ($P > 0.4$ d)	17–12	2000	75	0.001
Intermediate population II	Long period variables	15–10	700	25	0.005
Disc population	Planetary nebulae, novae bright red giants	12–2	400	18	0.01–0.02
Old population I	A stars, Me dwarfs classical cepheids	2–0.1	160	10	0.02
Young population I	Gas, dust, supergiants, T Tau stars	0.1	120		0.03–0.04

Fig. 17.11. The distribution of various objects in the galactic plane. Three condensations can be discerned: the Sagittarius arm (*lowest*), the local arm near the Sun and (*outermost*) the Perseus arm

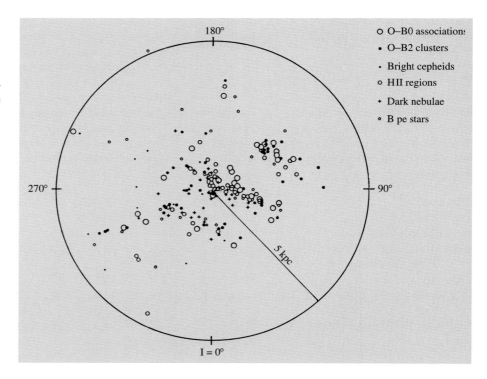

○ O–B0 associations
● O–B2 clusters
· Bright cepheids
○ HII regions
✦ Dark nebulae
○ B pe stars

Old objects, particularly the globular clusters, have an almost spherical distribution about the centre of the Milky Way (Fig. 17.12). The space density of old objects increases towards the galactic centre. They can be used to determine the distance of the Sun from the galactic centre; the value of this distance is about 8.5 kpc.

Stellar Populations. Studies of the motions of the stars in the Milky Way have revealed that the orbits of stars moving in the galactic plane are almost circular. These stars are also usually young, a few hundred million years at the most. They also contain a relatively large amount of heavy elements, about 2–4%. The interstel-

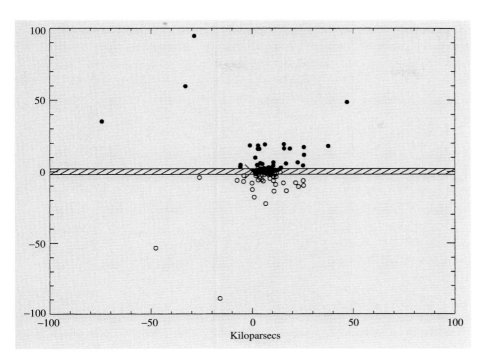

Fig. 17.12. The distribution of globular clusters. (From S.R. Majewski, *Stellar populations and the Milky Way*, in C. Martínez Roger, I. Perez Fournón, F. Sánchez (Eds.) *Globular Clusters*, Cambridge University Press, 1999)

lar material similarly moves in the galactic plane in almost circular orbits. On the basis of their motions and their chemical composition, the interstellar medium and the youngest stars are collectively referred to as *population I*.

Outside the plane of the Milky Way, an almost spherically symmetric halo extends out to 50 kpc and beyond. The stellar density is largest near the galactic centre and decreases outwards. The halo contains very little interstellar matter, and its stars are old, perhaps up to 15×10^9 years. These stars are also very metal-poor. Their orbits may be very eccentric and show no preference for the galactic plane. On the basis of these criteria, one defines stars of *population II*. Typical population II objects are the globular clusters, and the RR Lyrae and W Virginis stars.

The stars of population II have large velocities with respect to the local standard of rest, up to more than $300 \, \mathrm{km \, s^{-1}}$. In reality their velocities at the solar distance from the galactic centre are quite small and may sometimes be opposite to the direction of motion of the LSR. The large relative velocities only reflect the motion of the LSR with a velocity of about $220 \, \mathrm{km \, s^{-1}}$ round the galactic centre.

Between these two extremes, there is a sequence of intermediate populations. In addition to populations I and II, one generally also speaks of a disc population, including the Sun, for instance. The typical motions, chemical composition and age of the various populations (Table 17.1) contain information about the evolution of our Galaxy and about the formation of its stars.

17.3 The Rotation of the Milky Way

Differential Rotation. Oort's Formulas. The flatness of the Milky Way is already suggestive of a general rotation about an axis normal to the galactic plane. Observations of the motions both of stars and of interstellar gas have confirmed this rotation and shown it to be differential. This means that the angular velocity of rotation depends on the distance from the galactic centre (Fig. 17.13). Thus the Milky Way does not rotate like a rigid body. Near the Sun, the rotational velocity decreases with radius.

The observable effects of the galactic rotation were derived by the Dutch astronomer *Jan H. Oort*. Let us suppose the stars are moving in circular orbits about

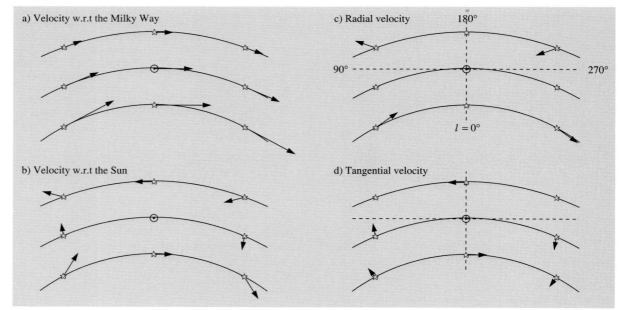

Fig. 17.13a–d. The effect of differential rotation on the radial velocities and proper motions of stars. (**a**) Near the Sun the orbital velocities of stars decrease outwards in the Galaxy. (**b**) The relative velocity with respect to the Sun is obtained by subtracting the solar velocity from the velocity vectors in (**a**). (**c**) The radial components of the velocities with respect to the Sun. This component vanishes for stars on the same orbit as the Sun. (**d**) The tangential components of the velocities

the galactic centre (Fig. 17.14). This approximation is acceptable for population I stars and gas. The star S, seen from the Sun $\odot$ at galactic longitude l at distance r, has circular velocity V at a distance R from the centre. Similarly for the Sun the galactic radius and velocity are R_0 and V_0. The relative radial velocity v_r of the star with respect to the Sun is the difference between the projections of the circular velocities on the line of sight:

$$v_r = V \cos \alpha - V_0 \sin l , \qquad (17.6)$$

where α is the angle between the velocity vector of the star and the line of sight. From Fig. 17.14 the angle $CS\odot = \alpha + 90°$. By applying the sine theorem to the triangle $CS\odot$ one obtains

$$\frac{\sin(\alpha + 90°)}{\sin l} = \frac{R_0}{R}$$

or

$$\cos \alpha = \frac{R_0}{R} \sin l . \qquad (17.7)$$

Denoting the angular velocity of the star by $\omega = V/R$ and that of the Sun by $\omega_0 = V_0/R_0$, one obtains the observable radial velocity in the form

$$v_r = R_0(\omega - \omega_0) \sin l . \qquad (17.8)$$

The tangential component of the relative velocity of the Sun and the star is obtained as follows. From Fig. 17.14,

$$v_t = V \sin \alpha - V_0 \cos l = R\omega \sin \alpha - R_0\omega_0 \cos l .$$

The triangle $\odot CP$ gives

$$R \sin \alpha = R_0 \cos l - r ,$$

and hence

$$v_t = R_0(\omega - \omega_0) \cos l - \omega r . \qquad (17.9)$$

Oort noted that in the close neighbourhood of the Sun ($r \ll R_0$), the difference of the angular velocities will be very small. Therefore a good approximation for the exact equations (17.8) and (17.9) is obtained by keeping only the first term of the Taylor series of $\omega - \omega_0$ in the neighbourhood of $R = R_0$:

$$\omega - \omega_0 = \left(\frac{d\omega}{dR}\right)_{R=R_0} (R - R_0) + \dots .$$

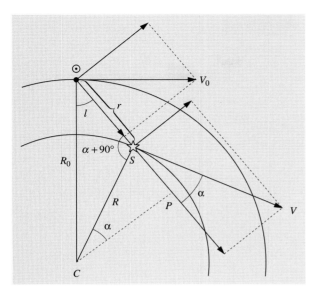

Fig. 17.14. In order to derive Oort's formulas, the velocity vectors of the Sun and the star S are divided into components along the line $\odot S$ and normal to it

Using $\omega = V/R$ and $V(R_0) = V_0$, one finds

$$\omega - \omega_0 \approx \frac{1}{R_0^2}\left[R_0\left(\frac{\mathrm{d}V}{\mathrm{d}R}\right)_{R=R_0} - V_0\right](R - R_0) \, .$$

For $R \approx R_0 \gg r$, the difference $R - R_0 \approx -r\cos l$. One thus obtains an approximate form

$$v_{\mathrm{r}} \approx \left[\frac{V_0}{R_0} - \left(\frac{\mathrm{d}V}{\mathrm{d}R}\right)_{R=R_0}\right] r\cos l \sin l$$

or

$$v_{\mathrm{r}} \approx A r \sin 2l \, , \tag{17.10}$$

where A is a characteristic parameter of the solar neighbourhood of the Galaxy, the *first Oort constant*:

$$A = \frac{1}{2}\left[\frac{V_0}{R_0} - \left(\frac{\mathrm{d}V}{\mathrm{d}R}\right)_{R=R_0}\right] \, . \tag{17.11}$$

For the tangential relative velocity, one similarly obtains, since $\omega r \approx \omega_0 r$:

$$v_{\mathrm{t}} \approx \left[\frac{V_0}{R_0} - \left(\frac{\mathrm{d}V}{\mathrm{d}R}\right)_{R=R_0}\right] r\cos^2 l - \omega_0 r \, .$$

Because $2\cos^2 l = 1 + \cos 2l$, this may be written

$$v_{\mathrm{t}} \approx A r \cos 2l + B r \, , \tag{17.12}$$

where A is the same as before and B, the *second Oort constant*, is

$$B = -\frac{1}{2}\left[\frac{V_0}{R_0} + \left(\frac{\mathrm{d}V}{\mathrm{d}R}\right)_{R=R_0}\right] \, . \tag{17.13}$$

The proper motion $\mu = v_{\mathrm{t}}/r$ is then given by the expression

$$\mu \approx A\cos 2l + B \, . \tag{17.14}$$

Equation (17.10) says that the observed radial velocities of stars at the same distance should be a double sine curve as a function of galactic longitude. This has been confirmed by observations (Fig. 17.15a). If the distance to the stars involved is known, the amplitude of the curve determines the value of the Oort constant A.

Independently of distance, the proper motions of the stars form a double sine wave as a function of galactic

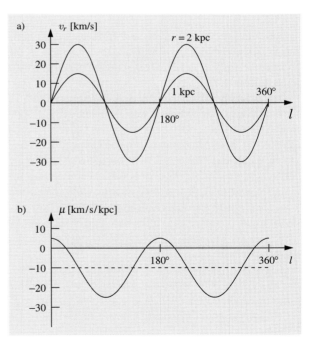

Fig. 17.15a,b. The velocity components due to differential rotation according to Oort's formulas as functions of galactic longitude. (**a**) Radial velocities for objects at a distance of 1 and 2 kpc. (Compare with Fig. 17.13.) Strictly, the longitude at which the radial velocity vanishes depends on the distance. Oort's formulas are valid only in the close vicinity of the Sun. (**b**) Proper motions

longitude, as seen in Fig. 17.15b. The amplitude of the curve is A and its mean value, B.

In 1927 on the basis of this kind of analysis, Oort established that the observed motions of the stars indicated a differential rotation of the Milky Way. Taking into account an extensive set of observational data, the International Astronomical Union IAU has confirmed the present recommended values for the Oort constants:

$$A = 15 \,\mathrm{km\,s^{-1}\,kpc^{-1}}, \quad B = -10 \,\mathrm{km\,s^{-1}\,kpc^{-1}} .$$

The Oort constants obey some interesting relations. By subtracting (17.13) from (17.11), one obtains

$$A - B = \frac{V_0}{R_0} = \omega_0 . \tag{17.15}$$

Adding (17.13) and (17.11) gives

$$A + B = -\left(\frac{\mathrm{d}V}{\mathrm{d}R}\right)_{R=R_0} . \tag{17.16}$$

Knowing the values of A and B, one can calculate the angular velocity $\omega_0 = 0.0053''/\text{year}$, which is the angular velocity of the local standard of rest around the galactic centre.

The circular velocity of the Sun and the LSR can be measured in an independent way by using extragalactic objects as a reference. In this way a value of about $220 \,\mathrm{km\,s^{-1}}$ has been obtained for V_0. Using (17.15) one can now calculate the distance of the galactic centre R_0. The result is about 8.5 kpc, in good agreement with the distance to the centre of the globular cluster system. The direction to the galactic centre obtained from the distribution of radial velocities and proper motions by means of (17.10) and (17.14) also agrees with other measurements.

The orbital period of the Sun in the Galaxy according to these results is about 2.5×10^8 years. Since the Sun's age is nearly 5×10^9 years, it has made about 20 revolutions around the galactic centre. At the end of the previous revolution, the Carboniferous period had ended on Earth and the first mammals would soon appear.

The Distribution of Interstellar Matter. Radio radiation from interstellar gas, in particular that of neutral hydrogen, is not strongly absorbed or scattered by interstellar dust. It can therefore be used to map the structure

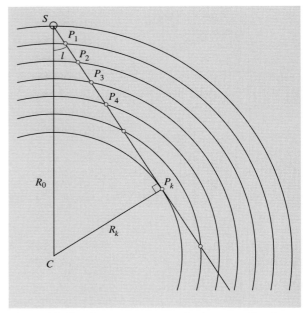

Fig. 17.16. Clouds P_1, P_2, ... seen in the same direction at various distances

of the Milky Way on large scales. Radio signals can be detected even from the opposite edge of the Milky Way.

The position of a radio source, for example an H I cloud, in the Galaxy cannot be directly determined. However, an indirect method exists, based on the differential rotation of the Galaxy.

Figure 17.16 is a schematic view of a situation in which gas clouds on the circles P_1, P_2, ... are observed in the direction l ($-90° < l < 90°$). The angular velocity increases inwards, and therefore the greatest angular velocity along the line of sight is obtained at the point P_k, where the line of sight is tangent to a circle. This means that the radial velocity of the clouds in a fixed direction grows with distance up to the maximum velocity at cloud P_k:

$$v_{\mathrm{r,max}} = R_k(\omega - \omega_0) , \tag{17.17}$$

where $R_k = R_0 \sin l$. The distance of cloud P_k from the Sun is $r = R_0 \cos l$. When r increases further, v_r decreases monotonically. Figure 17.17 shows how the observed radial velocity in a given direction varies with distance r, if the gas moves in circular orbits and the angular velocity decreases outwards.

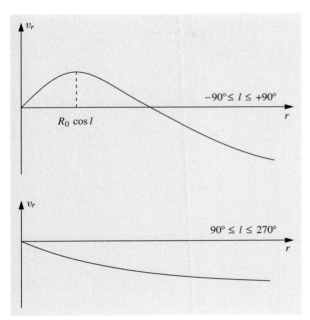

Fig. 17.17. The radial velocity as a function of distance (shown schematically)

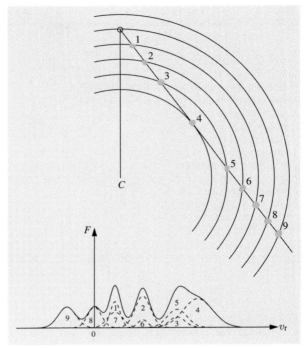

Fig. 17.18. Clouds at different distances have different velocities and therefore give rise to emission lines with different Doppler shifts. The observed flux density profile (*continuous curve*) is the sum of the line profiles of all the individual line profiles (*dashed curves*). The numbers of the line profiles correspond to the clouds in the upper picture

The neutral hydrogen 21 cm line has been particularly important for mapping the Milky Way. Figure 17.18 gives a schematic view of how the hydrogen spectral line is made up of the radiation of many individual concentrations of neutral hydrogen, clouds or spiral arms. The line component produced by each cloud has a wavelength which depends on the radial velocity of the cloud and a strength depending on its mass and density. The total emission is the sum of these contributions.

By making observations at various galactic longitudes and assuming that the clouds form at least partly continuous spiral arms, the distribution of neutral hydrogen in the galactic plane can be mapped. Figure 15.17 shows a map of the Milky Way obtained from 21 cm line observations of neutral hydrogen. It appears that the neutral hydrogen is concentrated in spiral arms. However, interpretation of the details is difficult because of the uncertainties of the map. In order to obtain the distances to the gas clouds, one has to know the *rotation curve*, the circular velocity as a function of the galactic radius. This is determined from the same radial velocity observations and involves assumptions concerning the density and rotation of the gas. The interpretation of

the spiral structure obtained from radio observations is also still uncertain. For example, it is difficult to fit the radio spiral structure to the one obtained near the Sun from optical observations of young stars and associations.

The Rotation, Mass Distribution and Total Mass of the Milky Way. In (17.17) the galactic longitude l gives the galactic radius R_k of the clouds with maximum radial velocity. By making observations at different longitudes, one can therefore use (17.17) to determine the angular velocity of the gas for various distances from the galactic centre. (Circular motions must be assumed.) In this way, the rotation curve $\omega = \omega(R)$ and the corresponding velocity curve $V = V(R) (= \omega R)$ are obtained.

Figure 17.19 shows the rotation curve of the Milky Way. Its central part rotates like a rigid body, i.e. the

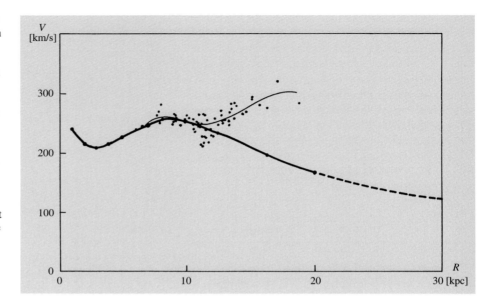

Fig. 17.19. Rotation curve of the Milky Way based on the motions of hydrogen clouds. Each point represents one cloud. The thick line represents the rotation curve determined by Maarten Schmidt in 1965. If all mass were concentrated within the radius 20 kpc, the curve would continue according to Kepler's third law (*broken line*). The rotation curve determined by Leo Blitz on the basis of more recent observations begins to rise again at 12 kpc

angular velocity is independent of the radius. Outside this region, the velocity first drops and then begins to rise gradually. A maximum velocity is reached at about 8 kpc from the centre. Near the Sun, about 8.5 kpc from the centre, the rotational velocity is about $220 \, \text{km s}^{-1}$. According to earlier opinions, the velocity continues to decrease outwards. This would mean that most of the mass is inside the solar radius. This mass could then be determined from Kepler's third law. According to (6.34),

$$M = R_0 V_0^2 / G \,.$$

Using the values $R_0 = 8.5 \, \text{kpc}$ and $V_0 = 220 \, \text{km s}^{-1}$, one obtains

$$M = 1.9 \times 10^{41} \, \text{kg} = 1.0 \times 10^{11} \, M_\odot \,.$$

The escape velocity at radius R is

$$V_e = \sqrt{\frac{2GM}{R}} = V\sqrt{2} \,. \tag{17.18}$$

This gives an escape velocity near the Sun $V_e = 310 \, \text{km s}^{-1}$. One therefore should not see many stars moving in the direction of galactic rotation, $l = 90°$, with velocities larger than $90 \, \text{km s}^{-1}$ with respect to the local standard of rest, since the velocity of

such stars would exceed the escape velocity. This has been confirmed by observations.

The preceding considerations have been based on the assumption that near the Sun, the whole mass of the Galaxy can be taken to be concentrated in a central point. If this were true, the rotation curve should be of the Keplerian form, $V \propto R^{-1/2}$. That this is not the case can be established from the values of the Oort constants.

The derivative of the Keplerian relation

$$V = \sqrt{\frac{GM}{R}} = \sqrt{GM} \, R^{-1/2}$$

yields

$$\frac{dV}{dR} = -\frac{1}{2}\sqrt{GM} \, R^{-3/2} = -\frac{1}{2}\frac{V}{R} \,.$$

Using the properties (17.15) and (17.16) of the Oort constants, one finds

$$(A - B)/(A + B) = 2 \tag{17.19}$$

for a Keplerian rotation curve. This disagrees with the observed value and thus the assumed Keplerian law does not apply.

The mass distribution in the Milky Way can be studied on the basis of the rotation curve. One looks for a suitable mass distribution, such that the correct rotation curve is reproduced. Recently distant globular clusters have been discovered, showing that the Milky Way is larger than expected. Also, observations of the rotation curve outside the solar circle suggest that the rotational velocity might begin to rise again. These results suggest that the mass of the Galaxy might be as much as ten times larger than had been thought.

17.4 The Structure and Evolution of the Milky Way

Spiral Structure. As mentioned earlier, the Milky Way appears to be a *spiral galaxy*, but there is no general agreement on the detailed form of the spiral pattern. For example, in 1976 *Y.M. Georgelin* and *Y.P. Georgelin* determined the distances of H II regions by radio and optical observations. In the optical region their method is independent of assumptions about the galactic rotation law. They then fitted four spiral arms through the H II regions (Fig. 17.20).

The cause of the spiral structure is a long-standing problem. A small perturbation in the disc will quickly be stretched into a spiral shape by differential rotation. However, such a spiral would disappear in a few galactic revolutions, a few hundred million years.

An important step forward in the study of the spiral structure was the *density wave theory* developed by *C. C. Lin* and *Frank H. Shu* in the 1960's. The spiral structure is taken to be a wavelike variation of the density of the disc. The spiral pattern rotates as a solid body with an angular velocity smaller than that of the galactic rotation, while the stars and gas in the disc pass through the wave.

The density wave theory explains in a natural way why young objects, like molecular clouds, H II regions and bright young stars are found in the spiral arms. As gas passes through the wave, it is strongly compressed. The internal gravity of the gas clouds then becomes more important and causes them to collapse and form stars.

It takes about 10^7 years for the material to pass through a spiral arm. By that time, the hot, bright stars have finished their evolution, their ultraviolet radiation has ceased and the H II regions have disappeared. The less massive stars formed in the spiral arms are spread out in the disc by their peculiar velocities.

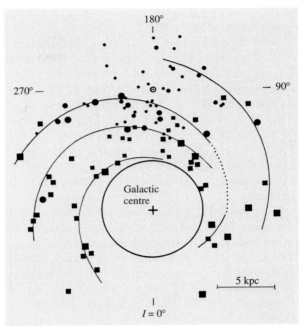

Fig. 17.20. The positions of H II regions according to optical (*circles*) and radio (*squares*) observations. Four spiral arms have been fitted to the data

It is not yet clear what gives rise to the spiral wave. For some further remarks, see Sect. 18.3.

The Galactic Centre. Our knowledge of the centre of the Milky Way is mostly based on radio and infrared observations. In the optical region the view to the centre is blocked by the dark clouds in the Sagittarius spiral arm about 2 kpc from us. The galactic centre is interesting because it may be a small-scale version of the much more violently active nuclei of some external galaxies (see Sect. 18.7). It therefore provides opportunities to study at close hand phenomena related to active galaxies. Since active galactic nuclei are thought to contain black holes with masses larger than $10^7 M_\odot$, there may also be a large black hole at the galactic centre.

As one approaches the galactic centre the stellar density continues to rise towards a sharp central peak. In contrast, the galactic gas disc has a central hole of radius about 3 kpc. According to one model the central bulge of the Milky Way is bar-shaped. The effect of this bar is to channel gas into the galactic nucleus leaving a gas-free zone at larger radii.

Inside the central hole is a dense nuclear gas disc. Its radius is about 1.5 kpc in neutral hydrogen, but most of its mass is molecular and concentrated within 300 pc of the nucleus. In this region the mass of molecular gas is about $10^8\,M_\odot$, or 5% of the total molecular mass of the Milky Way. The molecular clouds are probably confined by the pressure from surrounding very hot ($T \approx 10^8$ K) gas. This hot gas may then expand vertically, forming a galactic wind. Gas lost to a wind or to star formation is replenished with infalling gas from larger radii.

The central 10 pc are dominated by the radio continuum source Sgr A and a dense star cluster observed in the infrared. There is also molecular gas with complex motions and signs of star formation activity. Within Sgr A there is a unique point-like radio continuum source known as Sgr A*. The position of Sgr A* agrees to within $1''$ with the centre of a cluster of stars that is much denser than anything observed in the galactic disc. If the galactic centre contains a large black hole, Sgr A* is the natural candidate.

The luminosity of the galactic centre could be provided by the central star cluster, but there are still reasons to expect that a large black hole may be present. The central mass distribution can be estimated by modelling the observed motions of stars and gas (cf. Sect. 18.2). The best fit with the observations is obtained with models with an extended stellar mass distribution together with a point-mass of a few times $10^6\,M_\odot$. The size of Sgr A* measured by very long baseline interferometry is less than 10 AU. The most plausible explanation of this very compact structure is that Sgr A* is indeed a black hole with mass of a few million solar masses.

17.5 Examples

Example 17.1 Show that if the stars are uniformly distributed in space and there is no interstellar extinction, the number of stars brighter than apparent magnitude m is

$$N_0(m) = N_0(0) \times 10^{0.6m} .$$

Let us suppose first that all stars have the same absolute magnitude M. The distance in parsecs of those stars having the apparent magnitude m is

$$r = 10 \times 10^{0.2(m-M)} .$$

In order to appear brighter than m, a star has to be within a sphere of radius r. Because the stellar density is constant, the number of such stars is proportional to the volume:

$$N_0(m) \propto r^3 \propto 10^{0.6m} .$$

The result does not depend on the absolute magnitudes of the stars, so that the same result still applies when the magnitudes are not equal, as long as the luminosity function does not depend on distance. Thus the equation is generally valid under the stated conditions.

Example 17.2 *The Estimation of Distances by Means of Oort's Formulas*

An object in the galactic plane at longitude $l = 45°$ has the radial velocity of 30 km s^{-1} with respect to the LSR. What is its distance?

According to (17.10),

$$v_r = Ar \sin 2l .$$

Thus

$$r = \frac{v_r}{A \sin 2l} = \frac{30\ \mathrm{km\,s^{-1}}}{15\ \mathrm{km\,s^{-1}\,kpc^{-1}}} = 2\ \mathrm{kpc} .$$

In practice, the peculiar velocities are so large that this method cannot be used for distance determination. Oort's formulas are mostly suitable only for statistical studies.

Example 17.3 *Discussion of the Gravitational Field of a Uniform Disc*

It can be shown that the gravitational field of a homogeneous infinite thin disc is constant and directed towards the plane of the disc. If the mass per unit area of the disc is σ, the gravitational field is

$$g = 2\pi G\sigma .$$

A test particle located outside the plane will therefore get a constant acceleration towards the plane. Taking a numerical example, assume a mass of $10^{11}\,M_\odot$ distributed uniformly on a circular disc 20 kpc in diameter.

The mass per unit area will be a

$$\sigma = \frac{10^{11} \times 2 \times 10^{30} \text{ kg}}{\pi (10^4 \times 3.086 \times 10^{16} \text{ m})^2}$$

$$= 0.67 \text{ kg m}^{-2} .$$

The corresponding gravitational field is

$$g = 2.8 \times 10^{-10} \text{ m s}^{-2} .$$

Let a star be located at $d = 1$ kpc above the plane, initially at rest (not very near the edge of the disk, in order to keep our approximation valid). The disc will pull it towards the plane, and when the star crosses the plane, it has acquired a velocity given by

$$v = \sqrt{2gd} = 130 \text{ km s}^{-1} .$$

The time required to reach the plane is

$$t = v/g = 15 \times 10^6 \text{ a} .$$

17.6 Exercises

Exercise 17.1 Assume that the Sun and a star move around the Galaxy at the same speed in the same circular orbit in the galactic plane. Show that the proper motion of the star is independent of its distance. How big is this proper motion?

Exercise 17.2 a) A cepheid has a radial velocity of 80 km s^{-1}, and its galactic longitude is $145°$. What is the distance of the cepheid?

b) The period of the cepheid is 3.16 d, and the apparent visual magnitude is 12.3. What is the distance derived from this information? Are the distances consistent?

Exercise 17.3 a) How many of the nearest stars (Table C.15) are also among the brightest stars (Table C.16)? Explain.

b) If the stellar density were constant, how many stars would there be within the distance of Canopus?

Exercise 17.4 a) Assume that the Galaxy is a homogeneous disk and the Sun lies in the central plane of the disk. The absolute magnitude of a star is M, galactic latitude b, and distance from the central plane z. What is the apparent magnitude of the star, if the extinction inside the Galaxy is $a \text{ mag kpc}^{-1}$?

b) Assume that the thickness of the galactic disk is 200 pc. Find the apparent magnitude of a star with $M = 0.0$, $b = 30°$, distance $r = 1$ kpc, and $a = 1 \text{ mag kpc}^{-1}$.

18. Galaxies

The galaxies are the fundamental building blocks of the Universe. Some of them are very simple in structure, containing only normal stars and showing no particular individual features. There are also galaxies that are almost entirely made of neutral gas. On the other hand, others are complex systems, built up from many separate components — stars, neutral and ionized gas, dust, molecular clouds, magnetic fields, cosmic rays .... The galaxies may form small groups or large clusters in space. At the centre of many galaxies, there is a compact nucleus that may sometimes be so bright that it overwhelms all the normal radiation of the galaxy.

The luminosity of the brightest normal galaxies may correspond to 10^{12} solar luminosities, but most of them are much fainter — the smallest ones that have been discovered are about $10^5 L_\odot$. Since galaxies do not have a sharp outer edge, to some extent their masses and radii depend on how these quantities are defined. If only the bright central parts are included, a giant galaxy may typically have a mass of about $10^{13} M_\odot$, and a radius of 30 kpc, and a dwarf, correspondingly, $10^7 M_\odot$, and 0.5 kpc. In addition, it seems that the outer parts of most galaxies contain large quantities of non-luminous matter that might increase galaxy masses by an order of magnitude.

The density of matter may be very different in different galaxies and in different parts of the same galaxy. Thus the evolution of a galaxy will be the result of processes occurring on vastly different time and energy scales, and no generally accepted comprehensive picture of it exists as yet. In the following, the most important observed properties of galaxies will be presented. Many of them still await an explanation in current theories of galaxy evolution.

18.1 The Classification of Galaxies

A useful first step towards an understanding of galaxies is a classification based on their various forms. Although such a morphological classification must always be to some extent subjective, it provides a framework within which the quantitative properties of galaxies can be discussed in a systematic fashion. However, it should always be remembered that the picture thus obtained will be limited to those galaxies that are large and bright enough to be easily visible in the sky. An idea of the consequent limitations can be obtained from Fig. 18.1, showing the radii and magnitudes of normal galaxies. One sees that only within a narrow region of this diagram can galaxies be easily found. If a galaxy has too large a radius for its magnitude (small surface brightness), it will disappear in the background light from the

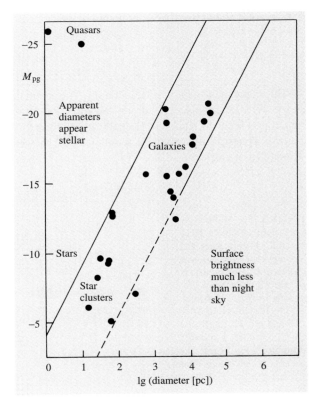

Fig. 18.1. Magnitudes and diameters of observable extragalactic objects. Objects to the upper left look like stars. The quasars in this region have been discovered on the basis of their spectra. Objects to the lower right have a surface brightness much smaller than that of the night sky. In recent years large numbers of low surface brightness galaxies have been discovered in this region. (Arp, H. (1965): Astrophys. J. **142**, 402)

night sky. On the other hand, if its radius is too small, it looks like a star and is not noticed on a photographic plate. In the following, we shall mainly be concerned with bright galaxies that fit within these limits.

If a classification is to be useful, it should at least roughly correspond to important physical properties of the galaxies. Most classifications accord in their main features with the one put forward by *Edwin Hubble* in 1926. Hubble's own version of the *Hubble sequence* is shown in Fig. 18.2. The various types of galaxies are ordered in a sequence from early to late types. There are three main types: *elliptical*, *lenticular*, and *spiral* galaxies. The spirals are divided into two sequences, *normal* and *barred* spirals. In addition, Hubble included a class of *irregular galaxies*.

The elliptical galaxies appear in the sky as elliptical concentrations of stars, in which the density falls off in a regular fashion as one goes outwards. Usually there are no signs of interstellar matter (dark bands of dust, bright young stars). The ellipticals differ from each other only in shape and on this basis they are classified as E0, E1, ..., E7. If the major and minor axes of an elliptical galaxy are *a* and *b*, its type is defined to be E*n*, where

$$n = 10 \left(1 - \frac{b}{a} \right) . \tag{18.1}$$

An E0 galaxy thus looks circular in the sky. The apparent shape of an E galaxy depends on the direction from which it is seen. In reality an E0 galaxy may therefore be truly spherical or it may be a circular disc viewed directly from above.

A later addition to the Hubble sequence is a class of *giant elliptical galaxies* denoted cD. These are generally found in the middle of clusters of galaxies. They consist of a central part looking like a normal elliptical surrounded by an extended fainter halo of stars.

In the Hubble sequence the lenticulars or S0 galaxies are placed between the elliptical and the spiral types. Like the ellipticals they contain only little interstellar matter and show no signs of spiral structure. However, in addition to the usual elliptical stellar component, they also contain a flat disc made up of stars. In this respect they are like spiral galaxies (Figs. 18.3, 18.4).

The characteristic feature of spiral galaxies is a more or less well-defined spiral pattern in the disc. Spiral galaxies consist of a central *bulge*, which is structurally similar to an E galaxy, and of a stellar disc, like in an S0 galaxy. In addition to these, there is a thin disc of gas and other interstellar matter, where young stars are being born, forming the spiral pattern. There are two sequences of spirals, normal Sa–Sb–Sc, and barred SBa–SBb–SBc spirals. In the barred spirals the spiral pattern ends at a central bar, whereas in the normal spirals the spiral pattern may end at an inner ring or continue all the way to the centre. The position of a galaxy within the spiral sequence is determined on the basis of three criteria (which are not always in agreement): later types have a smaller central bulge, more narrow spiral arms and a more open spiral pattern. The Milky Way Galaxy is thought to be of type SABbc (intermediate between Sb and Sc, and between normal and barred spirals).

The classical Hubble sequence is essentially based on bright galaxies; faint galaxies have been less easy to

Fig. 18.2. The Hubble sequence in Hubble's 1936 version. At this stage the existence of type S0 was still doubtful. Photographs of the Hubble types are shown in Figs. 18.6 and 18.15 (E); 18.3 and 18.4 (S0 and S); 18.12 (S and Irr II); 18.5 (Irr I and dE). (Hubble, E.P. (1936): *The Realm of the Nebulae* (Yale University Press, New Haven))

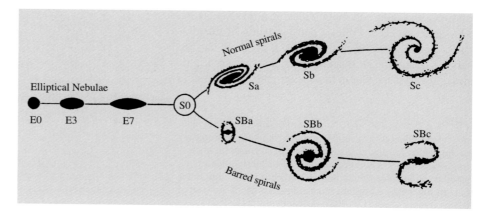

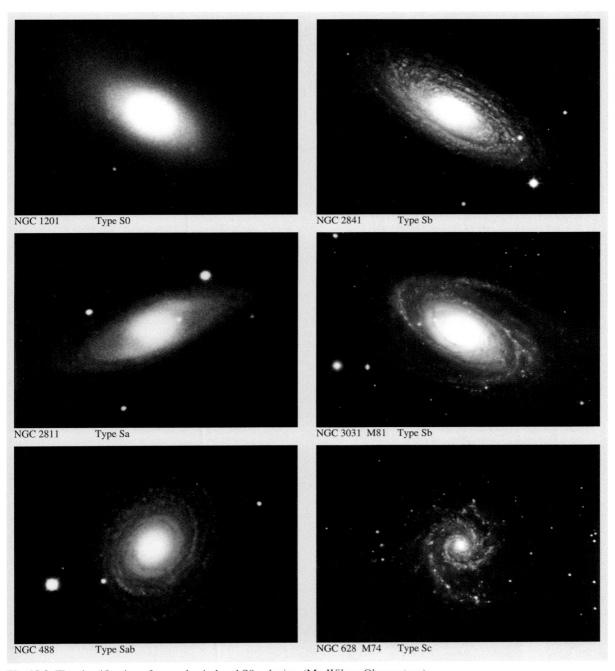

Fig. 18.3. The classification of normal spiral and S0 galaxies. (Mt. Wilson Observatory)

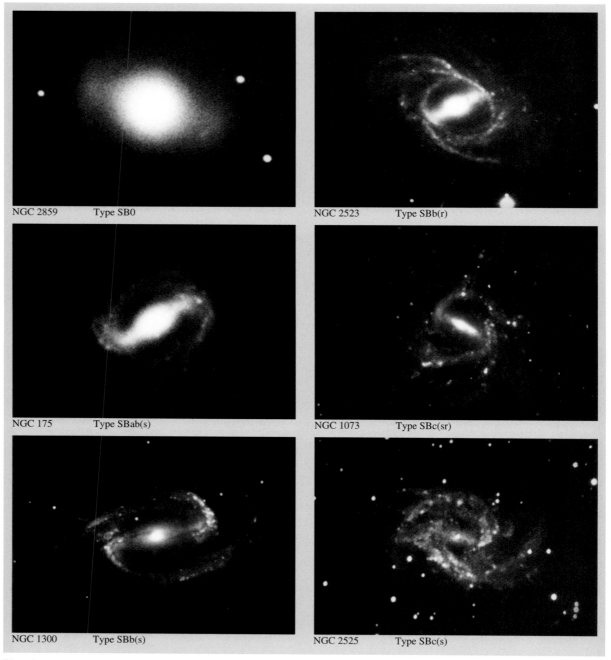

NGC 2859   Type SB0

NGC 2523   Type SBb(r)

NGC 175   Type SBab(s)

NGC 1073   Type SBc(sr)

NGC 1300   Type SBb(s)

NGC 2525   Type SBc(s)

Fig. 18.4. Different types of SB0 and SB galaxies. The type (r) or (s) depends on whether the galaxy has a central ring or not. (Mt. Wilson Observatory)

Fig. 18.5. *Above*: The Small Magellanic Cloud (Hubble type Irr I), a dwarf companion of the Milky Way. (Royal Observatory, Edinburgh). *Below*: The Sculptor Galaxy, a dE dwarf spheroidal. (ESO)

Fig. 18.6. M32 (type E2), a small elliptical companion of the Andromeda Galaxy. (NOAO/Kitt Peak National Observatory)

fit into it (Fig. 18.5). For example, the irregular galaxies of the original Hubble sequence can be divided into the classes Irr I and Irr II. The Irr I galaxies form a continuation of the Hubble sequence towards later types beyond the Sc galaxies. They are rich in gas and contain many young stars. Type Irr II are dusty, somewhat irregular small ellipticals. Other types of dwarf galaxies are often introduced. One example is the *dwarf spheroidal* type dE, similar to the ellipticals, but with a much less centrally concentrated star distribution. Another is the *blue compact galaxies* (also called extragalactic H II regions), in which essentially all the light comes from a small region of bright, newly formed stars.

18.2 Luminosities and Masses

Distances. In order to determine the absolute luminosities and linear dimensions of galaxies one needs to know their distances. Distances are also needed in order to estimate the masses of galaxies, because these estimates depend on the absolute linear size. Distances within the Local Group can be measured by the same methods as inside the Milky Way, most importantly by means of variable stars. On the very large scale (beyond 50 Mpc), the distances can be deduced on the basis of the expansion of the Universe (see Sect. 19.1). In order to connect these two regions one needs methods of distance determination based on the properties of individual galaxies.

To some extent local distances can be determined using structural components of galaxies, such as the sizes of H II regions or the magnitudes of globular clusters. However, to measure distances of tens of megaparsecs, one needs a distance-independent method to determine the absolute luminosities of entire galaxies. Several such methods have been proposed. For example, a luminosity classification has been introduced for late spiral types by *Sidney van den Bergh*. This is based on a correlation between the luminosity of a galaxy and the prominence of its spiral pattern.

Other distance indicators are obtained if there is some intrinsic property of the galaxy, which is correlated with its total luminosity, and which can be measured independently of the distance. Such properties are the colour, the surface brightness and the internal velocities in galax-

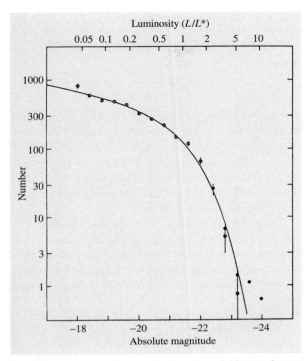

Fig. 18.7. Compound luminosity function of thirteen clusters of galaxies. The open symbols have been obtained by omitting the cD galaxies. The distribution is then well described by (18.2). The cD galaxies (filled symbols) cause a deviation at the bright end. (Schechter, P. (1976): Astrophys. J. **203**, 297)

ies. All of these have been used to measure distances to both spiral and elliptical galaxies. For example, the absolute luminosity of a galaxy should depend on its mass. The mass, in turn, will be reflected in the velocities of stars and gas in the galaxy. Accordingly there is a relationship between the absolute luminosity and the velocity dispersion (in ellipticals) and the rotational velocity (in spirals). Since rotational velocities can be measured very accurately from the width of the hydrogen 21-cm line, the latter relationship (known as the *Tully–Fisher relation*) is perhaps the best distance indicator currently available.

The luminosity of the brightest galaxies in clusters has been found to be reasonably constant. This fact can be used to measure even larger distances, providing a method which is important in cosmology.

Luminosities. The definition of the total luminosity of a galaxy is to some extent arbitrary, since galaxies do not have a sharp outer edge. The usual convention is to measure the luminosity of a galaxy out to a given value of the surface brightness, e.g. to 26.5 mag/sq.arcsec. For a given Hubble type, the total luminosity L may vary widely.

As in the case of stars, the distribution of galaxy luminosities is described by the *luminosity function* $\Phi(L)$. This is defined so that the space density of galaxies with luminosities between L and $L + dL$ is $\Phi(L)\,dL$. It can be determined from the observed magnitudes of galaxies, once their distances have been estimated in some way. In practice, one assumes some suitable functional form for $\Phi(L)$, which is then fitted to the observations. One common form is *Schechter's luminosity function*,

$$\Phi(L)\,dL = \Phi^* \left(\frac{L}{L^*}\right)^\alpha e^{-L/L^*} d\left(\frac{L}{L^*}\right) . \qquad (18.2)$$

The values of the parameters Φ^*, L^*, α are observationally determined for different types of objects; in general, they will be functions of position.

The shape of the luminosity function is described by the parameters α and L^*. The relative number of faint galaxies is described by α. Since its observed value is about -1.1, the density of galaxies grows monotonically as one goes towards fainter luminosities. The luminosity function falls off steeply above the luminosity L^*, which therefore represents a characteristic luminosity of bright galaxies. The observed L^* corresponds to an absolute magnitude $M^* = -21.0$ mag. The corresponding magnitude for the Milky Way Galaxy is probably -20.2 mag. The cD giant galaxies do not obey this brightness distribution; their magnitudes may be -24 mag and even brighter.

The parameter Φ^* is proportional to the space density of galaxies and is therefore a strong function of position. Since the total number density of galaxies predicted by relation (18.2) is infinite, we define $n^* =$ density of galaxies with luminosity $> L^*$. The observed average value of n^* over a large volume of space is $n^* = 3.5 \times 10^{-3}$ Mpc^{-3}. The mean separation between galaxies corresponding to this density is 4 Mpc. Since most galaxies are fainter than L^*, and since, in addition, they often belong to groups, we see that the distances between normal galaxies are generally not much larger than their diameters.

Masses. The distribution of mass in galaxies is a crucial quantity, both for cosmology and for theories of the origin and evolution of galaxies. Observationally it is determined from the velocities of the stars and interstellar gas. Total masses of galaxies can also be derived from their motions in clusters of galaxies. The results are usually given in terms of the corresponding mass-luminosity ratio M/L, using the solar mass and luminosity as units. The value measured in the solar neighbourhood of the Milky Way is $M/L = 3$. If M/L were constant, the mass distribution could be determined from the observed luminosity distribution by multiplying with M/L.

The masses of eliptical galaxies may be obtained from the stellar velocity dispersion given by the broadening of spectral lines. The method is based on the virial theorem (see Sect. 6.10), which says that in a system in equilibrium, the kinetic energy T and the potential energy U are related according to the equation

$$2T + U = 0 . \tag{18.3}$$

Since ellipticals rotate slowly, the kinetic energy of the stars may be written

$$T = Mv^2/2 , \tag{18.4}$$

where M is the total mass of the galaxy and v the velocity width of the spectral lines. The potential energy is

$$U = -GM^2/2R , \tag{18.5}$$

where R is a suitable average radius of the galaxy that can be estimated or calculated from the light distribution. Introducing (18.4) and (18.5) into (18.3) we obtain:

$$M = 2v^2R/G . \tag{18.6}$$

From this formula the mass of an elliptical galaxy can be calculated when v^2 and R are known. Some observations of velocities in elliptical galaxies are given in Fig. 18.8. These will be further discussed in Sect. 18.4. The value of M/L derived from such observations is about 10 within a radius of 10 kpc. The mass of a bright elliptical might thus be up to $10^{13}\,M_\odot$.

The masses of spiral galaxies are obtained from their *rotation curve* $v(R)$, which gives the variation of their rotational velocity with radius. Assuming that most of the mass is in the almost spherical bulge, the mass within radius R, $M(R)$, can be estimated from Kepler's third law:

$$M(R) = Rv(R)^2/G . \tag{18.7}$$

Some typical rotation curves are shown in Fig. 18.9. In the outer parts of many spirals, $v(R)$ does not depend on R. This means that $M(R)$ is directly proportional to the radius – the further out one goes, the larger the interior mass is. Since the outer parts of spirals are very faint, at large radii the value of M/L is directly proportional to the radius. For the disc, one finds that $M/L = 8$ for early and $M/L = 4$ for late spiral types. The largest measured total mass is $2 \times 10^{12}\,M_\odot$.

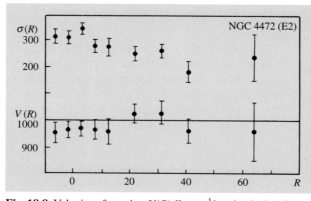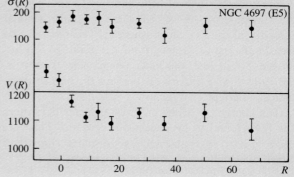

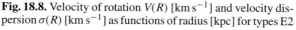

Fig. 18.8. Velocity of rotation $V(R)$ [km s^{-1}] and velocity dispersion $\sigma(R)$ [km s^{-1}] as functions of radius [kpc] for types E2 and E5. The latter galaxy is rotating, the former is not. (Davies, R. L. (1981): Mon. Not. R. Astron. Soc. **194**, 879)

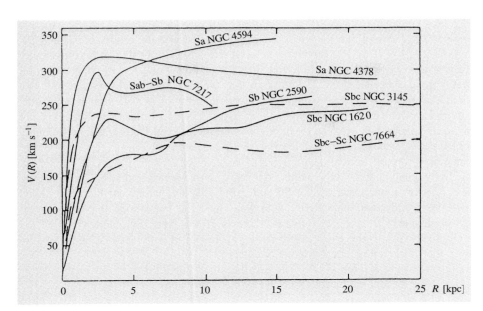

Fig. 18.9. Rotation curves for seven spiral galaxies. (Rubin, V.C., Ford, W.K., Thonnard, N. (1978): Astrophys. J. (Lett.) **225**, L107)

In order to measure the mass at even larger radii where no emission can be detected, motions in systems of galaxies have to be used. One possibility is to use pairs of galaxies. In principle, the method is the same as for binary stars. However, because the orbital period of a binary galaxy is about 10^9 years, only statistical information can be obtained in this way. The results are still uncertain, but seem to indicate values of $M/L = 20$–30 at pair separations of about 50 kpc.

A fourth method to determine galaxy masses is to apply the virial theorem to clusters of galaxies, assuming that these are in equilibrium. The kinetic energy T in (18.4) can then be calculated from the observed redshifts and the potential energy U, from the separations between cluster galaxies. If it is assumed that the masses of galaxies are proportional to their luminosities, it is found that M/L is about 200 within 1 Mpc of the cluster centre. However, there is a large variation from cluster to cluster.

Present results suggest that as one samples larger volumes of space, one obtains larger values for the mass-luminosity ratio. Thus a large fraction of the total mass of galaxies must be in an invisible and unknown form, mostly found in the outer parts. This is known as the *missing mass problem*, and is one of the central unsolved questions of extragalactic astronomy.

18.3 Galactic Structures

Ellipticals and Bulges. In all galaxies the oldest stars have a more or less round distribution. In the Milky Way this component is represented by the population II stars. Its inner parts are called the bulge, and its outer parts are often referred to as the halo. There does not appear to be any physically significant difference between the bulge and the halo. The population of old stars can be best studied in ellipticals, which only contain this component. The bulges of spiral and S0 galaxies are very similar to ellipticals of the same size.

The surface brightness distribution in elliptical galaxies essentially depends only on the distance from the centre and the orientation of the major and minor axis. If r is the radius along the major axis, the surface brightness $I(r)$ is well described by *de Vaucouleurs' law*:

$$\log \frac{I(r)}{I_e} = -3.33 \left[\left(\frac{r}{r_e} \right)^{1/4} - 1 \right]. \tag{18.8}$$

The constants in (18.8) have been chosen so that half of the total light of the galaxy is radiated from within the radius r_e and the surface brightness at that radius is I_e. The parameters r_e and I_e are determined by fitting (18.8) to observed brightness profiles. Typical values for elliptical, normal spiral and S0 galaxies are in the ranges $r_e = 1$–$10\,\text{kpc}$ and I_e corresponds to 20–23 magnitudes per square arc second.

Although de Vaucouleurs' law is a purely empirical relation, it still gives a remarkably good representation of the observed light distribution. However, in the outer regions of elliptical galaxies, departures may often occur: the surface brightness of dwarf spheroidals often falls off more rapidly than (18.8), perhaps because the outer parts of these galaxies have been torn off in tidal encounters with other galaxies. In the giant galaxies of type cD, the surface brightness falls off more slowly (see Fig. 18.10). It is thought that this is connected with their central position in clusters of galaxies.

Although the isophotes in elliptical galaxies are ellipses to a good approximation, their ellipticities and the orientation of their major axes may vary as a function of radius. Different galaxies differ widely in this respect, indicating that the structure of ellipticals is not as simple as it might appear. In particular, the fact that the direction of the major axis sometimes changes within a galaxy suggests that some ellipticals may not be axially symmetric in shape.

From the distribution of surface brightness, the three-dimensional structure of a galaxy may be inferred as explained in *Three-Dimensional Shape of Galaxies.

The relation (18.8) gives a brightness profile which is very strongly peaked towards the centre. The real distribution of axial ratios for ellipticals can be statistically inferred from the observed one. On the (questionable) assumption that they are rotationally symmetric, one obtains a broad distribution with a maximum corresponding to types E3–E4. If the true shape is not axisymmetric, it cannot even statistically be uniquely determined from the observations.

Discs. A bright, massive stellar disc is characteristic for S0 and spiral galaxies, which are therefore called *disc galaxies*. There are indications that in some ellipticals there is also a faint disc hidden behind the bright bulge. In the Milky Way the disc is formed by population I stars.

The distribution of surface brightness in the disc is described by the expression

$$I(r) = I_0 \, e^{-r/r_0} \, . \tag{18.9}$$

Figure 18.11 shows how the observed radial brightness distribution can be decomposed into a sum of two components: a centrally dominant bulge and a disc contributing significantly at larger radii. The central surface brightness I_0 typically corresponds to 21–22 mag./sq.arcsec, and the radial scale

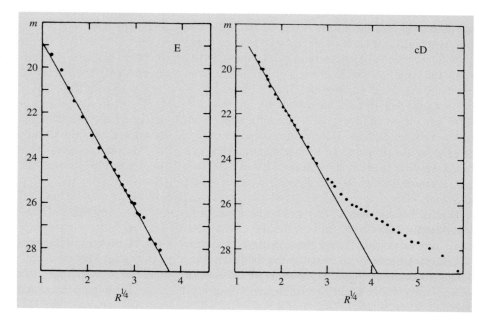

Fig. 18.10. The distribution of surface brightness in E and cD galaxies. Ordinate: surface magnitude, mag/sq.arcsec; abscissa: (radius [kpc])$^{1/4}$. Equation (18.8) corresponds to a straight line in this representation. It fits well with an E galaxy, but for type cD the luminosity falls off more slowly in the outer regions. Comparison with Fig. 18.11 shows that the brightness distribution in S0 galaxies behaves in a similar fashion. cD galaxies have often been erroneously classified as S0. (Thuan, T.X., Romanishin, W. (1981): Astrophys. J. **248**, 439)

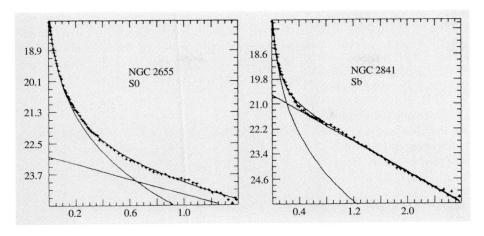

Fig. 18.11. The distribution of surface brightness in types S0 and Sb. Ordinate: mag/sq.arc sec; abscissa: radius [arc sec]. The observed surface brightness has been decomposed into a sum of bulge and disc contributions. Note the larger disc component in type Sb. (Boroson, T. (1981): Astrophys. J. Suppl. **46**, 177)

length $r_0 = 1$–5 kpc. In Sc galaxies the total brightness of the bulge is generally slightly smaller than that of the disc, whereas in earlier Hubble types the bulge has a larger total brightness. The thickness of the disc, measured in galaxies that are seen edge-on, may typically be about 1.2 kpc. Sometimes the disc has a sharp outer edge at about $4r_0$.

The Interstellar Medium. Elliptical and S0 galaxies contain very little interstellar gas. However, in some ellipticals neutral hydrogen amounting to about 0.1% of the total mass has been detected, and in the same galaxies there are also often signs of recent star formation. In some S0 galaxies much larger gas masses have been observed, but the relative amount of gas is very variable from one galaxy to another. The lack of gas in these galaxies is rather unexpected, since during their evolution the stars release much more gas than is observed.

The relative amount of neutral hydrogen in spiral galaxies is correlated with their Hubble type. Thus Sa spirals contain about 2%, Sc spirals 10%, and Irr I galaxies up to 30% or more.

The distribution of neutral atomic hydrogen has been mapped in detail in nearby galaxies by means of radio observations. In the inner parts of galaxies the gas forms a thin disc with a fairly constant thickness of about 200 pc, sometimes with a central hole of a few kpc diameter. The gas disc may continue far outside the optical disc, becoming thicker and often warped from the central disc plane.

Most of the interstellar gas in spiral galaxies is in the form of molecular hydrogen. The hydrogen molecule cannot be observed directly, but the distribution of carbon monoxide has been mapped by radio observations. The distribution of molecular hydrogen can then be derived by assuming that the ratio between the densities of CO and H_2 is everywhere the same, although this may not always be true. It is found that the distribution obeys a similar exponential law as the young stars and H II regions, although in some galaxies (such as the Milky Way) there is a central density minimum. The surface density of molecular gas may be five times larger than that of H I, but because of its strong central concentration its total mass is only perhaps two times larger.

The distribution of cosmic rays and magnetic fields in galaxies can be mapped by means of radio observations of the synchrotron radiation from relativistic electrons. The strength of the magnetic field deduced in this way is typically 0.5–1 nT. The observed emission is polarized, showing that the magnetic field is fairly well-ordered on large scales. Since the plane of polarization is perpendicular to the magnetic field, the large-scale structure of the magnetic field can be mapped. However, the plane of polarization is changed by Faraday rotation, and for this reason observations at several wavelengths are needed in order to determine the direction of the field. The results show that the field is generally strongest in the plane of the disc, and is directed along the spiral arms in the plane. The field is thought to have been produced by the combined action of rising elements of gas, perhaps produced by supernova explosions, and the differential rotation, in principle in the same way as the production of solar magnetic fields was explained in Chapter 12.

* Three-Dimensional Shape of Galaxies

Equations (18.8) and (18.9) describe the distribution of galactic light projected on the plane of the sky. The actual three-dimensional luminosity distribution in a galaxy is obtained by inverting the projection. This is easiest for spherical galaxies.

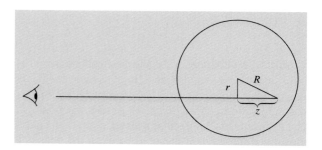

Let us suppose that a spherical galaxy has the projected luminosity distribution $I(r)$ (e. g. as in (18.8)). With coordinates chosen according to the figure, $I(r)$ is given in terms of the three-dimensional luminosity distribution $\rho(R)$ by

$$I(r) = \int_{-\infty}^{\infty} \rho(R)\, dz .$$

Since $z^2 = R^2 - r^2$, a change of the variable of integration yields

$$I(r) = 2 \int_{r}^{\infty} \frac{\rho(R) R\, dR}{\sqrt{R^2 - r^2}} .$$

This is known as an Abel integral equation for $\rho(R)$, and has the solution

$$\rho(r) = -\frac{1}{\pi R} \frac{d}{dR} \int_{R}^{\infty} \frac{I(r) r\, dr}{\sqrt{r^2 - R^2}}$$

$$= -\frac{1}{\pi} \int_{R}^{\infty} \frac{(dI/dr)\, dr}{\sqrt{r^2 - R^2}} .$$

Introducing the observed $I(r)$ into this expression, one obtains the actual luminosity distribution $\rho(R)$. In the figure the solid curve shows the three-dimensional luminosity distribution obtained from the Vancouleurs' law (the dashed line).

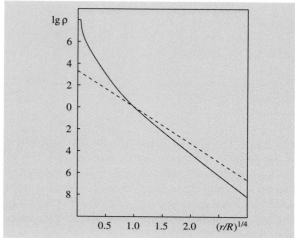

If the galaxy is not spherical, its three-dimensional shape can only be determined if its inclination with respect to the line of sight is known. Since galactic discs are thin and of constant thickness, the inclination i of a disc galaxy is obtained directly from the axis ratio of its projected image: $\sin i = b/a$.

When the inclination is known, the real axis ratio of the bulge q_0 can be determined from the projected value q. For a rotationally symmetric bulge the relation between q and q_0 is

$$\cos^2 i = \frac{1 - q^2}{1 - q_0^2} .$$

The flattenings of disc galaxy bulges obtained from this relation lie in the range $q_0 = 0.3$–0.5. Since the inclinations of ellipticals are generally unknown, only the statistical distribution of q can be determined from that of q_0.

18.4 Dynamics of Galaxies

We have seen how the masses of galaxies can be derived from observed velocities of stars and gas. The same observations can be used to study the internal distribution of mass in more detail.

Slowly Rotating Systems. The dynamics of elliptical galaxies and disc galaxy bulges are studied by means of the Doppler shifts and broadenings of stellar absorption lines. Since a given absorption line is the sum of contributions from many individual stars, its Doppler shift gives their mean velocity, while its broadening is increased by an amount depending on the dispersion of stellar velocities around the mean. By observing how the wavelengths and widths of spectral lines behave as functions of the radius, one can get some insight into the distribution of mass in the galaxy.

Examples of the observed radial dependence of the rotational velocity and velocity dispersion derived for some ellipticals were given in Fig. 18.8. The observed rotational velocities are often small ($< 100 \, \mathrm{km \, s^{-1}}$), while the velocity dispersion may typically be about $200 \, \mathrm{km \, s^{-1}}$. If elliptical galaxies were in fact ellipsoids of revolution, there should be a statistical relation (when projection effects have been taken into account) between flatness, rotational velocity and velocity dispersion. Such a relationship has been observed for fainter ellipticals and for disc galaxy bulges. However, some of the brightest ellipticals rotate very slowly. Therefore their flattening cannot be due to rotation.

The radial dependence of the velocity dispersion gives information on the distribution of mass within the galaxy. Since it also depends on how the shapes of stellar orbits in the galaxy are distributed, its interpretation requires detailed dynamical models.

Rotation Curves. In spiral galaxies the distribution of mass can be studied directly using the observed rotational velocities of the interstellar gas. This can be observed either at optical wavelengths from the emission lines of ionised gas in H II regions or at radio wavelengths from the hydrogen 21 cm line. Typical galactic rotation curves were shown in Fig. 18.9.

The qualitative behaviour of the rotation curve in all spiral galaxies is similar to the rotation curve of the Milky Way: there is a central portion, where the rotational velocity is directly proportional to the radius, corresponding to solid body rotation. At a few kpc radius the curve turns over and becomes flat, i. e. the rotational velocity does not depend on the radius. In early Hubble types, the rotation curve rises more steeply near the centre and reaches larger velocities in the flat region (Sa about $300 \, \mathrm{km \, s^{-1}}$, Sc about $200 \, \mathrm{km \, s^{-1}}$). A higher rotational velocity indicates a larger mass according to (18.7), and thus Sa types must have a larger mass density near the centre. This is not unexpected, since a more massive bulge is one of the defining properties of early type spirals.

A decrease of the rotational velocity at large radii would be an indication that most of the mass is inside that radius. In some galaxies such a decrease has been detected, in others the rotational velocity remains constant as far out as the observations can reach.

Spiral Structure. Spiral galaxies are relatively bright objects. Some have a well-defined, large-scale two-armed spiral pattern, whereas in others the spiral structure is made up of a large number of short filamentary arms. From galaxies where the pattern is seen in front of the central bulge, it has been deduced that the sense of winding of the spiral is trailing with respect to the rotation of the galaxy.

The spiral structure is most clearly seen in the interstellar dust, H II regions, and the OB associations formed by young stars. The dust often forms thin lanes along the inner edge of the spiral arms, with star forming regions on their outside. Enhanced synchrotron radio emission associated with spiral arms has also been detected.

The spiral pattern is generally thought to be a wave in the density of the stellar disc, as discussed in Sect. 17.4. As the interstellar gas streams through the density wave a shock, marked by the dust lanes, is formed as the interstellar gas is compressed, leading to the collapse of molecular clouds and the formation of stars. The density wave theory predicts characteristic streaming motions within the arm, which have been detected in some galaxies by observations of the H I 21 cm line.

It is not known how the spiral wave was produced. In multiarmed galaxies the spiral arms may be short-lived, constantly forming and disappearing, but extensive, regular, two-armed patterns have to be more long-lived. In barred spirals the bar can drive a spiral wave in the

Fig. 18.12. *Above*: A spiral galaxy from above: M51 (type Sc). The interacting companion is NGC 5195 (type Irr II). (Lick Observatory). *Below*: A spiral galaxy from the side: the Sb spiral NGC 4565. (NOAO/Kitt Peak National Observatory)

gas. Some normal spirals may have been produced by the tidal force from another galaxy passing nearby. Finally, in some galaxies a two-armed spiral may have been spontaneously generated by an instability of the disc.

18.5 Stellar Ages and Element Abundances in Galaxies

From the Milky Way we know that stars of populations I and II are different not only in respect to their spatial distribution, but also in respect to their ages and heavy element abundances. This fact gives important evidence about the formation of the Milky Way, and it is therefore of interest if a similar connection can be found in other galaxies.

The indicators of composition most easily measured are the variations of colour indices inside galaxies and between different galaxies. Two regularities have been discovered in these variations: First, according to the *colour–luminosity relation* for elliptical and S0 galaxies, brighter galaxies are redder. Secondly, there is a *colour–aperture effect*, so that the central parts of galaxies are redder. For spirals this relationship is due to the presence of young, massive stars in the disc, but it has also been observed for elliptical and S0 galaxies.

Galactic spectra are composed of the spectra of all their stars added together. Thus the colours depend both on the ages of the stars (young stars are bluer) and on the heavy element abundance Z (stars with larger Z are redder). The interpretation of the observational results thus has to be based on detailed modelling of the stellar composition of galaxies or *population synthesis*.

Stars of different spectral classes contribute different characteristic absorption features to the galaxy spectrum. By observing the strength of various spectral features, one can find out about the masses, ages and chemical composition of the stars that make up the galaxy. For this purpose, a large number of characteristic properties of the spectrum, strengths of absorption lines and broad-band colours are measured. One then attempts to reproduce these data, using a representative collection of stellar spectra. If no satisfactory solution can be found, more stars have to be added to the model. The final result is a population model, giving the stellar composition of the galaxy. Combining this with theoretical stellar evolution calculations, the evolution of the light of the galaxy can also be computed.

Population synthesis of E galaxies show that practically all their stars were formed simultaneously about 15×10^9 years ago. Most of their light comes from red giants, whereas most of their mass resides in lower main sequence stars of less than one solar mass. Since all stars have roughly the same age, the colours of elliptical galaxies are directly related to their metallicities. Thus the colour–luminosity relation indicates that Z in giant ellipticals may be double that in the solar neighbourhood, while it may be smaller by a factor 100 in dwarfs. Similarly, the radial dependence of the colours can be explained if the value of Z at the centre is an order of magnitude larger than it is at larger radii.

The stellar composition of disc galaxy bulges is generally similar to that of ellipticals. The element abundances in the gas in spirals can be studied by means of the emission lines from H II regions ionised by newly formed stars. In this case too, the metallicity increases towards the centre, but the size of the variation varies in different galaxies, and is not yet well understood.

18.6 Systems of Galaxies

The galaxies are not smoothly distributed in space; rather, they form systems of all sizes: galaxy pairs, small *groups*, large *clusters* and *superclusters* formed from several groups and clusters. The larger a given system, the less its density exceeds the mean density of the Universe. On the average, the density is twice the background density for systems of radius 5 Mpc and 10% above the background at radius 20 Mpc.

Interactions of Galaxies. The frequency of different types of galaxies varies in the various kinds of groups. This could either be because certain types of galaxies are formed preferentially in certain environments, or because interactions between galaxies have changed their shapes. There are many observed interacting systems where strong tidal forces have produced striking distortions, "bridges" and "tails" in the member galaxies.

The interactions between galaxies are not always dramatic. For example, the Milky Way has two satellites,

Fig. 18.13. *Above*: The irregular Virgo Cluster of galaxies. *Below*: The regular Coma Cluster (ESO and Karl-Schwarzschild-Observatorium)

the Large and Small Magellanic Clouds (see Fig. 18.5), which are Irr I type dwarf galaxies at about 60 kpc distance. It is thought that approximately 5×10^8 years ago, these passed the Milky Way at a distance of about 10–15 kpc, leaving behind the *Magellanic Stream*, a 180° long thin stream of neutral hydrogen clouds. Systems of this type, where a giant galaxy is surrounded by a few small companions, are quite common. Computations show that in many such cases the tidal interactions are so strong that the companions will merge with the parent galaxy at the next close approach. This is likely to happen to the Magellanic Clouds.

During earlier epochs in the Universe, when the density was larger, interactions between galaxies must have been much more common than at present. Thus it has been proposed that a large fraction of bright galaxies have undergone major mergers at some stage in their history. In particular, there are good reasons to believe that the slowly rotating, non-axisymmetric giant ellipticals may have formed by the merger of disc galaxies.

Groups. The most common type of galaxy systems are small, irregular groups of a few tens of galaxies. A typical example is the *Local Group*, which contains two larger galaxies in addition to the Milky Way – the Andromeda Galaxy M31, an Sb spiral of about the same size as the Milky Way with two dwarf companions, and the smaller Sc spiral M33. The rest of the about 35 members of the Local Group are dwarfs; about 20 are of type dE and 10 of type Irr I. The diameter of the Local Group is about 1.2 Mpc.

Clusters. A system of galaxies may be defined to be a cluster if it contains a larger number (at least 50) of bright galaxies. The number of members and the size of a cluster depend on how they are defined. One way of doing this is to fit the observed distribution of galaxies within a cluster with an expression of the form (18.8). In this way a characteristic cluster radius of about 2–5 Mpc is obtained. The number of members depends both on the cluster radius and on the limiting magnitude. A large cluster may contain several hundred galaxies that are less than two magnitudes fainter than the characteristic luminosity L^* of (18.2).

Clusters of galaxies can be ordered in a sequence from extended, low-density, irregular systems (some-times called clouds of galaxies) to denser and more regular structures (Fig. 18.13). The galaxy type composition also varies along this sequence in the sense that in the loose irregular clusters, the bright galaxies are predominantly spirals, whereas the members of dense clusters are almost exclusively type E and S0. The nearest cluster of galaxies is the Virgo Cluster at a distance of about 15 Mpc. It is a relatively irregular cluster, where a denser central region containing early galaxy types is surrounded by a more extended distribution of mainly spiral galaxies. The nearest regular cluster is the Coma Cluster, roughly 90 Mpc away. In the Coma Cluster a central pair of giant ellipticals is surrounded by a flattened (axis ratio about 2 : 1) system of early type galaxies.

X-ray emission from hot gas has been detected in many clusters. In irregular clusters the gas temperature is about 10^7 K and the emission is generally concentrated near individual galaxies; in the regular ones the gas is hotter, 10^8 K, and the emission tends to be more evenly distributed over the whole cluster area. By and large, the X-ray emission follows the distribution of the galaxies. The amount of gas needed to explain this emission is about equal to the mass of the galaxies – thus it cannot solve the missing mass problem. X-ray emission lines from multiply ionised iron have also been observed. On basis of these lines, it has been concluded that the metal abundance of intergalactic gas is roughly equal to that of the Sun. For this reason it is most likely that the gas has been ejected from the galaxies in the cluster.

Superclusters. Groups and clusters of galaxies may form even larger systems, superclusters. For example, the Local Group belongs to the *Local Supercluster*, a flattened system whose centre is the Virgo Cluster, containing tens of smaller groups and clouds of galaxies. The Coma Cluster is part of another supercluster. The diameters of superclusters are 10–20 Mpc. However, on this scale, it is no longer clear whether one can reasonably speak of individual systems. Perhaps it would be more accurate to think of the distribution of galaxies as a continuous network, where the large clusters are connected by walls and strings formed by smaller systems. Between these there remain empty regions containing very few galaxies, which can be up to 50 Mpc in diameter (Fig. 18.14).

Fig. 18.14. Distribution of 9325 galaxies. The upper sector covers the region $8.5° \le \delta \le 55.5°$, $8\,h \le \alpha \le 17\,h$ in the Northern sky, the lower sector the region $-40° \le \delta \le -2.5°$, $20.8\,h \le \alpha \le 4\,h$ in the Southern sky. The radial coordinate is the measured velocity. If this were due only to the expansion of the Universe, the velocity would be directly proportional to the distance (Sect. 19.1). In reality clusters are stretched along the line of sight by the peculiar motions of galaxies. (da Costa et al. (1994): Astrophys. J. **424**, L1)

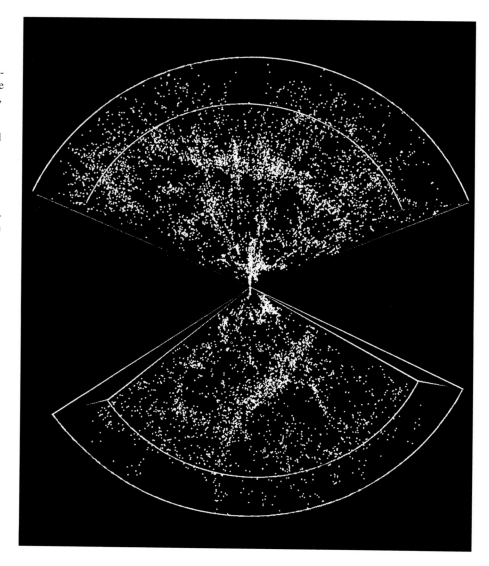

18.7 Active Galaxies and Quasars

So far in this chapter we have been concerned with the properties of normal galaxies. In some galaxies, however, the normal galaxy is overshadowed by violent activity. This activity is produced in the nucleus, which is then called an *active galactic nucleus (AGN)*.

The luminosities of active galactic nuclei may be extremely large, sometimes much larger than that of the rest of the galaxy. It seems unlikely that a galaxy could maintain such a large power output for long. For this reason it is thought that active galaxies do not form a separate class of galaxy, but rather represent a passing stage in the evolution of normal galaxies.

Activity appears in many different forms. Some galaxies have an exceptionally bright nucleus similar to a large region of ionised hydrogen. These may be young galaxies, where near the centre large numbers of stars are forming and evolving into supernovae (starburst nuclei). In other nuclei the radiation cannot have been produced by stars, and the most plausible source of energy in these nuclei is the gravitational energy of a supermassive black hole (mass $> 10^8 \, M_\odot$). In some galaxies, the spectral lines are unusually broad, indicating large internal velocities. These may be either rotational velocities near a black hole or due to explosive events in the nucleus. In some galaxies, jets are seen coming out of the nucleus. Many active galaxies radiate a nonthermal spectrum, apparently synchrotron radiation produced by fast electrons in a magnetic field.

The classification of active galaxies has been developed rather unsystematically, since many of them have been discovered only recently, and have not been completely studied. For example, the *Markarian galaxies* catalogued by *Benyamin Yerishevich Markarian* in the early 1970's are defined by strong ultraviolet emission. Many Markarian galaxies are Seyfert galaxies; others are galaxies undergoing a burst of star formation. The N galaxies form another class closely similar to the Seyfert galaxies.

Two natural basic classes of active galaxies are the *Seyfert galaxies* and the *radio galaxies*. The former are spirals; the latter are ellipticals. Some astronomers think that the Seyfert galaxies represent the active stage of normal spiral galaxies and the radio galaxies that of ellipticals.

Seyfert Galaxies. The Seyfert galaxies are named after *Carl Seyfert*, who discovered them in 1943. Their most important characteristics are a bright, pointlike central nucleus and a spectrum showing broad emission lines. The continuous spectrum has a nonthermal component, which is most prominent in the ultraviolet. The emission lines are thought to be produced in gas clouds moving close to the nucleus with large velocities.

On the basis of the spectrum, Seyfert galaxies are classified as type 1 or 2. In a type 1 spectrum, the allowed lines are broad (corresponding to a velocity of $10^4 \, \mathrm{km \, s^{-1}}$), much broader than the forbidden lines. In type 2, all lines are similar and narrower ($< 10^3 \, \mathrm{km \, s^{-1}}$). Transitions between these types and intermediate cases have sometimes been observed. The reason for the difference is thought to be that the allowed lines are formed in denser gas near the nucleus, and the forbidden lines in more diffuse gas further out. In type 2 Seyfert galaxies, the denser gas is missing or obscured.

Almost all Seyfert galaxies with known Hubble types are spirals; the possible exceptions are of type 2. They are strong infrared sources. Type 1 galaxies often show strong X-ray emission.

The true Seyfert galaxies are relatively weak radio sources. However, there are compact radio galaxies with an optical spectrum that is essentially the same as for Seyfert galaxies. These should probably be classified with the Seyfert galaxies. In general, the stronger radio emission seems to come with a type 2 spectrum.

It is estimated that about 1% of all bright spiral galaxies are Seyfert galaxies. The luminosities of their nuclei are about 10^{36}–10^{38} W, of the same order as all the rest of the galaxy. Brightness variations are common.

Radio Galaxies. By definition, radio galaxies are galaxies that are powerful radio sources. The radio emission of a radio galaxy is non-thermal synchrotron radiation. The radio luminosity of radio galaxies is typically 10^{33}–10^{38} W, and may thus be as large as the total luminosity of a normal galaxy. The main problem in explaining radio emission is to understand how the electrons and magnetic fields are produced, and above all, where the electrons get their energy.

The forms and sizes of the radio emitting regions of radio galaxies have been studied ever since the 1950's, when radio interferometers achieved the resolution of optical telescopes. The characteristic feature of a strong radio galaxy is a double structure: there are two large radio emitting regions on opposite sides of the observed galaxy. The radio emitting regions of some radio galaxies are as far apart as 6 Mpc, almost ten times the distance between the Milky Way and Andromeda galaxies. One of the smallest double radio sources is the galaxy M87 (Fig. 18.15), whose two components are only a few kpc distant from each other.

The double structure of radio galaxies appears to be produced by ejections from the nucleus. However, the electrons in the radio lobes cannot be coming from the centre of the galaxy, because they would lose all their energy during such a long transit. Therefore electrons have to be continuously accelerated within the radio-emitting regions. Within the radio lobes there are almost point-like regions, hot spots. These are generally symmetrically placed with respect to the nucleus, and are apparently consequences of nuclear ejections.

Fig. 18.15. *Above*: The active galaxy M87. In the lower right-hand corner a short exposure of the core region has been inserted (same scale as in the main photograph). One sees a blue jet coming out of the nucleus of a normal E0 galaxy. (NOAO/Kitt Peak National Observatory). *Below*: In the radio map made using the VLA the jet is observed to be two-sided. The area shown is much smaller than in the upper picture. (Owen, F.N., Hardee, P.E., Bignell, R.C. (1980): Astrophys. J. (Lett.) **239**, L11)

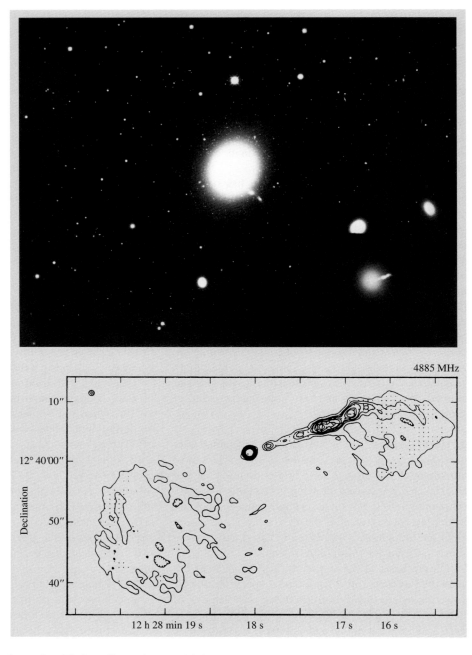

"Tailed" radio sources also exist. Their radio emission mainly comes from one side of the galaxy, forming a curved tail, which is often tens of times longer than the diameter of the galaxy. The best examples are NGC 1265 in the Perseus cluster of galaxies and 3C129, which appears to be in an elliptical orbit around a companion galaxy. The tail is interpreted as the trail left by the radio galaxy in intergalactic space.

Another special feature revealed by the radio maps is the presence of *jets*, narrow lines of radio emission, usually starting in the nucleus and stretching far outside the galaxy. The best known may be the M87 jet, which has also been observed as an optical and X-ray jet. The optically observed jet is surrounded by a radio source. A similar radio source is seen on the opposite side of the nucleus, where no optical jet is seen. Our nearest radio galaxy Centaurus A also has a jet extending from the nucleus to near the edge of the galaxy. VLBI observations of radio jets have also revealed *superluminal motions*: in many compact sources the components appear to be separating faster than the speed of light. Since such velocities are impossible according to the theory of relativity, the observed velocities can only be apparent, and several models have been proposed to account for them.

Quasars. The first quasar was discovered in 1963, when *Maarten Schmidt* interpreted the optical emission lines of the known radio source 3C273 as hydrogen Balmer lines redshifted by 16%. Such large redshifts are the most remarkable characteristics of the quasars. Properly speaking, the word quasar is an abbreviation for quasistellar radio source, and some astronomers prefer to use the designation QSO (quasistellar object), since not all quasars emit radio radiation.

Optically the quasars appear almost as point sources, although improved observational techniques have revealed an increasing number of quasars located inside more or less normal galaxies (Fig. 18.16). Although the first quasars where discovered by radio observations, only a small fraction of all optically identified quasars are bright radio sources. Most radio quasars are point sources, but some have a double structure like the radio galaxies. Satellite X-ray pictures also show the quasars to be pointlike.

In the visible region the quasar spectra are dominated by spectral lines with rest wavelengths in the ultraviolet. The first observed quasar redshifts were $z = 0.16$ and 0.37, and later searches have continued to turn up ever larger redshifts. The present record is 6.3. The light left the quasar when the Universe was less than one-tenth of its present age. The large inferred distances of the quasars mean that their luminosities have to be extremely large. Typical values lie in the range of 10^{38}–10^{41} W. The brightness of quasars may vary rapidly, within a few days or less. Thus the emitting region can be no larger than a few light-days, i.e. about 100 AU.

The quasars often have both emission and absorption lines in their spectra. The emission lines are very broad and are probably produced in the quasar itself. Much of the absorption spectrum consists of densely distributed narrow lines that are thought to be hydrogen Lyman α lines formed in gas clouds along the line of sight to the quasar. The clouds producing this *"Lyman α forest"* are young galaxies or protogalaxies, and they therefore provide important evidence about the formation of galaxies.

Some astronomers have questioned the cosmological interpretation of the redshift. Thus Halton Arp has discovered small systems of quasars and galaxies where some of the components have widely discrepant redshifts. For this reason, Arp thinks that the quasar redshifts are produced by some unknown process. This claim is highly controversial.

Unified Models. Although the forms of galactic activity may at first sight appear diverse, they can be unified within a fairly widely accepted schematic model. According to this model, most galaxies contain a compact central nucleus, usually supposed to be a black hole, with mass 10^7–$10^9 M_\odot$, surrounded by a disc or ring of gas. The source of energy is the gravitational energy released as gas is accreted into the black hole. The disc may also give rise to a jet, where some of the energy is converted into perpendicular motions along the rotational axis. Thus active galactic nuclei are similar to the nucleus of the Milky Way, although the masses of both the black hole and the gas disc may be much larger.

A first characteristic parameter of the model described is obviously the total luminosity. For example, the only essential difference between Seyfert 1 galaxies and radio-quiet quasars is the larger luminosity of quasars. A second basic parameter is the radio brightness, which may be related to the strength of a jet. On the basis of their radio luminosity one can connect Seyfert galaxies and radio-quiet quasars on one hand, and radio galaxies and radio quasars on the other.

The third important parameter of unified models is the angle from which we happen to view the nuclear disc. For example, if the disc is seen edge-on, the actual nucleus is obscured by the disc. This could explain the difference between Seyfert types 1 and 2: in type 2 we do not see the broad emission lines formed near the black hole, but only the narrower lines from the disc. Similarly

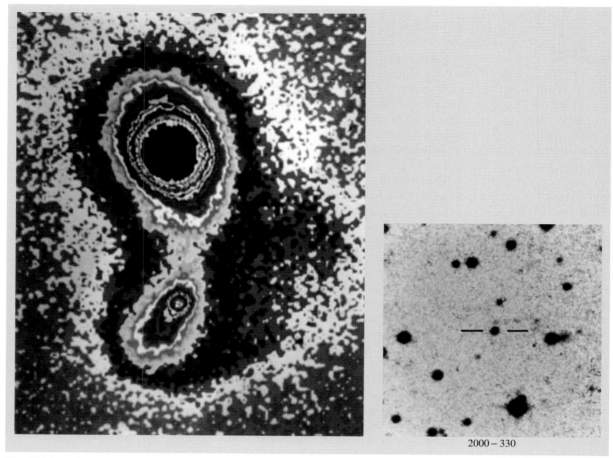

Fig. 18.16. *Left*: QSO 0351+026, a quasar–galaxy interacting pair. The components have the same redshift. *Right*: The distant quasar PKS 2000–330 with the redshift of $z = 3.78$.

(NOAO/Kitt Peak National Observatory; Jet Propulsion Laboratory)

a galaxy that looks like a double radio source when seen edge-on, would look like a radio quasar if the disc were seen face-on. In the latter case there is a possibility that we may be seeing an object directly along the jet. It will then appear as a *blazar*, an object with rapid and violent variations in brightness and polarization, and very weak or invisible emission lines. If the jet is almost relativistic, its transverse velocity may appear larger than the speed of light, and thus superluminal motions can also be understood. One prediction of the unified model is that there should be a large number of quasars where the nucleus is obscured by the disc as in

the Seyfert 2 galaxies. Some possible candidates have been found.

Gravitational Lenses. An interesting phenomenon first discovered in connection with quasars are gravitational lenses. Since light rays are bent by gravitational fields, a mass (e.g. a galaxy) placed between a distant quasar and the observer will distort the image of the quasar. The first example of this effect was discovered in 1979, when it was found that two quasars, 5.7″ apart in the sky, had essentially identical spectra. It was concluded that the "pair" was really a double image of a single

Fig. 18.17. The components of the Einstein cross are gravitationally lensed images of the same quasar. (ESA/NASA)

quasar. Since then several other gravitationally lensed quasars have been discovered.

Gravitational lenses have also been discovered in clusters of galaxies. Here the gravitational field of the cluster distorts the images of distant galaxies into arcs around the cluster centre. In addition in 1993 microlensing was observed in the Milky Way, where the brightness of a star is momentarily increased by the lensing effect from a mass passing in front of the star. Thus the study of gravitational lens effects offers a new promising method of obtaining information on the distribution of mass in the Universe.

18.8 The Origin and Evolution of Galaxies

Our present ideas about the origin and evolution of galaxies are still rather preliminary. However, there is a fairly wide consensus about the general framework within which more detailed questions can be investigated. In the currently dominant model most of the mass is in the form of cold (non-relativistic) dark matter (CDM). Overdense regions of dark matter will fall together under their own gravity. The galaxies then form from gas collapsing into the dark matter concentrations. Elliptical galaxies and disc galaxy bulges were formed

in the first infall of gas about 15×10^9 years ago. In some galaxies, some of the gas was left to settle into a disc before turning into stars. Within some of these galaxies (spirals) star formation still continues in the disc; in others (S0s) all the gas has been consumed.

By means of numerical simulations of the collapse of gas clouds and star formation in them, one can try to compute the evolution of the spectrum and the chemical abundances under various assumptions. The results of the models can be compared with observational data on galaxy spectra and the distribution of elements in galaxies.

The density distribution of dark matter is expected to be very irregular, containing numerous small-scale clumps. The collapse will therefore be highly inhomogeneous, and subsequent mergers between smaller systems should be common. There are additional complicating factors. Gas may be expelled from the galaxy, or there may be an influx of fresh gas. Interactions with the surroundings may radically alter the course of evolution – in dense systems they may lead to the complete merging of the individual galaxies into one giant elliptical. Much remains to be learned about how the formation of stars is affected by the general dynamical state of the galaxy and of how an active nucleus may influence the formation process.

18.9 Exercises

Exercise 18.1 The galaxy NGC 772 is an Sb spiral, similar to M31. Its angular diameter is $7'$ and apparent magnitude 12.0. The corresponding values of M31 are $3.0°$ and 5.0. Find the ratio of the distances of the galaxies

a) assuming their sizes are equal,
b) assuming they are equally bright.

Exercise 18.2 The brightness of the quasar 3C279 has shown changes with the time scale of one week. Estimate the size of the region producing the radiation. The apparent magnitude is 18. If the distance of the quasar is 2000 Mpc, what is its absolute magnitude and luminosity? How much energy is produced per AU^3?

19. Cosmology

After the demise of the Aristotelian world picture, it took hundreds of years of astronomical observations and physical theories to reach a level at which a satisfactory modern scientific picture of the physical universe could be formed. The decisive steps in the development were the clarification of the nature of the galaxies in the 1920's and the general theory of relativity developed by Einstein in the 1910's. Research in cosmology tries to answer questions such as: How large and how old is the Universe? How is matter distributed? How were the elements formed? What will be the future of the Universe? The central tenet of modern cosmology is the model of the expanding universe. On the basis of this model, it has been possible to approach these questions.

19.1 Cosmological Observations

The Olbers Paradox. The simplest cosmological observation may be that the sky is dark at night. This fact was first noted by *Johannes Kepler*, who, in 1610, used it as evidence for a finite universe. As the idea of an infinite space filled with stars like the Sun became widespread in consequence of the Copernican revolution, the question of the dark night sky remained a problem. In the 18th and 19th centuries *Edmond Halley*, *Loys de Chéseaux* and *Heinrich Olbers* considered it in their writings. It has become known as the *Olbers paradox* (Fig. 19.1).

The paradox is the following: Let us suppose the Universe is infinite and that the stars are uniformly distributed in space. No matter in what direction one looks, sooner or later the line of sight will encounter the surface of a star. Since the surface brightness does not depend on distance, each point in the sky should appear to be as bright as the surface of the Sun. This clearly is not true. The modern explanation of the paradox is that the stars have only existed for a finite time, so that the light from very distant stars has not yet reached us. Rather than proving the world to be finite in space, the Olbers paradox has shown it to be of a finite age.

Extragalactic Space. In 1923 *Edwin Hubble* showed that the Andromeda Galaxy M31 was far outside the Milky Way, thus settling a long-standing controversy concerning the relationship between the nebulae and the Milky Way. The numerous galaxies seen in photographs form an extragalactic space vastly larger than the dimensions of the Milky Way. It is important for

Fig. 19.1. The Olbers paradox. If the stars were uniformly distributed in an unending, unchanging space, the sky should be as bright as the surface of the Sun, since each line of sight would eventually meet the surface of a star. A two-dimensional analogy can be found in an optically thick pine forest where the line of sight meets a trunk wherever one looks. (Photo M. Poutanen and H. Karttunen)

cosmology that the distribution and motions of the basic components of extragalactic space, the galaxies and clusters of galaxies, should everywhere be the same as in our local part of the Universe.

Galaxies generally occur in various systems, ranging from small groups to clusters of galaxies and even larger superclusters. The largest structures observed are about 100 Mpc in size (see Sect. 18.6). They are thus significantly smaller than the volume of space (a few thousand Mpc in size) in which the distribution of galaxies has been investigated. One way of studying the large-scale homogeneity of the galaxy distribution is to count the number of galaxies brighter than some limiting magnitude m. If the galaxies are uniformly distributed in space, this number should be proportional to $10^{0.6m}$ (see Fig. 19.2 and Example 17.1). For example, the galaxy counts made by Hubble in 1934, which included 44,000 galaxies, were consistent with a galaxy distribution independent of position (homogeneity) and

of direction (isotropy). Hubble found no "edge" of the Universe, nor have later galaxy counts found one.

Similar counts have been made for extragalactic radio sources. (Instead of magnitudes, flux densities are used. If F is the flux density, then because $m = -2.5 \lg(F/F_0)$, the number count will be proportional to $F^{-3/2}$.) These counts mainly involve very distant radio galaxies and quasars (Fig. 19.3). The results seem to indicate that the radio sources were either much brighter or much more common at earlier epochs than at present. This constitutes evidence in favour of an evolving, expanding universe.

Hubble's Law (Fig. 19.4). In the late 1920's, Hubble discovered that the spectral lines of galaxies were shifted towards the red by an amount proportional to their distances. If the redshift is due to the Doppler effect, this means that the galaxies move away from each other with velocities proportional to their separations, i.e. that the Universe is expanding as a whole.

In terms of the redshift $z = (\lambda - \lambda_0)/\lambda_0$, Hubble's law can be written as

$$z = (H/c)r , \qquad (19.1)$$

where c is the speed of light, H is the *Hubble constant* and r the distance of the galaxy. For small velocities ($V \ll c$) the Doppler redshift $z = V/c$, and hence

$$V = Hr , \qquad (19.2)$$

which is the most commonly used form of Hubble's law.

For a set of observed "standard candles", i.e. galaxies whose absolute magnitudes are close to some mean M_0, Hubble's law corresponds to a linear relationship between the apparent magnitude m and the logarithm of the redshift, $\lg z$. This is because a galaxy at distance r has an apparent magnitude $m = M_0 + 5 \lg(r/10 \, \text{pc})$, and hence Hubble's law yields

$$m = M_0 + 5 \lg \left(\frac{cz}{H \times 10 \, \text{pc}} \right) = 5 \lg z + C , \quad (19.3)$$

where the constant C depends on H and M_0. Suitable standard candles are e.g. the brightest galaxies in clusters and Sc galaxies of a known luminosity class. Some

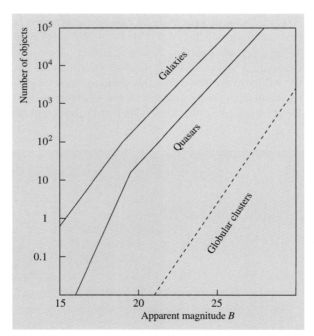

Fig. 19.2. The number of galaxies as a function of limiting magnitude obeys the $10^{0.6m}$ law of a uniform distribution down to the apparent magnitude $B = 20$. The flattening of the relation at fainter magnitudes can be explained by the curvature and expansion of the Universe

Fig. 19.3. The quasar 3C295 and its spectrum. The quasars are among the most distant cosmological objects. (Photograph Palomar Observatory)

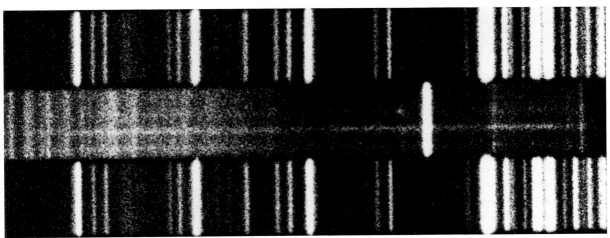

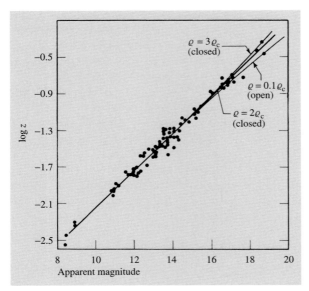

Fig. 19.4. Hubble's law for the brightest galaxies in clusters and the predictions of the Friedmann models. A choice of model cannot be made on the basis of these observations. Here ρ_c is the critical density defined in Sect. 19.4

velocity towards the centre of the Local Supercluster (the Virgo Cluster). Because the Virgo Cluster is often used to determine the value of H, neglecting this peculiar velocity leads to a large error in H. The size of the peculiar velocity is not yet well known, but it is probably about $250\,\mathrm{km\,s^{-1}}$.

The most ambitious recent project for determining H used the Hubble Space Telescope in order to measure cepheid distances to a set of nearby galaxies. These distances were then used to calibrate

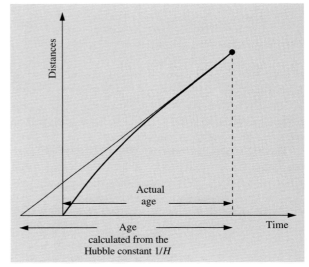

Fig. 19.5. Because the expansion of the Universe is slowing down, the inverse Hubble constant gives an upper limit of the age. The real age depends on the value of the deceleration parameter

other methods of distance determination for galaxies were discussed in Sect. 18.2. Most recently, type Ia supernovae (Sect. 13.3) in distant galaxies have been used to determine distances out to the redshift $z = 1$, where departures from Hubble's law are already detectable.

If the Universe is expanding, the galaxies were once much nearer to each other. If the rate of expansion had been unchanging, the inverse of the Hubble constant, $T = H^{-1}$, would represent the age of the Universe. Since the expansion is thought to be gradually slowing down, the inverse Hubble constant gives an upper limit on the age of the Universe (Fig. 19.5). According to present estimates, $60\,\mathrm{km\,s^{-1}\,Mpc^{-1}} < H < 80\,\mathrm{km\,s^{-1}\,Mpc^{-1}}$, corresponding to $11\,\mathrm{Ga} < T < 17\,\mathrm{Ga}$.

One reason for the difficulty in determining the value of the Hubble constant is the uncertainty in extragalactic distances. A second problem is that the measured values of the velocity V, corrected to take into account the motion of the Sun within the Local Group, contain a significant component due to the peculiar motions of the galaxies. These peculiar velocities are caused by local mass concentrations like groups and clusters of galaxies. It is possible that the Local Group has a significant

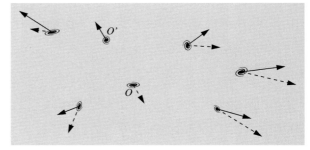

Fig. 19.6. A regular expansion according with Hubble's law does not mean that the Milky Way (O) is the centre of the Universe. Observers at any other galaxy (O') will see the same Hubble flow (*dashed lines*)

other distance indicators, such as the Tully–Fisher relation and type Ia supernovae. The final result was $H = (72 \pm 8) \, \text{km s}^{-1} \, \text{Mpc}^{-1}$. The largest remaining source of error in this result is the distance to the Large Magellanic Cloud, used for calibrating the cepheid luminosity.

The form of Hubble's law might give the impression that the Milky Way is the centre of the expansion, in apparent contradiction with the Copernican principle. Figure 19.6 shows that, in fact, the same Hubble's law is valid at each point in a regularly expanding universe. There is no particular centre of expansion.

The Thermal Microwave Background Radiation. The most important cosmological discovery since Hubble's law was made in 1965. In that year *Arno Penzias* and *Robert Wilson* discovered that there is a universal microwave radiation, with a spectrum corresponding to that of blackbody radiation (see Sect. 5.6) at a temperature of about 3 K (Fig. 19.7). For their discovery, they received the Nobel prize in physics in 1979.

The existence of a thermal cosmic radiation background had been predicted in the late 1940's by *George Gamow*, who was one of the first to study the initial phases of expansion of the Universe. According to Gamow, the Universe at that time was filled with extremely hot radiation. As it expanded, the radiation cooled, until at present, its temperature would be a few kelvins. After its discovery by Penzias and Wilson, the cosmic background radiation has been studied at wavelengths from 50 cm to 0.5 cm. The most recent measurements, made from the COBE (Cosmic Background Explorer) satellite show that it corresponds closely to a Planck spectrum at 2.73 ± 0.06 K.

The existence of the thermal cosmic microwave background (CMB) gives strong support to the belief that the Universe was extremely hot in its early stages. The background is very nearly isotropic, which supports the isotropic and homogeneous models of the Universe. The COBE satellite also detected temperature variations of a relative amplitude 6×10^{-6} in the background. These fluctuations are interpreted as a gravitational redshift of the background produced by the mass concentrations that would later give rise to the observed structures in the Universe. They are the direct traces of initial irregularities in the big bang, and provide important constraints for theories of galaxy formation. Perhaps even more importantly, the amplitude of the fluctuations on different angular scales have provided crucial constraints on the cosmological model.

The Isotropy of Matter and Radiation. Apart from the CMB, several other phenomena confirm the isotropy of the Universe. The distribution of radio sources, the X-ray background, and faint distant galaxies, as well as Hubble's law are all isotropic. The observed isotropy is also evidence that the Universe is homogeneous, since a large-scale inhomogeneity would be seen as an anisotropy.

The Age of the Universe. Estimates of the ages of the Earth, the Sun and of star clusters are important cosmological observations that do not depend on specific cosmological models. From the decay of radioactive isotopes, the age of the Earth is estimated to be 4600 million years. The age of the Sun is thought to be slightly larger than this. The ages of the oldest star clusters in the Milky Way are 10–15 Ga.

The values thus obtained give a lower limit to the age of the Universe. In an expanding universe, the inverse Hubble constant gives an upper limit to that age. It is most remarkable that the directly determined ages of cosmic objects are so close to the upper limit given by the Hubble constant. This is strong evidence that Hubble's law really is due to the expansion of the Universe.

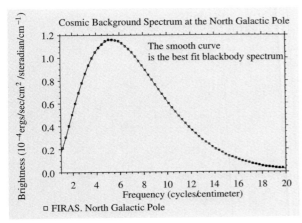

Fig. 19.7. Observations of the cosmic microwave background radiation made by the COBE satellite in 1990 are in agreement with a blackbody law at 2.7 K

It also shows that the oldest star clusters formed very early in the history of the Universe.

The Relative Helium Abundance. A cosmological theory should also give an acceptable account of the origin and abundances of the elements. Even the abundance of the elementary particles and the lack of antimatter are cosmological problems that have begun to be investigated in the context of theories of the early Universe.

Observations show that the oldest objects contain about 25% by mass of helium, the most abundant element after hydrogen. The amount of helium produced is sensitive to the temperature of the Universe, which is related to that of the background radiation. The computations made for the standard models of the expanding Universe (the Friedmann models) yield a helium abundance of exactly the right size.

19.2 The Cosmological Principle

One hopes that as ever larger volumes of the Universe are observed, its average properties will become simple and well defined. Figure 19.8 attempts to show this.

It shows a distribution of galaxies in the plane. As the circle surrounding the observer O becomes larger, the mean density inside the circle becomes practically independent of its size. The same behaviour occurs, regardless of the position of the centre of O: at close distances, the density varies randomly (Fig. 19.9), but in a large enough volume, the average density is constant. This is an example of the *cosmological principle:* apart from local irregularities, the Universe looks the same from all positions in space.

The cosmological principle is a far-reaching assumption, which has been invoked in order to put constraints on the large variety of possible cosmological theories. If in addition to the cosmological principle one also assumes that the Universe is isotropic, then the only possible cosmic flow is a global expansion. In that case, the local velocity difference V between two nearby points has to be directly proportional to their separation ($V = Hr$); i.e. Hubble's law must apply.

The plane universe of Fig. 19.8 is homogeneous and isotropic, apart from local irregularities. Isotropy at each point implies homogeneity, but homogeneity does not require isotropy. An example of an anisotropic, homogeneous universe would be a model containing a constant

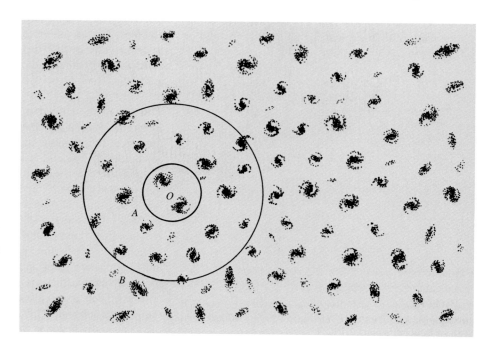

Fig. 19.8. The cosmological principle. In the small circle (A) about the observer (O) the distribution of galaxies does not yet represent the large-scale distribution. In the larger circle (B) the distribution is already uniform on the average

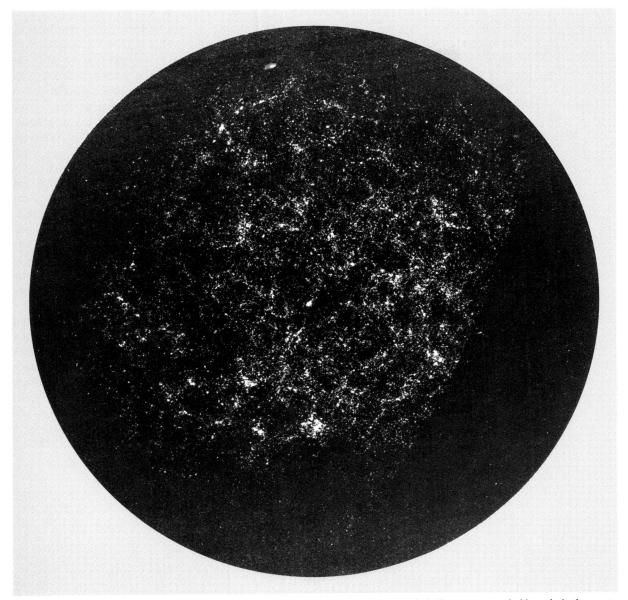

Fig. 19.9. The galaxies seem to be distributed in a "foamlike" way. Dense strings and shells are surrounded by relatively empty regions. (Seldner, M. et al. (1977): Astron. J. **82**, 249)

magnetic field: because the field has a fixed direction, space cannot be isotropic.

We have already seen that astronomical observations support the homogeneity and isotropy of our observable neighbourhood, the *metagalaxy*. On the grounds of the cosmological principle, these properties may be extended to the whole of the Universe.

The cosmological principle is closely related to the Copernican principle that our position in the Universe is in no way special. From this principle, it is only a short step to assume that on a large enough scale, the local properties of the metagalaxy are the same as the global properties of the Universe.

Homogeneity and isotropy are important simplifying assumptions when trying to construct *cosmological models* which can be compared with local observations. They may therefore reasonably be adopted, at least as a preliminary hypothesis.

19.3 Homogeneous and Isotropic Universes

Under general conditions, space and time coordinates in a universe may be chosen so that the values of the space coordinates of observers moving with the matter are constant. It can be shown that in a homogeneous and isotropic universe, the line element (Appendix B) then takes the form

$$ds^2 = -c^2 dt^2 + R^2(t)$$
$$\times \left[\frac{dr^2}{1 - kr^2} + r^2(d\theta^2 + \cos^2\theta d\phi^2) \right], \quad (19.4)$$

known as the *Robertson–Walker line element*. (The radial coordinate r is defined to be dimensionless.) $R(t)$ is a time-dependent quantity representing the scale of the Universe. If R increases with time, all distances, including those between galaxies, will grow. The coefficient k may be $+1$, 0 or -1, corresponding to the three possible geometries of space, the *elliptic* or *closed*, the *parabolic* and the *hyperbolic* or *open* model.

The space described by these models need not be Euclidean, but can have positive or negative curvature.

Depending on the curvature, the volume of the universe may be finite or infinite. In neither case does it have a visible edge.

The two-dimensional analogy to elliptical ($k = +1$) geometry is the surface of a sphere (Fig. 19.10): its surface area is finite, but has no edge. The scale factor $R(t)$ represents the size of the sphere. When R changes, the distances between points on the surface change in the same way. Similarly, a three-dimensional "spherical surface", or the space of elliptical geometry, has a finite volume, but no edge. Starting off in an arbitrary direction and going on for long enough, one always returns to the initial point.

When $k = 0$, space is flat or Euclidean, and the expression for the line element (19.4) is almost the same as in the Minkowski space. The only difference is the scale factor $R(t)$. All distances in a Euclidean space change with time. The two-dimensional analogue of this space is a plane.

The volume of space in the hyperbolic geometry ($k = -1$) is also infinite. A two-dimensional idea of the geometry in this case is given by a saddle surface.

In a homogeneous and isotropic universe, all physical quantities will depend on time through the scale factor $R(t)$. For example, from the form of the line element, it is evident that all distances will be proportional to R (Fig. 19.11). Thus, if the distance to a galaxy is r at time t, then at time t_0 (in cosmology, the subscript 0 refers to the present value) it will be

$$\frac{R(t_0)}{R(t)} r . \quad (19.5)$$

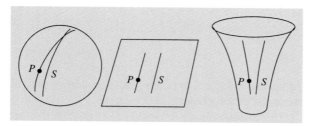

Fig. 19.10. The two-dimensional analogues of the Friedmann models: A spherical surface, a plane and a saddle surface

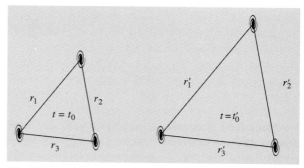

Fig. 19.11. When space expands, all galaxy separations grow with the scale factor R: $r' = [R(t_0')/R(t_0)]r$

Similarly, all volumes will be proportional to R^3. From this it follows that the density of any conserved quantity (e. g. mass) will behave as R^{-3}.

It can be shown that the wavelength of radiation in an expanding universe is proportional to R, like all other lengths. If the wavelength at the time of emission, corresponding to the scale factor R, is λ, then it will be λ_0 when the scale factor has increased to R_0:

$$\frac{\lambda_0}{\lambda} = \frac{R_0}{R} \,. \tag{19.6}$$

The redshift is $z = (\lambda_0 - \lambda)/\lambda$, and hence

$$1 + z = \frac{R_0}{R} \,; \tag{19.7}$$

i. e. the redshift of a galaxy expresses how much the scale factor has changed since the light was emitted. For example, the light from a quasar with $z = 1$ was emitted at a time when all distances were half their present values.

For small values of the redshift, (19.7) approaches the usual form of Hubble's law. This can be seen as follows. When z is small, the change in R during the propagation of a light signal will also be small and proportional to the light travel time t. Because $t = r/c$ approximately, where r is the distance of the source, the redshift will be proportional to r. If the constant of proportionality is denoted by H/c, one has

$$z = Hr/c \,. \tag{19.8}$$

This is formally identical to Hubble's law (19.1). However, the redshift is now interpreted in the sense of (19.7).

As the universe expands, the photons in the background radiation will also be redshifted. The energy of each photon is inversely proportional to its wavelength, and will therefore behave as R^{-1}. It can be shown that the number of photons will be conserved, and thus their number density will behave as R^{-3}. Combining these two results, one finds that the energy density of the background radiation is proportional to R^{-4}. The energy density of blackbody radiation is proportional to T^4, where T is the temperature. Thus the temperature of cosmic background radiation will vary as R^{-1}.

19.4 The Friedmann Models

The results of the preceding section are valid in any homogeneous and isotropic universe. In order to determine the precise time-dependence of the scale factor $R(t)$ a theory of gravity is required.

In 1917 *Albert Einstein* presented a model of the Universe based on his general theory of relativity. It described a geometrically symmetric (spherical) space with finite volume but no boundary. In accordance with the cosmological principle, the model was homogeneous and isotropic. It was also *static*: the volume of space did not change.

In order to obtain a static model, Einstein had to introduce a new repulsive force, the *cosmological term*, in his equations. The size of this cosmological term is given by the *cosmological constant Λ*. Einstein presented his model before the redshifts of the galaxies were known, and taking the Universe to be static was then reasonable. When the expansion of the Universe was discovered, this argument in favour of the cosmological constant vanished. Einstein himself later called it the biggest blunder of his life. Nevertheless, the most recent observations now seem to indicate that a non-zero cosmological constant has to be present.

The St. Petersburg physicist *Alexander Friedmann* and later, independently, the Belgian *Georges Lemaître* studied the cosmological solutions of Einstein's equations. If $\Lambda = 0$, only evolving, expanding or contracting models of the Universe are possible. From the Friedmann models exact formulas for the redshift and Hubble's law may be derived.

The general relativistic derivation of the law of expansion for the Friedmann models will not be given here. It is interesting that the existence of three types of models and their law of expansion can be derived from purely Newtonian considerations, with results in complete agreement with the relativistic treatment. The detailed derivation is given on p. 386, but the essential character of the motion can be obtained from a simple energy argument.

Let us consider a small expanding spherical region in the Universe. In a spherical distribution of matter, the gravitational force on a given spherical shell depends only on the mass inside that shell. We shall here assume $\Lambda = 0$.

We can now consider the motion of a galaxy of mass m at the edge of our spherical region. According to Hubble's law, its velocity will be $V = Hr$ and the corresponding kinetic energy,

$$T = mV^2/2 .\qquad(19.9)$$

The potential energy at the edge of a sphere of mass M is $U = -GMm/r$. Thus the total energy is

$$E = T + U = mV^2/2 - GMm/r ,\qquad(19.10)$$

which has to be constant. If the mean density of the Universe is ρ, the mass is $M = (4\pi r^3/3)\rho$. The value of ρ corresponding to $E = 0$ is called the *critical density*, ρ_c. We have

$$
\begin{aligned}
E &= \frac{1}{2}mH^2 r^2 - \frac{GMm}{r} \\
&= \frac{1}{2}mH^2 r^2 - Gm\frac{4\pi}{3}\frac{r^3 \rho_c}{r} \\
&= mr^2\left(\frac{1}{2}H^2 - \frac{4}{3}\pi G\rho_c\right) = 0 ,
\end{aligned}
\qquad(19.11)
$$

whence

$$\rho_c = \frac{3H^2}{8\pi G} .\qquad(19.12)$$

The expansion of the Universe can be compared to the motion of a mass launched vertically from the surface of a celestial body. The form of the orbit depends on the initial energy. In order to compute the complete orbit, the mass M of the main body and the initial velocity have to be known. In cosmology, the corresponding parameters are the mean density and the Hubble constant.

The $E = 0$ model corresponds to the Euclidean Friedmann model, the *Einstein–de Sitter* model. If the density exceeds the critical density, the expansion of any spherical region will turn to a contraction and it will collapse to a point. This corresponds to the closed Friedmann model. Finally, if $\rho < \rho_c$, the ever expanding hyperbolic model is obtained. The behaviour of the scale factor in these three cases is shown in Fig. 19.12.

These three models of the universe are called the *standard models*. They are the simplest relativistic cosmological models for $\Lambda = 0$. Models with $\Lambda \neq 0$ are mathematically more complicated, but show the same general behaviour.

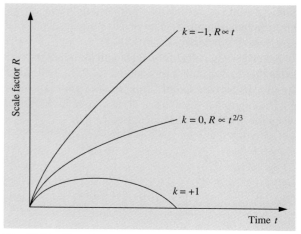

Fig. 19.12. The time dependence of the scale factor for different values of k. The cosmological constant $\Lambda = 0$

The simple Newtonian treatment of the expansion problem is possible because Newtonian mechanics is approximately valid in small regions of the Universe. However, although the resulting equations are formally similar, the interpretation of the quantities involved (e. g. the parameter k) is not the same as in the relativistic context. The global geometry of the Friedmann models can only be understood within the general theory of relativity.

Consider two points at a separation r. Let their relative velocity be V. Then

$$r = \frac{R(t)}{R(t_0)}r_0 \quad \text{and} \quad V = \dot r = \frac{\dot R(t)}{R(t_0)}r_0 ,\qquad(19.13)$$

and thus the Hubble constant is

$$H = \frac{V}{r} = \frac{\dot R(t)}{R(t)} .\qquad(19.14)$$

The deceleration of the expansion is described by the *deceleration parameter q*, defined as

$$q = -R\ddot R/\dot R^2 .\qquad(19.15)$$

The deceleration parameter describes the change of the rate of the expansion $\dot R$. The additional factors have been included in order to make it dimensionless, i. e. independent of the choice of units of length and time.

The value of the deceleration parameter can be expressed in terms of the mean density. The *density parameter* Ω is defined as $\Omega = \rho/\rho_c$, so that $\Omega = 1$ corresponds to the Einstein–de Sitter model. From the conservation of mass, it follows that $\rho_0 R_0^3 = \rho R^3$. Using expression (19.12) for the critical density, one then obtains

$$\Omega = \frac{8\pi G}{3}\frac{\rho_0 R_0^3}{R^3 H^2}\,. \tag{19.16}$$

On the other hand, using (19.27), q can be written

$$q = \frac{4\pi G}{3}\frac{\rho_0 R_0^3}{R^3 H^2}\,. \tag{19.17}$$

Thus there is a simple relation between Ω and q:

$$\Omega = 2q\,. \tag{19.18}$$

The value $q = 1/2$ of the deceleration parameter corresponds to the critical density $\Omega = 1$. Both quantities are in common use in cosmology. It should be noted that the density and the deceleration can be observed independently. The validity of (19.18) is thus a test for the correctness of general relativity with $\Lambda = 0$.

19.5 Cosmological Tests

A central cosmological problem is the question of which Friedmann model best represents the real Universe. Different models make different observational predictions. Recently there has been considerable progress in the determination of the cosmological parameters, and for the first time there is now a set of parameters that appear capable of accounting for all observations. In the following, some possible tests will be considered. These tests are related to the average properties of the Universe. Further cosmological constraints can be obtained from the observed structures. These will be discussed in Sect. 19.7.

The Critical Density. If the average density ρ is larger than the critical density ρ_c, the Universe is closed. For the Hubble constant $H = 100\ \mathrm{km\,s^{-1}\,Mpc^{-1}}$, the value of $\rho_c = 1.9 \times 10^{-26}\ \mathrm{kg\,m^{-3}}$, corresponding to roughly ten hydrogen atoms per cubic metre. Mass determinations for individual galaxies lead to smaller values for the density, favouring an open model. However, the density determined in this way is a lower limit, since there

may be significant amounts of invisible mass outside the main parts of the galaxies.

If most of the mass of clusters of galaxies is dark and invisible, it will increase the mean density nearer the critical value. Using the virial masses of X-ray clusters of galaxies (Sect. 18.2), one finds $\Omega_0 = 0.3$. Considerations of the observed velocities of clusters of galaxies indicate that the relative amount of dark matter does not increase further on even larger scales.

It may be that the neutrino has a small mass (about 10^{-4} electron mass). A large neutrino background should have been produced in the big bang. In spite of the small suggested mass of the neutrino, it would still be large enough to make neutrinos the dominant form of mass in the Universe, and would probably make the density larger than critical. Laboratory measurement of the mass of the neutrino is very difficult, and the claims for a measured nonzero mass have not won general acceptance. Instead much recent work in cosmology has been based on the hypothesis of *Cold Dark Matter* (CDM), i.e. the idea that a significant part of the mass of the Universe is in the form of non-relativistic particles of an unknown kind.

The Magnitude-Redshift Test. Although for small redshifts, all models predict the Hubble relationship $m = 5\lg z + C$ for standard candles, for larger redshifts there are differences, depending on the deceleration parameter q. This provides a way of measuring q.

The models predict that galaxies at a given redshift look brighter in the closed models than in the open ones (see Fig. 19.4). Measurements of type Ia supernovae out to redshifts $z = 1$ using the Hubble Space Telescope have now shown that the observed q is inconsistent with models having $\Lambda = 0$. Assuming $\Omega_0 = 0.3$ these observations require $\Omega_\Lambda = 0.7$, where Ω_Λ is defined below ((19.31)).

The Angular Diameter–Redshift Test. Along with the magnitude-redshift test, the relation between angular diameter and redshift has been used as a cosmological test. Let us first consider how the angular diameter θ of a standard object varies with distance in static models with different geometries. In a Euclidean geometry, the angular diameter is inversely proportional to the distance. In an elliptical geometry, θ decreases more slowly with distance, and even begins to increase beyond a certain point. The reason for this can be understood by thinking of the surface of a sphere. For an observer at the pole, the angular diameter is the angle between two meridians marking the edges of his standard object. This angle is smallest when the object is at the equator, and grows without limit towards the opposite pole. In a hyperbolic geometry, the angle θ decreases more rapidly with distance than in the Euclidean case.

In an expanding closed universe the angular diameter should begin to increase at a redshift of about one. This effect has been looked for in the diameters of radio galaxies and quasars. No turnover has been observed, but this may also be due to evolution of the radio sources or to the selection of the observational data. At smaller redshifts, the use of the diameters of clusters of galaxies has yielded equally inconclusive results.

Basically the same idea can be applied to the angular scale of the strongest fluctuation in the cosmic microwave background. The linear size of these depends only weakly on the cosmological model and can therefore be treated as a standard measuring rod. Their redshift is determined by the decoupling of matter and radiation (see Sect. 19.6). Observations of their angular size have provided strong evidence that $\Omega_0 + \Omega_\Lambda = 1$, i.e. the Universe is flat.

Primordial Nucleosynthesis. The standard model predicts that 25% of the mass of the Universe turned into helium in the "big bang". This amount is not sensitive to the density and thus does not provide a strong cosmological test. However, the amount of deuterium left over from helium synthesis does depend strongly on the density. Almost all deuterons formed in the big bang unite into helium nuclei. For a larger density the collisions destroying deuterium were more frequent. Thus a small present deuterium abundance indicates a high cosmological density. Similar arguments apply to the amounts of ^{3}He and ^{7}Li produced in the big bang. The interpretation of the observed abundances is difficult, since they have been changed by later nuclear processes. Still, present results for the abundances of these nuclei are consistent with each other and with a density corresponding to Ω_0 about 0.04. Note that this number only refers to the mass in the form of baryons, i.e. protons and neutrons. Since the virial masses of clusters of galaxies indicate that Ω_0 is

about 0.3, this has stimulated models such as the CDM model, where most of the mass is not in the form of baryons.

Ages. The ages of different Friedmann models can be compared with known ages of various systems. For a Friedmann model with Hubble constant H and density parameter Ω, the age may be written

$$t_0 = \frac{f(\Omega)}{H} , \qquad (19.19)$$

where the known function $f(\Omega) < 1$. For the critical density $f(\Omega = 1) = 2/3$. Thus if $H = 75\,\mathrm{km\,s^{-1}\,Mpc^{-1}}$, the age is $t_0 = 9\,\mathrm{Ga}$. Larger values of Ω give smaller ages, whereas positive values of Λ lead to larger ages. If there are systems of the known age t_k, one has to demand that $t_0 > t_k$ or

$$f(\Omega) > t_k H . \qquad (19.20)$$

In principle, this equation gives an upper limit for the allowed density in a model. It has been a source of embarrassment that the best values of H have tended to give an age for the Universe only marginally consistent with the ages of the oldest astronomical objects. With the introduction of a positive cosmological constant and a slight downward revision of stellar ages this problem has disappeared. The best current parameter values give 13–15 Ga for the age of the Universe.

The "Concordance" Model. In summary there has been a remarkable recent convergence between different cosmological tests. The resulting model has a positive cosmological constant, and most of the matter is cold and dark. It is thus referred to as the ΛCDM model. The best parameter values are $H = 70\,\mathrm{km\,s^{-1}\,Mpc^{-1}}$, $\Omega_\Lambda = 0.7$, $\Omega_0 = 0.3$, with cold dark matter making up 85% of the total density.

The concordance model is by no means definitive. In particular the reason for the cosmological constant is a major puzzle. Even if some alternative mechanism can produce the same effect as a non-zero Λ, finding at least one set of acceptable parameters is an important step forward.

These parameter values have now been confirmed to great accuracy by the first results from the WMAP satellite mapping the CMB.

19.6 History of the Universe

We have seen how the density of matter and of radiation energy and temperature can be computed as functions of the scale factor R. Since the scale factor is known as a function of time, these quantities can be calculated backwards in time.

At the earliest times, densities and temperatures were so immense that all theories about the physical processes taking place are highly conjectural. Nevertheless first attempts have been made at understanding the most fundamental properties of the Universe on the basis of modern theories of particle physics. For example, no indications of significant amounts of antimatter in the Universe have been discovered. Thus, for some reason, the number of matter particles must have exceeded that of antimatter particles by a factor of 1.000000001. Because of this *symmetry breaking*, when 99.9999999% of the hadrons were annihilated, 10^{-7}% was left later to form galaxies and everything else. It has been speculated that the broken symmetry originated in particle processes about 10^{-35} s after the initial time.

The breaking of fundamental symmetries in the early Universe may lead to what is known as *inflation* of the Universe. In consequence of symmetry breaking, the dominant energy density may be the zero-point energy of a quantum field. This energy density will lead to inflation, a strongly accelerated expansion, which will dilute irregularities and drive the density very close to the critical value. One may thus understand how the present homogeneity, isotropy and flatness of the Universe have come about. In the inflationary picture the Universe has to be very nearly flat, $\Omega_0 + \Omega_\Lambda = 1$. The inflationary models also make specific predictions for the form of the irregularities in the CMB. These predictions are in general agreement with what has been observed.

As the Universe expanded, the density and temperature decreased (Fig. 19.13) and conditions became such that known physical principles can be applied. During the hot early stages, photons and massive particles were continually changing into each other: high-energy photons collided to produce particle-antiparticle pairs, which then were annihilated and produced photons. As the Universe cooled, the photon energies became too small to maintain this equilibrium. There is a *threshold temperature* below which particles of a given type are

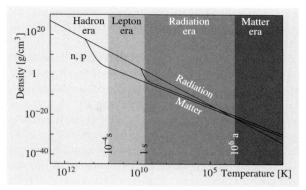

Fig. 19.13. The energy densities of matter and radiation decrease as the Universe expands. Nucleon–antinucleon pairs annihilate at 10^{-4} s; electron–positron pairs at 1 s

no longer produced. For example, the threshold temperature for *hadrons* (protons, neutrons and mesons) is $T = 10^{12}$ K, reached at the time $t = 10^{-4}$ s. Thus the present building blocks of the atomic nuclei, protons and neutrons, are relics from the time 10^{-8}–10^{-4} s, known as the *hadron era*.

The Lepton Era. In the time period 10^{-4}–1 s, the *lepton era*, the photon energies were large enough to produce light particles, such as electron–positron pairs. Because of matter-antimatter symmetry breaking, some of the electrons were left over to produce present astronomical bodies. During the lepton era *neutrino decoupling* took place. Previously the neutrinos had been kept in equilibrium with other particles by fast particle reactions. As the density and temperature decreased, so did the reaction rates, and finally they could no longer keep the neutrinos in equilibrium. The neutrinos decoupled from other matter and were left to propagate through space without any appreciable interactions. It has been calculated that there are at present 600 such cosmological neutrinos per cubic centimetre, but their negligible interactions make them extremely difficult to observe.

The Radiation Era. After the end of the lepton era, about 1 s after the initial time, the most important form of energy was electromagnetic radiation. This stage is called the *radiation era*. At its beginning the temperature was about 10^{10} K and at its end, about one million years later, when the radiation energy density had dropped to that of the particles, it had fallen to a few thousand degrees. At the very beginning of the radiation era within a few hundred seconds helium was produced.

Just before the epoch of helium synthesis, the number ratio of free protons and neutrons was changing because of the decay of the free neutrons. After about 100 s the temperature had dropped to about 10^9 K, which is low enough for deuterons to be formed. All remaining neutrons were then incorporated in deuterons; these, in turn, were almost entirely consumed to produce helium nuclei. Thus the amount of helium synthesized was determined by the number ratio of protons and neutrons at the time of deuterium production $t = 100$ s. Calculations show that this ratio was about 14 : 2. Thus, out of 16 nucleons, 2 protons and 2 neutrons were incorporated in a helium nucleus. Consequently $4/16 = 25\%$ of the mass turned into helium. This is remarkably close to the measured primordial helium abundance.

Only the isotopes ^{2}H, ^{3}He, ^{4}He and ^{7}Li were produced in appreciable numbers by nuclear processes in the big bang. The heavier elements have formed later in stellar interiors, in supernova explosions and perhaps in energetic events in galactic nuclei.

Radiation Decoupling. The Matter Era. As we have seen, the mass density of radiation (obtained from the formula $E = mc^2$) behaves as R^{-4}, whereas that of ordinary matter behaves as R^{-3}. Thus the radiation mass density decreases more rapidly. At the end of the radiation era it became smaller than the ordinary mass density. The *matter era* began, bringing with it the formation of galaxies, stars, planets and human life. At present, the mass density of radiation is much smaller than that of matter. Therefore the dynamics of the Universe is completely determined by the density of massive particles.

Soon after the end of the radiation era, radiation decoupled from matter. This happened when the temperature had dropped to a few thousand degrees, and the protons and electrons combined to form hydrogen atoms. It was the beginning of the *"dark ages"* at redshifts $z = 1000$–100, before stars and galaxies had formed, when the Universe only contained dark matter, blackbody radiation, and slowly cooling neutral gas.

At present, light can propagate freely through space. The world is transparent to radiation: the light from distant galaxies is weakened only by the r^{-2} law and by the redshift. Since there is no certain detection of absorp-

tion by neutral gas, there must have been a *reionisation* of the Universe. It is thought that this occurred around $z = 5$–10.

19.7 The Formation of Structure

As we go backward in time from the present, the distances between galaxies and clusters of galaxies become smaller. For example, the typical separation between galaxies is 100 times their diameter. At the redshift $z = 101$ most galaxies must have been practically in contact. For this reason galaxies in their present form cannot have existed much earlier than about $z = 100$. Since the stars were presumably formed after the galaxies, all present astronomical systems must have formed later than this.

It is thought that all observed structures in the Universe have arisen by the gravitational collapse of small overdensities. Whereas the presently observable galaxies have undergone considerable evolution, which makes it difficult to deduce their initial state, on larger scales the density variations should still be small and easier to study.

CMB Fluctuations. Density variations on the largest scale can be detected as fluctuations in the cosmic microwave background. In Sect. 19.5 we noted that the scale of the strongest fluctuations show the Universe to be flat. Other features in the *power spectrum* describing the scale dependence of the background fluctuations provide further constraints on the cosmological model as well as information on the character of the initial perturbations of the Universe.

Very Large Scale Structure. Structures on scales larger than about 10 Mpc have not had time to undergo significant evolution. By studying the statistics of the galaxy distribution in extensive redshift surveys, such as the one shown in Fig. 18.14, one can get additional constraints on the cosmological parameters.

The tightest cosmological constraints can be obtained by combining the CMB observations with information from the large scale structure. Remarkably, this has yielded values for the cosmological parameters in good agreement with those obtained by the independent methods described in Sect. 19.5.

The Formation of Galaxies. The frequency of regions of higher density, with masses corresponding to galaxies and clusters of galaxies, can be estimated by extrapolating from the strength of the density variations observed on larger scales. On this basis one can connect the characteristics of the structures formed by the collapse of these regions with density perturbations in the early Universe.

Overdensities in the dark matter component can start growing as soon as matter becomes the dominant form of energy at a redshift about 10^4. For the ordinary matter the situation is different.

The minimum mass of a collapsing gas cloud is given by the *Jeans mass* M_J:

$$M_J \approx \frac{P^{3/2}}{G^{3/2}\rho^2} \,, \tag{19.21}$$

where ρ and P are the density and pressure in the cloud (see Sect. 6.11). The value of M_J before decoupling was

$$M_J = 10^{18}\, M_\odot \tag{19.22}$$

and after decoupling

$$M_J = 10^5\, M_\odot \,. \tag{19.23}$$

The reason for the large difference is that before decoupling, matter feels the large radiation pressure ($P = u/3$, see *Gas Pressure and Radiation Pressure, p. 230). After decoupling, this pressure no longer affects the gas.

The large Jeans mass before decoupling means that overdense regions of normal gas cannot start growing before $z = 1000$. After decoupling a large range of masses become Jeans unstable. By then density perturbations of dark matter have had time to grow, and the gas will therefore rapidly fall into heir potential wells.

Because the expansion of the Universe works against the collapse, the density of Jeans unstable regions grows rather slowly. In order to produce the observed systems, the density perturbations at decoupling cannot be too small. In models without dark matter the variations in the CMB predicted on this basis tended to be too large. In the CDM model the predicted variations are of the expected amplitude.

The way galaxies and clusters form in the CDM model is described as the *hierarchical* scenario. In this picture systems of all masses above $10^5\, M_\odot$ begin forming after decoupling. Because smaller systems will collapse more rapidly, they are the first to form, at redshifts about 30. Once the first sources of light, starbursts or AGNs, had formed, they could reionise the gas. This marked the ending of the dark ages at redshifts $z = 10\text{--}5$.

One finally has to ask where the initial perturbations came from. An attractive feature of the inflationary model is that it makes specific predictions for these initial perturbations, deriving them from quantum effects at very early times. In this way the observed properties of the largest astronomical systems contain information about the earliest stages of our Universe.

19.8 The Future of the Universe

The standard models allow two alternative prospects for the future development of the Universe. Expansion may either go on forever, or it may reverse to a contraction, where everything is eventually squeezed back to a point. In the final squeeze, the early history of the Universe would be repeated backwards: in turn, galaxies, stars, atoms and nucleons would be broken up. Finally the state of the Universe could no longer be treated by present-day physics.

In the open models the future is quite different. The evolution of the stars may lead to one of four end results: a white dwarf, a neutron star or a black hole may be formed, or the star may be completely disrupted. After about 10^{11} years, all present stars will have used up their nuclear fuel and reached one of these four final states.

Some of the stars will be ejected from their galaxies; others will form a dense cluster at the centre. In about 10^{27} years the central star clusters will become so dense that a black hole is formed. Similarly the galaxies in large clusters will collide and form very massive black holes.

Not even black holes last forever. By a quantum mechanical tunnelling process, mass can cross the event horizon and escape to infinity – the black hole is said to "evaporate". The rate of this phenomenon, known as the *Hawking process*, is inversely proportional to the mass of the hole. For a galactic-mass black hole, the evaporation time is roughly 10^{98} years. After this time, almost all black holes will have disappeared.

The ever expanding space now contains black dwarfs, neutron stars and planet-size bodies (unless the predictions of a finite proton lifetime of about 10^{31} years are confirmed; in that case all these systems will have been destroyed by proton decay). The temperature of the cosmic background radiation will have dropped to 10^{-20} K.

Even further in the future, other quantum phenomena come into play. By a tunnelling process, black dwarfs can change into neutron stars and these, in turn, into black holes. In this way, all stars are turned into black holes, which will then evaporate. The time required has been estimated to be $10^{10^{26}}$ years! At the end, only radiation cooling towards absolute zero will remain.

It is of course highly doubtful whether our current cosmological theories really are secure enough to allow such far-reaching predictions. New theories and observations may completely change our present cosmological ideas.

* Newtonian Derivation of a Differential Equation for the Scale Factor R(t)

Let us consider a galaxy at the edge of a massive sphere (see figure). It will be affected by a central force due to gravity and the cosmological force

$$m\ddot{r} = -\frac{4\pi G}{3}\frac{r^3 \rho m}{r^2} + \frac{1}{3}m\Lambda r,$$

or

$$\ddot{r} = -\frac{4\pi}{3}G\rho r + \frac{1}{3}\Lambda r. \tag{19.24}$$

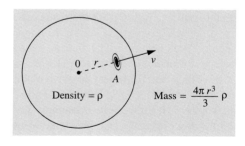

Density $= \rho$ Mass $= \dfrac{4\pi r^3}{3}\rho$

In these equations, the radius r and the density ρ are changing with time. They may be expressed in terms of the scale factor R:

$$r = (R/R_0)r_0, \tag{19.25}$$

where R is defined to be R_0 when the radius $r = r_0$.

$$\rho = (R_0/R)^3 \rho_0, \tag{19.26}$$

where the density $\rho = \rho_0$ when $R = R_0$. Introducing (19.25) and (19.26) in (19.24), one obtains

$$\ddot{R} = -\frac{a}{R^2} + \frac{1}{3}\Lambda R, \tag{19.27}$$

where $a = 4\pi G R_0^3 \rho_0/3$. If (19.27) is multiplied on both sides by $\dot{R}$, the left-hand side yields

$$\dot{R}\ddot{R} = \frac{1}{2}\frac{d(\dot{R}^2)}{dt},$$

and thus (19.27) takes the form

$$d(\dot{R}^2) = -\frac{2a}{R^2}dR + \frac{2}{3}\Lambda R dR. \tag{19.28}$$

Let us define $R_0 = R(t_0)$. Integrating (19.28) from t_0 to t gives

$$\dot{R}^2 - \dot{R}_0^2 = 2a\left(\frac{1}{R} - \frac{1}{R_0}\right) + \frac{1}{3}\Lambda(R^2 - R_0^2). \tag{19.29}$$

The constants $\dot{R}_0$ and a can be eliminated in favour of the Hubble constant H_0 and the density parameter Ω_0. Because $\rho_c = 3H_0^2/8\pi G$,

$$2a = 8\pi G R_0^3 \rho_0/3$$
$$= H_0^2 R_0^3 \rho_0/\rho_c = H_0^2 R_0^3 \Omega_0, \tag{19.30}$$

where $\Omega_0 = \rho_0/\rho_c$. Using expression (19.30) and $\dot{R}_0 = H_0 R_0$ in (19.29), and defining $\Omega_\Lambda = \Lambda/(3H_0^2)$, one obtains

$$\frac{\dot{R}^2}{H_0^2 R_0^2} = \Omega_0\frac{R_0}{R} + \Omega_\Lambda\left(\frac{R}{R_0}\right)^2 + 1 - \Omega_0 - \Omega_\Lambda \tag{19.31}$$

as the basic differential equation governing $R(t)$.

For simplicity we now set $\Omega_\Lambda = 0$. Then the time behaviour of the scale factor R depends on the value of the density parameter Ω_0. Because $\dot{R}^2 > 0$ always, according to (19.31)

$$\Omega_0\frac{R_0}{R} - \Omega_0 + 1 \geq 0,$$

or

$$\frac{R_0}{R} \geq \frac{\Omega_0 - 1}{\Omega_0}. \tag{19.32}$$

If $\Omega_0 > 1$, this means that

$$R \leq R_0 \frac{\Omega_0}{\Omega_0 - 1} \equiv R_{max} .$$

When the scale factor reaches its maximum value R_{max}, then according to (19.31), $\dot{R} = 0$, and the expansion turns into contraction. If $\Omega_0 < 1$, the right-hand side of (19.31) is always positive and the expansion continues forever.

The equation for the time dependence of the scale factor in the general theory of relativity contains the constant k which determines the geometry of space:

$$\dot{R}^2 = \frac{8\pi G R_0^3 \rho_0}{3R} - kc^2 . \tag{19.33}$$

Equations (19.33) and (19.29) (or (19.31)) can be made identical if one chooses

$$H_0^2 R_0^2 (\Omega_0 - 1) = kc^2 .$$

Thus, complete agreement between the Newtonian and the relativistic equation for R is obtained. The values of the geometrical constant $k = +1, 0, -1$ correspond respectively to $\Omega_0 > 1, = 1$ and < 1. More generally, the condition for a flat model, $k = 0$, corresponds to $\Omega_0 + \Omega_\Lambda = 1$.

When $k = 0$, the time dependence of the expansion is very simple. Setting $\Omega_0 = 1$ and using (19.33) and (19.30), one obtains

$$\dot{R}^2 = \frac{H_0^2 R_0^3}{R} .$$

The solution of this equation is

$$R = \left(\frac{3H_0 t}{2}\right)^{2/3} R_0 . \tag{19.34}$$

It is also easy to calculate the time from the beginning of the expansion: $R = R_0$ at the time

$$t_0 = \frac{2}{3} \frac{1}{H_0} .$$

This is the age of the Universe in the Einstein–de Sitter model.

* Three Redshifts

The redshift of a distant galaxy is the result of three different mechanisms acting together. The first one is the peculiar velocity of the observer with respect to the mean expansion: the Earth moves about the Sun, the Sun about the centre of the Milky Way, and the Milky Way and the Local Group of galaxies is falling towards the Virgo Cluster. The apparatus measuring the light from a distant galaxy is not at rest; the velocity of the instrument gives rise to a Doppler shift that has to be corrected for. Usually the velocities are much smaller than the speed of light. The Doppler shift is then

$$z_D = v/c . \tag{19.35}$$

For large velocities the relativistic formula has to be used:

$$z_D = \sqrt{\frac{c+v}{c-v}} - 1 . \tag{19.36}$$

The redshift appearing in Hubble's law is the *cosmological redshift* z_c. It only depends on the values of the scale factor at the times of emission and detection of the radiation (R and R_0) according to

$$z_c = R_0/R - 1 . \tag{19.37}$$

The third type of redshift is the *gravitational redshift* z_g. According to general relativity, light will be redshifted by a gravitational field. For example, the redshift of radiation from the surface of a star of radius R and mass M will be

$$z_g = \frac{1}{\sqrt{1 - R_S/R}} - 1 , \tag{19.38}$$

where $R_S = 2GM/c^2$ is the Schwarzschild radius of the star. The gravitational redshift of the radiation from galaxies is normally insignificant.

The combined effect of the redshifts can be calculated as follows. If the rest wavelength λ_0 is redshifted by the amounts z_1 and z_2 by two different processes, so that

$$z = \frac{\lambda_2 - \lambda_0}{\lambda_0} = \frac{\lambda_2}{\lambda_0} - 1 = \frac{\lambda_2}{\lambda_1}\frac{\lambda_1}{\lambda_0} - 1 ,$$

or

$$(1+z) = (1+z_1)(1+z_2) .$$

Similarly, the three redshifts z_D, z_c and z_g will combine to give an observed redshift z, according to

$$1+z = (1+z_D)(1+z_c)(1+z_g) . \qquad (19.39)$$

19.9 Examples

Example 19.1 a) In a forest there are n trees per hectare, evenly spaced. The thickness of each trunk is D. What is the distance of the wood not seen for the trees? (Find the probability that the line of sight will hit a trunk within a distance x.) b) How is this related to the Olbers paradox?

a) Imagine a circle with radius x around the observer. A fraction $s(x)$, $0 \le s(x) \le 1$, is covered by trees. Then we'll move a distance dx outward, and draw another circle. There are $2\pi nx\, dx$ trees growing in the annulus limited by these two circles. They hide a distance $2\pi xnD\, dx$ or a fraction $nD\, dx$ of the perimeter of the circle. Since a fraction $s(x)$ was already hidden, the contribution is only $(1-s(x))nD\, dx$. We get

$$s(x+dx) = s(x) + (1-s(x))nD\, dx ,$$

which gives a differential equation for s:

$$\frac{ds(x)}{dx} = (1-s(x))nD .$$

This is a separable equation, which can be integrated:

$$\int_0^s \frac{ds}{1-s} = \int_0^x nD\, dx .$$

This yields the solution

$$s(x) = 1 - e^{-nDx} .$$

This is the probability that in a random direction we can see at most to a distance x. This function s is a cumulative probability distribution. The corresponding probability density is its derivative ds/dx. The mean

free path λ is the expectation of this distribution:

$$\lambda = \int_0^\infty x \left(\frac{ds(x)}{dx} \right) dx = \frac{1}{nD} .$$

For example, if there are 2000 trees per hectare, and each trunk is 10 cm thick, we can see to a distance of 50 m, on the average.

b) The result can easily be generalized into three dimensions. Assume there are n stars per unit volume, and each has a diameter D and surface $A = \pi D^2$ perpendicular to the line of sight. Then we have

$$s(x) = 1 - e^{-nAx} ,$$

where

$$\lambda = 1/nA .$$

For example, if there were one sun per cubic parsec, the mean free path would be 1.6×10^4 parsecs. If the universe were infinitely old and infinite in size, the line of sight would eventually meet a stellar surface in any direction, although we could see very far indeed.

Example 19.2 Find the photon density of the 2.7 K background radiation.

The intensity of the radiation is

$$B_\nu = \frac{2h\nu^3}{c^2} \frac{1}{e^{h\nu/(kT)} - 1}$$

and the energy density

$$u_\nu = \frac{4\pi}{c} B_\nu = \frac{8\pi h\nu^3}{c^3} \frac{1}{e^{h\nu/(kT)} - 1} .$$

The number of photons per unit volume is found by dividing the energy density by the energy of a single photon, and integrating over all frequencies:

$$N = \int_0^\infty \frac{u_\nu d\nu}{h\nu} = \frac{8\pi}{c^3} \int_0^\infty \nu^2 \frac{d\nu}{e^{h\nu/(kT)} - 1} .$$

We substitute $h\nu/kT = x$ and $d\nu = (kT/h)dx$:

$$N = 8\pi \left(\frac{kT}{hc} \right)^3 \int_0^\infty \frac{x^2\, dx}{e^x - 1} .$$

The integral cannot be expressed in terms of elementary functions (however, it can be expressed as an infinite sum $2 \sum_{n=0}^{\infty} (1/n^3)$), but it can be evaluated numerically. Its value is 2.4041. Thus the photon density at 2.7 K is

$$N = 16\pi \left(\frac{1.3805 \times 10^{-23} \text{ J K}^{-1} \times 2.7 \text{ K}}{6.6256 \times 10^{-34} \text{ Js} \times 2.9979 \times 10^8 \text{ m s}^{-1}} \right)^3$$
$$\times 1.20206$$
$$= 3.99 \times 10^8 \text{ m}^{-3} \approx 400 \text{ cm}^{-3} \, .$$

19.10 Exercises

Exercise 19.1   The apparent diameter of the galaxy NGC 3159 is 1.3′, apparent magnitude 14.4, and radial velocity with respect to the Milky Way 6940 km s^{-1}. Find the distance, diameter and absolute magnitude of the galaxy. What potential sources of error can you think of?

Exercise 19.2   The radial velocity of NGC 772 is 2562 km s^{-1}. Compute the distance obtained from this information and compare the result with Exercise 18.1.

Exercise 19.3   If the neutrinos have nonzero mass, the universe can be closed. What is the minimum mass needed for this? Assume that $\Lambda = 0$, the density of neutrinos is 600 cm^{-3}, and the density of other matter is one tenth of the critical density.

Appendices

A. Mathematics

A.1 Geometry

Units of Angle and Solid Angle. *Radian* is the angular unit most suitable for theoretical studies. One radian is the angle subtended by a circular arc whose length equals the radius. If r is the radius of a circle and s the length of an arc, the arc subtends an angle

$$\alpha = s/r \, .$$

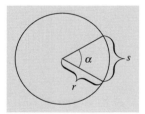

Since the circumference of the circle is $2\pi r$, we have

$$2\pi \text{ rad} = 360° \quad \text{or} \quad 1 \text{ rad} = 180°/\pi \, .$$

In an analogous way we can define a *steradian*, a unit of solid angle, as the solid angle subtended by a unit area on the surface of a unit sphere as seen from the centre. An area A on the surface of a sphere with radius r subtends a solid angle

$$\omega = A/r^2 \, .$$

Since the area of the sphere is $4\pi r^2$, a full solid angle equals 4π steradians.

Circle

Area $A = \pi r^2$.

Area of a sector $A_s = \dfrac{1}{2}\alpha r^2$.

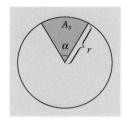

Sphere

Area $A = 4\pi r^2$.

Volume $V = \dfrac{4}{3}\pi r^3$.

Volume of a sector $V_s = \dfrac{2}{3}\pi r^2 h = \dfrac{2}{3}\pi r^3 (1 - \cos\alpha)$

$\qquad = V_{\text{sphere}} \text{ hav } \alpha$.

Area of a segment $A_s = 2\pi r h = 2\pi r^2 (1 - \cos\alpha)$

$\qquad = A_{\text{sphere}} \text{ hav } \alpha$.

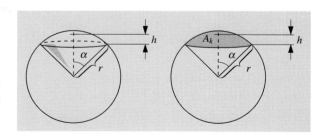

A.2 Conic Sections

As the name already says, conic sections are curves obtained by intersecting circular cones with planes.

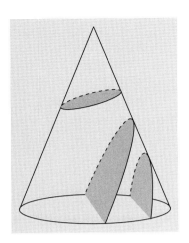

Hyperbola. Equations in rectangular and polar coordinates

$$\frac{x^2}{a^2} - \frac{y^2}{b^2} = 1, \quad r = \frac{p}{1 + e \cos f}.$$

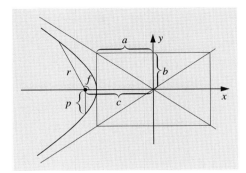

Eccentricity $e > 1$.

Semi-minor axis $b = a\sqrt{e^2 - 1}$.

Parameter $p = a(e^2 - 1)$.

Asymptotes $y = \pm \frac{b}{a} x$.

Ellipse. Equation in rectangular coordinates

$$\frac{x^2}{a^2} + \frac{y^2}{b^2} = 1.$$

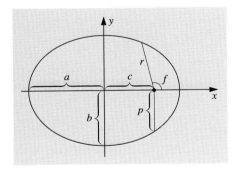

$a = $ the semimajor axis,

$b = $ the semiminor axis $b = a\sqrt{1 - e^2}$,

$e = $ eccentricity $0 \le e < 1$.

Distance of the foci from the centre $c = ea$.

Parameter (semilatus rectum) $p = a(1 - e^2)$.

Area $A = \pi ab$.

Equation in polar coordinates

$$r = \frac{p}{1 + e \cos f},$$

where the distance r is measured from one focus, not from the centre.

When $e = 0$, the curve becomes a circle.

Parabola. Parabola is a limiting case between the previous ones; its eccentricity is $e = 1$.

Equations

$$x = -ay^2, \quad r = \frac{p}{1 + \cos f}.$$

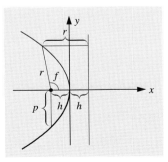

Distance of the focus from the apex $h = 1/4\,a$.

Parameter $p = 1/2\,a$.

A.3 Taylor Series

Let us consider a differentiable real-valued function of one variable $f : R \rightarrow R$. The tangent to the graph of the function at x_0 is

$$y = f(x_0) + f'(x_0)(x - x_0) ,$$

where $f'(x_0)$ is the derivative of f at x_0. Now, if x is close to x_0, the graph of the tangent at x will not be very far from the graph of the function itself. Thus, we can approximate the function by

$$f(x) \approx f(x_0) + f'(x_0)(x - x_0) .$$

The approximation becomes worse, the more the derivative f' varies in the interval $[x_0, x]$. The rate of change of f' is described by the second derivative f'', and so on. To improve accuracy, we have to also include higher derivatives. It can be shown that the value of the function f at x is (assuming that the derivatives exist)

$$f(x) = f(x_0) + f'(x_0)(x - x_0)$$

$$+ \frac{1}{2} f''(x_0)(x - x_0)^2 + \dots$$

$$+ \frac{1}{n!} f^{(n)}(x_0)(x - x_0)^n + \dots ,$$

where $f^{(n)}(x_0)$ is the nth derivative at x_0 and $n!$ is the n-factorial, $n! = 1 \cdot 2 \cdot 3 \cdot \dots \cdot n$. This expansion is called the *Taylor series* of the function at x_0.

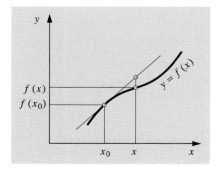

The following list gives some useful Taylor series (in all these cases we have $x_0 = 0$):

$$\frac{1}{1+x} = 1 - x + x^2 - x^3 + \dots$$
$$\text{converges if } |x| < 1$$

$$\frac{1}{1-x} = 1 + x + x^2 + x^3 + \dots$$

$$\sqrt{1+x} = 1 + \frac{1}{2}x - \frac{1}{8}x^2 + \frac{1}{16}x^3 - \dots$$

$$\sqrt{1-x} = 1 - \frac{1}{2}x - \frac{1}{8}x^2 - \frac{1}{16}x^3 - \dots$$

$$\frac{1}{\sqrt{1+x}} = 1 - \frac{1}{2}x + \frac{3}{8}x^2 - \frac{5}{16}x^3 - \dots$$

$$\frac{1}{\sqrt{1-x}} = 1 + \frac{1}{2}x + \frac{3}{8}x^2 + \frac{5}{16}x^3 + \dots$$

$$e^x = 1 + x + \frac{1}{2!}x^2 + \frac{1}{3!}x^3 + \dots + \frac{1}{n!}x^n + \dots$$
$$\text{converges for all } x$$

$$\ln(1+x) = x - \frac{1}{2}x^2 + \frac{1}{3}x^3 - \frac{1}{4}x^4 + \dots$$
$$x \in (-1, 1]$$

$$\sin x = x - \frac{1}{3!}x^3 + \frac{1}{5!}x^5 - \dots \quad \text{for all } x$$

$$\cos x = 1 - \frac{1}{2!}x^2 + \frac{1}{4!}x^4 - \dots \quad \text{for all } x$$

$$\tan x = x + \frac{1}{3}x^3 + \frac{2}{15}x^5 + \dots \quad |x| < \frac{\pi}{2} .$$

Many problems involve small perturbations, in which case it is usually possible to find expressions having very rapidly converging Taylor expansions. The great advantage of this is the reduction of complicated functions to simple polynomials. Particularly useful are linear approximations, such as

$$\sqrt{1+x} \approx 1 + \frac{1}{2}x , \quad \frac{1}{\sqrt{1+x}} \approx 1 - \frac{1}{2}x , \quad \text{etc.}$$

A.4 Vector Calculus

A *vector* is an entity with two essential properties: *magnitude* and *direction*. Vectors are usually denoted by boldface letters *a*, *b*, *A*, *B* etc. The *sum* of the vectors

A and *B* can be determined graphically by moving the origin of *B* to the tip of *A* and connecting the origin of *A* to the tip of *B*. The vector $-A$ has the same magnitude as *A*, is parallel to *A*, but points in the opposite direction. The *difference* $A - B$ is defined as $A + (-B)$.

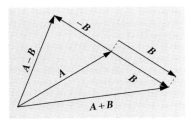

Addition of vectors satisfies the ordinary rules of commutativity and associativity,

$$A + B = B + A ,$$
$$A + (B + C) = (A + B) + C .$$

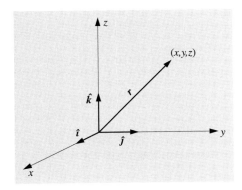

A point in a coordinate frame can be specified by giving its *position* or *radius vector*, which extends from the origin of the frame to the point. The position vector *r* can be expressed in terms of *basis vectors*, which are usually *unit vectors*, i.e. have a length of one distance unit. In a rectangular *xyz*-frame, we denote the basis vectors parallel to the coordinate axes by $\hat{i}$, $\hat{j}$ and $\hat{k}$. The position vector corresponding to the point (x, y, z) is then

$$r = x\hat{i} + y\hat{j} + z\hat{k} .$$

The numbers x, y and z are the *components* of *r*. Vectors can be added by adding their components. For example, the sum of

$$A = a_x\hat{i} + a_y\hat{j} + a_z\hat{k} ,$$
$$B = b_x\hat{i} + b_y\hat{j} + b_z\hat{k} ,$$

is

$$A + B = (a_x + b_x)\hat{i} + (a_y + b_y)\hat{j} + (a_z + b_z)\hat{k} .$$

The magnitude of a vector *r* in terms of its components is

$$r = |r| = \sqrt{x^2 + y^2 + z^2} .$$

The *scalar product* of two vectors *A* and *B* is a real number (scalar)

$$A \cdot B = a_xb_x + a_yb_y + a_zb_z = |A||B|\cos(A, B) ,$$

where (A, B) is the angle between the vectors *A* and *B*. We can also think of the scalar product as the projection of, say, *A* in the direction of *B* multiplied by the length of *B*. If *A* and *B* are perpendicular, their scalar product vanishes. The magnitude of a vector expressed as a scalar product is $A = |A| = \sqrt{A \cdot A}$.

The *vector product* of the vectors *A* and *B* is a vector

$$A \times B = (a_yb_z - a_zb_y)\hat{i} + (a_zb_x - a_xb_z)\hat{j}$$
$$+ (a_xb_y - a_yb_x)\hat{k}$$
$$= \begin{vmatrix} \hat{i} & \hat{j} & \hat{k} \\ a_x & a_y & a_z \\ b_x & b_y & b_z \end{vmatrix} .$$

This is perpendicular to both *A* and *B*. Its length gives the area of the parallelogram spanned by *A* and *B*. The vector product of parallel vectors is a null vector. The vector product is anti-commutative:

$$A \times B = -B \times A .$$

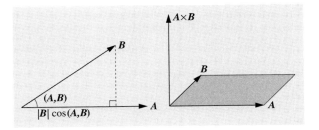

Scalar and vector products satisfy the laws of distributivity:

$$A \cdot (B+C) = A \cdot B + A \cdot C \, ,$$

$$A \times (B+C) = A \times B + A \times C \, ,$$

$$(A+B) \cdot C = A \cdot C + B \cdot C \, ,$$

$$(A+B) \times C = A \times C + B \times C \, .$$

A *scalar triple product* is a scalar

$$A \times B \cdot C = \begin{vmatrix} a_x & a_y & a_z \\ b_x & b_y & b_z \\ c_x & c_y & c_z \end{vmatrix} \, .$$

Here the cross and dot can be interchanged and the factors permuted cyclically without affecting the value of the product. For example $A \times B \cdot C = B \times C \cdot A = B \cdot C \times A$, but $A \times B \cdot C = -B \times A \cdot C$.

A *vector triple product* is a vector, which can be evaluated using one of the expansions

$$A \times (B \times C) = B(A \cdot C) - C(A \cdot B) \, ,$$

$$(A \times B) \times C = B(A \cdot C) - A(B \cdot C) \, .$$

In all these products, scalar factors can be moved around without affecting the product:

$$A \cdot kB = k(A \cdot B) \, ,$$

$$A \times (B \times kC) = k(A \times (B \times C)) \, .$$

The position vector of a particle is usually a function of time $r = r(t) = x(t)\hat{i} + y(t)\hat{j} + z(t)\hat{k}$. The velocity of the particle is a vector, tangent to the trajectory, obtained by taking the derivative of r with respect to time:

$$v = \frac{d}{dt}r(t) = \dot{r} = \dot{x}\hat{i} + \dot{y}\hat{j} + \dot{z}\hat{k} \, .$$

The acceleration is the second derivative, $\ddot{r}$.

Derivatives of the various products obey the same rules as derivatives of products of real-valued function:

$$\frac{d}{dt}(A \cdot B) = \dot{A} \cdot B + A \cdot \dot{B} \, ,$$

$$\frac{d}{dt}(A \times B) = \dot{A} \times B + A \times \dot{B} \, .$$

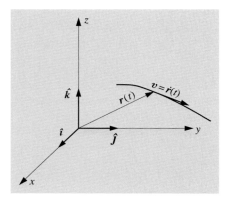

When computing a derivative of a vector product, one must be careful to retain the order of the factors, since the sign of the vector product changes if the factors are interchanged.

A.5 Matrices

Assume we have a vector x with components (x, y, z). We can calculate another vector $x' = (x', y', z')$, the components of which are linear combinations of the original components:

$$x' = a_{11}x + a_{12}y + a_{13}z \, ,$$

$$y' = a_{21}x + a_{22}y + a_{23}z \, ,$$

$$z' = a_{31}x + a_{32}y + a_{33}z \, .$$

This is a linear transform that maps the vector x to a vector x'.

We can collect the coefficients to an array, called a *matrix A*:

$$A = \begin{pmatrix} a_{11} & a_{12} & a_{13} \\ a_{21} & a_{22} & a_{23} \\ a_{31} & a_{32} & a_{33} \end{pmatrix} \, .$$

A general matrix can consist of an arbitrary number of rows and columns. In this book we need only matrices operating on vectors of a three-dimensional space, and they always have three rows and columns. Two subscripts refer to the different elements of the matrix, the first one giving the row and the second one the column.

When using matrix formalism it is convenient to write vectors in the form of column vectors:

$$A = \begin{pmatrix} x \\ y \\ z \end{pmatrix} .$$

We now define that the product of a matrix and a column vector

$$x' = Ax$$

or

$$\begin{pmatrix} x' \\ y' \\ z' \end{pmatrix} = \begin{pmatrix} a_{11} & a_{12} & a_{13} \\ a_{21} & a_{22} & a_{23} \\ a_{31} & a_{32} & a_{33} \end{pmatrix} \begin{pmatrix} x \\ y \\ z \end{pmatrix}$$

means just

$$x' = a_{11}x + a_{12}y + a_{13}z ,$$
$$y' = a_{21}x + a_{22}y + a_{23}z ,$$
$$z' = a_{31}x + a_{32}y + a_{33}z .$$

Comparing these equations we see, for example, that the first component of x' is obtained by taking the first row of the matrix, multiplying the components of the vector x by the corresponding components of that row, and finally adding the products.

This definition can easily be generalised to the product of two matrices. The elements of the matrix

$$C = AB$$

are

$$c_{ij} = \sum_k a_{ik}b_{kj} .$$

This is easy to remember by noting that we take the row i of the first factor A and the column j of the second factor B and evaluate the scalar product of the two vectors. For example

$$\begin{pmatrix} 1 & 1 & 1 \\ 0 & 1 & 2 \\ 1 & 2 & 3 \end{pmatrix} \begin{pmatrix} 1 & 2 & 0 \\ 2 & 1 & 1 \\ 1 & 3 & 2 \end{pmatrix}$$

$$= \begin{pmatrix} 1+2+1 & 2+1+3 & 0+1+2 \\ 0+2+2 & 0+1+6 & 0+1+4 \\ 1+4+3 & 2+2+9 & 0+2+6 \end{pmatrix}$$

$$= \begin{pmatrix} 4 & 6 & 3 \\ 4 & 7 & 5 \\ 8 & 13 & 8 \end{pmatrix} .$$

When multiplying matrices, we have to be careful with the order of the factors, because usually $AB \neq BA$. If we multiply the matrices of the previous example in the reverse order, we get quite a different result:

$$\begin{pmatrix} 1 & 2 & 0 \\ 2 & 1 & 1 \\ 1 & 3 & 2 \end{pmatrix} \begin{pmatrix} 1 & 1 & 1 \\ 0 & 1 & 2 \\ 1 & 2 & 3 \end{pmatrix} = \begin{pmatrix} 1 & 3 & 5 \\ 3 & 5 & 7 \\ 3 & 8 & 13 \end{pmatrix} .$$

A *unit matrix* I is a matrix, which has ones on its diagonal and zeros elsewhere:

$$I = \begin{pmatrix} 1 & 0 & 0 \\ 0 & 1 & 0 \\ 0 & 0 & 1 \end{pmatrix} .$$

If a vector or a matrix is multiplied by a unit matrix, it will remain unchanged.

If the product of two matrices is a unit matrix, the two matrices are *inverse matrices* of each others. The inverse matrix of A is denoted by A^{-1}. It satisfies the equations

$$A^{-1}A = AA^{-1} = I .$$

In spherical astronomy we need mainly *rotation matrices*, describing the rotation of a coordinate frame. The following matrices correspond to rotations around the x, y and z axes, respectively:

$$R_x(\alpha) = \begin{pmatrix} 1 & 0 & 0 \\ 0 & \cos\alpha & \sin\alpha \\ 0 & -\sin\alpha & \cos\alpha \end{pmatrix} ,$$

$$R_y(\alpha) = \begin{pmatrix} \cos\alpha & 0 & \sin\alpha \\ 0 & 1 & 0 \\ -\sin\alpha & 0 & \cos\alpha \end{pmatrix} ,$$

$$R_z(\alpha) = \begin{pmatrix} \cos\alpha & \sin\alpha & 0 \\ -\sin\alpha & \cos\alpha & 0 \\ 0 & 0 & 1 \end{pmatrix} .$$

If the angle is $\alpha = 0$, only a unit matrix remains.

The elements of a rotation matrix can easily be determined. For example, a rotation around the x axis will leave the x coordinate unaffected, and thus the first row and column must be zeroes, except for the diagonal element, which must be one. This will leave four elements. When the angle is zero, the matrix has to reduce to a unit matrix; thus the diagonal elements must be cosines and the other ones sines. The only problem is to decide, which of the sines will get the minus sign. This is most easily done by testing the effect of the matrix on some basis vector.

The inverse matrix of a rotation matrix corresponds to a rotation in the opposite direction. Thus it is obtained from the original matrix by replacing the angle α by $-\alpha$. The only change in the matrix is that the signs of the sines are changed.

For example, the *precession matrix* is a product of three rotation matrices. Since the matrix product is not commutative, these rotations must be carried out in the correct order.

A.6 Multiple Integrals

An integral of a function f over a surface A

$$I = \int_A f \, dA$$

can be evaluated as a double integral by expressing the surface element dA in terms of coordinate differentials. In rectangular coordinates,

$$dA = dx \, dy$$

and in polar coordinates

$$dA = r \, dr \, d\varphi .$$

The integration limits of the innermost integral may depend on the other integration variable. For example, the function xe^y integrated over the shaded area is

$$I = \int_A xe^y \, dA = \int_{x=0}^{1} \int_{y=0}^{2x} xe^y \, dx \, dy$$

$$= \int_0^1 \left[\Big|_0^{2x} xe^y \right] dx = \int_0^1 (xe^{2x} - x) \, dx$$

$$= \Big|_0^1 \frac{1}{2} xe^{2x} - \frac{1}{4} e^{2x} - \frac{1}{2} x^2 = \frac{1}{4} (e^2 - 1) .$$

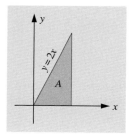

The surface need not be confined to a plane. For example, the area of a sphere is

$$A = \int_S dS ,$$

where the integration is extended over the surface S of the sphere. In this case the surface element is

$$dS = R^2 \cos \theta \, d\varphi \, d\theta ,$$

and the area is

$$A = \int_{\varphi=0}^{2\pi} \int_{\theta=-\pi/2}^{\pi/2} R^2 \cos \theta \, d\varphi \, d\theta$$

$$= \int_0^{2\pi} \left[\Big|_{-\pi/2}^{\pi/2} R^2 \sin \theta \right] d\varphi$$

$$= \int_0^{2\pi} 2R^2 \, d\varphi = 4\pi R^2 .$$

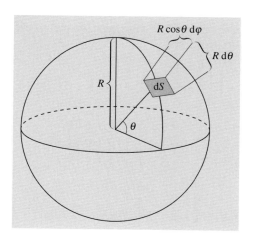

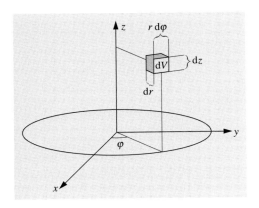

Similarly, a volume integral

$$I = \int_V f \, dV$$

can be evaluated as a triple integral. In rectangular coordinates, the volume element dV is

$$dV = dx \, dy \, dz \,;$$

in cylindrical coordinates

$$dV = r \, dr \, d\varphi \, dz \,,$$

and in spherical coordinates

$$dV = r^2 \cos\theta \, dr \, d\varphi \, d\theta$$

(θ measured from the xy plane)

or

$$dV = r^2 \sin\theta \, dr \, d\varphi \, d\theta$$

(θ measured from the z axis) .

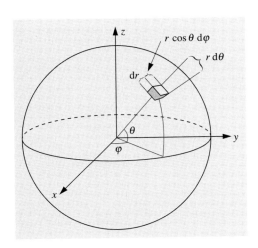

For example, the volume of a sphere with radius R is

$$V = \int_V dV$$

$$= \int_{r=0}^{R} \int_{\varphi=0}^{2\pi} \int_{\theta=-\pi/2}^{\pi/2} r^2 \cos\theta \, dr \, d\varphi \, d\theta$$

$$= \int_0^R \int_0^{2\pi} \left[\Big|_{-\pi/2}^{\pi/2} r^2 \sin\theta \right] dr \, d\varphi$$

$$= \int_0^R \int_0^{2\pi} 2r^2 \, dr \, d\varphi$$

$$= \int_0^R 4\pi r^2 \, dr = \Big|_0^R \frac{4\pi r^3}{3} = \frac{4}{3}\pi R^3 \,.$$

A.7 Numerical Solution of an Equation

We frequently meet equations defying analytical solutions. Kepler's equation is a typical example. If we cannot do anything else, we can always apply some numerical method. Next we shall present two very simple methods, the first of which is particularly suitable for calculators.

Method 1: Direct Iteration. We shall write the equation as $f(x) = x$. Next we have to find an initial value x_0 for the solution. This can be done, for example, graphically or by just guessing. Then we compute a succession of new iterates $x_1 = f(x_0)$, $x_2 = f(x_1)$, and so on, until the difference of successive solutions becomes smaller than some preset limit. The last iterate x_i is the solution. After computing a few x_i's, it is easy to see if they are going to converge. If not, we rewrite the equation as $f^{-1}(x) = x$ and try again. (f^{-1} is the inverse function of f.)

As an example, let us solve the equation $x = -\ln x$. We guess $x_0 = 0.5$ and find

$$x_1 = -\ln 0.5 = 0.69 \,, \quad x_2 = 0.37 \,, \quad x_3 = 1.00 \,.$$

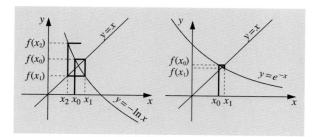
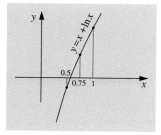

This already shows that something is wrong. Therefore we change our equation to $x = e^{-x}$ and start again:

$$x_0 = 0.5 \,,$$
$$x_1 = e^{-0.5} = 0.61 \,,$$
$$x_2 = 0.55 \,,$$
$$x_3 = 0.58 \,,$$
$$x_4 = 0.56 \,,$$
$$x_5 = 0.57 \,,$$
$$x_6 = 0.57 \,.$$

Thus the solution, accurate to two decimal places, is 0.57.

Method 2. Interval Halving. In some pathological cases the previous method may refuse to converge. In such situations we can use the foolproof method of interval halving. If the function is continuous (as most functions of classical physics are) and we manage to find two points x_1 and x_2 such that $f(x_1) > 0$ and $f(x_2) < 0$, we know that somewhere between x_1 and x_2 there must be a point x in which $f(x) = 0$. Now we find the sign of f in the midpoint of the interval, and select the half of the interval in which f changes sign. We repeat this procedure until the interval containing the solution is small enough.

We shall try also this method on our equation $x = -\ln x$, which is now written as $f(x) = 0$, where $f(x) = x + \ln x$. Because $f(x) \to -\infty$, when $x \to 0$ and $f(1) > 0$, the solution must be in the range $(0, 1)$. Since $f(0.5) < 0$, we know that $x \in (0.5, 1)$. We continue in this way:

$$f(0.75) > 0 \;\Rightarrow\; x \in (0.5, 0.75) \,,$$

$$f(0.625) > 0 \;\Rightarrow\; x \in (0.5, 0.625) \,,$$

$$f(0.563) < 0 \;\Rightarrow\; x \in (0.563, 0.625) \,,$$

$$f(0.594) > 0 \;\Rightarrow\; x \in (0.563, 0.594) \,.$$

The convergence is slow but certain. Each iteration restricts the solution to an interval which is half as large as the previous one, thus improving the solution by one binary digit.

B. Theory of Relativity

Albert Einstein published his *special theory of relativity* in 1905 and the *general theory of relativity* ten years later. Especially the general theory, which is essentially a gravitation theory, has turned out to be very important for the theories of the evolution of the Universe. Therefore it is appropriate to consider here some basic principles of relativity theory. A more detailed discussion would require sophisticated mathematics, and is beyond the scope of this very elementary book.

B.1 Basic Concepts

Everyone knows the famous Pythagorean theorem

$$\Delta s^2 = \Delta x^2 + \Delta y^2 \,,$$

where Δs is the length of the hypotenuse of a right-angle triangle, and Δx and Δy are the lengths of the other two sides. (For brevity, we have denoted $\Delta s^2 = (\Delta s)^2$.) This is easily generalized to three dimensions:

$$\Delta s^2 = \Delta x^2 + \Delta y^2 + \Delta z^2 \,.$$

This equation describes the *metric* of an ordinary rectangular frame in a Euclidean space, i.e. tells us how to measure distances.

Generally the expression for the distance between two points depends on the exact location of the points. In such a case the metric must be expressed in terms of infinitesimal distances in order to correctly take into account the curvature of the coordinate curves. (A coordinate curve is a curve along which one coordinate changes while all the others remain constant.) An infinitesimal distance ds is called the *line element*. In a rectangular frame of a Euclidean space it is

$$ds^2 = dx^2 + dy^2 + dz^2 \,,$$

and in spherical coordinates

$$ds^2 = dr^2 + r^2(d\theta^2 + \cos^2\theta \, d\phi^2) \,.$$

Generally ds^2 can be expressed as

$$ds^2 = \sum_{i,j} g_{ij} \, dx_i \, dx_j \,,$$

where the x_i's are arbitrary coordinates, and the coefficients g_{ij} are components of the *metric tensor*. These can be functions of the coordinates, as in the case of the spherical coordinates. The metric tensor of an n-dimensional space can be expressed as an $n \times n$ matrix. Since $dx_i \, dx_j = dx_j dx_i$, the metric tensor is symmetric, i.e. $g_{ij} = g_{ji}$. If all coordinate curves intersect perpendicularly, the coordinate frame is *orthogonal*. In an orthogonal frame, $g_{ij} = 0$ for all $i \neq j$. For example, the spherical coordinates form an orthogonal frame, the metric tensor of which is

$$(g_{ij}) = \begin{pmatrix} 1 & 0 & 0 \\ 0 & r^2 & 0 \\ 0 & 0 & r^2\cos^2\theta \end{pmatrix} \,.$$

If it is possible to find a frame in which all the components of g are constant, the space is *flat*. In the rectangular frame of a Euclidean space, we have $g_{11} = g_{22} = g_{33} = 1$; hence the space is flat. The spherical coordinates show that even in a flat space, we can use frames in which the components of the metric tensor are not constant. The line element of a two-dimensional spherical surface is obtained from the metric of the spherical coordinate frame by assigning to r some fixed value R:

$$ds^2 = R^2(d\theta^2 + \cos^2\theta \, d\phi^2) \,.$$

The metric tensor is

$$(g_{ij}) = \begin{pmatrix} 1 & 0 \\ 0 & \cos^2\phi \end{pmatrix} \,.$$

This cannot be transformed to a constant tensor. Thus the surface of a sphere is a *curved space*.

If we know the metric tensor in some frame, we can compute a fourth-order tensor, the *curvature tensor* R_{ijkl}, which tells us whether the space is curved or flat. Unfortunately the calculations are slightly too laborious to be presented here.

The metric tensor is needed for all computations involving distances, magnitudes of vectors, areas, and so on. Also, to evaluate a scalar product, we must know the metric. In fact the components of the metric tensor can be expressed as scalar products of the basis vectors:

$$g_{ij} = \hat{e}_i \cdot \hat{e}_j .$$

If A and B are two arbitrary vectors

$$A = \sum_i a^i \hat{e}_i, \quad B = \sum_i b^i \hat{e}_i ,$$

their scalar product is

$$A \cdot B = \sum_i \sum_j a^i b^j \hat{e}_i \cdot \hat{e}_j = \sum_i \sum_j g_{ij} a^i b^j .$$

B.2 Lorentz Transformation. Minkowski Space

The special theory of relativity abandoned the absolute Newtonian time, supposed to flow at the same rate for every observer. Instead it required that the speed of light must have the same value c in all coordinate frames. The constancy of the speed of light follows immediately from the basic assumptions of the special theory of relativity. This is possible only if different observers measure time intervals differently.

Let us send a beam of light from the origin. It travels along a straight line at the speed c. Thus, at the moment t its space and time coordinates satisfy the equation

$$x^2 + y^2 + z^2 = c^2 t^2 . \tag{B.1}$$

Next we study what the situation looks like in another frame, $x'y'z'$, moving at a velocity v with respect to the xyz frame. Let us select the new frame so that it coincides with the xyz frame at $t = 0$. Also, let the time coordinate t' of the $x'y'z'$ frame be $t' = 0$ when $t = 0$. And finally we assume that the $x'y'z'$ frame moves in the direction of the positive x axis. Since the beam of light must also travel at the speed c in the $x'y'z'$ frame, we must have

$$x'^2 + y'^2 + z'^2 = c^2 t'^2 .$$

If we require that the new (dashed) coordinates be obtained from the old ones by a linear transformation, and also that the inverse transformation be obtained simply by replacing v by $-v$, we find that the transformation

must be

$$
\begin{aligned}
x' &= \frac{x - vt}{\sqrt{1 - v^2/c^2}} , \\
y' &= y , \\
z' &= z , \\
t' &= \frac{t - vx/c^2}{\sqrt{1 - v^2/c^2}} .
\end{aligned}
\tag{B.2}
$$

This transformation between frames moving at a constant speed with respect to each other is called the *Lorentz transformation*.

Because the Lorentz transformation is derived assuming the invariance of (B.1), it is obvious that the *interval*

$$\Delta s^2 = -c^2 \Delta t^2 + \Delta x^2 + \Delta y^2 + \Delta z^2$$

of any two events remains invariant in all Lorentz transformations. This interval defines a metric in four-dimensional spacetime. A space having such a metric is called the *Minkowski space* or Lorentz space. The components of the metric tensor g are

$$
(g_{ij}) = \begin{pmatrix} -c^2 & 0 & 0 & 0 \\ 0 & 1 & 0 & 0 \\ 0 & 0 & 1 & 0 \\ 0 & 0 & 0 & 1 \end{pmatrix} .
$$

Since this is constant, the space is flat. But it is no longer an ordinary Euclidean space, because the sign of the time component differs from the sign of the space components. In the older literature, a variable ict is often used instead of time, i being the imaginary unit. Then the metric looks Euclidean, which is misleading; the properties of the space cannot be changed just by changing notation.

In a Minkowski space position, velocity, momentum and other vector quantities are described by *four-vectors*, which have one time and three space components. The components of a four-vector obey the Lorentz transformation when we transform them from one frame to another, both moving along a straight line at a constant speed.

According to classical physics the distance of two events depends on the motion of the observer, but the time interval between the events is the same for all observers. The world of special relativity is more anarchistic: even time intervals have different values for different observers.

B.3   General Relativity

The Equivalence Principle. Newton's laws relate the acceleration, a, of a particle and the applied force F by

$$F = m_i a \, ,$$

where m_i is the *inertial mass* of the particle, resisting the force trying to move the particle. The gravitational force felt by the particle is

$$F = m_g f \, ,$$

where m_g is the *gravitational mass* of the particle, and f is a factor depending only on other masses. The masses m_i and m_g appear as coefficients related to totally different phenomena. There is no physical reason to assume that the two masses should have anything in common. However, already the experiments made by Galilei showed that evidently $m_i = m_g$. This has been verified later with very high accuracy.

The *weak equivalence principle*, which states that $m_i = m_g$, can therefore be accepted as a physical axiom. The *strong equivalence principle* generalises this: if we restrict our observations to a sufficiently small region of spacetime, there is no way to tell whether we are subject to a gravitational field or are in uniformly accelerated motion. The strong equivalence principle is one of the fundamental postulates of general relativity.

Curvature of Space. General relativity describes gravitation as a geometric property of the spacetime. The equivalence principle is an obvious consequence of this idea. Particles moving through spacetime follow the shortest possible paths, *geodesics*. The projection of a geodesic onto a three-dimensional space need not be the shortest way between the points.

The geometry of spacetime is determined by the mass and energy distribution. If this distribution is known, we can write the *field equations*, which are partial differential equations connecting the metric to the mass and energy distribution and describing the curvature of spacetime.

In the case of a single-point mass, the field equations yield the Schwarzschild metric, the line element of which is

$$ds^2 = -\left(1 - \frac{2GM}{c^2 r}\right) c^2 \, dt^2 + \frac{dr^2}{1 - 2GM/c^2 r} + r^2 (d\theta^2 + \cos^2 \theta \, d\phi^2) \, .$$

Here M is the mass of the point; r, θ and ϕ are ordinary spherical coordinates. It can be shown that the components cannot be transformed to constants simultaneously: spacetime must be curved due to the mass.

If we study a very small region of spacetime, the curvature has little effect. Locally the space is always a Minkowski space where special relativity can be used. Locality means not only a limited spatial volume but also a limited interval in time.

Unlike in the Newtonian formalism in general relativity there are no equations of motion describing the motion of a particle. It is possible to use the positions of the bodies to compute the metric tensor of the space and then the geodesics representing the orbits of the bodies. When the bodies are moving, the metric of the space keeps changing, and thus this method is very laborious. The *PPN* (Parametrized Post-Newtonian) formalism has been developed for practical calculations, yielding approximate equations of motion. It contains ten constants, the values of which are specific for Einstein's theory of gravity. Alternative theories give different values for these constants. The formalism therefore provides a framework for experimental tests of general relativity.

The PPN formalism is an approximation that can be used if velocities are well below the speed of light and the gravitational field is weak and hence the curvature of the space is low.

B.4   Tests of General Relativity

General relativity gives predictions different from classical physics. Although the differences are usually very small, there are some phenomena in which the deviation can be measured and used to test the validity of general relativity. At presents five different astronomical tests have verified the theory.

First of all the orbit of a planet is no longer a closed Keplerian eclipse. The effect is strongest for the innermost planets, whose perihelia should turn little by little.

Most of the motion of the perihelion of Mercury is predicted by Newtonian mechanics; only a small excess of 43 arc seconds per century remains unexplained. And it so happens that this is exactly the correction suggested by general relativity.

Secondly a beam of light should bend when it travels close to the Sun. For a beam grazing the surface, the deviation should be about $1.75''$. Such effects have been observed during total solar eclipses and also by observing pointlike radio sources, just before and after occultation by the Sun.

The third classical way of testing general relativity is to measure the redshift of a photon climbing in a gravitational field. We can understand the redshift as an energy loss while the photon does work against the gravitational potential. Or we can think of the redshift as caused by the metric only: a distant observer finds that near a mass, time runs slower, and the frequency of the radiation is lower. The time dilation in the Schwarzschild metric is described by the coefficient of dt; thus it is not surprising that the radiation emitted at a frequency ν from a distance r from a mass M has the frequency

$$\nu_\infty = \nu\sqrt{1 - \frac{2GM}{c^2 r}} \, , \tag{B.3}$$

if observed very far from the source. This gravitational redshift has been verified by laboratory experiments.

The fourth test employs the slowing down of the speed of light in a gravitational field near the Sun. This has been verified by radar experiments.

The previous tests concern the solar system. Outside the solar system binary pulsars have been used to test general relativity. An asymmetric system in accelerated motion (like a binary star) loses energy as it radiates *gravitational waves*. It follows that the components approach each other, and the period decreases. Usually the gravitational waves play a minor role, but in the case of a compact source the effect may be strong enough to be observed. The first source to show this predicted shortening of period was the binary pulsar PSR 1913+16.

C. Tables

Table C.1. SI basic units

Quantity	Symbol	Unit	Abbr.	Definition
Length	$l, s, \ldots$	Metre	m	The length of the path travelled by light in vacuum during a time interval of $1/299{,}792{,}458$ of a second.
Mass	m, M	Kilogram	kg	Equal to the mass of the international prototype of the kilogram.
Time	t	Second	s	The duration of $9{,}192{,}631{,}770$ periods of the radiation corresponding to the transition between two hyperfine levels of the ground state of the caesium-133 atom.
Electric current	I	Ampere	A	That constant current which, if maintained in two straight parallel conductors of infinite length, of negligible circular cross section and placed 1 metre apart in a vacuum, would produce a force equal to 2×10^{-7} Newton per metre of length between these conductors.
Temperature	T	Kelvin	K	The fraction $1/273.16$ of the thermodynamic temperature of the triple point of water.
Amount of substance	n	Mole	mol	The amount of substance of a system which contains as many elementary entities as there. are atoms in 0.012 kg of ^{12}C
Luminous intensity	I	Candela	cd	The luminous intensity in a given direction of a source which emits monochromatic radiation of frequency 540×10^{12} Hz and of which the radiant intensity in that direction is $1/683$ Watt per steradian.

Table C.2. Prefixes for orders of ten

Prefix	Symbol	Multiple	Prefix	Symbol	Multiple
yocto	y	10^{-24}	deca	da	10^{1}
zepto	z	10^{-21}	hecto	h	10^{2}
atto	a	10^{-18}	kilo	k	10^{3}
femto	f	10^{-15}	Mega	M	10^{6}
pico	p	10^{-12}	Giga	G	10^{9}
nano	n	10^{-9}	Tera	T	10^{12}
micro	μ	10^{-6}	Peta	P	10^{15}
milli	m	10^{-3}	Exa	E	10^{18}
centi	c	10^{-2}	Zetta	Z	10^{21}
deci	d	10^{-1}	Yotta	Y	10^{24}

Table C.3. Constants and units

Radian	1 rad	$= 180°/\pi = 57.2957795°$
		$= 206,264.8''$
Degree	$1°$	$= 0.01745329$ rad
Arc second	$1''$	$= 0.000004848$ rad
Velocity of light	c	$= 299,792,458$ m s^{-1}
Gravitational constant	G	$= 6.673 \times 10^{-11}$ m^3 kg^{-1} s^{-2}
		$= 4\pi^2$ AU^3 $M_\odot^{-1}$ a^{-2}
		$= 3,986,005 \times 10^8$ m^3 $M_\oplus^{-1}$ s^{-2}
Planck constant	h	$= 6.6261 \times 10^{-34}$ J s
	$\hbar$	$= h/2\pi = 1.0546 \times 10^{-34}$ J s
Boltzmann constant	k	$= 1.3807 \times 10^{-23}$ J K^{-1}
Radiation density constant	a	$= 7.5659 \times 10^{-16}$ J m^{-3} K^{-4}
Stefan-Boltzmann constant	σ	$= ac/4$
		$= 5.6705 \times 10^{-8}$ W m^{-2} K^{-4}
Atomic mass unit	amu	$= 1.6605 \times 10^{-27}$ kg
Electron volt	eV	$= 1.6022 \times 10^{-19}$ J
Electron charge	e	$= 1.6022 \times 10^{-19}$ C
Mass of electron	m_e	$= 9.1094 \times 10^{-31}$ kg
		$= 0.511$ MeV
Mass of proton	m_p	$= 1.6726 \times 10^{-27}$ kg
		$= 938.3$ MeV
Mass of neutron	m_n	$= 1.6749 \times 10^{-27}$ kg
		$= 939.6$ MeV
Mass of ^{1}H atom	m_H	$= 1.6735 \times 10^{-27}$ kg
		$= 1.0078$ amu
Mass of $^4_2$He atom	m_{He}	$= 6.6465 \times 10^{-27}$ kg
		$= 4.0026$ amu
Rydberg constant for ^{1}H	R_H	$= 1.0968 \times 10^7$ m^{-1}
Rydberg consnat for ∞ mass	R_∞	$= 1.0974 \times 10^7$ m^{-1}
Gas constant	R	$= 8.3145$ J K^{-1} mol^{-1}
Normal atmospheric pressure	atm	$= 101,325$ Pa $= 1013$ mbar
		$= 760$ mmHg
Astronomical unit	AU	$= 1.49597870 \times 10^{11}$ m
Parsec	pc	$= 3.0857 \times 10^{16}$ m $= 206,265$ AU
		$= 3.26$ ly
Light-year	ly	$= 0.9461 \times 10^{16}$ m $= 0.3066$ pc

Table C.4. Units of time

Unit	Equivalent to
Sidereal year	365.2564 d (with respect to fixed stars)
Tropical year	365.2422 d (equinox to equinox)
Anomalistic year	365.2596 d (perihelion to perihelion)
Gregorian calendar year	365.2425 d
Julian year	365.25 d
Julian century	36,525 d
Eclipse year	346.6200 d (with respect to the ascending node of the Moon)
Lunar year	354.367 d = 12 synodical months
Synodical month	29.5306 d (newmoon to newmoon)
Sidereal month	27.3217 d (with respect to fixed stars)
Tropical month	27.3216 d (with respect to the vernal equinox)
Anomalistic month	27.5546 d (perigee to perigee)
Draconic month	27.2122 d (node to node)
Mean solar day	24 h mean solar time
	$= 24$ h 03 min 56.56 s siderel time
	$= 1.00273791$ sidereal days
Sidereal day	24 h sidereal time
	$= 23$ h 56 min 04.09 s mean solar time
	$= 0.99726957$ mean solar days
Rotation period of the Earth (referred to fixed stars)	1.000000097 sidereal days
	$= 23$ h 56 min 04.10 s mean solar time

Table C.5. The Greek alphabet

A, α	B, β	Γ, γ	Δ, δ	E, ϵ, ε	Z, ζ	H, η	Θ, θ, ϑ
alpha	beta	gamma	delta	epsilon	zeta	eta	theta
I, ι	K, κ	Λ, λ	M, μ	N, ν	Ξ, ξ	O, o	Π, π, ϖ
iota	kappa	lambda	mu	nu	xi	omicron	pi
P, ρ	Σ, σ, ς	T, τ	Υ, υ	Φ, ϕ, φ	X, χ	Ψ, ψ	Ω, ω
rho	sigma	tau	upsilon	phi	chi	psi	omega

Table C.6. The Sun

Property	Symbol	Numerical value
Mass	$M_\odot$	1.989×10^{30} kg
Radius	$R_\odot$	6.960×10^8 m
		$= 0.00465$ AU
Effective temperature	T_e	5785 K
Luminosity	$L_\odot$	3.9×10^{26} W
Apparent visual magnitude	V	-26.78
Colour indices	$B - V$	0.62
	$U - B$	0.10
Absolute visual magnitude	M_V	4.79
Absolute bolometric magnitude	M_{bol}	4.72
Inclination of equator to ecliptic		$7°15'$
Equatorial horizontal parallax	$\pi_\odot$	$8.794''$
Motion: direction of apex		$\alpha = 270°$
		$\delta = 30°$
velocity in LSR		19.7 km s^{-1}
Distance from galactic centre		8.5 kpc

Table C.7. The Earth

Property	Symbol	Numerical value
Mass	$M_\oplus$	$= M_\odot/332{,}946$
		$= 5.974 \times 10^{24}$ kg
Mass, Earth+Moon	$M_\oplus + M_{\mathbb{C}}$	$= M_\odot/328{,}900.5$
		$= 6.048 \times 10^{24}$ kg
Equatorial radius	R_e	$= 6{,}378{,}137$ m
Polar radius	R_p	$= 6{,}356{,}752$ m
Flattening	f	$= (R_e - R_p)/R_e$
		$= 1/298.257$
Surface gravity	g	$= 9.81$ m s^{-2}

Table C.8. The Moon

Property	Symbol	Numerical value
Mass	$M_{\mathbb{C}}$	$= M_\oplus/81.30$
		$= 7.348 \times 10^{22}$ kg
Radius	$R_{\mathbb{C}}$	$= 1738$ km
Surface gravity	$g_{\mathbb{C}}$	$= 1.62$ m s^{-2}
		$= 0.17\, g$
Mean equatorial horizontal parallax	$\pi_{\mathbb{C}}$	$= 57'$
Semimajor axis of the orbit	a	$= 384{,}400$ km
Smallest distance from Earth	r_{min}	$= 356{,}400$ km
Greatest distance from Earth	r_{max}	$= 406{,}700$ km
Mean inclination of orbit to ecliptic	ι	$= 5.145°$

Table C.9. Planets. R_e = equatorial radius, ρ = mean density, τ_{sid} = sidereal rotation period (R indicates retrograde rotation), ε = inclination of the equator with respect to the ecliptic (at the beginning of 2000), f = flattening, T = surface temperature, p = geometric albedo, V_0 = mean opposition magnitude. The data are based mostly on the *Astronomical Almanac*

Name	R_e	Mass			ρ	Number of known satellites
		planet	planet + satellites			
	[km]	[kg]	[$M_{\oplus}$]	[$M_{\odot}$]	[g/cm^3]	
Mercury	2,440	3.30×10^{23}	0.0553	1/6,023,600	5.4	–
Venus	6,052	4.87×10^{24}	0.8150	1/408,523.5	5.2	–
Earth	6,378	5.97×10^{24}	1.0123	1/328,900.5	5.5	1
Mars	3,397	6.42×10^{23}	0.1074	1/3,098,710	3.9	2
Jupiter	71,492	1.90×10^{27}	317.89	1/1,047.355	1.3	39
Saturn	60,268	5.69×10^{26}	95.17	1/3,498.5	0.7	30
Uranus	25,559	8.66×10^{25}	14.56	1/22,869	1.3	20
Neptune	24,764	1.03×10^{26}	17.24	1/19,314	1.8	8
Pluto	1,195	1.5×10^{22}	0.003		1.1	1

Name	τ_{sid}	ε [deg]	f	T [K]	p	V_0
Mercury	58.646 d	0.0	0	130–615	0.106	–
Venus	243.019 d R	177.4	0	750	0.65	–
Earth	23 h 56 min 04.1 s	23.4	0.003353	300	0.367	–
Mars	24 h 37 min 22.6 s	25.2	0.006476	220	0.150	– 2.01
Jupiter	9 h 55 min 30 s	3.1	0.06487	140	0.52	– 2.70
Saturn	10 h 39 min 22 s	26.7	0.09796	100	0.47	+ 0.67
Uranus	17 h 14 min 24 s R	97.8	0.02293	65	0.51	+ 5.52
Neptune	16 h 06 min 36 s	28.3	0.01708	55	0.41	+ 7.84
Pluto	6.387 d R	122.5	0	45	0.3	+15.12

Table C.10. Magnitudes of the planets. The table gives the expressions used in the *Astronomical Almanac*. They give the magnitudes as functions of the phase angle α. Interior planets can be observed at relatively large phase angles, and thus several terms are needed to describe their phase curves. Phase angles of the exterior planets are always small, and thus very simple expressions are sufficient. The magnitude of Saturn means the magnitude of the planet only; the total magnitude also depends on the orientation of the rings

	$V(1, 0)$	$V(1, \alpha)$
Mercury	−0.36	$V(1, 0) + 3.80(\alpha/100°)$ $-2.73(\alpha/100°)^2 + 2.00(\alpha/100°)^3$
Venus	−4.29	$V(1, 0) + 0.09(\alpha/100°)$ $+2.39(\alpha/100°)^2 - 0.65(\alpha/100°)^3$
Mars	−1.52	$V(1, 0) + 1.60(\alpha/100°)$
Jupiter	−9.25	$V(1, 0) + 0.5(\alpha/100°)$
Saturn	−8.88	
Uranus	−7.19	$V(1, 0) + 0.28(\alpha/100°)$
Neptune	−6.87	$V(1, 0)$
Pluto	−1.01	$V(1, 0) + 4.1(\alpha/100°)$

Table C.11. Osculating elements of planetary orbits on JD 2451600.5 (Feb 24, 2000). a = semimajor axis, e = eccentricity i = inclination, Ω = longitude of ascending node, ϖ = longitude of perihelion, L = men longitude, P_{sid} = mean sidereal orbital period (here 1 a means a Julian year, or 365.25 days), P_{syn} = mean synodic period

	a [AU]	[10^6 km]	e	i [°]	Ω [°]	ϖ [°]	L [°]	P_{sid} [a]	[d]	P_{syn} [d]
Mercury	0.387	57.9	0.2056	7.01	48.3	77.5	119.4	0.2408	87.97	115.9
Venus	0.723	108.2	0.0068	3.39	76.7	131.9	270.9	0.6152	224.7	583.9
Earth	1.000	149.6	0.0167	0.00	143.9	102.9	155.2	1.0000	365.3	–
Mars	1.524	228.0	0.0934	1.85	49.6	336.1	24.5	1.8807	686.9	779.9
Jupiter	5.204	778.6	0.0488	1.30	100.5	15.5	39.0	11.8565	4,330.6	398.9
Saturn	9.582	1433.4	0.0558	2.49	113.6	89.9	51.9	29.4235	10,747	378.1
Uranus	19.224	2875.8	0.0447	0.77	74.0	170.3	314.1	83.747	30,589	369.7
Neptune	30.092	4501.7	0.0112	1.77	131.8	39.5	305.5	163.723	59,800	367.5
Pluto	39.257	5872.8	0.2446	17.15	110.3	224.0	239.3	248.02	90,590	366.7

Table C.12. Mean elements of planets with respect to the equator and equinox of J2000.0. The variable t is the time in days since J2000.0 and T is the same time in Julian centuries: $t = J - 2451545.0$, $T = t/36525$. L is the mean longitude, $L = M + \varpi$. The elements do not contain periodic terms, and the accuracy of the positions computed from them is of the order of a few minutes of arc. The values are from the *Explanatory Supplement to the Astronomical Almanac*. The elements of the Earth describe the orbit of the barycentre of the Earth–Moon system

Mercury	$a = 0.38709893 + 0.00000066T$	$e = 0.20563069 + 0.00002527T$
	$i = 7.00487° - 23.51''T$	$\Omega = 48.33167° - 446.30''T$
	$\varpi = 77.45645° + 573.57''T$	$L = 252.25084° + 4.09233880°t$
Venus	$a = 0.72333199 + 0.00000092T$	$e = 0.00677323 - 0.00004938T$
	$i = 3.39471° - 2.86''T$	$\Omega = 76.68069° - 996.89''T$
	$\varpi = 131.53298° - 108.80''T$	$L = 181.97973° + 1.60213047°t$
Earth + Moon	$a = 1.00000011 - 0.00000005T$	$e = 0.01671022 - 0.00003804T$
	$i = 0.00005° - 46.94''T$	$\Omega = -11.26064° - 18,228.25''T$
	$\varpi = 102.94719° + 1198.28''T$	$L = 100.46435° + 0.98560910°t$
Mars	$a = 1.52366231 - 0.00007221T$	$e = 0.09341233 + 0.00011902T$
	$i = 1.85061° - 25.47''T$	$\Omega = 49.57854° - 1020.19''T$
	$\varpi = 336.04084° + 1560.78''T$	$L = 355.45332° + 0.52403304°t$
Jupiter	$a = 5.20336301 + 0.00060737T$	$e = 0.04839266 - 0.00012880T$
	$i = 1.30530° - 4.15''T$	$\Omega = 100.55615° + 1217.17''T$
	$\varpi = 14.75385° + 839.93''T$	$L = 34.40438° + 0.08308676°t$
Saturn	$a = 9.53707032 - 0.00301530T$	$e = 0.05415060 - 0.00036762T$
	$i = 2.48446° + 6.11''T$	$\Omega = 113.71504° - 1591.05''T$
	$\varpi = 92.43194° - 1948.89''T$	$L = 49.94432° + 0.03346063°t$
Uranus	$a = 19.19126393 + 0.00152025T$	$e = 0.04716771 - 0.00019150T$
	$i = 0.76986° - 2.09''T$	$\Omega = 74.22988° + 1681.40''T$
	$\varpi = 170.96424° + 1312.56''T$	$L = 313.23218° + 0.01173129°t$
Neptune	$a = 30.06896348 - 0.00125196T$	$e = 0.00858587 + 0.00002514T$
	$i = 1.76917° - 3.64''T$	$\Omega = 131.72169° - 151.25''T$
	$\varpi = 44.97135° - 844.43''T$	$L = 304.88003° + 0.00598106°t$
Pluto	$a = 39.48168677 - 0.00076912T$	$e = 0.24880766 + 0.00006465T$
	$i = 17.14175° + 11.07''T$	$\Omega = 110.30347° - 37.33''T$
	$\varpi = 224.06676° - 132.25''T$	$L = 238.92881° + 0.00397557°t$

Table C.13. Largest satellites of the planets. The giant planets have a large number of satellites, and new ones are found frequently. Also the distinction between a large ring particle and a small satellite is somewhat arbitrary. Therefore this table is not complete. $a =$ semimajor axis, $P_{sid} =$ sidereal period (tropical for Saturn's moons), R means that the motion is retrograde, $e =$ eccentricity, $i =$ inclination with respect to the equator (E = ecliptic), $r =$ radius $M =$ mass in planetary masses, $\rho =$ density (calculated), $p =$ geometric albedo and V_0 mean opposition magnitude

	Discoverer	Year of discovery	a [10^3 km]	a [R_p]	P_{sid} [d]	e	i [°]	r [km]	M [M_{plan}]	ρ [g/cm³]	p	V_0
Earth												
Moon			384.4	60.27	27.3217	0.055	18.28–28.58	1737	0.0123	3.34	0.12	−12.74
Mars												
Phobos	Hall	1877	9.38	2.76	0.3189	0.015	1.0	13×11×9	1.7×10^{-8}	2.0	0.07	11.3
Deimos	Hall	1877	23.46	6.91	1.2624	0.0005	0.9–2.7	7×6×5	3.7×10^{-9}	2.7	0.08	12.4
Jupiter												
XVI Metis	Synott	1979	128	1.79	0.295			20	5×10^{-11}	2.8	0.05	17.5
XV Adrastea	Jewitt Danielson Synott	1979	129	1.80	0.298			13×10×8	1×10^{-11}	4.4	0.05	19.1
V Amalthea	Barnard	1892	181	2.53	0.498	0.003	0.40	131×73×67	3.8×10^{-9}	2.7	0.07	14.1
XIV Thebe	Synott	1979	222	3.11	0.674	0.015	0.8	55×45	4×10^{-10}	1.3	0.04	15.7
I Io	Galilei	1610	422	5.90	1.769	0.004	0.04	1830×1819×1815	4.7×10^{-5}	3.5	0.63	5.0
II Europa	Galilei	1610	671	9.39	3.551	0.009	0.47	1565	2.5×10^{-5}	3.0	0.67	5.3
III Ganymedes	Galilei	1610	1,070	15.0	7.155	0.002	0.21	2634	7.8×10^{-5}	1.9	0.44	4.6
IV Callisto	Galilei	1610	1,883	26.3	16.689	0.007	0.51	2403	5.7×10^{-5}	1.9	0.20	5.6
XIII Leda	Kowal	1974	11,094	155.2	238.72	0.148	26.07	5	3×10^{-12}		0.07	20.2
VI Himalia	Perrine	1905	11,480	160.6	250.566	0.158	27.63	85	5×10^{-9}		0.03	14.8
X Lysithea	Nicholson	1938	11,720	163.9	259.22	0.107	29.02	12	4×10^{-11}		0.06	18.4
VII Elara	Perrine	1905	11,737	164.2	259.653	0.207	24.77	40	4×10^{-10}		0.03	16.8
XII Ananke	Nicholson	1951	21,200	297	631 R	0.169	147	10	2×10^{-11}		0.06	18.9
XI Carme	Nicholson	1938	22,600	316	692 R	0.207	164	15	5×10^{-11}		0.06	18.0
VIII Pasiphae	Melotte	1908	23,500	329	735 R	0.378	145	18	1×10^{-10}		0.10	17.0
IX Sinope	Nicholson	1914	23,700	332	758 R	0.275	153	14	4×10^{-11}		0.05	18.3
Saturn												
XVIII Pan	Showalter	1990	133.58	2.22	0.575	0.000	0.3	10			0.5	18
XV Atlas	Terrile	1980	137.67	2.28	0.602	0.003	0.0	18×17×13			0.8	18
XVI Prometheus	Collins Carlson	1980	139.35	2.31	0.613	0.004	0.0	74×50×34			0.5	16
XVII Pandora	Collins Carlson	1980	141.70	2.35	0.629	0.004	0	55×44×31			0.7	16
XI Epimetheus	Cruikshank	1980	151.42	2.51	0.694	0.009	0.34	69×55×55	9.5×10^{-10}	0.6	0.8	15
X Janus	Pascu	1980	151.47	2.51	0.695	0.007	0.14	97×95×77	3.4×10^{-9}	0.7	0.9	14
I Mimas	W. Herschel	1789	185.52	3.08	0.942	0.020	1.53	209×196×191	6.6×10^{-8}	1.1	0.5	12.9
II Enceladus	W. Herschel	1789	238.02	3.95	1.370	0.005	0.00	256×247×245	1×10^{-7}	0.9	1.0	11.7
XIII Telesto	Smith Larson Reitsema	1980	294.66	4.89	1.888			15×12×7			1.0	18.5
III Tethys	Cassini	1684	294.66	4.89	1.888	0.000	1.86	536×528×526	1.1×10^{-6}	1.0	0.9	10.2

		Discoverer	Year of discovery	a [10^3 km]	a [R_p]	P_{sid} [d]	e	i [°]	r [km]	M [M_{plan}]	ρ [g/cm^3]	p	V_0
XIV	Calypso	Pascu Seidelmann Baum Currie	1980	294.66	4.89	1.888			$15 \times 8 \times 8$			1.0	18.7
IV	Dione	Cassini	1684	377.40	6.26	2.737	0.002	0.02	560	1.9×10^{-6}	1.5	0.7	10.4
XII	Helene	Laques Lecacheux	1980	377.40	6.26	2.737	0.005	0.0	$18 \times 16 \times 15$			0.7	18
V	Rhea	Cassini	1672	527.04	8.74	4.517	0.001	0.35	764	4.1×10^{-6}	1.2	0.7	9.7
VI	Titan	Huygens	1665	1,221.83	20.3	15.945	0.03	0.33	2575	2.4×10^{-4}	1.9	0.22	8.3
VII	Hyperion	Bond	1848	1,481.1	24.6	21.277	0.10	0.43	$180 \times 140 \times 113$	4×10^{-8}	1.4	0.3	14.2
VIII	Iapetus	Cassini	1671	3,561.3	59.1	79.330	0.028	14.72	718	2.8×10^{-6}	1.0	0.5/0.05	11.1
IX	Phoebe	Pickering	1898	12,952	215	550.48 R	0.163	177 E	110	7×10^{-10}?		0.06	16.4
Uranus													
VI	Cordelia	Voyager 2	1986	49.77	1.95	0.335	0.00	0.08	13			0.07	24.1
VII	Ophelia	Voyager 2	1986	53.79	2.10	0.376	0.01	0.10	15			0.07	23.8
VIII	Bianca	Voyager 2	1986	59.17	2.32	0.435	0.00	0.19	21			0.07	23.0
IX	Cressida	Voyager 2	1986	61.78	2.42	0.464	0.00	0.01	31			0.07	22.2
X	Desdemona	Voyager 2	1986	62.68	2.45	0.474	0.00	0.11	27			0.07	22.5
XI	Juliet	Voyager 2	1986	64.35	2.52	0.493	0.00	0.07	42			0.07	21.5
XII	Portia	Voyager 2	1986	66.09	2.59	0.513	0.00	0.06	54			0.07	21.0
XIII	Rosalind	Voyager 2	1986	69.94	2.74	0.558	0.00	0.28	27			0.07	22.5
XIV	Belinda	Voyager 2	1986	75.26	2.94	0.624	0.00	0.03	33			0.07	22.1
XVIII	S/1986U10	Karkoschka	1999	77.30	3.0	0.637			20			0.07	
XV	Puck	Voyager 2	1985	86.01	3.37	0.762	0.00	0.32	77			0.07	20.2
V	Miranda	Kuiper	1948	129.39	5.06	1.413	0.003	4.2	$240 \times 234 \times 233$	8×10^{-7}	1.3	0.27	16.3
I	Ariel	Lassell	1851	191.02	7.47	2.520	0.003	0.3	$581 \times 578 \times 578$	1.6×10^{-5}	1.7	0.35	14.2
II	Umbriel	Lassell	1851	266.30	10.42	4.144	0.005	0.36	585	1.4×10^{-5}	1.4	0.19	14.8
III	Titania	W. Herschel	1787	435.91	17.06	8.706	0.002	0.14	789	4.1×10^{-5}	1.7	0.28	13.7
IV	Oberon	W. Herschel	1787	583.52	22.83	13.463	0.001	0.10	761	3.5×10^{-5}	1.6	0.25	13.9
XVI	Caliban	Nicholson	1997	7,169	281	579 R	0.08	140 E	30			0.07	22.4
XVII	Sycorax	Nicholson	1997	12,214	477	1,289 R	0.5	153 E	60			0.07	20.9
Neptune													
III	Naiad	Voyager 2	1989	48.23	1.95	0.294	0.000	4.74	29			0.06	24.7
IV	Thalassa	Voyager 2	1989	50.07	2.02	0.311	0.000	0.21	40			0.06	23.8
V	Despina	Voyager 2	1989	52.53	2.12	0.335	0.000	0.07	74			0.06	22.6
VI	Galatea	Voyager 2	1989	61.95	2.50	0.429	0.000	0.05	79			0.06	22.3
VII	Larissa	Voyager 2	1989	73.55	2.97	0.555	0.001	0.20	104×89			0.06	22.0
VIII	Proteus	Voyager 2	1989	117.65	4.75	1.122	0.000	0.55	$218 \times 208 \times 201$			0.06	20.3
I	Triton	Lassell	1846	354.76	14.33	5.877R	0.000	157.35	1353	2.1×10^{-4}	2.1	0.77	13.5
II	Nereid	Kuiper	1949	5,513.4	222.6	360.136	0.751	27.6	170	2×10^{-7}	1.0	0.4	18.7
Pluto													
I	Charon	Christy	1978	19.6	17	6.387	0.00	99	593	0.125	2.1	0.5	16.8

Table C.14. Some well-known asteroids. a = semimajor axis, e = eccentricity, i = incination, d = diameter, τ_{sid} = sidereal period of rotation, p = geometric albedo, V_0 = mean opposition magnitude

	Asteroidi	Discoverer	Year of discovery	a [AU]	e	i [°]	d [km]	τ_{sid} [h]	p	V_0	Type
1	Ceres	Piazzi	1801	2.77	0.08	10.6	946	9.08	0.07	7.9	C
2	Pallas	Olbers	1802	2.77	0.23	34.8	583	7.88	0.09	8.5	U
3	Juno	Harding	1804	2.67	0.26	13.0	249	7.21	0.16	9.8	S
4	Vesta	Olbers	1807	2.36	0.09	7.1	555	5.34	0.26	6.8	U
5	Astraea	Hencke	1845	2.58	0.19	5.3	116	16.81	0.13	11.2	S
6	Hebe	Hencke	1847	2.42	0.20	14.8	206	7.27	0.16	9.7	S
7	Iris	Hind	1847	2.39	0.23	5.5	222	7.14	0.20	9.4	S
8	Flora	Hind	1847	2.20	0.16	5.9	160	13.60	0.13	9.8	S
9	Metis	Graham	1848	2.39	0.12	5.6	168	5.06	0.12	10.4	S
10	Hygiea	DeGasparis	1849	3.14	0.12	3.8	443	18.00	0.05	10.6	C
243	Ida	Palisa	1884	2.86	0.04	1.1	32	4.63	0.16		
433	Eros	Witt	1898	1.46	0.22	10.8	20	5.27	0.18	11.5	S
588	Achilles	Wolf	1906	5.18	0.15	10.3	70	?	?	16.4	U
624	Hektor	Kopff	1907	5.16	0.03	18.3	230	6.92	0.03	15.3	U
944	Hidalgo	Baade	1920	5.85	0.66	42.4	30	10.06	?	19.2	MEU
951	Gaspra	Neujmin	1916	2.21	0.17	4.1	19	7.04	0.15		
1221	Amor	Delporte	1932	1.92	0.43	11.9	5	?	?	20.4	?
1566	Icarus	Baade	1949	1.08	0.83	22.9	2	2.27	?	12.3	U
1862	Apollo	Reinmuth	1932	1.47	0.56	6.4	2	?	?	16.3	?
2060	Chiron	Kowal	1977	13.64	0.38	6.9	320	?	?	17.3	?
5145	Pholus	Rabinowitz	1992	20.46	0.58	24.7	190				

Table C.15. Principal meteor showers

Shower	Period of visibility	Maximum	Radiant		Meteors per hour	Comet
			α	δ		
Quadrantids	Jan 1–5	Jan 3–4	15.5 h	+50°	30–40	
Lyrids	Apr 19–25	Apr 22	18.2 h	+34°	10	Thatcher
η Aquarids	May 1–12	May 5	22.4 h	− 1°	5–10	Halley
Perseids	Jul 20–Aug 18	Aug 12	3.1 h	+58°	40–50	Swift-Tuttle
κ Cygnids	Aug 17–24	Aug 20	19.1 h	+59°	5	
Orionids	Oct 17–26	Oct 21	6.3 h	+16°	10–15	Halley
Taurids	Oct 10–Dec 5	Nov 1	3.8 h	+14°, +22°	5	Encke
Leonids	Nov 14–20	Nov 17	10.2 h	+22°	10	Tempel-Tuttle
Geminids	Dec 7–15	Dec 13–14	7.5 h	+33°	40–50	
Ursids	Dec 17–24	Dec 22	13.5 h	+78°	5	

Table C.16. Periodic comets with several perihelion passages observed. N = number of passages observed, τ = time of perihelion passage, P = sidereal period, q = perihelion distance, e = eccentricity, ω = argument of perihelion (1950.0), Ω = longitude of ascending node (1950.0), i = inclination, l = longitude of perihelion, defined here as $l = \Omega + \arctan(\tan \omega \cos i)$, b = latitude of perihelion ($\sin b = \sin \omega \sin i$), Q = aphelion distance. The elements are affected by planetary perturbations as well as reaction forces due to evaporating material the amount of which is difficult to predict

Comet	N	τ	P [a]	q [AU]	e	ω [°]	Ω [°]	i [°]	l [°]	b [°]	Q [AU]
Encke	56	Feb 9, 1994	3.28	0.331	0.850	186.3	334.0	11.9	160.2	− 1.3	4.09
Grigg-Skjellerup	16	Jul 24, 1992	5.10	0.995	0.664	359.3	212.6	21.1	212.0	− 0.3	4.93
Honda-Mrkos-Pajdušáková	8	Sep 12, 1990	5.30	0.541	0.822	325.8	88.6	4.2	54.5	− 2.4	5.54
Tuttle-Giacobini-Kresák	8	Feb 8, 1990	5.46	1.068	0.656	61.6	140.9	9.2	202.1	8.1	5.14
Tempel 2	19	Mar 16, 1994	5.48	1.484	0.522	194.9	117.6	12.0	312.1	− 3.1	4.73
Wirtanen	8	Sep 21, 1991	5.50	1.083	0.652	356.2	81.6	11.7	77.9	− 0.8	5.15
Clark	8	Nov 28, 1989	5.51	1.556	0.501	208.9	59.1	9.5	267.7	− 4.6	4.68
Forbes	8	Mar 15, 1993	6.14	1.450	0.568	310.6	333.6	7.2	284.5	− 5.4	5.25
Pons-Winnecke	20	Aug 19, 1989	6.38	1.261	0.634	172.3	92.8	22.3	265.6	2.9	5.62
d'Arrest	15	Feb 4, 1989	6.39	1.292	0.625	177.1	138.8	19.4	316.0	1.0	5.59
Schwassmann-Wachmann 2	11	Jan 24, 1994	6.39	2.070	0.399	358.2	125.6	3.8	123.8	− 0.1	4.82
Wolf-Harrington	8	Apr 4, 1991	6.51	1.608	0.539	187.0	254.2	18.5	80.8	− 2.2	5.37
Ciacobini-Zinner	12	Apr 14, 1992	6.61	1.034	0.707	172.5	194.7	31.8	8.3	3.9	6.01
Reinmuth 2	8	Jun 29, 1994	6.64	1.893	0.464	45.9	295.4	7.0	341.1	5.0	5.17
Perrine-Mrkos	8	Mar 1, 1989	6.78	1.298	0.638	166.6	239.4	17.8	46.6	4.1	5.87
Arend-Rigaux	7	Oct 3, 1991	6.82	1.438	0.600	329.1	121.4	17.9	91.7	− 9.1	5.75
Borrelly	11	Dec 18, 1987	6.86	1.357	0.624	353.3	74.8	30.3	69.0	− 3.4	5.86
Brooks 2	14	Sep 1, 1994	6.89	1.843	0.491	198.0	176.2	5.5	14.1	− 1.7	5.40
Finlay	11	Jun 5, 1988	6.95	1.094	0.700	322.2	41.7	3.6	4.0	− 2.2	6.19
Johnson	7	Nov 19, 1990	6.97	2.313	0.366	208.3	116.7	13.7	324.3	− 6.4	4.98
Daniel	8	Aug 31, 1992	7.06	1.650	0.552	11.0	68.4	20.1	78.7	3.8	5.71
Holmes	8	Apr 10, 1993	7.09	2.177	0.410	23.2	327.3	19.2	349.4	7.4	5.21
Reinmuth 1	8	May 10, 1988	7.29	1.869	0.503	13.0	119.2	8.1	132.0	1.8	5.65
Faye	19	Nov 15, 1991	7.34	1.593	0.578	204.0	198.9	9.1	42.6	− 3.7	5.96
Ashbrook-Jackson	7	Jul 13, 1993	7.49	2.316	0.395	348.7	2.0	12.5	350.9	− 2.4	5.34
Schaumasse	10	Mar 5, 1993	8.22	1.202	0.705	57.5	80.4	11.8	137.3	10.0	6.94
Wolf	14	Aug 28, 1992	8.25	2.428	0.406	162.3	203.4	27.5	7.6	8.1	5.74
Whipple	9	Dec 22, 1994	8.53	3.094	0.259	201.9	181.8	9.9	23.4	− 3.7	5.25
Comas Solá	8	Aug 18, 1987	8.78	1.830	0.570	45.5	60.4	13.0	105.2	9.2	6.68
Väisälä 1	6	Apr 29, 1993	10.8	1.783	0.635	47.4	134.4	11.6	181.2	8.5	7.98
Tuttle	11	Jun 27, 1994	13.5	0.998	0.824	206.7	269.8	54.7	106.1	−21.5	10.3
Halley	30	Feb 9, 1986	76.0	0.587	0.967	111.8	58.1	162.2	305.3	16.4	35.3

Table C.17. Nearest stars. V = apparent visual magnitude, $B - V$ = colour index, r = distance, μ = proper motion, v_{r} = radial velocity, (positive for receding objects)

Name	α_{2000} [h]	[min]	δ_{2000} [°]	[']	V	$B - V$	Spectrum	r [pc]	μ ["/a]	v_{r} [km/s]
Sun					−26.8	0.6	G2V			
α Cen C (Proxima)	14	29.7	−62	41	11.0	2.0	M5eV	1.30	3.9	− 16
α Cen A	14	39.6	−60	50	− 0.0	0.7	G2V	1.33	3.7	− 22
α Cen B	14	39.6	−60	50	1.3	0.9	K1V	1.33	3.7	− 22
Barnard's star	17	57.8	4	42	9.5	1.7	M5V	1.83	10.3	−108
Wolf 359	10	56.5	7	01	13.5	2.0	M6eV	2.39	4.7	+ 13
BD+36°2147	11	03.3	35	58	7.5	1.5	M2V	2.54	4.8	− 86
α CMa (Sirius) A	6	45.1	−16	43	− 1.5	0.0	A1V	2.66	1.3	− 8
α CMa (Sirius) B	6	45.1	−16	43	8.4		wdA	2.66	1.3	− 8
Luyten 726–8 A	1	39.0	−17	57	12.5		M6eV	2.66	3.3	+ 29
Luyten 726–8 B (UV Cet)	1	39.0	−17	57	13.0		M6eV	2.66	3.3	+ 32
Ross 154	18	49.8	−23	50	10.4		M4eV	2.92	0.7	− 4
Ross 248	23	41.9	44	11	12.2	1.9	M5eV	3.13	1.6	− 81
ε Eri	3	32.9	− 9	27	3.7	0.9	K2V	3.26	1.0	+ 16
Ross 128	11	47.7	0	48	11.1	1.8	M5V	3.31	1.4	− 13
Luyten 789–6 A	22	38.6	−15	17	12.8	2.0	M5eV	3.40	3.3	− 60
Luyten 789–6 B	22	38.6	−15	17	13.3			3.40	3.3	− 60
BD+43°44 A	0	18.4	44	01	8.1	1.6	M3V	3.44	2.9	+ 13
BD+43°44 B	0	18.4	44	01	11.0	1.8	M6V	3.44	2.9	+ 20
ε Ind	22	03.4	−56	47	4.7	1.1	K5V	3.45	4.7	− 40
BD+59°1915 A	18	42.8	59	38	8.9	1.5	M4V	3.45	2.3	0
BD+59°1915 B	18	42.8	59	38	9.7	1.6	M4V	3.45	2.3	+ 10
61 Cyg A	21	06.9	38	45	5.2	1.2	K5V	3.46	5.2	− 64
61 Cyg B	21	06.9	38	45	6.0	1.4	K7V	3.46	5.2	− 64
τ Cet	1	43.1	−15	56	3.5	0.7	G8V	3.48	1.9	− 16
CD−36°15693	23	05.9	−35	51	7.4	1.5	M2V	3.51	6.9	+ 10
α CMi (Procyon) A	7	39.3	5	14	0.4	0.4	F5IV	3.51	1.3	− 3
α CMi (Procyon) B	7	39.3	5	14	10.7		wdF	3.51	1.3	
G 51–15	8	29.8	26	47	14.8		M7V	3.62	1.3	
BD+5°1668	7	27.4	5	13	9.8	1.6	M4V	3.76	3.8	+ 26
Luyten 725–32	1	12.6	−17	00	11.8		M6eV	3.77	1.4	
Kapteyn's star	5	11.7	−45	01	8.8	1.6	M1VI	3.85	8.8	+245
CD−39°14192	21	17.2	−38	52	6.7	1.4	M0eV	3.85	3.5	+ 21
Krüger 60 A	22	28.0	57	42	9.9	1.6	M3V	3.95	0.9	− 26
Krüger 60 B	22	28.0	57	42	11.5	1.8	M4eV	3.95	0.9	− 26
Ross 614 A	6	29.4	− 2	49	11.2	1.7	M4eV	4.13	1.0	+ 24
Ross 614 B	6	29.4	− 2	49	14.8			4.13	1.0	+ 24
BD−12°4523	16	30.3	−12	40	10.2	1.6	M5V	4.15	1.2	− 13
Wolf 424 A	12	33.3	9	01	13.2	1.8	M6V	4.29	1.8	− 5
Wolf 424 B	12	33.3	9	01	13.2			4.29	1.8	− 5
van Maanen's star	0	49.2	5	23	12.4	0.6	wdG	4.33	3.0	+ 54
Luyten 1159–16	2	00.2	13	03	12.2		M5eV	4.48	2.1	
CD−37°15492	0	05.4	−37	21	8.6	1.5	M3V	4.48	6.1	+ 23
Luyten 143–23	10	44.5	−61	12	13.9		dM	4.48	1.7	
CD−46°11540	17	28.7	−46	54	9.4	1.5	M3	4.52	1.1	
LP 731–58	10	48.2	−11	20	15.6		M7V	4.55	1.6	
Luyten 145–141	11	45.7	−64	50	11.4	0.2	wdA	4.57	2.7	
BD+68°946	17	36.4	68	20	9.1	1.5	M3V	4.63	1.3	− 22
CD−49°13515	21	33.6	−49	01	8.7	1.5	M2V	4.63	0.8	+ 8
BD+50°1725	10	11.3	49	27	6.6	1.4	K2V	4.67	1.5	− 26
G 158–27	0	06.7	−07	32	13.7		M5V	4.67	2.1	
BD−15°6290	22	53.3	−14	18	10.2	1.6	M4V	4.69	1.1	+ 9
CD−44°11909	17	37.1	−44	19	11.0		M5V	4.72	1.1	

Table C.17 (continued)

Name	α_{2000} [h]	[min]	δ_{2000} [°]	[']	V	B − V	Spectrum	r [pc]	μ ["/a]	v_r [km/s]
G 208–44/45 A	19	53.9	44	25	13.4		M6eV	4.72	0.7	
G 208–44/45 B	19	53.9	44	25	14.3		dM	4.72	0.7	
G 208–44/45 C	19	53.9	44	25	15.5		dM	4.72	0.7	
o^2 Eri A	4	15.3	− 7	39	4.4	0.8	K0V	4.76	4.0	− 43
o^2 Eri B	4	15.3	− 7	39	9.5	0.0	wdA	4.76	4.0	− 21
o^2 Eri C	4	15.3	− 7	39	11.2	1.7	M4eV	4.76	4.0	− 45
BD+20°2465	10	19.6	19	52	9.4	1.5	M4V	4.88	0.5	+ 11
70 Oph A	18	05.5	2	30	4.2	0.9	K0V	4.98	1.1	− 7
70 Oph B	18	05.5	2	30	6.0		K5V	4.98	1.1	− 10
BD+44°2051 A	11	05.5	43	32	8.7		M2V	5.00	4.5	+ 65
BD+44°2051 B	11	05.5	43	32	14.4		M5eV	5.00	4.5	+ 65
α Aql (Altair)	19	50.8	8	52	0.8	0.2	A7V	5.08	0.7	− 26

Table C.18. Brightest stars ($V \leq 2$). V = apparent visual magnitude, $B - V$ = colour index, r = distance. Remarks: b=binary, sb=spectroscopic binary, v=variable

Name		α_{2000} [h]	[min]	δ_{2000} [°]	[']	V	$B - V$	Spectrum	r [pc]	Remarks
α CMa	Sirius	6	45.2	−16	43	−1.5	0.0	A1V,wdA	2.7	b
α Car	Canopus	6	24.0	−52	42	−0.7	0.2	A9II	60	
α Cen	Rigil Kentaurus	14	39.6	−60	50	−0.3	0.7	G2V,K1V	1.3	b, Proxima 2.2° apart
α Boo	Arcturus	14	15.7	19	11	−0.0	1.2	K2IIIp	11	
α Lyr	Vega	18	36.9	38	47	0.0	0.0	A0V	8	
α Aur	Capella	5	16.7	46	00	0.1	0.8	G2III,G6III	14	b
β Ori	Rigel	5	14.5	− 8	12	0.1	−0.0	B8Ia	90	b
α CMi	Procyon	7	39.3	5	14	0.4	0.4	F5IV,wdF	3.5	b
α Eri	Achernar	1	37.7	−57	14	0.5	−0.2	B3Vp	40	
α Ori	Betelgeuze	5	55.2	7	24	0.5	1.9	M2I	200	v 0.4 − 1.3, sb
β Cen	Hadar	14	03.8	−60	22	0.6	−0.2	B1III	60	b
α Aql	Altair	19	50.8	8	52	0.8	0.2	A7V	5.1	
α Cru	Acrux	12	26.6	−63	06	0.8	−0.3	B0.5IV,B1V	120	b 1.6+2.1
α Tau	Aldebaran	4	35.9	16	31	0.9	1.5	K5III	20	b, v
α Vir	Spica	13	25.2	−11	10	1.0	−0.2	B1IV	50	sb, several comp.
α Sco	Antares	16	29.4	−26	26	1.0	1.8	M1.5I,B2.5V	50	v 0.9 − 1.8
β Gem	Pollux	7	45.3	28	02	1.2	1.1	K0III	11	
α PsA	Fomalhaut	22	57.6	−29	37	1.2	0.1	A3V	7.0	
α Cyg	Deneb	20	41.4	45	17	1.3	0.1	A2Ia	500	
β Cru	Mimosa	12	47.7	−59	41	1.3	−0.2	B0.5III	150	v, sb
α Leo	Regulus	10	08.4	11	58	1.4	−0.1	B7V	26	b
ϵ CMa	Adhara	6	58.6	−28	58	1.5	−0.2	B2II	170	b
α Gem	Castor	7	34.6	31	53	1.6	0.0	A1V,A2V	14	b
γ Cru	Gacrux	12	31.2	−57	07	1.6	1.6	M3.5III	40	v
λ Sco	Shaula	17	33.6	−37	06	1.6	−0.2	B1.5IV		v
γ Ori	Bellatrix	5	25.1	6	21	1.6	−0.2	B2III	40	
β Tau	Elnath	5	26.3	28	36	1.7	−0.1	B7III	55	
β Car	Miaplacidus	9	13.2	−69	43	1.7	0.0	A1III	30	
ϵ Ori	Alnilam	5	36.2	− 1	12	1.7	−0.2	B0Ia		
α Gru	Al Na'ir	22	08.2	−46	58	1.7	−0.1	B7IV	20	
ϵ UMa	Alioth	12	54.0	55	58	1.8	−0.0	A0IVp	120	v
γ Vel	Regor	8	09.5	−47	20	1.8	−0.2	WC8,B1IV		b 1.8 + 4.3, each sb
α Per	Mirfak	3	24.3	49	52	1.8	0.5	F5Ib	35	
α UMa	Dubhe	11	03.7	61	45	1.8	1.1	K0III	30	b
ϵ Sgr	Kaus Australis	18	24.2	−34	23	1.9	−0.0	A0II	70	
δ CMa	Wezen	7	08.4	−26	23	1.9	0.7	F8Ia		
ϵ Car	Avior	8	22.5	−59	31	1.9	1.3	K3III,B2V	25	sb
η UMa	Alkaid	13	47.5	49	19	1.9	−0.2	B3V		
θ Sco	Girtab	17	37.3	−43	00	1.9	0.4	F0II	50	
β Aur	Menkalinan	5	59.6	44	57	1.9	0.0	A1IV	30	
ζ Ori	Alnitak	5	40.8	− 1	57	1.9	−0.2	O9.5Ib, B0III	45	b 2.1 + 4.2
α TrA	Atria	16	48.7	−69	02	1.9	1.4	K2II–III	40	
γ Gem	Alhena	6	37.7	16	24	1.9	0.0	A1IV	30	
α Pav	Peacock	20	25.7	−56	44	1.9	−0.2	B3V		
δ Vel		8	44.7	−54	43	2.0	0.0	A1V	20	
β CMa	Mirzam	6	22.7	−17	57	2.0	−0.2	B1II–III	70	v
α Hya	Alphard	9	27.6	− 8	40	2.0	1.4	K3II–III	60	
α Ari	Hamal	2	07.2	23	28	2.0	1.2	K2III	25	

Table C.19. Some double stars. Magnitudes of the components are m_1 and m_2, and the angular separation d; r is the distance of the star

Name		α_{2000} [h]	[min]	δ_{2000} [°]	[']	m_1	m_2	Spectrum		d ["]	r [pc]
η Cas	Achird	0	49.1	57	49	3.7	7.5	G0V	M0	12	6
γ Ari	Mesarthim	1	53.5	19	18	4.8	4.9	A1p	B9V	8	40
α Psc	Alrescha	2	02.0	2	46	4.3	5.3	A0p	A3m	2	60
γ And	Alamak	2	03.9	42	20	2.4	5.1	K3IIb	B8V,A0V	10	100
δ Ori	Mintaka	5	32.0	− 0	18	2.5	7.0	B0III,O9V	B2V	52	70
λ Ori	Meissa	5	35.1	9	56	3.7	5.7	O8e	B0.5V	4	140
ζ Ori	Alnitak	5	40.8	− 1	56	2.1	4.2	O9.5Ibe	B0III	2	40
α Gem	Castor	7	34.6	31	53	2.0	3.0	A1V	A5Vm	3	15
γ Leo	Algieba	10	20.0	19	50	2.6	3.8	K1III	G7III	4	80
ξ UMa	Alula Australis	11	18.2	31	32	4.4	4.9	G0V	G0V	1	7
α Cru	Acrux	12	26.6	−63	06	1.6	2.1	B0.5IV	B1V	4	120
γ Vir	Porrima	12	41.7	− 1	27	3.7	3.7	F0V	F0V	3	10
α CVn	Cor Caroli	12	56.1	38	18	2.9	5.5	A0p	F0V	20	40
ζ UMa	Mizar	13	23.9	54	56	2.4	4.1	A1Vp	A1m	14	21
α Cen	Rigil Kentaurus	14	39.6	−60	50	0.0	1.3	G2V	K1V	21	1.3
ε Boo	Izar	14	45.0	27	04	2.7	5.3	K0II-III	A2V	3	60
δ Ser		15	34.8	10	32	4.2	5.3	F0IV	F0IV	4	50
β Sco	Graffias	16	05.4	−19	48	2.6	4.9	B1V	B2V	14	110
α Her	Rasalgethi	17	14.6	14	23	3.0–4.0	5.7	M5Ib-II	G5III,F2V	5	120
ρ Her		17	23.7	37	08	4.5	5.5	B9.5III	A0V	4	
70 Oph		18	05.5	2	30	4.3	6.1	K0V	K5V	2	5
ε Lyr		18	44.3	39	40	4.8	4.4	A4V,F1V	A8V,F0V	208	50
ε¹ Lyr		18	44.3	39	40	5.1	6.2	A4V	F1V	3	50
ε² Lyr		18	44.4	39	37	5.1	5.3	A8V	F0V	2	50
ζ Lyr		18	44.8	37	36	4.3	5.7	Am	F0IV	44	30
θ Ser	Alya	18	56.2	4	12	4.5	4.9	A5V	A5V	22	30
γ Del		20	46.7	16	07	4.5	5.4	K1IV	F7V	10	40
ζ Aqr		22	28.8	− 0	01	4.4	4.6	F3V	F6IV	2	30
δ Cep		22	29.1	58	24	3.5–4.3	7.5	F5Ib–G2Ib	B7IV	41	90

Table C.20. Milky Way Galaxy

Property	Value
Mass	$> 2 \times 10^{11}\ M_\odot$
Disc diameter	30 kpc
Disc thickness (stars)	1 kpc
Disc thickness (gas and dust)	200 pc
Halo diameter	50 kpc
Sun's distance from the centre	8.5 kpc
Sun's orbital velocity	220 km s^{-1}
Sun's period	240×10^6 a
Direction of the centre (2000.0)	$\alpha = 17$ h 45.7 min $\delta = -29°00'$
Direction of the north pole (2000.0)	$\alpha = 12$ h 51.4 min $\delta = +27°08'$
Galactic coordinates of the celestial north pole	$l = 123°00'$ $b = +27°08'$

Table C.21. Members of the Local Group of Galaxies. V = apparent visual magnitude, M_V = absolute visual magnitude, r = distance

	α_{2000}		δ_{2000}		Type	V	M_V	r
	[h]	[min]	[°]	[']				[kpc]
Milky Way	17	45.7	−29	00	Sbc		−20.9	8
NGC 224 = M31	00	42.7	41	16	Sb	3.2	−21.2	760
NGC 598 = M33	01	33.8	30	30	Sc	5.6	−18.9	790
Large Magellanic Cloud	05	19.6	−69	27	Irr	0.0	−18.5	50
Small Magellanic Cloud	00	52.6	−72	48	Irr	1.8	−17.1	60
NGC 221 = M32	00	42.7	40	52	E2	7.9	−16.5	760
NGC 205	00	40.4	41	41	dE5	8.0	−16.4	760
IC 10	00	20.4	59	17	Irr	7.8	−16.3	660
NGC 6822	19	44.9	−14	48	Irr	7.5	−16.0	500
NGC 185	00	39.0	48	20	dE3	8.5	−15.6	660
IC 1613	01	04.8	02	08	Irr	9.0	−15.3	720
NGC 147	00	33.2	48	30	dE4	9.0	−15.1	660
WLM	00	02.0	−15	28	Irr	10.4	−14.4	930
Sagittarius	18	55.1	−30	29	dE7	3.1	−13.8	24
Fornax	02	39.9	−34	30	dE3	7.6	−13.1	140
Pegasus	23	28.6	14	45	Irr	12.1	−12.3	760
Leo I	10	08.4	12	18	dE3	10.1	−11.9	250
And II	01	16.5	33	26	dE3	12.4	−11.8	700
And I	00	45.7	38	00	dE0	12.7	−11.8	810
Leo A	09	59.4	30	45	Irr	12.7	−11.5	690
Aquarius	20	46.9	−12	51	Irr	13.7	−11.3	1020
SagDIG	19	30.0	−17	41	Irr	15.0	−10.7	1400
Pegasus II = And VI	23	51.7	24	36	dE	14.0	−10.6	830
Pisces = LGS 3	01	03.9	21	54	Irr	14.1	−10.4	810
And III	00	35.3	36	30	dE6	14.2	−10.2	760
And V	01	10.3	47	38	dE	14.3	−10.2	810
Leo II	11	13.5	22	10	dE0	11.5	−10.1	210
Cetus	00	26.1	−11	02	dE	14.3	−10.1	780
Sculptor	01	00.1	−33	43	dE3	10.0	− 9.8	90
Phoenix	01	51.1	−44	27	Irr	13.2	− 9.8	400
Tucana	22	41.8	−64	25	dE5	15.1	− 9.6	870
Sextans	10	13.0	−01	37	dE4	10.3	− 9.5	90
Cassiopeia = And VII	23	26.5	50	42	dE	14.7	− 9.5	690
Carina	06	41.6	−50	58	dE4	10.6	− 9.4	100
Ursa Minor	15	08.8	67	07	dE5	10.0	− 8.9	60
Draco	17	20.3	57	55	dE3	10.9	− 8.6	80

Table C.22. Optically brightest galaxies. $B =$ apparent blue magnitude, $d =$ apparent diameter, $r =$ distance

Name	α_{2000} [h]	[min]	δ_{2000} [°]	[']	Type	B	d ["]	r [Mpc]
NGC 55	0	15.1	−39	13	Sc/Irr	7.9	30 × 5	2.3
NGC 205	0	40.4	41	41	E6	8.9	12 × 6	0.7
NGC 221 = M32	0	42.7	40	52	E2	9.1	3.4 × 2.9	0.7
NGC 224 = M31	0	42.8	41	16	Sb	4.3	163 × 42	0.7
NGC 247	0	47.2	−20	46	S	9.5	21 × 8	2.3
NGC 253	0	47.6	−25	17	Sc	7.0	22 × 5	2.3
Small Magellanic Cloud	0	52.6	−72	48	Irr	2.9	216 × 216	0.06
NGC 300	0	54.9	−37	41	Sc	8.7	22 × 16	2.3
NGC 598 = M33	1	33.9	30	39	Sc	6.2	61 × 42	0.7
Fornax	2	39.9	−34	32	dE	9.1	50 × 35	0.2
Large Magellanic Cloud	5	23.6	−69	45	Irr/Sc	0.9	432 × 432	0.05
NGC 2403	7	36.9	65	36	Sc	8.8	22 × 12	2.0
NGC 2903	9	32.1	21	30	Sb	9.5	16 × 7	5.8
NGC 3031 = M81	9	55.6	69	04	Sb	7.8	25 × 12	2.0
NGC 3034 = M82	9	55.9	69	41	Sc	9.2	10 × 1.5	2.0
NGC 4258 = M106	12	19.0	47	18	Sb	8.9	19 × 7	4.3
NGC 4472 = M49	12	29.8	8	00	E4	9.3	10 × 7	11
NGC 4594 = M104	12	40.0	−11	37	Sb	9.2	8 × 5	11
NGC 4736 = M94	12	50.9	41	07	Sb	8.9	13 × 12	4.3
NGC 4826 = M64	12	56.8	21	41		9.3	10 × 4	3.7
NGC 4945	13	05.4	−49	28	Sb	8.0	20 × 4	3.9
NGC 5055 = M63	13	15.8	42	02	Sb	9.3	8 × 3	4.3
NGC 5128 = Cen A	13	25.5	−43	01	E0	7.9	23 × 20	3.9
NGC 5194 = M51	13	29.9	47	12	Sc	8.9	11 × 6	4.3
NGC 5236 = M83	13	37.0	−29	52	Sc	7.0	13 × 12	2.4
NGC 5457 = M101	14	03.2	54	21	Sc	8.2	23 × 21	4.3
NGC 6822	19	45.0	−14	48	Irr	9.2	20 × 10	0.7

Table C.23. Constellations. The first column gives the abbreviation of the Latin name used to form star names

Abbreviation	Latin name	Genitive	English name
And	Andromeda	Andromedae	Andromeda
Ant	Antlia	Antliae	Air Pump
Aps	Apus	Apodis	Bird of Paradise
Aql	Aquila	Aquilae	Eagle
Aqr	Aquarius	Aquarii	Water-bearer
Ara	Ara	Arae	Altar
Ari	Aries	Arietis	Ram
Aur	Auriga	Aurigae	Charioteer
Boo	Boötes	Boötis	Herdsman
Cae	Caelum	Caeli	Chisel
Cam	Camelopardalis	Camelopardalis	Giraffe
Cnc	Cancer	Cancri	Crab
CMa	Canis Major	Canis Majoris	Great Dog
CMi	Canis Minor	Canis Minoris	Little Dog
Cap	Capricornus	Capricorni	Sea-goat
Car	Carina	Carinae	Keel
Cas	Cassiopeia	Cassiopeiae	Cassiopeia
Cen	Centaurus	Centauri	Centaurus
Cep	Cepheus	Cephei	Cepheus
Cet	Cetus	Ceti	Whale
Cha	Chamaeleon	Chamaeleontis	Chameleon
Cir	Circinus	Circini	Compasses
Col	Columba	Columbae	Dove
Com	Coma Berenices	Comae Berenices	Berenice's Hair
CrA	Corona Austrina	Coronae Austrinae	Southern Crown
CrB	Corona Borealis	Coronae Borealis	Northern Crown
Crv	Corvus	Corvi	Crow
Crt	Crater	Crateris	Cup
Cru	Crux	Crucis	Southern Cross
CVn	Canes Venatici	Canum Venaticorum	Hunting Dogs
Cyg	Cygnus	Cygni	Swan
Del	Delphinus	Delphini	Dolphin
Dor	Dorado	Doradus	Swordfish
Dra	Draco	Draconis	Dragon
Equ	Equuleus	Equulei	Little Horse
Eri	Eridanus	Eridani	Eridanus
For	Fornax	Fornacis	Furnace
Gem	Gemini	Geminorum	Twins
Gru	Grus	Gruis	Crane
Her	Hercules	Herculis	Hercules
Hor	Horologium	Horologii	Clock
Hya	Hydra	Hydrae	Water Serpent
Hyi	Hydrus	Hydri	Water Snake
Ind	Indus	Indi	Indian
Lac	Lacerta	Lacertae	Lizard
Leo	Leo	Leonis	Lion
Lep	Lepus	Leporis	Hare
Lib	Libra	Librae	Scales
LMi	Leo Minor	Leonis Minoris	Little Lion
Lup	Lupus	Lupi	Wolf
Lyn	Lynx	Lyncis	Lynx
Lyr	Lyra	Lyrae	Lyre
Men	Mensa	Mensae	Table Mountain
Mic	Microscopium	Microscopii	Microscpe
Mon	Monoceros	Monocerotis	Unicorn
Mus	Musca	Muscae	Fly

Table C.23 (continued)

Abbreviation	Latin name	Genitive	English name
Nor	Norma	Normae	Square
Per	Perseus	Persei	Perseus
Phe	Phoenix	Phoenicis	Phoenix
Pic	Pictor	Pictoris	Painter
PsA	Piscis Austrinus	Piscis Austrini	Southern Fish
Psc	Pisces	Piscium	Fishes
Pup	Puppis	Puppis	Poop
Pyx	Pyxis	Pyxidis	Compass
Ret	Reticulum	Reticuli	Net
Scl	Sculptor	Sculptoris	Sculptor
Sco	Scorpius	Scorpii	Scorpion
Sct	Scutum	Scuti	Sobieski's Shield
Ser	Serpens	Serpentis	Serpent
Sex	Sextans	Sextantis	Sextant
Sge	Sagitta	Sagittae	Arrow
Sgr	Sagittarius	Sagittarii	Archer
Tau	Taurus	Tauri	Bull
Tel	Telescopium	Telescopii	Telescope
TrA	Triangulum Australe	Trianguli Australis	Southern Triangle
Tri	Triangulum	Trianguli	Triangle
Tuc	Tucana	Tucanae	Toucan
UMa	Ursa Major	Ursae Majoris	Great Bear
UMi	Ursa Minor	Ursae Minoris	Little Bear
Vel	Vela	Velorum	Sails
Vir	Virgo	Virginis	Virgin
Vol	Volans	Volantis	Flying Fish
Vul	Vulpecula	Vulpeculae	Fox
Oct	Octans	Octantis	Octant
Oph	Ophiuchus	Ophiuchi	Serpent-bearer
Ori	Orion	Orionis	Orion
Pav	Pavo	Pavonis	Peacock
Peg	Pegasus	Pegasi	Pegasus

Table C.24. Largest optical telescopes. $D =$ diameter of the mirror

Telescope	Location	Completion year	D [m]
William M. Keck Telescope I	Mauna Kea, Hawaii	1992	10
William M. Keck Telescope II	Mauna Kea, Hawaii	1996	10
Hobby-Eberle Telescope	Mt. Fowlkes, Texas	1997	9.2
Subaru Telescope	Mauna Kea, Hawaii	1999	8.3
Antu Telescope (VLT 1)	Cerro Paranal, Chile	1998	8.2
Kueyen Telescope (VLT 2)	Cerro Paranal, Chile	1999	8.2
Melipal Telescope (VLT 3)	Cerro Paranal, Chile	2000	8.2
Yepun Telescope (VLT 4)	Cerro Paranal, Chile	2000	8.2
Gemini North Telescope	Mauna Kea, Hawaii	1999	8.1
Gemini South Telescope	Cerro Pachon, Chile	2000	8.1
Multi-Mirror Telescope	Mt. Hopkins, Arizona	1999	6.5
Magellan 1 Telescope	Las Campanas, Chile	2000	6.5
BTA	Zelentšukskaja, Russia	1975	6.0
Hale Telescope	Mt. Palomar, USA	1948	5.0
William Herschel Telescope	La Palma, Canary Islands	1987	4.2

Table C.25. Largest parabolic radio telescopes. D = diameter of the antenna, λ_{min} = shortest wavelength

		Completion year	D [m]	λ_{min} [cm]	Remarks
Arecibo	Puerto Rico, USA	1963	305	5	Fixed disk; limited tracking
Green Bank	West Virginia, USA	2001	100×110	0.3	The largest fully steerable telescope
Effelsberg	Bonn, Germany	1973	100	0.8	
Jodrell Bank	Macclesfield, Great Britain	1957	76.2	10–20	First large paraboloid antenna
Jevpatoria	Crimea	1979	70	1.5	
Parkes	Australia	1961	64	2.5	Innermost 17 m of dish can be used down to 3 mm wavelengths
Goldstone	California, USA		64	1.5	Belongs to NASA deep space network
Tidbinbilla	Australia		64	1.3	NASA
Madrid	Spain		64	1.3	NASA

Table C.26. Millimetre and submillimetre telescopes and interferometers. $h =$ altitude above sea level, $D =$ diameter of the antenna, $\lambda_{min} =$ shortest wavelength

Institute	Location	h [m]	D [m]	λ_{min} [mm]	Remarks; operational since
NRAO, VLA	New Mexico, USA	2124	25	7	27 antennas $d_{max} = 36.6$ km 1976
NRAO, VLBA	USA	16–3720	25	13	10 antennas 1988–1993
Max-Planck-Institut für Radioastronomie& University of Arizona	Mt. Graham, USA	3250	10	0.3	1994
California Institute of Technology	Mauna Kea, Hawaii	4100	10.4	0.3	1986
Science Research Council England & Holland	Mauna Kea, Hawaii	4100	15.0	0.5	The James Clerk Maxwell Telescope 1986
California Institute of Technology	Owens Valley, USA	1220	10.4	0.5	3 antenna interferometer 1980
Sweden-ESO Southern Hemisphere Millimeter Antenna (SEST)	La Silla, Chile	2400	15.0	0.6	1987
Institut de Radioastronomie Millimetrique (IRAM), France & Germany	Plateau de Bure, France	2550	15.0	0.6	3 antenna interferometer 1990; fourth antenna 1993
IRAM	Pico Veleta, Spain	2850	30.0	0.9	1984
National Radio Astronomy Observatory (NRAO)	Kitt Peak, USA	1940	12.0	0.9	1983 (1969)
University of Massachusetts	New Salem, USA	300	13.7	1.9	radom 1978
University of California, Berkeley	Hat Creek Observatory	1040	6.1	2	3 antenna interferometer 1968
Purple Mountain Observatory	Nanjing, China	3000	13.7	2	radom 1987
Daeduk Radio Astronomy Observatory	Söul, South-Korea	300	13.7	2	radom 1987
University of Tokyo	Nobeyama, Japan	1350	45.0	2.6	1982
University of Tokyo	Nobeyama, Japan	1350	10.0	2.6	5 antenna interferometer 1984
Chalmers University of Technology	Onsala, Sweden	10	20.0	2.6	radom 1976

Table C.27. Some important astronomical satellites and space probes 1980–2002

Satellite		Launch date	Target
Solar Max	USA	Feb 14, 1980	Sun
Venera 13	SU	Oct 30, 1981	Venus
Venera 14	SU	Nov 4, 1981	Venus
IRAS	USA	Jan 25, 1983	infrared
Astron	SU	Mar 23, 1983	ultraviolet
Venera 15	SU	Jun 2, 1983	Venus
Venera 16	SU	Jun 7, 1983	Venus
Exosat	ESA/USA	May 26, 1983	X-ray
Vega 1	SU	Dec 15 1984	Venus/Halley
Vega 2	SU	Dec 21, 1984	Venus/Halley
Giotto	ESA	Jul 2, 1985	Halley
Suisei	Japan	Aug 18, 1985	Halley
Ginga	Japan	Feb 5, 1987	X-ray
Magellan	USA	May 4, 1989	Venus
Hipparcos	ESA	Aug 8, 1989	astrometry
COBE	USA	Nov 18, 1989	cosmic background radiation
Galileo	USA	Oct 18, 1989	Jupiter etc.
Granat	SU	Dec 1, 1989	gamma ray
Hubble	USA/ESA	Apr 24, 1990	UV, visible
Rosat	Germany	Jun 1, 1990	X-ray
Gamma	SU	Jul 11, 1990	gamma ray
Ulysses	ESA	Oct 6, 1990	Sun
Compton	USA	Apr 5, 1991	gamma ray
EUVE	USA	Jun 7, 1992	extreme UV
Asuka	Japan	Feb 20, 1993	X-ray
Clementine	USA	Jan 25, 1994	Moon
ISO	ESA	Nov 17, 1995	infrred
SOHO	ESA	Dec 2, 1995	Sun
Near-Shoemaker	USA	Feb 17, 1996	Mathilde, Eros
BeppoSAX	Italy	Apr 30, 1996	X-ray
Mars Global Surveyor	USA	Nov 7, 1996	Mars
Cassini/Huygens	USA/ESA	Oct 15, 1997	Saturn, Titan
Mars Pathfinder/Sojourner	USA	Dec 4, 1996	Mars
Lunar Prospector	USA	Jan 6, 1998	Moon
Nozomi	Japan	Jul 4, 1998	Mars
Deep Space 1	USA	Oct 24, 1998	Braille, Borrelly
Stardust	USA	Feb 7, 1999	Wild 2
Chandra	USA	Jul 23, 1999	X-ray
XMM-Newton	ESA	Dec 10, 1999	X-ray
Hete 2	USA	Oct 9, 2000	gamma ray
Mars Odyssey	USA	Apr 7, 2001	Mars
MAP	USA	Jun 30, 2001	cosmic background radiation
Genesis	USA	Aug 8, 2001	solar particles
RHESSI	USA	Feb 5, 2002	Sun

Answers to Exercises

Chapter 2

2.1 The distance is ≈ 7640 km, the northernmost point is 79°N, 45°W, in North Greenland, 1250 km from the North Pole.

2.2 The star can culminate south or north of zenith. In the former case we get $\delta = 65°$, $\phi = 70°$, and in the latter $\delta = 70°$, $\phi = 65°$.

2.3 a) $\phi > 58°7'$. If refraction is taken into account, the limit is $57°24'$. b) $\phi = \delta = 7°24'$. c) $-59°10' \leq \phi \leq -0°50'$.

2.4 Pretty bad.

2.5 $\lambda_\odot = 70°22'$, $\beta_\odot = 0°0'$, $\lambda_\oplus = 250°22'$, $\beta_\oplus = 0°0'$.

2.6 c) $\Theta_0 = 18$ h.

2.8 $\alpha = 6$ h 45 min 9 s, $\delta = -16°43'$.

2.9 $v_t = 16.7$ km s^{-1}, $v = 18.5$ km s^{-1}, after $\approx$ 61,000 years. $\mu = 1.62''$ per year, parallax $0.42''$.

Chapter 3

3.1 a) The flux density in the focal plane as well as the exposure time are proportional to $(D/f)^2$. Thus the required exposure is 3.2 s. b) 1.35 cm and 1.80 cm. c) 60 and 80.

3.2 a) $0.001''$ (note that the aperture is a line rather than circular; therefore the coefficient 1.22 should not be used). b) 140 m.

Chapter 4

4.1 0.9.

4.2 The absolute magnitude will be -17.5 and apparent 6.7.

4.3 $N(m+1)/N(m) = 10^{3/5} = 3.98$.

4.4 $r = 2.1$ kpc, $E_{B-V} = 0.7$, and $(B-V)_0 = 0.9$.

4.5 a) $\Delta m = 1.06$ mag, $m = 2.42$. b) $\tau = -\ln 0.85^6 \approx 0.98$.

Chapter 5

5.2 $n = 166$, which corresponds to $\lambda = 21.04$ cm. Such transitions would keep the population of the state $n = 166$ very high, resulting in downward transitions also from this state. Such transitions have not been detected. Hence, the line is produced by some other process.

5.3 If we express the intensity as something per unit wavelength, we get $\lambda_{max} = 1.1$ mm. If the intensity is given per unit frequency, we have $\lambda_{max} = 1.9$ mm. The total intensity is 2.6×10^{13} W m^{-2} sterad^{-1}. At 550 nm the intensity is practically zero.

5.4 a) $L = 1.35 \times 10^{29}$ W. The flux in the given interval can be found by integrating Planck's law numerically. Using the Wien approximation a rather complicated expression can be derived. Both methods give the result that 3.3% of the radiation is in the visual range, thus $L_V = 4.45 \times 10^{27}$ W. b) At a distance of 10 pc the observed flux density is 3.7×10^{-9} W m^{-2}. b) 10.3 km.

5.5 $M_{bol} = 0.87$, whence $R = 2.0\,R_\odot$.

5.6 $T = 1380$ K. There are several strong absorption lines in this spectral region, reducing the brightness temperature.

5.8 $v_{rms} \approx 6700$ km s^{-1}.

Chapter 6

6.1 $v_a/v_p = (1-e)/(1+e)$. For the Earth this is 0.97.

6.2 $a = 1.4581$ AU, $v \approx 23.6$ km s^{-1}.

6.3   The period must equal the sidereal rotation period of the Earth. $r = 42{,}339$ km $= 6.64 R_\oplus$. Areas within 8.6° from the poles cannot be seen from geostationary satellites. The hidden area is 1.1% of the total surface area.

6.4 $\rho = 3\pi/(G P^2 (\alpha/2)^3) \approx 1400$ kg m^{-3}.

6.5 $M = 90°$, $E = 90.96°$, $f = 91.91°$.

6.6   The orbit is hyperbolic, $a = 3.55 \times 10^7$ AU, $e = 1 + 3.97 \times 10^{-16}$, $r_\mathrm{p} = 2.1$ km. The comet will hit the Sun.

6.7   The orbital elements of the Earth calculated from Table C.12 are $a = 1.0000$, $e = 0.0167$, $i = 0.0004°$, $\Omega = -11.13°$, $\varpi = 102.9°$, $L = 219.5°$. The geocentric radius vector of the Sun in the ecliptic coordinates is

$$r = \begin{pmatrix} 0.7583 \\ 0.6673 \\ 0.0 \end{pmatrix}.$$

The corresponding equatorial radius vector is

$$r = \begin{pmatrix} 0.7583 \\ 0.6089 \\ 0.2640 \end{pmatrix},$$

which gives $\alpha \approx 2$ h 35 min 3 s, $\delta \approx 15.19°$. The exact direction is $\alpha = 2$ h 34 min 53 s, $\delta = 15.17°$.

Chapter 7

7.1   Assuming the orbits are circular, the greatest elongation is $\arcsin(a/1 \text{ AU})$. For Mercury this is 23° and for Venus 46°. The elongation of a superior planet can be anything up to 180°. The sky revolves about 15° per hour, and thus corresponding times for Mercury and Venus are 1 h 30 min and 3 h 5 min, respectively. In opposition Mars is visible the whole night. These values, however, depend on the actual declinations of the planets.

7.2   a) 8.7°. b) The Earth must be 90° from the ascending node of Venus, which is the situation about 13 days before vernal and autumnal equinoxes, around March 8 and September 10.

7.3 $P_\mathrm{sid} = 11.9$ a, $a = 5.20$ AU, $d = 144{,}000$ km. Obviously the planet is Jupiter.

7.4   a) Hint: If there is a synodic period P there must be integers p and q such that $(n_2 - n_1)P = 2\pi p$ and $(n_3 - n_1)P = 2\pi q$. Sometimes one can see claims that the configuration of the whole planetary system will recur after a certain period. Such claims are obviously nonsense. b) 7.06 d.

7.5   a) If the radii of the orbits are a_1 and a_2, the angular velocity of the retrograde motion is

$$\frac{d\lambda}{dt} = \frac{\sqrt{GM}}{\sqrt{a_1 a_2}(\sqrt{a_1} + \sqrt{a_2})}.$$

b) In six days Pluto moves about 0.128° corresponding to 4 mm. For a main belt asteroid the displacement is almost 4 cm.

7.6   If the orbital velocity of the planet is v the deviation in radians is

$$\alpha = \frac{v}{c} = \frac{1}{c}\sqrt{\frac{GM_\odot}{a}}.$$

This is greatest for Mercury, $\alpha = 0.00016$ rad $= 33''$. This planetary aberration must be taken into account when computing accurate ephemerides. The deviation is largest when the planet is in conjunction or opposition and moves almost perpendicularly to the line of sight.

7.7 $p = 0.11$, $q = 2$, and $A = 0.2$. In reality the Moon reflects most of the light directly backwards (opposition effect), and thus q and A are much smaller.

7.8 $\Delta m = 0.9$. The surface brightness remains constant.

7.9   The absolute magnitude is $V(1, 0) = 23$. a) $m = 18.7$. b) $m = 14.2$. At least a 15 cm telescope is needed to detect the asteroid even one day before the collision.

7.10   Assuming the comet rotates slowly the distances are 1.4 AU and 0.8 AU.

Chapter 8

8.1   c g a d f e b; the actual spectral classes from top to bottom are A0, M5, O6, F2, K5, G2, B3.

Chapter 9

9.1   The period is $P = 1/\sqrt{2}$ years and the relative velocity $42,100 \text{ m s}^{-1}$. The maximum separation of the lines is 0.061 nm.

9.3   Substituting the values to the equation of Example 9.1 we get an equation for a. The solution is $a = 4.4$ AU. The mass of the planet is $0.0015 \, M_\odot$.

Chapter 10

10.1   10.5.

10.2   a) $9.5 \times 10^{37} \text{ s}^{-1}$. b) The neutrino production rate is $1.9 \times 10^{38} \text{ s}^{-1}$, and each second 9×10^{28} neutrinos hit the Earth.

10.3   The mean free path is $1/\kappa\rho \approx 42,000$ AU.

Chapter 11

11.1 $t_{ff} = 6.7 \times 10^5$ a. Stars are born at the rate $0.75 \, M_\odot$ per year.

11.2 $t_t \approx 400,000$ a, $t_n \approx 3 \times 10^8$ a.

11.3   About 900 million years.

Chapter 12

12.1   a) $6.3 \times 10^7 \text{ W m}^{-2}$. b) 16 m^2.

12.2 807 W m^{-2}.

Chapter 13

13.1 $dr/r = -0.46 \, dM = 0.14$.

13.2   a) $T = 3570$ K. b) $R_{\min}/R_{\max} = 0.63$.

13.3   a) 1300 pc. b) 860 years ago; due to inaccuracies a safe estimate would be 860 ± 100 years. Actually, the explosion was observed in 1054. c) -7.4.

Chapter 14

14.1 $L = 2.3 \times 10^{40} \text{ kg m}^2 \text{ s}^{-1}$. $dR = 45$ m.

14.2   a) $M = 0.5 \, M_\odot$, $a = 0.49 \times 10^9 \text{ m s}^{-2} \approx 4 \times 10^7 g$. b) A standing astronaut will be subject to a stretching tidal force. The gravitational acceleration felt by the feet is $3479 \text{ m s}^{-2} \approx 355 \, g$ larger than that felt by the head. If the astronaut lies tangentially, (s)he will experience a compressing force of $177 \, g$.

14.3 $v = v_e \left(1 - GM/(Rc^2)\right)$. If $\Delta v/v$ is small, we have also $\Delta\lambda = (GM/Rc^2)\lambda_e$. A photon emitted from the Sun is reddened by $2.1 \times 10^{-6}\lambda_e$. In yellow light (550 nm) the change is 0.0012 nm.

Chapter 15

15.1   2.6 kpc and 0.9 kpc, $a = 1.5$ mag/kpc.

15.2 7 km s^{-1}.

15.3   The velocity of the proton is $v = 0.0462c = 1.38 \times 10^7 \text{ m s}^{-1}$. The radius of the orbit is $r = mv/qB = 0.01$ AU.

Chapter 16

16.1   7.3.

16.2   The potential energy is approximately $U = -G(m^2 n^2/(2R))$, where m is the mass of one star, n the number of stars (there are $n(n-1)/2 \approx n^2/2$ pairs), and R the radius of the cluster. The average velocity is $\approx \sqrt{Gmn/(2R)} = 0.5 \text{ km s}^{-1}$.

Chapter 17

17.1 $\mu = 0.0055'' \, a^{-1}$.

17.2 a) 5.7 kpc. b) 11 kpc. Possible reasons for the discrepancy include: 1) The distance is so large that the approximations used for deriving Oort's formulae are not very good. 2) Taking into account the interstellar extinction will reduce the distance in b). 3) The peculiar velocity of the star was neglected.

17.3 a) 3 (and the Sun). b) The number is of the order of 100,000. This is a typical selection effect: bright stars are rare but they are visible overt long distances.

17.4 a) If the thickness of the disk is H, the light has to to travel a distance $s = \min\{r, (H/2)\sec b\}$ in the interstellar medium. Thus the magnitude will be $m = M + 5 \lg(r/10 \, \mathrm{pc}) + as$. b) $s = 200 \, \mathrm{pc}$, $m = 10.2$.

Chapter 18

18.1 a) 26, b) 25.

18.2 The diameter must be of the order one one light-week ≈ 1200 AU. $M = -23.5$. If this is the bolometric magnitude, the luminosity is $L \approx 2 \times 10^{11} \, L_\odot$, corresponding to $210 \, L_\odot \mathrm{AU}^{-3}$.

Chapter 19

19.1 $r = v/H = 93$ Mpc (if $H = 75$), diameter is 35 kpc, $M = -20.4$. Potential sources of error include: 1) inaccuracy of the Hubble constant, 2) peculiar velocity of the galaxy, 3) intergalactic extinction, 4) only 2-dimensional projection is observed, and the edge depends on the limiting magnitude used.

19.2 If $H = 50 \, \mathrm{km \, s^{-1} Mpc^{-1}}$, $r = 51$ Mpc, or 74 times the distance of M31. If H is doubled, the distance is reduced to half of that, but is still higher than the value obtained in Exercise 18.1. A possible explanation is the peculiar velocity of the galaxy.

19.3 $m_\nu = 1.5 \times 10^{-35}$ kg, or 0.00002 times the mass of the electron.

Further Reading

The following list of references is not intended as a complete bibliography. It gives a number of intermediate level or more advanced works, which can serve as starting points for those who wish to learn about some specific topic in more detail or depth.

General Reference Works

Cox: *Allen's Astrophysical Quantities*, Springer 2000.
Harwit: *Astrophysical Concepts*, John Wiley & Sons 1973.
Lang: *Astrophysical Formulae*, Springer 1974.
Maran (ed.): *The Astronomy and Astrophysics Encyclopedia*, Van Nostrand – Cambridge University Press 1992.
Meeus: *Astronomical Algorithms*, Willman-Bell 1991.
Schaifers, Voigt (eds.): *Landolt-Börnstein Astronomy and Astrophysics*, Springer 1981–82.
Shu: *The Physical Universe*, University Science Books 1982.
Unsöld: *The New Cosmos*, Springer 1969, 3rd ed. 1983.

Chapter 2. Spherical Astronomy

Green: *Spherical Astronomy*, Cambridge University Press 1985.
Seidelmann (ed.): *Explanatory Supplement to the Astronomical Almanac*, University Science Books 1992.
Smart: *Spherical Astronomy*, Cambridge University Press 1931.

Chapter 3. Observations and Instruments

Evans: *Observation in Modern Astronomy*, English Universities Press 1968.
Hecht, Zajac (1974): *Optics*, Addison-Wesley 1974.
Howell (ed.): *Astronomical CCD Observing and Reduction Techniques*, ASP Conference Series 23, 1992.
King: *The History of the Telescope*, Charles Griffin & Co. 1955, Dover 1979.
Kitchin: *Astrophysical Techniques*, Hilger 1984.
Roth (ed.): *Compendium of Practical Astronomy 1–3*, Springer 1994.
Rutten, van Venrooij: *Telescope Optics*, Willman-Bell 1988.

Chapters 4 and 5. Photometry and Radiation Mechanisms

Chandrasekhar: *Radiative Transfer*, Dover 1960.
Rybicki, Lightman: *Radiative Processes in Astrophysics*, Wiley 1979.

Chapter 6. Celestial Mechanics

Brouwer, Clemence: *Methods of Celestial Mechanics*, Academic Press 1960.
Danby: *Fundamentals of Celestial Mechanics*, MacMillan 1962; 2nd ed. Willman-Bell, 3rd revised and enlarged printing 1992.
Goldstein: *Classical Mechanics*, Addison-Wesley 1950.
Roy: *Orbital Motion*, John Wiley & Sons 1978; 3rd ed. Institute of Physics Publishing 1988, reprinted 1991, 1994.

Chapter 7. The Solar System

Atreya, Pollack, Matthews (eds.): *Origin and Evolution of Planetary and Satellite Atmospheres*, 1989;
Bergstrahl, Miner, Matthews (eds.): *Uranus*, 1991;
Binzel, Gehrels, Matthews (eds.): *Asteroids II*, 1989;
Burns, Matthews (ed.): *Satellites*, 1986;
Cruikshank: *Neptune and Triton*, 1996;
Gehrels (ed.): *Jupiter*, 1976;
Gehrels (ed.): *Asteroids*, 1979;
Gehrels (ed.): *Saturn*, 1984;
Gehrels (ed.): *Hazards due to Comets and Asteroids*, 1994;
Greenberg, Brahic (eds.): *Planetary Rings*, 1984;
Hunten, Colin, Donahue, Moroz (eds.): *Venus*, 1983;
Kieffer, Jakosky, Snyder, Matthews (eds.): *Mars*, 1992;
Lewis, Matthews, Guerreri (eds.): *Resources of Near-Earth Space*, 1993;
Morrison (ed.): *Satellites of Jupiter*, 1982;
Vilas, Chapman, Matthews (eds.): *Mercury*, 1988;
Wilkening (ed.): *Comets*, 1982;

The previous items belong to a series of books on planetary astronomy published by Arizona University Press.

Beatty, Chaikin (eds.): *The New Solar System*, Sky Publishing, 3rd ed. 1990.

Encrenaz, Bibring, Blanc: *The Solar System*, Springer 1990.

Heiken, Vaniman, French (eds.): *Lunar Sourcebook*, Cambridge University Press 1991.

Lewis: *Physics and Chemistry of the Solar System*, Academic Press, Revised Edition 1997.

Minnaert: *Light and Color in the Outdoors*, Springer 1993.

Schmadel: *Dictionary of Minor Planet Names*, Springer 1992.

Chapter 8. Stellar Spectra

Böhm-Vitense: *Introduction to Stellar Astrophysics*, Cambridge University Press, Vol. 1–3, 1989–1992.

Gray: *Lectures on Spectral-line Analysis: F, G, and K stars*, Arva, Ontario, 1988.

Gray: *The Observation and Analysis of Stellar Photospheres*, 2nd edition, Cambridge University Press 1992.

Mihalas: *Stellar Atmospheres*, Freeman 1978.

Novotny: *Introduction to Stellar Atmospheres and Interiors*, Oxford University Press 1973.

Chapter 9. Binary Stars

Aitken: *The Binary Stars*, Dover 1935, 1964.

Heinz: *Double Stars*, Reidel 1978.

Sahade, Wood: *Interacting Binary Stars*, Pergamon Press 1978.

Chapters 10 and 11. Stellar Structure and Stellar Evolution

Bowers, Deeming: *Astrophysics I: Stars*, Jones and Bartlett Publishers 1984.

Böhm-Vitense: *Introduction to Stellar Astrophysics*; Cambridge University Press, Vol. 1–3 1989–1992.

Clayton: *Principles of Stellar Evolution and Nucleosynthesis*, McGraw-Hill 1968.

Hansen, Kawaler: *Stellar interiors, Physical principles, structure and evolution*, Springer 1994.

Harpaz: *Stellar Evolution*, Peters Wellesley 1994.

Kippenhahn, Weigert: *Stellar Structure and Evolution*, Springer, 2nd ed. 1994.

Phillips: *The physics of Stars*, Manchester Phys. Ser. 1994, 2nd ed. 1999.

Taylor: *The Stars: their structure and evolution*, Cambridge University Press 1994.

Chapter 12. The Sun

Golub, Pasachoff: *The Solar Corona*, Cambridge university Press 1997.

Priest: *Solar Magnetohydrodynamics*, Reidel 1982.

Stix: *The Sun – An Introduction*, Springer 1989.

Taylor: *The Sun as a Star*, Cambridge University Press 1997.

Chapter 13. Variable Stars

Cox: *Theory of Stellar Pulsations*, Princeton University Press 1980.

Glasby: *Variable Stars*, Constable 1968.

Chapter 14. Compact Stars

Chandrasekhar: *The Mathematical Theory of Black Holes*, Oxford University Press 1983.

Frank, King, Raine: *Accretion Power in Astrophysics*, Cambridge Astrophysics series 21, 2nd ed., Cambridge University Press 1992.

Glendenning: *Compact stars, Nuclear physics, Particle physics and General relativity*, Springer 1997.

Lewin, van Paradijs, van den Heuvel (eds.), *X-ray Binaries*, Cambridge University Press 1995.

Manchester, Taylor: *Pulsars*, Freeman 1977.

Smith: *Pulsars*, Cambridge University Press 1977.

Poutanen, Svensson: *High energy processes in accreting black holes*, Astronomical Society of Pacific Conference series Vol. 161, 1999.

Shapiro, Teukolsky: *Black Holes, White Dwarfs and Neutron Stars*, Wiley 1983.

Chapter 15. Interstellar Matter

Dyson, Williams: *The Physics of the Interstellar Medium*, Manchester University Press 1980.

Longair: *High Energy Astrophysics*, Cambridge University Press, 2nd ed. Vol. 1–2 1992, 1994.

Spitzer: *Physical Processes in the Interstellar Medium*, Wiley 1978.

Chapter 16. Stellar Clusters

Hanes, Madore (eds.): *Globular Clusters*, Cambridge University Press 1980.

Spitzer: *Dynamical Evolution of Globular Clusters*, Princeton University Press 1987.

Chapter 17. The Milky Way

Binney, Merrifield: *Galactic Astronomy*, Princeton University Press 1998.

Bok, Bok: *Milky Way*, Harvard University Press, 5. painos 1982.

Gilmore, King, van der Kruit: *The Milky Way as a Galaxy*, University Science Books 1990

Mihalas, Binney: *Galactic Astronomy*, Freeman 1981.

Scheffler, Elsässer: *Physics of the Galaxy and the Interstellar Matter*, Springer 1988.

Chapter 18. Galaxies

Binney, Tremaine: *Galactic Dynamics*, Princeton University Press 1987.

Combes, Boissé, Mazure, Blanchard: *Galaxies and Cosmology*, Springer 1995.

Frank, King, Raine: *Accretion Power in Astrophysics*, Cambridge Astrophysics series 21, 2nd ed., Cambridge University Press 1992.

Krolik: *Active Galactic Nuclei*, Princeton University Press 1999.

Sandage: *The Hubble Atlas of Galaxies*, Carnegie Institution 1961.

Sparke, Gallagher: *Galaxies in the Universe*, Cambridge University Press 2000.

Chapter 19. Cosmology

Harrison: *Cosmology*, Cambridge University Press 1981.

Kolb, Turner: *The Early Universe*, Perseus Books 1993.

Peebles: *Physical Cosmology*, Princeton University Press 1971.

Peebles: *Principles of Physical Cosmology*, Princeton University Press 1993.

Raine: *The Isotropic Universe*, Hilger 1981.

Roos: *Introduction to Cosmology*, Wiley 1994, 2nd ed. 1997.

Weinberg: *Gravitation and Cosmology*, Wiley 1972.

Physics

Feynman, Leighton, Sands: *The Feynman Lectures on Physics I-III*, Addison-Wesley 1963.

Shu: *The Physics of Astrophysics I–II*, University Science Books 1991

Taylor, Wheeler: *Spacetime Physics*, Freeman 1963.

Misner, Thorne, Wheeler: *Gravitation*, Freeman 1970.

Maps and Catalogues

Burnham: *Burnham's Celestial Handbook I, II, III*, Dover 1966, 2nd ed. 1978.

de Vaucouleurs et al.: *Reference Catalogue of Bright Galaxies*, University of Texas Press 1964, 2nd catalogue 1976.

Hirshfeld, Sinnott: *Sky Catalogue 2000.0*, Sky Publishing 1985.

Hoffleit: *Bright Star Catalogue*, Yale University Observatory 1982.

Kholopov (ed.): *Obshij katalog peremennyh zvezd*, Nauka, 4th edition 1985.

Luginbuhl, Skiff: *Observing Handbook and Catalogue of Deep-sky Objects*, Cambridge University Press 1989.

Ridpath: *Norton's 2000.0*, Longman 1989.

Rükl: *Atlas of the Moon*, Hamlyn 1991.

Ruprecht, Baláz, White: *Catalogue of Star Clusters and Associations*, Akadémiai Kiadó (Budapest) 1981.

Tirion: *Sky Atlas 2000.0*, Sky Publishing 1981.

Tirion, Rappaport, Lovi: *Uranometria 2000.0*, Willman-Bell 1987.

Greeley, Batson: *The NASA Atlas of the Solar System*, Cambridge University Press 1997.

Yearbooks

The Astronomical Almanac, Her Majesty's Stationery Office.

Photograph Credits

We are grateful to the following institutions who kindly gave us permission to use illustrations (abbreviated references in the figure captions)

Anglo-Australian Observatory, photograph by David R. Malin

Arecibo Observatory, National Astronomy and Ionosphere Center, Cornell University

Arp, Halton C., Mount Wilson and Las Campanas Observatories (colour representation of plate by Jean Lorre)

Big Bear Solar Observatory, California Institute of Technology

Catalina Observatory, Lunar and Planetary Laboratory

CSIRO (Commonwealth Scientific and Industrial Research Organisation), Division of Radiophysics, Sydney, Australia

ESA (copyright Max-Planck-Institut für Astronomie, Lindau, Harz, FRG)

European Southern Observatory (ESO)

Helsinki University Observatory

High Altitude Observatory, National Center for Atmospheric Research, National Science Foundation

Karl-Schwarzschild-Observatory Tautenburg of the Central Institute of Astrophysics of the Academy of Sciences of the GDR (Archives)

Lick Observatory, University of California at Santa Cruz

Lowell Observatory

Lund Observatory

Mount Wilson and Las Campanas Observatories, Carnegie Institution of Washington

NASA (National Aeronautics and Space Administration)

National Optical Astronomy Observatories, operated by the Association of Universities for Research in Astronomy (AURA), Inc., under contract to the National Science Foundation (Plate 27: observers Richard J. Tuffs, Richard A. Perley, Martin T. Brown, Stephen F. Gull; Plate 31: observers Farhad Yusef-Zadeh, Mark R. Morris, Don R. Chance; Plate 33: observers Frazer N. Owen, Richard A. White; Plate 36: observers Peter A.G. Scheuer, Robert A. Laing, Richard A. Perley)

NRL-ROG (Space Research Laboratory, University of Groningen)

Palomar Observatory, California Institute of Technology

Yerkes Observatory, University of Chicago

Name and Subject Index

Colour Supplement

Photograph on previous page. NGC 6872 and IC 4970 are a pair of interacting galaxies far south in the constellation Pavo (the Peacock). NGC 6872 is one of the largest known barred spiral galaxies; its diameter is over 7 arcmin corresponding about 200,000 pc. It is accompanied by a smaller galaxy IC 4970 of type S0 (just above the centre). Their distance is about 90 million pc. North is to the upper right. (Photograph European Southern Observatory)

Chapter 2. Spherical Astronomy

Plate 1. A long exposure shows the apparent motion of the celestial sphere. In photographs the colours of the stars are seen more clearly than with the naked eye. The brightest object is the planet Venus. (Photograph Pekka Parviainen)

Plate 2. Atmospheric refraction has distorted and discoloured the image of the setting Sun. A rare green segment can be seen above the Sun, and a mirage has formed below it. (Photograph Pekka Parviainen)

Chapter 3. Observations and Instruments

Plate 3. Astronomical observatories are usually built in high, dry places with a clear sky. The picture shows the forest of domes at Kitt Peak in Arizona in the United States. At the right is the building of the 4 m telescope. (National Optical Astronomy Observatories)

Plate 4. The most powerful radio telescope in the world is the VLA or Very Large Array in New Mexico in the United States. It is composed of 27 movable 25 m antennas. (Photograph Hannu Karttunen)

Chapter 7. The Solar System

Plate 5. Mercury can easily be seen with the naked eye in the morning or evening sky, if only one looks at the right time. The picture shows Venus (in the *upper part*) and Mercury (*lower*) in March 1985. (Photograph Hannu Heiskanen)

Plate 6. Eclipses are astronomical phenomena that tend to attract the attention of the general public. The picture shows a total eclipse of the Moon in January 1982. (Photograph Juhani Laurila)

Plate 7. The first colour pictures from the surface of Venus were obtained in 1982. Panoramic views from Venera 13 (*above*) and Venera 14 (*below*). At the edges, the camera is looking at the horizon and in the centre, at the ground around the base of the space probe. Parts of the base, the protective covers of the camera and a colour map can be seen.

Plate 8. The surface of the Earth as seen from space. The picture shows a part of Egypt with the Nile in the middle and the Red Sea in the background. (Photograph Gemini 12/NASA)

Plate 9. The surface of the Moon was visited by twelve astronauts in 1969–1972. On the last visits, a lunar vehicle was used to extend the domain of exploration trips. (Photograph Apollo 15/NASA)

Plate 10. The red desert of Mars in the Chryse Plain. (Viking 1/NASA)

Plate 11. The surface of Jupiter shows many atmospheric structures. The picture was taken by Voyager 2 from a distance of 20 million kilometres. Io is seen over the Red Spot, and Europa on the right. (NASA)

Plate 12. The strangest moon of Jupiter is Io, where the Voyager space probes observed several volcanic eruptions. One of the largest active areas is the Pele region. (Voyager 1/NASA)

Plate 13. An approach picture of Saturn taken by Voyager 2 shows the planet and its rings from a distance of 40 million kilometres. The colours in the picture have been artificially enhanced in order to bring out the atmospheric details more clearly. (NASA)

Plate 14. Neptune was imaged by Voyager 2 in August 1989. The bright spots are high-altitude methane clouds, and the red hue is due to semitransparent haze covering the planet. (NASA)

Plate 15. The nucleus of Halley's comet was imaged by the Halley Multicolour camera on board ESA's Giotto spacecraft in March 1986. The nucleus is seen as a dark elongated area in the centre of the photograph. The bright areas are jets of dust emanating from the comet. The longest diameter of the comet is about 13 km. (ESA/MPI for Aeronomie. Courtesy Dr. H.U. Keller)

Plate 16. Objective prism spectra of Regulus (*above*) and Saturn (*below*). The spectrum of Saturn is essentially the reflected spectrum of the Sun (spectral class G2V). The spectral class of Regulus is B7V.

Plate 17. A part of the solar spectrum in the green light (about 520–540 nm). The dark absorption lines on the left are due to magnesium. (High Altitude Observatory)

Chapter 11. Stellar Evolution

Plate 18. M16 is a star cluster surrounded by interstellar matter. New stars are possibly born from the protruding condensates. (Photograph NASA/Hubble Space Telescope)

Chapter 12. The Sun

Plate 19. X-ray image of the Sun showing active regions (*bright areas*) and dark coronal holes. The picture was obtained by the Skylab X-ray telescope, which is sensitive to the wavelengths 0.2–3.2 nm and 4.4–5.4 nm. (NASA)

Plate 20. In 1980, images of the Sun at visual and ultraviolet wavelengths were obtained by the Solar Maximum Mission satellite. This false colour picture shows the outward density decrease in the corona. The streamer going to the upper right extends 1.5 million kilometres from the surface. (NASA)

Chapter 15. The Interstellar Medium

Plate 21. The region around ρ Ophiuchi as seen by the Infrared Astronomical Satellite (IRAS). The same region is seen in optical wavelengths in Fig. 15.8. The cool dust shows up in this far-infrared picture in emission. (NLR-ROG)

Plate 22. The interstellar dust appears in visible light as dark clouds or as reflection nebulae. The photograph shows the Trifid Nebula M20, which is a reddish gas (emission) nebula. Dark streaks of dust can be seen in front of it and a bluish reflection nebula below it. (National Optical Astronomy Observatories)

Plate 23. Different forms of interstellar matter are found in the Eagle Nebula M16 in Serpens: red emission nebulae and dark streaks of interstellar dust. (National Optical Astronomy Observatories)

Plate 24. The brightest infrared source in the sky is the nebula around η Carinae. The red emission is from hydrogen gas and the dark areas are dust. (European Southern Observatory)

Plate 25. The Dumbbell Nebula M27 in Vulpecula. A hot star at the centre of the nebula ionizes the gas. The greenish colour is from twice ionized oxygen; the reddish hue of the outer edges, from ionized nitrogen and from the hydrogen Balmer lines. (European Southern Observatory)

Plate 26. Another beautiful example of a planetary nebula: the Ring Nebula M57 in Lyra. (European Southern Observatory)

Plate 27. The radio source Cassiopeia A is the remnant of a supernova that exploded approximately 300 years ago. The source was observed by the Very Large Array (VLA) in 1982–1983 at a wavelength of 6 cm. The field of view is 6×6 arcmin. (National Radio Astronomy Observatory)

Plate 28. The Vela supernova remnant. The picture is a composite of three black-and-white photographs taken with the ESO 1 m Schmidt telescope. The filamentary gas in the supernova remnant is visible over a six-degree area. (European Southern Observatory)

Plate 29. The structure of the galactic magnetic field is apparent in this whole-sky radio map. The red areas are bright at radio wavelengths; the blue ones are faint. The interstellar medium is mainly confined to the galactic plane, but filaments along the magnetic field lines may extend quite a long way out of the plane. The map is based on measurements at 73 cm wavelength made at Bonn, Jodrell Bank and Parkes (Max-Planck-Institut für Radioastronomie)

Chapter 17. The Milky Way

Plate 30. The image of the centre of our galaxy was produced from observations by the IRAS satellite. The yellow and green knots and blobs scattered along the band of the Milky Way are giant clouds of interstellar gas and dust heated by nearby stars. (NASA)

Plate 31. The bright radio source Sagittarius A is located at the galactic centre. This high-resolution map was made with the VLA in 1982–1984. The wavelength was 20 cm and the region shown covers approximately as large an area as the full moon. The yellow dot in the middle is a radio point source at the exact centre of the Milky Way. Its nature is unknown. In the upper part of the picture lies the "continuum arc", where narrow filamentary structures can be seen. The side of the picture corresponds to about 70 pc. (National Radio Astronomy Observatory)

Chapter 18. Galaxies

Plate 32. The Andromeda Galaxy M31 is our nearest large galaxy. Its central parts can be seen with the naked eye. A satellite trail is visible across the image. (National Optical Astronomy Observatories)

Plate 33. False colour CCD image of the central galaxy (3C75) in the cluster of galaxies Abell 400. The two fuzzy images are the twin nuclei of the galaxy. The two objects with spikes are bright stars. Observations were made with the Kitt Peak 0.9 m telescope. (National Radio Astronomy Observatory)

Plate 34. Enhanced colour image of the galaxy M81 in Ursa Major, showing the blue H II regions in the spiral arms and in the red, old stars in the central bulge. (Halton C. Arp)

Plate 35. The Southern radio galaxy Centaurus A is a peculiar galaxy that radiates at all wavelengths. Visually it looks like an old, round elliptical galaxy crossed by a strong dust lane. (National Optical Astronomy Observatories)

Plate 36. Cygnus A (3C405), a typical radio galaxy. The actual galaxy is located at the bright point in the middle, between the two large radio lobes on both sides. The image was obtained with the VLA at 6–20 cm wavelength. (National Radio Astronomy Observatory)

Plate 1

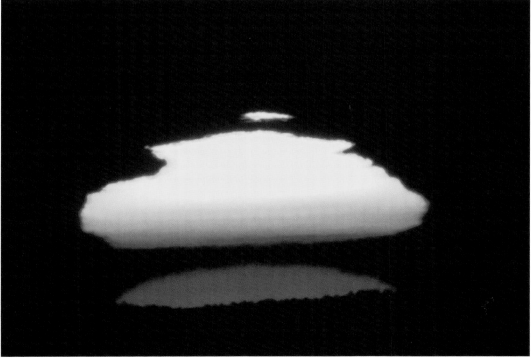

Plate 2

Plate 3

Plate 4

Plate 6

Plate 5

Plate 7

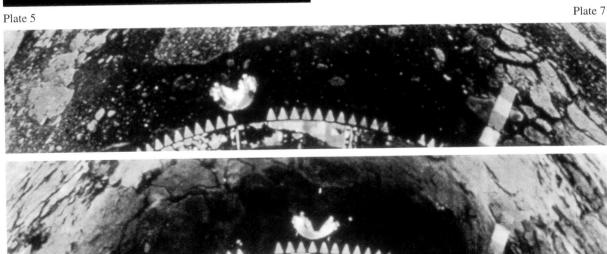

Plate 8

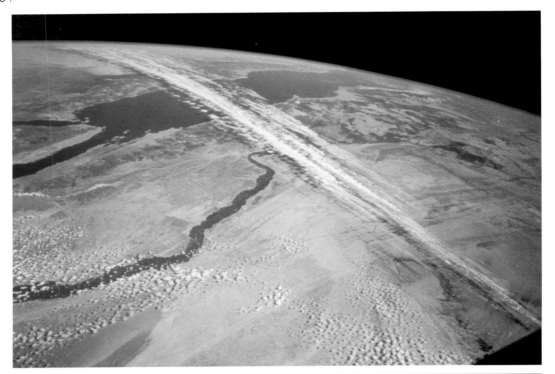

Plate 9

Plate 10

Plate 11

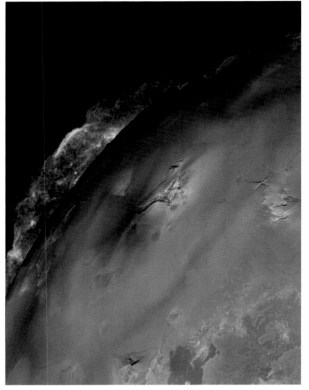

Plate 12

Plate 13

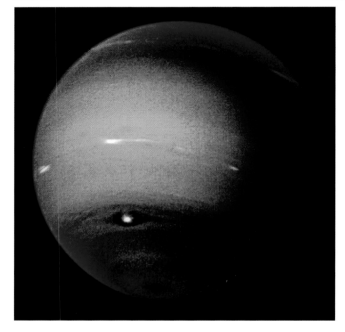

Plate 14

Plate 15

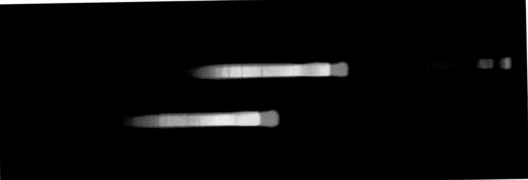

Plate 16

Plate 17

Plate 18

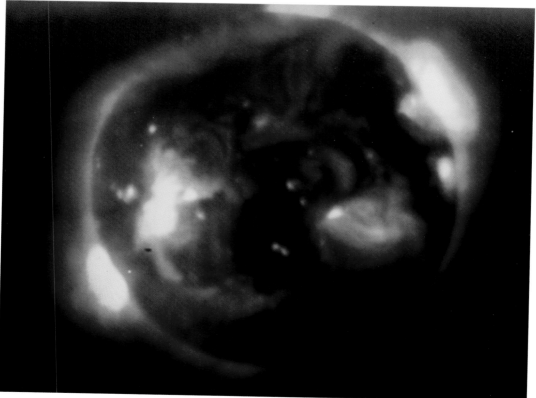

Plate 19

Plate 20

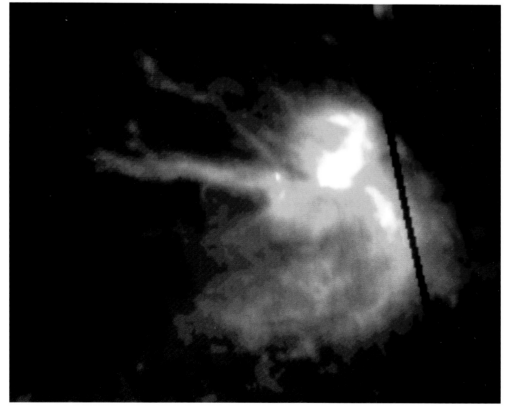

Plate 21

Plate 22

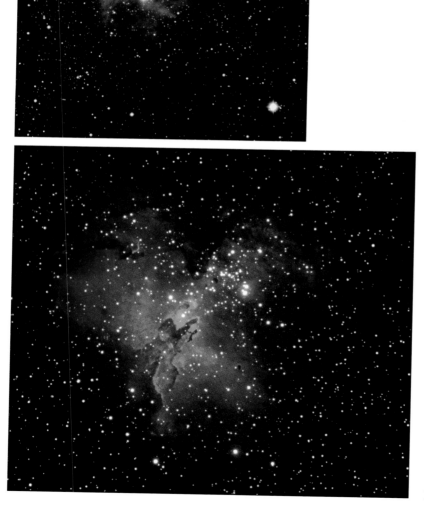

Plate 23

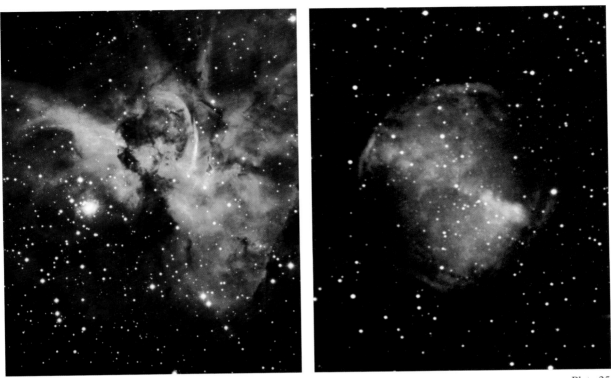

Plate 24

Plate 25

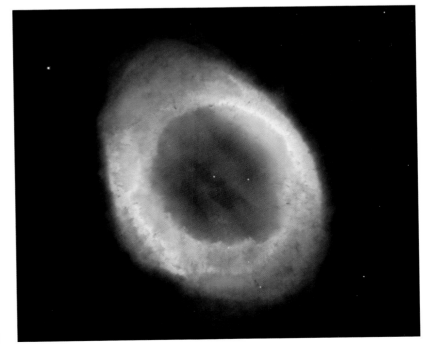

Plate 26

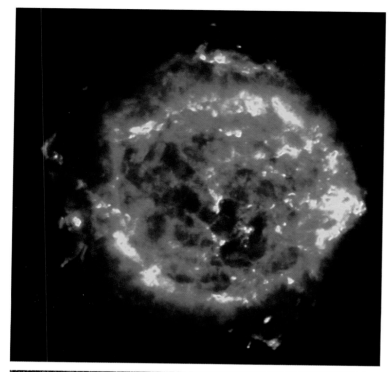

Plate 27

Plate 28

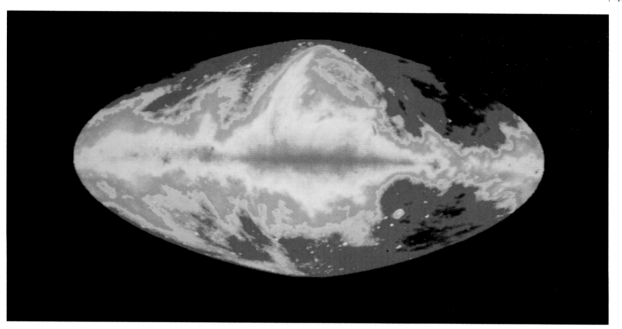

Plate 29
Plate 30

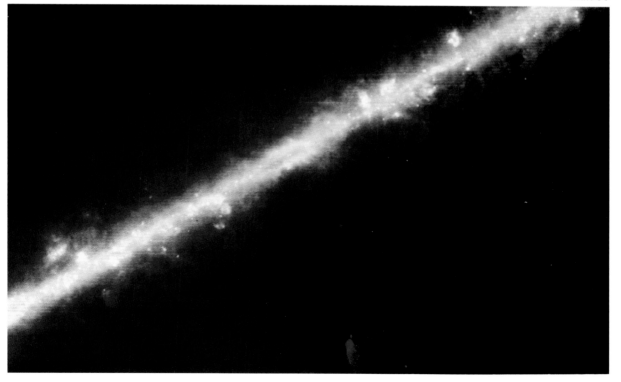

Plate 31

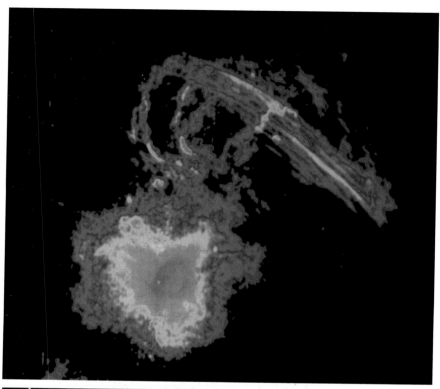

Plate 32

Plate 33

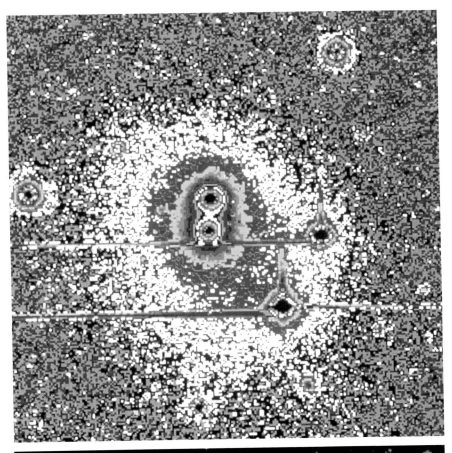

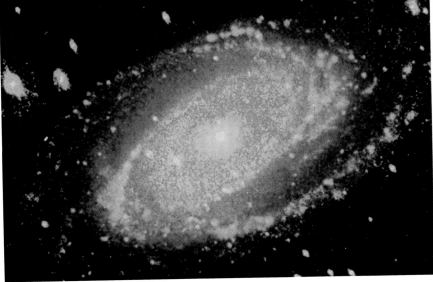

Plate 34

Plate 35

Plate 36

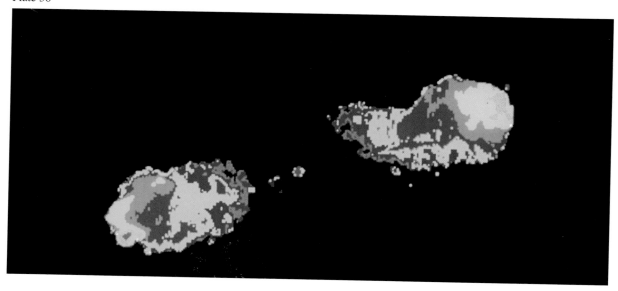